Applications of the theory of distributions in mechanics

W. Kecs * P. P. Teodorescu

Applications of the theory of distributions in mechanics

Editura Academiei
Bucureşti
România

Abacus Press
Tunbridge Wells, Kent
England

1974

Revised and enlarged translation of the
Romanian language edition of
"Aplicaţii ale teoriei distribuţiilor în mecanică",
first published
by
EDITURA ACADEMIEI ROMÂNE, Bucharest, 1970.

Translated by Radu Georgescu
Translation edited by John Hammel, Ph D
The English language edition first published in 1974 under
the joint imprints of
EDITURA ACADEMIEI ROMÂNE
str. Gutenberg 3 bis, Bucharest
and
ABACUS PRESS, TUNBRIDGE WELLS, KENT

ISBN 0 85626 006 1
© 1974

PRINTED IN ROMANIA

Preface to the Romanian edition

Both authors have been eminent students of the Faculty of Mathematics at the University of Bucharest. P. P. Teodorescu is already a well-respected mathematician and is expected to attain further achievements.

Wilhelm Kecs has recently submitted a very interesting thesis for his doctoral degree. We have admired his perseverence to master a vast modern area of mathematical analysis, the theory of distributions, in order to explain certain phenomena in mechanics which are inexplicable by classical analysis without resort to various conventions and contrivances.

The restrictions in the study of mechanics, and many mechanical phenomena, imposed by the limits of applicability of classical mathematical analysis disappear with the systematic application of the theory of distributions.

A brief survey of the contents demonstrates the extension of the area of research treated by the authors. The mathematical models of Newtonian mechanics (Chapter 4); the theory of concentrated loads (Chapter 5 — this is the most natural application of the theory of distributions); bars and strings (Chapter 8); plane and space problems of the theory of elasticity (Chapters 9 and 10).

A particular merit of the authors is the endeavour to confer the monograph an autonomous status by the introduction of a chapter devoted to the theory of distributions, expounded with a minimum of mathematical background, as well as a chapter dealing with the applications of the theory of distributions to functional equations.

Undoubtedly, the teaching of general mechanics and of the mechanics of deformable solids in the faculties of mathematics and also in polytechnics, will suffer deep mutations, as in the teaching of analysis, algebra and geometry.

This monograph may represent the beginning of such a change.

15 June 1970.

MIRON NICOLESCU
President of the Romanian Academy

Contents

Part 3

APPLICATIONS OF THE THEORY OF DISTRIBUTIONS IN MECHANICS OF SOLIDS

Foreword

The solution of many theoretical and practical problems is closely connected to the method applied and the mathematical tool which is used. In the mathematical description of physical phenomena and the solution of the corresponding boundary-value problems difficulties may appear owing to the additional conditions resulting sometimes from the limited range of applicability of the mathematical tool used; in general, such conditions may be neither necessary nor connected to the physical phenomenon considered.

The methods of classical mathematical analysis are usually employed but their applicability is often limited. Thus, the fact that not all continuous functions possess derivatives is a severe restriction imposed on the mathematical tool; it affects the unity and the generalization of the results. For example, it may lead to the conclusion of the non-existence of the velocity of a particle at any moment of its motion, a conclusion which obviously is not true.

On the other hand, the development of theoretical physics, and particularly of modern quantum mechanics, the study of various phenomena of electromagnetism, optics, wave propagation and the solution of certain boundary-value problems have brought about the introduction of new concepts and computations which cannot be justified within the frame of classical mathematical analysis.

In this way P. A. M. Dirac introduced in 1926 the delta function (denoted by δ), which from a physical point of view represents the density of a load equal to unity located at one point. A formalism has been worked out about that function and its use which justifies and simplifies different results.

With the exception of a small number of incipient investigations, the theory of distributions has been elaborated as a new chapter of functional analysis in the sixth decade of the century. The theory represents a mathematical instrument applicable to a large class of problems which cannot be solved with the aid of the classical analysis; it eliminates the restrictions which are not imposed by the physical phenomenon and justifies procedures and results which can thus be stated in a unitary and general form.

The present monograph presents, in a systematic manner, a large number of applications of the theory of distributions to problems of general Newtonian mechanics as well as of problems pertaining to the mechanics of deformable solids; special stress is laid upon the introduction of corresponding mathematical models.

Some notions and theorems of Newtonian mechanics are stated in a generalized form; in that way the effect of discontinuities on the motion of a particle and its mechanical interpretation is emphasized. This permits the study in a unitary form of collision as well as the study of bodies of variable mass by using the methods applied to the study of other mechanical phenomena; thus they will no longer appear as special phenomena.

Particular stress is laid upon the correct mathematical representation of concentrated and distributed loads; in this way the solution of the problems encountered in the mechanics of deformable solids may be obtained in a unitary form.

Newton's fundamental equation, the equations of equilibrium and of motion of the theory of elasticity are presented in a modified form which includes the boundary and the initial conditions. In that case, the Fourier and the Laplace transforms may be easily applied to obtain the fundamental solutions of the corresponding differential equations; the use of the convolution product permits them to express the solution for an arbitrary load.

We remark that in the treatment of various problems of mechanics we have placed no emphasis on the mechanical phenomenon although (besides the presentation, in a rigorous mathematical form of known results) a large number of new results are given. We considered that the most important point was to present a general method of approaching the problems, to exemplify it, to point out its different aspects and also the difficulties that may occur.

With respect to the theory of distributions only the results and the principal theorems, which are useful for the study of the mechanical problems considered, are given; however, some mathematical results of interest by themselves are also given.

The bibliography has been placed at the end of the volume but no references will be found throughout the book which is based mostly on the original works of the authors. The reference list is far from exhaustive but stress has been laid on the work accomplished in this direction in Romania.

The aim of the book is to draw attention on the possibility of applying modern mathematical methods to the study of mechanical phenomena and to be useful to mathematicians, engineers and researchers in the field of mechanics.

March 19, 1970

The authors

Foreword to the English edition

In this English edition the presentation of some of the problems and the proof of certain theorems have been improved.

New topics are included such as the study of bilocal problems in the case of differential equations, a more detailed study of the elements of the calculus of variations and their connection to certain differential operators, etc.; this has led, in addition, to an interesting mathematical basis of the calculus of variations in distributions and has pointed out different possible applications.

Particular stress has been laid on the notion of fundamental solution in the sense of the theory of distributions and various results are given concerning problems of three-dimensional elasticity. In this connection we mention the study of the dynamic problem of the elastic half-space.

The bibliography includes forty new references.

October 24, 1972   *The authors*

0

Introduction

0.1. Brief historical survey

The development of theoretical physics and particularly of modern quantum mechanics as well as the treatment of various optical, wave propagation, and electromagnetic phenomena, or the solution of boundary-value problems have brought about the introduction of new notions outside the realm of classical mathematical analysis, wherein they can no longer be justified.

We shall present briefly some of the problems which have led to the *notion of distribution* as well as some of the fundamental research work related to it.

0.1.1. Problems leading to the notion of distribution

For the purpose of simplifying the study of certain physical phenomena and in order to present the results in a unitary form, it has been necessary to introduce point masses, point charges and point dipoles, the expression of which cannot be obtained with the usual functions as well as the notion of density, expressed by a continuous function and corresponding to a continuously distributed mass. This has led to the introduction of new mathematical notions, which were also called functions initially.

0.1.1.1. The "delta" function

In order to retain the definition of the term density in the case of point charges, P. A. M. Dirac [21] introduced, in 1926, in quantum mechanics the *delta function* (denoted by δ); from a physical point of view, this represents the density of a charge equal to unity and located at the origin of the co-ordinate axis. If the punctual charge at the origin is m, then the linear density $\rho(x)$ is expressed by the relation

$$\rho(x) = m\delta(x). \tag{0.1.1}$$

From this definition, it follows that the function $\delta(x)$ vanishes everywhere except at the origin, where it has the value ∞ and satisfies the relation

$$\int_{-\infty}^{\infty} \delta(x)\,dx = 1. \tag{0.1.2}$$

The above properties show that $\delta(x)$ is not a function in the usual sense of the word since, from the viewpoint of classical analysis, no function satisfying these properties exists. However, by setting certain *formal rules* concerning the use of the *delta* function, it has been possible to simplify and justify various results and to describe phenomena of discontinuity encountered both in physics and in mathematics applied to different branches of techniques.

0.1.1.2. Other functions

Besides the Dirac function $\delta(x)$, quantum mechanics currently uses also the *delta* functions *introduced by W. Heisenberg* [*] and defined by the relations

$$\delta_+(x) = \frac{1}{2}\delta(x) + \frac{i}{2\pi x}, \quad \delta_-(x) = \frac{1}{2}\delta(x) - \frac{i}{2\pi x}, \tag{0.1.3}$$

from which

$$\delta(x) = \delta_+(x) + \delta_-(x). \tag{0.1.4}$$

The *formalism of the delta function* was used not only in quantum mechanics but also in the symbolical calculus introduced by O. Heaviside [35] for the study of certain physical phenomena. The Heaviside calculus also uses derivatives of Heaviside functions, since the delta function can be shown to be the derivative of the *Heaviside function* $\theta(x)$ (which vanishes for $x < 0$ and is equal to unity for $x \geqslant 0$).

The connection between the delta and the Heaviside functions and other relations derived from it cannot be justified with the help of classical mathematical analysis; checking the correctness of their application has been achieved indirectly by considering the physical significance of differential equations and their solutions.

In different fields of mathematics, new ideas, singular functions and generalizations or extensions of certain operations have been introduced, such as the idea of *principal value*, introduced by A. L. Cauchy for $\dfrac{1}{x}$, the *pseudofunction Pfx^m*, introduced by J. Hadamard [32], the consideration of the finite part of a divergent integral (the *regularization* of such an integral), etc.

0.1.2. Fundamental investigations in the theory of distributions

The different above-mentioned functions, whose introduction was imposed by a great number of physical phenomena which could not be studied otherwise, have led to fundamental research of a mathematical nature to give the *theory of distributions*.

---

[*] E.g. see [94].

0.1.2.1. First mathematical works in the theory of distributions

In a study of partial differential equations of the hyperbolic type, S. L. Sobolev [106] introduced for the first time the notion of a *generalized function*, which he considered as a generalized solution of such differential equations.

The study of the Fourier transform and the Fourier integral have led S. Bochner [8] to introduce derivatives of continuous functions which, in general, have no derivatives.

The variety of new ideas, singular functions introduced in different fields, calculation rules, and methods which were incorrect from the viewpoint of classical analysis have brought about the elaboration of a new theory which unifies, simplifies, and justifies them from a mathematical point of view, thus leading to the *theory of distributions* (or of *generalized functions*), developed as *a part of functional analysis*.

The theory of distributions is the natural frame for the general treatment of the Fourier and Laplace transforms, since it offers an adequate mathematical tool for solving a large class of problems which could not be handled by means of classical mathematical analysis; at the same time, it eliminates restrictions resulting not from the phenomena investigated but from the limited applicability of classical mathematical analysis.

Thus, the fact that not any continuous function is differentiable constitutes a restriction, resulting from the mathematical tool, and affects the unity and generality of the results; from such a restriction there would follow, for example, the non-existence of velocity at any time of the motion of a particle, a conclusion which obviously contradicts the physical reality. This restriction is eliminated in the theory of distributions.

The theory not only justifies procedures and results which are currently used, but also allows their formulation in a unitary and general form.

0.1.2.2. Methods of introducing the notion of distribution

Among the founders of the theory of distributions, L. Schwartz must be mentioned; in two volumes [97], published in 1950 and 1951, the theory is presented in a systematic and general form, using the *methods of functional analysis*. The study of certain difficult problems concerning partial differential equations and of problems related to other fields have led I. M. Gelfand and G. E. Šilov [29], L. Hörmander [36], and others to develop considerably the theory elaborated by L. Schwartz.

A different viewpoint has been introduced by I. Mikusiński and R. Sikorski [77], who use the *method of fundamental sequences*, which is closer to classical analysis for defining distributions. This method of defining the distributions is similar to the cut-number method applied to the set of rational numbers for defining the set of real numbers. Owing to its similarity with the methods of classical analysis, the method of fundamental sequences has been widely used in the study of many physical phenomena, where it has proved to be a useful tool*.

Other authors, such as H. König [64], J. Korevaar [63], J. Sebastiano e Silva [99], R. Sikorski [100], and I. Halperin [34], have defined distributions by various classical methods (*method* of fundamental sequences, *of derivatives*, etc.).

In the study of problems of general mechanics or of mechanics of deformable bodies, we shall use throughout this monograph the theory of distributions based on the method of linear functionals; the method is more applicable to the structure of the problems encountered in mechanics and, at the same time, it provides the means for presenting various results in a general and unitary form.

0.2. Investigated problems

This monograph deals only with such results of the theory of distributions that are strictly required for the study of problems encountered in mechanics; at the same time, it presents some new unpublished results, as well as proofs of the more important theorems. A systematic treatment of the theory of distributions may be found in the handbooks of L. Schwartz [97], [98], I. M. Gelfand and G. E. Šilov [29], J. Mikusiński and R. Sikorski [77], and A. H. Zemanian [169], [170]. The Romanian literature in this field is represented by the monographs of G. Marinescu [74], R. Cristescu [15], and R. Cristescu and G. Marinescu [16].

0.2.1. Concentrated loads

One of the ideas used in mechanics is that of concentrated load, which is represented by means of bound vectors in the case of deformable bodies; a corresponding mathematical representation has been introduced, where use is made of the notion of *wrench* and of *the properties of the representative δ sequences*.

0.2.1.1. Forces. Moments. Dipoles

We remark that the *concentrated force*, considered as a bound vector, is represented in many works by means of the Dirac distribution, but quite often in a form which is incorrect from a mathematical viewpoint and without a proper justification. In this monograph we shall try to present this mechanical notion in a correct form.

Since the action of a concentrated force upon a deformable solid depends on its point of application as well as on its magnitude, line of action, and direction,

---

* E.g. see [9].

it follows that the limit of a group of concentrated forces will depend essentially on the directions in which the passage to the limit of the points of application takes place. Such limits are meaningless in the usual sense, unless they are considered in *the sense of the theory of distributions*.

For this reason, in the case of deformable solids, we define several types of moments, with particular emphasis on *directed concentrated moments, rotational concentrated moments* (centres of rotation), and *concentrated moments of dipole type*, whose specific properties and corresponding representations are expressed by distributions.

0.2.1.2. Methods of introducing concentrated loads

The results concerning the definition, classification, and representation of concentrated loads are based on the authors' papers [40], [42], [43], [45], [46], [51], [55], [121], [123], [126] — [132]. Evidently, there are also other points of view with respect to this problem.

Thus, I. Mikusiński [76] uses methods of operational calculus for the representation of particular forces and moments; he gives examples related to the bending of straight beams and the theory of structures. In fact, this represents a different form of the use of the theory of distributions.

In most monographs treating the mechanics of deformable solids, concentrated loads (concentrated forces, concentrated moments, etc.) are obtained by a process of passage to the limit, in the classical sense, starting generally from distributed loads; these procedures are justified and have a meaning in the theory of distributions if one may introduce certain particular representative δ sequences corresponding to the distributed loads mentioned above.

The notion of *concentrated moment of the nth order* is introduced starting from the equivalent load of a concentrated moment of the first order; these concentrated loads, too, may be expressed by means of distributions. The usefulness of these magnitudes becomes apparent when treating such problems as the *bending of straight bars* or the *equivalence of the action of forces*. Thus, it is found that the action of a concentrated force applied at a point of a deformable solid is equivalent to the action of the same force applied at another point, assuming that, at the second point, concentrated moments of different orders are also applied; the number of the latter is determined by the structure of the deformable solid.

0.2.2. Problems of general mechanics

The various notions of the theory of distributions provide an interesting application in certain problems of general mechanics, such as the geometry of masses and the general theorems (principles) of mechanics.

0.2.2.1. Stieltjes integral.
Elements of the geometry of masses

The introduction of the *Stieltjes integral in distributions* emphasizes in a general form, which also includes the cases of discontinuity, the different elements of the geometry of masses (density, total mass, static moment, moment of inertia, etc.). Certain extensions of the Stieltjes integral are made and it is shown that, under sufficiently broad conditions, the calculation of the Stieltjes integral reduces to that of a Riemann integral if the notion of the differential is used in the sense of the theory of distributions; these considerations are applied to one-, two-, and three-dimensional cases.

By proceeding in this way, the moments of inertia and the other magnitudes mentioned may be expressed from a mathematical viewpoint in a unitary form, whether the masses are concentrated or distributed. One way of expressing concentrated masses and moments of inertia is given in [2] and [16]; the results obtained are particular cases which can be easily derived with the help of the Stieltjes integral extended to distributions.

0.2.2.2. General theorems of mechanics. Applications

The different mechanical magnitudes which are used currently can be easily extended to distributions; thus, *velocity, acceleration, impulse, angular impulse, linear momentum, angular momentum, work, kinetic energy*, etc. will be extended, with emphasis on their mechanical interpretation and the effect of the *moments of discontinuity* on the motion of a particle. We shall show also the forms which are taken by the *general theorems* (or *principles*) *of mechanics*.

The equations of Meščerski and Lagrange will be extended to the *particle with variable discontinuous mass* and to *systems of particles with variable discontinuous mass*, by using the *notion of generalized variation* of the calculus of variations. In this connection, the phenomenon of *collision* will be discussed. It will be shown that, from a mathematical point of view, collision phenomena do not represent a special case and that they can be studied in distributions, using the same methods as for any other mechanical phenomena.

A *space of plastic impact* will be defined, wherein the equation of Meščerski represents the result of the composition of two of its elements; in this way, the theorem of Carnot concerning the loss of kinetic energy can be extended also to the *energy of acceleration*.

Other results concerning the Lagrange, Appell and Mangeron equations and the *plane-parallel motion* are given in references [11] — [13], [150] — [156]. The representation of *hinged supports* with the help of the theory of distributions may be useful in solving problems related to the strength of materials, as has been shown in references [166] — [168]. Other problems concerning straight bars and plane plates are treated in [3], [4], [7], [17] — [19], [20], [79] — [85], [113].

0.2.3. Problems of the mechanics of deformable solids

In the introductory part of the monograph an appropriate part is devoted to the representative δ sequences as well as to the Fourier and Laplace transforms, which are useful in the treatment of problems related to the *mechanics of deformable solids* and in particular to the *theory of elasticity*. An important part is played also by the concept of *fundamental solution* and of *Green's function* and *distribution*.

0.2.3.1. Equations of the theory of elasticity. Fundamental solution

In order to apply the method of integral transforms, it is necessary to transpose in distributions the equations of the mechanics of deformable solids; as a matter of fact, we shall confine ourselves to the case of elastic bodies. The respective equations will occur in a modified form, which includes the boundary conditions in the static case and the initial conditions in the dynamic case; this simplifies considerably the method for solving the different problems encountered.

By using the notion of fundamental solution for the equations of the theory of elasticity, both in the static and the dynamic case, one may give generalized solutions (in distributions) to a large number of problems, such as *elastic plane*, *elastic half-plane*, *elastic quarter-plane*, *elastic space*, *elastic half-space*, etc. The solutions are expressed with the help of the convolution product; thus, new problems can be solved and the unity and generality of the solutions are emphasized.

0.2.3.2. Green's functions

As the importance of Green's functions for solving boundary-value problems is well known, we will emphasize the relation between these functions and the Dirac distribution as well as the fundamental solution. The functions are determined for various particular problems, such as heat conduction, equations of the Poisson type, and others.

With the help of the notion of *particular fundamental solution* E^+, the expression of Green's function is obtained for ordinary differential equations; the results are applied to the case of forced linear vibrations and to the bending of straight bars.

The fundamental solution in the case of a certain class of partial differential equations is obtained by a simple method based on the properties of the characteristic curves.

Investigations in this field, concerning the infinite space, in the elastic and viscoelastic, static, and dynamic case, are presented in references [109], [111], [112].

0.2.4. Conclusions

The results expounded in this monograph are based mainly on the authors' studies mentioned in the reference list *. However, some results, such as the Stieltjes integral in distributions, the sectorial moments of the dipole type, the fundamental solutions for the elastic plane, the elastic half-plane, the elastic space, the elastic half-space, etc. have not yet been published and are presented here for the first time.

It is obvious that, besides the problems of mechanics mentioned above, there are also problems in which the theory of distributions plays an important part, such as *wave propagation with surfaces of discontinuity, stratified bodies* (with discontinuous non-homogeneity), etc. and which are not treated in this monograph.

---

* The reference list gives also other titles concerning the theory of distributions, which have not been mentioned above, as well as other works, such as tables of integral transforms, etc. used throughout the work.

Part 1

Introduction to the theory of distributions

1

Fundamental concepts and formulae of the theory of distributions

1.1. Fundamental spaces. Distribution concept

The *distribution concept* may be introduced, as we haves een, in several ways. From the point of view of practical applications the *method of linear functionals* seems the most convenient and hence it will be applied in the following discussion.

In the study of certain physical phenomena, functionals are introduced which act on a definite class of functions and have definite properties; they are called *fundamental functions* and their set constitutes the *fundamental space*.

1.1.1. Fundamental spaces

The consideration of a definite fundamental space is governed by the structure of the phenomenon investigated, that is, by the functionals which describe different properties of the physical phenomenon. This results in the introduction of various *classes of fundamental spaces* on which distributions are defined.

In particular we mention the fundamental spaces K, S, K^m and Z^*.

1.1.1.1. Fundamental space K

By definition, the *fundamental space*, K, consists of the functions of a real variable $\varphi(x)$, possessing derivatives of all orders (of class C^∞) and which are null, together with all their derivatives outside bounded domains, which together with their boundaries determine the supports of these functions called *fundamental* functions. We note that a proper support $\Omega_\varphi = \operatorname{supp} \varphi$ corresponds to every fundamental function $\varphi(x)$. Also, the function $\varphi(x)$ may be defined on the n-dimensional Euclidean space (R^n) in which case $x \equiv (x_1, x_2, \ldots, x_n)$ represents a point of R^n and $x_1, x_2, \ldots, x_n$ its co-ordinates.

In general, by the support of a function $\varphi(x)$ we mean the smallest closed set which contains the set of points, x, for which $\varphi(x) \neq 0$. Thus, if x_0 is a point of

---

* The space Z will be introduced in section 2.1.1.1.

the support of the function $\varphi(x)$, then for any neighbourhood, V, of x_0 it exists a point, $x \in V$, for which $\varphi(x) \neq 0$.

The support of a function is compact if it is *bounded*. Thus, it follows that the space K consists of functions with a compact support and possessing derivatives of all orders.

Denoting the boundary and the support of the function $\varphi(x)$ of the space K b$_J$ F and Ω_φ respectively, we may write

$$\varphi^{(k)}(x) = 0 \text{ for } x \notin \Omega_\varphi \qquad (k = 0, 1, 2, ...), \qquad (1.1.1)$$

$$\varphi^{(k)}(F) = 0 \qquad (k = 0, 1, 2, ...), \qquad (1.1.1')$$

where the derivatives of an arbitrary order are pointed out; this means that the fundamental function, $\varphi(x)$, has a contact of infinite order at zero.

As an example of a fundamental function defined on the real axis, R, let us consider the function

$$\varphi(x; a, b) = \begin{cases} e^{\frac{|ab|}{(x-a)(x-b)}} & \text{for} \quad x \in (a, b) \\ 0 & \text{for} \quad x \notin (a, b). \end{cases} \qquad (1.1.2)$$

The support of this function is the compactum $[a, b]$. At the points $x = a$ and $x = b$ the graph of the function $\varphi(x; a, b)$ has a contact of infinite order with the Ox axis. In particular, taking $a = -c$ and $b = c$ we obtain the fundamental function

$$\varphi_1(x; c) = \begin{cases} e^{-\frac{c^2}{c^2 - x^2}} & \text{for} \quad x \in (-c, c) \\ 0 & \text{for} \quad x \notin (-c, c), \end{cases} \qquad (1.1.2')$$

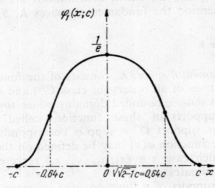

Fig. 1.1

the graph of which is shown in Figure 1.1.

Similarly, in the Euclidean space R^n, we may consider as an example of fundamental function, the function

$$\varphi(x) = \begin{cases} e^{-\frac{a^2}{a^2-r^2}} & \text{for} \quad r < a \\ 0 & \text{for} \quad r \geqslant a, \end{cases} \tag{1.1.3}$$

where

$$r = |x| = \sqrt{\sum_{i=1}^{n} x_i^2}. \tag{1.1.4}$$

The support of the function considered is $r \leqslant a$. The set of the fundamental functions $\varphi(x)$ whose supports are included in the set $r \leqslant a$ form a subspace of K which is denoted by $K(a)$; the union of these subspaces is K, i.e.

$$\bigcup_a K(a) = K. \tag{1.1.5}$$

The space K is a *vectorial space* since if $\varphi_1, \varphi_2 \in K$ and $\alpha_1, \alpha_2 \in R$ (or C) then we have

$$\alpha_1 \varphi_1, \alpha_2 \varphi_2 \in K \tag{1.1.6}$$

and

$$\alpha_1 \varphi_1 + \alpha_2 \varphi_2 \in K. \tag{1.1.6'}$$

Furthermore, the space K constitutes *an algebra* with respect to multiplication since from $\varphi_1, \varphi_2 \in K$ it follows that

$$\varphi_1 \varphi_2 \in K. \tag{1.1.7}$$

Thus, with the functions (1.1.2) we can construct the function

$$\varkappa(x_1, x_2, ..., x_n) = \varphi(x_1; a_1, b_1) \; \varphi(x_2; a_2, b_2) ... \varphi(x_n; a_n, b_n), \tag{1.1.8}$$

which is fundamental and defined on R^n, and has the support

$$\Omega_\varkappa = [a_1, b_1] \times [a_2, b_2] \times ... \times [a_n, b_n], \tag{1.1.8'}$$

where "$\times$" represents the symbol of the Cartesian product.

If $\psi(x)$ is a function possessing derivatives of all orders but without necessarily having a compact support (therefore without necessarily being a fundamental function), then the function $\psi(x)\varphi(x)$ is a fundamental function, the support of which is the intersection of the supports of the two functions. This remark permits us to construct fundamental functions, the respective property having a more general character than that specified by relation (1.1.7).

The function

$$\psi_1(x) = \begin{cases} x^2 & \text{for } x \in (-a, a) \\[2mm] 0 & \text{for } x \notin (-a, a) \end{cases} \tag{1.1.9}$$

has the compact support $[-a, a]$, although it vanishes at the origin, since in any neighbourhood of the origin there exists a point x_0 for which $\psi_1(x_0) \neq 0$. However,

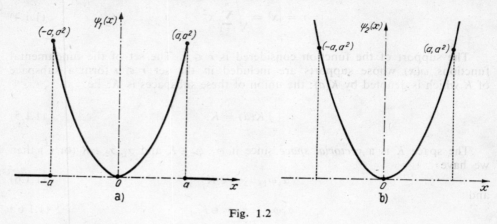

Fig. 1.2

the function considered is not fundamental, although it has a compact support, because it does not possess derivatives of all orders; indeed, the function does not possess a derivative at the points $x = \pm a$ (Figure 1.2a).

Unlike the function $\psi_1(x)$, the function

$$\psi_2(x) = x^2 \text{ for } x \in R \tag{1.1.9'}$$

possesses derivatives of all orders; yet it is not a fundamental function because its support is the real axis R (hence the support is not bounded) (Figure 1.2b).

If the function of a real variable $\varphi(x)$ is a fundamental function, then $\psi_2(x)\varphi(x)$ will also be a fundamental function whose support is the support of the function $\varphi(x)$.

Similarly, it may be seen that the function

$$\psi_3(x) = \sin x, \quad x \in R, \tag{1.1.10}$$

is of the class C^∞, but the function

$$\psi_4(x) = \sin |x|, \quad x \in R, \tag{1.1.10'}$$

is not of class C^∞ since it has no derivative at the origin. As has been pointed out before, such remarks permit us to construct fundamental functions.

We remark also that the function $\varphi(x) \equiv 0$, $x \in R$ belongs to the fundamental space K; it constitutes the *null element* of this space. But the function $\varphi(x) \equiv C \neq 0$, $x \in R$, is not a fundamental function.

In order that the vectorial space, K, may become a *topological space*, we shall define a *convergence in that space*.

If $K(a)$ represents the space of functions possessing derivatives of all orders, having the support included in $[-a, a]$ on the real axis R, then by the neighbourhood $V(m, \varepsilon)$ of the origin O we mean the set of all the points of this space, that is of the functions $\varphi^{(k)}(x) \in K(a)$, which satisfy the inequalities

$$|\varphi^{(k)}(x)| < \varepsilon \text{ for } \varepsilon > 0 \qquad (k = 0, 1, 2, ..., m), \tag{1.1.11}$$

where m is a given natural number.

Let us give the following

Definition 1.1.1. *The sequence* $\{\varphi_n(x) \in K = \bigcup_a K(a)\}$ *is said to converge to the function* $\varphi(x) \in K$, *if the functions* $\varphi_n^{(k)}(x)$ *have their supports contained in the same bounded domain, independent of n, and if, for any $k = 0, 1, 2, ...$ they converge uniformly to* $\varphi^{(k)}(x)$, *when $n \to \infty$; we shall write*

$$\lim_{n \to \infty} \varphi_n(x) = \varphi(x). \tag{1.1.12}$$

From this definition of convergence in the space K we deduce that the sequence $\{\varphi - \varphi_n\}$ belongs to the neighbourhood of O if φ, $\varphi_n \in K(a)$, that is

$$|\varphi^{(k)}(x) - \varphi_n^{(k)}(x)| < \varepsilon \qquad (k = 0, 1, 2, ..., m). \tag{1.1.13}$$

The system of neighbourhoods $V\left(m, \dfrac{1}{n}\right)$ of 0 constitues a *fundamental system of neighbourhoods* since $V_{n+1} \subset V_n$ and hence, all the neighbourhoods of 0 in $K(a)$ include a neighbourhood of the family $V\left(m, \dfrac{1}{n}\right)$.

The space K, wherein a convergence has been defined by means of the fundamental system of neighbourhoods of 0, is a *topological vectorial space*.

In the following discussion, we shall consider limits in the sense of the topology defined above in the space K.

In connection with the more important properties of the fundamental functions $\varphi \in K$, we mention the following

Theorem 1.1.1. *If E is a compact set in R^n and F is an open set which includes $E (F \supset E)$ then there exists a fundamental function $\varphi(x)$ equal to unity on E, null outside of F and comprised between 0 and 1 in the rest.*

Figure 1.3 illustrates graphically the theorem for the real axis R (the compactum $[a, b] = E$, $F = (a', b') \supset [a, b]$).

Based on the above theorem we can prove the following important property, i.e.

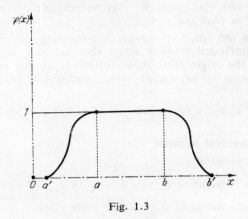

Fig. 1.3

Theorem 1.1.2. *If U_i $(i = 1, 2, \ldots)$ is a family of open sets which constitute a covering of R^n (any point of R^n belongs to a finite number of sets of U_i), then any fundamental function $\varphi(x)$ may be written in the form*

$$\varphi(x) = \sum_{i=1}^{\infty} \varphi_i(x), \qquad (1.1.14)$$

where $\varphi_i(x)$ are fundamental functions vanishing outside U_i.

Since fundamental functions have a compact support we have

$$\lim_{x \to \infty} \varphi(x) = 0, \qquad (1.1.15)$$

which states exactly the behaviour at infinity of these functions of the space K.

1.1.1.2. Fundamental space S

By an extension of the space K we reach another class of functions which determine the *fundamental space S*. The functions of this class also possess derivatives of all orders; for $|x| \to \infty$ they tend to zero together with their derivatives of any order more rapidly than any power of $\dfrac{1}{|x|}$. Hence we deduce that for any $x \in R$ the fundamental functions of the space S satisfy an inequality of the form

$$|x^k D^q \varphi| \leqslant C_{kq} \qquad (k, q = 0, 1, 2, \ldots), \qquad (1.1.16)$$

where we have introduced the differential operator

$$D^q = \frac{\partial^{q_1 + q_2 + \cdots + q_n}}{\partial x_1^{q_1} \partial x_2^{q_2} \ldots \partial x_n^{q_n}} .$$ (1.1.17)

These functions satisfy also relation (1.1.15).

Concerning *convergence in the space S* we shall use the following

Definition 1.1.2. *The sequence $\varphi_n(x) \in S$ is said to have the limit $\varphi(x) \in S$, for $n \to \infty$, if $\varphi_n^{(k)}(x)$ converges uniformly to $\varphi^{(k)}(x)$ on any compactum which satisfies relation (1.1.16), where C_{kq} is not dependent on n.*

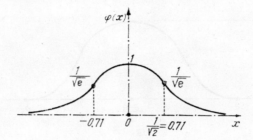

Fig. 1.4

It is easily seen that between the two fundamental spaces there exists the relation $K \subset S$.

Also, the convergence defined in K implies the convergence defined in S, which means that the convergence in the fundamental space K is stronger than the convergence in the fundamental space S.

The function

$$\varphi(x) = e^{-x^2}, \quad x \in R,$$ (1.1.18)

is an example of function of the class S (Figure 1.4).

1.1.1.3. The space K^m

Similarly, we introduce the *space K^m* which includes the functions with a compact support, having continuous derivatives up to and including the mth order. Convergence in this space is defined in the same manner as in the space K.

Between the space K^m and the space K we evidently have the relations

$$K^m \supset K, \quad \bigcap_{m=0}^{\infty} K^m = K.$$ (1.1.19)

For $m = 0$, we obtain the space K^0 which includes the continuous functions with compact support.

The function

$$\varphi(x) = \begin{cases} \sin^{m+1} \dfrac{x-a}{b-a}\,\pi & \text{for } x \in [a, b] \\[2mm] 0 & \text{in the rest} \end{cases} \tag{1.1.20}$$

is an example of a function of space K^m, with the support $[a, b]$ (Figure 1.5).

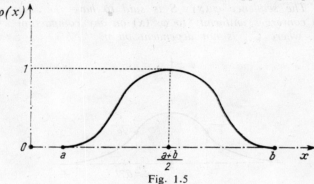

Fig. 1.5

We mention a useful property for the construction of functions belonging to the space K^m; the property extends a similar property which applies to the fundamental space K. If the function $\varphi(x) \in K^m$ and $\psi(x)$ is a function of class C^m, then the function $\psi(x)\varphi(x) \in K^m$ and its support is included in the intersection of the supports of the two functions.

1.1.2. The concept of a distribution. Properties

The *concept of a distribution* is strongly connected to that of *a functional*; therefore we shall begin by defining the latter concept.

1.1.2.1. The concept of a functional

Let X and Y be vectorial spaces relative to the field, R, of real numbers or to the field, C, of complex numbers. We denote elements of the two spaces by x and $y (x \in X, y \in Y)$. If, to every element $x \in X$, there corresponds a single element, $y \in Y$, we have defined an *operator* or a *mapping* U of X into Y and we write $y = Ux$. The set of all the operators which map X into Y is denoted by (X, Y).

Let $U \in (X, Y)$ be such an operator. The operator U is said to be linear if for any real or complex numbers α_1, α_2 we have the relation

$$U(\alpha_1 x_1 + \alpha_2 x_2) = \alpha_1 U(x_1) + \alpha_2 U(x_2). \tag{1.1.21}$$

In particular, by taking $\alpha_1 = \alpha_2 = 1$, we have

$$U(x_1 + x_2) = U(x_1) + U(x_2), \tag{1.1.21'}$$

which expresses the *property of additivity* of the operator. Similarly, by taking $\alpha_1 \neq 0$ and $\alpha_2 = 0$, we obtain

$$U(\alpha_1 x_1) = \alpha_1 U(x_1), \tag{1.1.21''}$$

which expresses the *property of homogeneity* of the operator. It is easily seen that when an operator satisfies relations (1.1.21') and (1.1.21''), i.e. when it is additive and homogeneous, it also satisfies relation (1.1.21), and is thus a linear operator.

The set (X, Y) of all the operators which map X into Y becomes *a vectorial* (linear) *space* if the operations of additive composition (addition) and multiplicative composition (multiplication) relative to the field of real numbers or to the field of complex numbers are defined by the formulae

$$(U_1 + U_2)(x) = U_1(x) + U_2(x), \tag{1.1.22}$$

$$(\alpha U)(x) = \alpha U(x). \tag{1.1.22'}$$

The null element is denoted by 0 and is defined by the formula

$$0(x) = 0 \text{ for any } x \in X. \tag{1.1.23}$$

Let $Y = K^0(a)$ be the space of continuous functions having the support $[-a, a]$, and $X = K^1(a)$ the space of functions with continuous derivatives of the first order and having the same support. Between the spaces X and Y we establish the following correspondence

$$x \to y = \frac{\mathrm{d}x}{\mathrm{d}t}, \tag{1.1.24}$$

which shows that the mapping so defined is the *differential operator* $U = \dfrac{\mathrm{d}}{\mathrm{d}t}$.

Obviously, this operator is linear.

If the vectorial spaces X and Y are topological, hence if we have defined a convergence in the two spaces, we can also define the concept of continuous operator.

Let us give the following

Definition 1.1.3. *The operator $U(X, Y)$ is said to be continuous at the point x if $y_n = U(x_n) \in Y$ converges to $y = U(x)$ when $x_n \to x$, x_n, $x \in X$.*

It may be shown that if U is continuous at the point 0, then it is continuous at any other point $x \in X$. Similarly, it is shown easily that the differential operator, defined by formula (1.1.24), is a continuous linear operator, since the uniform convergence $x_n \rightarrow x$ implies $\dfrac{\mathrm{d}x_n}{\mathrm{d}t} \rightarrow \dfrac{\mathrm{d}x}{\mathrm{d}t}$ owing to the convergence defined in the space K^m.

Continuous linear operators are encountered in theoretical and particularly in many practical problems.

The notion of a *linear functional* is obtained by particularizing the concept of the linear operator; we consider that the vectorial space, Y, is the field of real or complex numbers, and we give the following.

Definition 1.1.4. *The mapping of a vectorial space X (relative to Γ) into Γ is called a functional. If Γ is the field of real numbers R the functional is said to be a real functional.*

In the case of a real functional, to every element $x \in X$ there corresponds, in accordance with a certain law, a real number which shall be denoted by (f, x).

We assume that the functional is linear and write

$$(f, \alpha_1 x_1 + \alpha_2 x_2) = \alpha_1 f(x_1) + \alpha_2 f(x_2). \tag{1.1.25}$$

The set of real or complex linear functionals, defined on X, forms, as we have shown, a vectorial space which is called the *algebraic dual* of X and is denoted by X^*; here the operations of addition and multiplication are defined by the formulae

$$(f_1 + f_2, x) = (f_1, x) + (f_2, x), \tag{1.1.26}$$

$$(\alpha f, x) = \alpha(f, x), \quad \alpha \in R. \tag{1.1.26'}$$

If X is a topological space and the linear functional is continuous then the set of all these linear continuous functionals constitute the *topological dual* of X and is denoted by X'. In that case, besides relation (1.1.25), we have also the relation

$$(f, x_n) \rightarrow (f, x) \text{ for } x_n \rightarrow x \tag{1.1.27}$$

in the sense of the convergence defined in X.

1.1.2.2. Concept of distribution

We shall now introduce the *concept of distribution* by the following.

Definition 1.1.5. *A linear and continuous functional defined on a topological vectorial space X is called a distribution.*

As the functional is real or complex, the *distribution* is said to be *real* or *complex*. In the following we shall deal only with real distributions; in that case the mapping (f, x) is a real number.

The different types of distributions depend on the choice of the topological space X. Throughout the following we shall use only distributions defined on the vectorial topological spaces K, S and K^m.

From the preceding subsection, it follows that distributions coincide with the topological duals of the spaces on which they are defined. Therefore, the distributions defined on the spaces K, S and K^m coincide with the duals K', S' and $(K^m)'$. Between the algebraic and the topological duals of these spaces we have the obvious relations

$$K^* \supset K', \quad S^* \supset S', \quad (K^m)^* \supset (K^m)'. \tag{1.1.28}$$

The distributions defined on the spaces K, S and K^m are said to be of *infinite order*, *temperate* and of *finite order* $p \leqslant m$, respectively (because $K^m \subset K^p$, $m > p$, leads to $(K^p)' \subset (K^m)'$).

In particular the distributions defined on the space $K^0 (m = 0)$ of the continuous functions with compact support are called *measures;* hence a measure is an element of $(K^0)'$.

Since between the spaces K and S we have the relation $S \supset K$, it follows that between their topological duals we have the relation $S' \subset K'$ which shows that the distributions defined on the space S constitute a subset of the distributions defined on the space K.

It has been shown that by introducing the operations of addition and multiplication with numbers, operations which have been defined by relations (1.1.26), (1.1.26′), the topological duals become vectorial spaces; the topological duals become topological vectorial spaces if we define a convergence in these spaces. Thus, convergence in the distribution space K' is defined by the relation

$$\lim_{n \to \infty} (f_n, \varphi) = (f, \varphi); \tag{1.1.29}$$

this means that the sequence of distributions $f_n \in K'$ tends to the distribution $f \in K'$ if relation (1.1.29) is satisfied, with $\varphi \in K$. Convergence in the space S' which is a subset of the space K', is defined in a similar way.

Let $f(x)$ be a function defined on the real axis R. This function is said to be *absolutely integrable* in a finite domain $[a, b]$ of R if the integral

$$\int_a^b |f(x)| \, \mathrm{d}x \tag{1.1.30}$$

exists.

If the function $f(x)$ is absolutely integrable in any finite domain of R then $f(x)$ is said to be *locally integrable*. We remark that an absolutely integrable function is also *integrable*, i.e. the integral

$$\int_a^b f(x) \, \mathrm{d}x \tag{1.1.31}$$

exists.

Locally integrable functions generate an important class of distributions. To show this we assume that to any fundamental function $\varphi(x) \in K$ there corresponds a real number

$$(f, \varphi) = \int_{-\infty}^{\infty} f(x)\varphi(x)\,\mathrm{d}x = \int_R f(x)\varphi(x)\,\mathrm{d}x = \int_{a'}^{b'} f(x)\varphi(x)\,\mathrm{d}x, \qquad (1.1.32)$$

where $f(x)$ is a locally integrable function and $[a', b']$ is the support of $\varphi(x)$. It may be seen easily that the functional thus defined is linear and continuous. The functional defined on the space K, by means of the locally integrable function $f(x)$, defines a *distribution on the space K*, a distribution which will be denoted also by $f(x)$, like the generating function. Such distributions are called *regular distributions* or distributions of the *function type*.

Denoting the distribution by the same symbol as the corresponding locally integrable function $f(x)$ should cause no confusion since, by definition, in the case of a distribution we cannot refer to its value at a certain point as in the case of functions. Denoting the distribution by $f(x)$ marks the function which generated it, as well as the independent variable $x \in R$.

Assuming that $f(x)$ is a locally integrable function we define the functional

$$\varphi \in K \to \int_R f(x)|\varphi(x)|\,\mathrm{d}x. \qquad (1.1.33)$$

Since

$$|\varphi_1 + \varphi_2| \leqslant |\varphi_1| + |\varphi_2|, \qquad (1.1.33')$$

we deduce that the functional considered although continuous, is not linear; hence, it does not define a distribution.

Distributions on the space S are defined similarly. In general, any locally integrable function $f(x)$ which satisfies the condition

$$|f(x)| \leqslant A|x|^k \text{ for } |x| \to \infty, \qquad (1.1.34)$$

where A is a positive constant and k is a constant, defines a *temperate distribution*. Such distributions are defined for instance by the function $\mathrm{e}^{-a|x|}$, $a > 0$, the polynomial $P(x)$, etc.

Although the distributions defined with the help of locally integrable functions represent a large class, other distributions exist which are not of the function type and which play an important part.

1.1.2.3. Dirac distribution

Thus, if to any function $\varphi(x) \in K$ we attach its value at the origin, i.e. $\varphi(0)$ we find that the respective functional is linear and continuous, hence it is a distribution which is not of the function type; this is the *Dirac distribution* which will be denoted by the symbol $\delta(x)$. Hence we can write

$$(\delta(x), \varphi(x)) = \varphi(0). \qquad (1.1.35)$$

If the fundamental functions $\varphi \in K$ are defined in R^n we have

$$(\delta(x_1, x_2, ..., x_n), \varphi(x_1, x_2, ..., x_n)) = \varphi(0, 0, ..., 0). \tag{1.1.35'}$$

Since the Dirac distribution, $\delta(x)$, is not of the function type, it will be a *singular* distribution, which means that we can write

$$\int_R \delta(x)\varphi(x)\,\mathrm{d}x = \varphi(0) \tag{1.1.36}$$

in a symbolic way only.

We remark that in order to define this distribution it is sufficient that $\varphi(x) \in K^0$, which means that it is a distribution of the zeroth order and hence a measure.

It has been shown (see section 1.1.2.1) that in the distribution space K', S' or $(K^m)'$, addition is defined by the relation

$$(f_1 + f_2, \varphi) = (f_1, \varphi) + (f_2, \varphi). \tag{1.1.37}$$

Since besides singular distributions there exist also regular distributions generated by locally integrable functions the multiplication of distributions with a real number must be such as to include as a particular case the multiplication of ordinary functions with numbers. Thus, the multiplication of distributions with a real number α is given by the relation

$$(\alpha f, \varphi) = \alpha(f, \varphi) = (f, \alpha\varphi). \tag{1.1.38}$$

Now let $\alpha(x)$ be a function of class C^∞, and $\varphi(x) \in K$. As we have shown in section 1.1.1.1, we have $\alpha(x)\varphi(x) \in K$. Hence, we may define the multiplication of a distribution by functions $\alpha(x)$ possessing derivatives of any order, using the formula

$$(\alpha(x)f(x), \varphi(x)) = (f(x), \alpha(x)\,\varphi(x)); \tag{1.1.39}$$

it is obvious that this is a generalization of the multiplication by real numbers, stated in relation (1.1.38).

In some cases it is not necessary that the function $\alpha(x)$ is of class C^∞ in order that the product with a distribution may exist. For example, if the function $\alpha(x)$ is continuous at the origin, the product $\alpha(x)\delta(x)$ exists since we have

$$(\alpha(x)\delta(x), \varphi(x)) = (\delta(x), \alpha(x)\varphi(x)) = \alpha(0)\varphi(0). \tag{1.1.40}$$

In general, a *translated distribution* $f(x - x_0)$ is defined by the relation

$$(f(x - x_0), \varphi(x)) = (f(x), \varphi(x + x_0)), \quad x, x_0 \in R^n. \tag{1.1.41}$$

In particular, for $\delta(x - x_0)$ we may write

$$(\delta(x - x_0), \varphi(x)) = \varphi(x_0), \quad x, x_0 \in R^n. \tag{1.1.41'}$$

For distributions subjected to a *homothetic transformation* with respect to the independent variable, we shall use by definition, the formula

$$(f(\alpha x), \varphi(x)) = |\alpha|^{-n}\left(f(x), \varphi\left(\frac{1}{\alpha}x\right)\right), \quad x \in R^n. \tag{1.1.42}$$

For $\alpha = -1$ we find the property of *symmetry* and we may write

$$(f(-x), \varphi(x)) = (f(x), \varphi(-x)), \quad x \in R^n. \tag{1.1.43}$$

In particular, for the distribution $\delta(-x)$ we have

$$(\delta(-x), \varphi(x)) = (\delta(x), \varphi(-x)) = \varphi(0), \tag{1.1.43'}$$

which means that $\delta(-x) = \delta(x)$, i.e. $\delta(x)$ is an *even* distribution with respect to the independent variables $x \in R^n$.

Also, from formula (1.1.42) we obtain

$$\delta(\alpha x) = \frac{1}{|\alpha|}\,\delta(x), \quad x \in R, \tag{1.1.44}$$

$$\delta(\alpha x + \beta) = \frac{1}{|\alpha|^n}\,\delta\left(x + \frac{\beta}{\alpha}\right), \quad x \in R^n, \tag{1.1.44'}$$

and in general, we have

$$\delta(\alpha_1 x_1, \alpha_2 x_2, ..., \alpha_n x_n) = \frac{1}{|\alpha_1|\,|\alpha_2|\,...\,|\alpha_n|}\,\delta(x_1, x_2, ..., x_n). \tag{1.1.44''}$$

The definitions of distributions subjected to linear transformations with respect to the independent variable have been introduced so as to include, as a particular case, the distributions of the function type.

1.1.2.4. Heaviside distribution

The equality of two distributions $f(x)$ *and* $g(x)$ *is defined by the relation*

$$(f, \varphi) = (g, \varphi) \text{ for any } \varphi \in K; \tag{1.1.45}$$

hence, if the equality (1.1.45) is valid for any φ, we may write

$$f = g. \tag{1.1.45'}$$

If the distributions f and g are generated by continuous functions $f(x)$ and $g(x)$, the equality (1.1.45') occurs in the usual sense, i.e. punctual, since in that case the distributions f and g coincide everywhere with the functions f and g.

In the case when the functions $f(x)$ and $g(x)$ are locally integrable and coincide almost everywhere, the distributions generated by them will be equal and the relation (1.1.45') is valid.

Let us consider on the real axis the functions

$$\theta(x) = \begin{cases} 1 \text{ for } x \geqslant 0 \\ 0 \text{ for } x < 0, \end{cases} \tag{1.1.46}$$

$$\theta_1(x) = \begin{cases} 1 \text{ for } x > 0 \\ 2 \text{ for } x = 0 \\ 0 \text{ for } x < 0. \end{cases} \tag{1.1.46'}$$

The function θ is called the *Heaviside function* or the *unit function* (Figure 1.6); it coincides everywhere with θ_1 except at the origin. The distributions generated by these functions will be regular and coincident, i.e. we have the relation

$$\theta(x) = \theta_1(x), \tag{1.1.47}$$

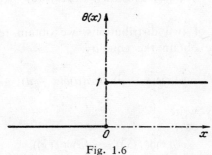

Fig. 1.6

since

$$\int_R \theta(x)\varphi(x)\,\mathrm{d}x = \int_R \theta_1(x)\varphi(x)\,\mathrm{d}x = \int_0^\infty \varphi(x)\,\mathrm{d}x, \tag{1.1.48}$$

i.e.

$$(\theta, \varphi) = (\theta_1, \varphi), \tag{1.1.48'}$$

for any $\varphi \in K$; the distribution generated by the Heaviside function will be called the *Heaviside distribution*.

The Heaviside function in R^3 is defined by

$$\theta(x_1, x_2, x_3) = \begin{cases} 1 \text{ for } x_1 \geqslant 0, \ x_2 \geqslant 0, \ x_3 \geqslant 0 \\ \\ 0 \text{ in the rest,} \end{cases} \tag{1.1.49}$$

and the distribution generated by it is defined by the relation

$$(\theta, \varphi) = \int_0^\infty \int_0^\infty \int^\infty \varphi(x_1, x_2, x_3) \, dx_1 \, dx_2 \, dx_3. \tag{1.1.49'}$$

It should be noted that the Heaviside distribution may be defined also on the fundamental space S, being thus a temperate distribution.

1.1.2.5. Other properties of distributions

If $\psi(x)$ is a function of class C^∞ we have the equality

$$\psi(x)\delta(x) = \psi(0)\delta(x); \tag{1.1.50}$$

indeed, we may write successively

$$(\psi(x)\delta(x), \ \varphi(x)) = (\delta(x), \ \psi(x)\varphi(x)) = \psi(0)\varphi(0) = (\psi(0)\delta(x), \ \varphi(x))$$

and from the equality of two distributions, we obtain relation (1.1.50).

By a translation we obtain the equality

$$\psi(x)\delta(x - a) = \psi(a)\delta(x - a); \tag{1.1.50'}$$

more generally we may write

$$\psi(x)\delta(f(x)) = \psi(a)\delta(f(x)), \tag{1.1.50''}$$

where $f(x)$ is a function of class C^∞, admitting one single simple root $x = a$.

In particular, considering $\psi(x) = x^n$, we have

$$x^n\delta(x) = 0, \tag{1.1.51}$$

which shows that the product of a distribution and a function possessing derivatives of any order may vanish without any of the factors being zero.

Another function which intervenes frequently in applications is

$$f(x) = \frac{1}{x} \text{ for } x \neq 0; \tag{1.1.52}$$

this function is not locally integrable, because the function is not integrable in the neighbourhood of the origin. The graph of this function (Figure 1.7) consists of

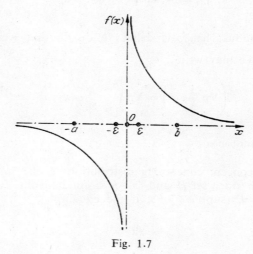

Fig. 1.7

the two branches of an equilateral hyperbola; the asymptotes are the co-ordinate axes.

The integral $\int_{-a}^{b} \frac{1}{x} \, dx$ with $a, b > 0$ does not exist in general. For this reason we shall introduce *the principal value of the integral in the sense of Cauchy*, and we shall write

$$\text{Vp} \int_{-a}^{b} \frac{1}{x} \, dx = \lim_{\varepsilon \to +0} \left(\int_{-a}^{-\varepsilon} \frac{1}{x} \, dx + \int_{\varepsilon}^{b} \frac{1}{x} \, dx \right), \quad \varepsilon > 0; \tag{1.1.53}$$

whence we have

$$\text{Vp} \int_{-a}^{b} \frac{1}{x} \, dx = \log \frac{b}{a}, \quad a, b > 0. \tag{1.1.53'}$$

In particular, we obtain

$$\text{Vp} \int_{-a}^{a} \frac{1}{x} \, dx = 0 \tag{1.1.54}$$

and

$$\text{Vp} \int_{-\infty}^{\infty} \frac{1}{x} \, dx = 0. \tag{1.1.54'}$$

It is easy to see that the functional $f(x)$ defined by the relation

$$(f, \varphi) = \text{Vp} \int_{-\infty}^{\infty} \frac{1}{x} \, \varphi(x) \, dx \tag{1.1.55}$$

is linear and continuous; hence it defines a distribution which is denoted by $\text{Vp} \dfrac{1}{x}$ or by $\dfrac{1}{x}$. We may write

$$\left(\text{Vp} \frac{1}{x}, \varphi \right) = \text{Vp} \int_{-\infty}^{\infty} \frac{1}{x} \, \varphi(x) \, dx. \tag{1.1.56}$$

1.1.2.6. Support of a distribution

Another important concept concerning distributions is the *support of a distribution*. Let there be an open set E and $\varphi \in K$ a fundamental function having the support included in E (supp $\varphi \subset E$); if the equality

$$(f, \varphi) = 0 \tag{1.1.57}$$

takes place for any φ having the support included in E, the distribution f is said to be null on the set E. Let us give the following

Definition 1.1.6. *The complement of the union of the open sets E on which a distribution f vanishes is said to be the support of this distribution (supp f). Therefore, the support of a distribution is a closed set.*

Although the concept of the value of a distribution at a point does not exist, we can nevertheless give a sense to the *concept of a null distribution at a point*. Let x_0 be a point and $\varphi(x) \in K$ a fundamental function having the support included in a neighbourhood of the point x_0. If relation (1.1.57) takes place for any φ having the support included in a neighbourhood of x_0, then the distribution is said to be zero in the neighbourhood of x_0, i.e. at x_0.

Now let us determine the support of the Dirac distribution $\delta(x)$; to this end we write

$$(\delta(x), \varphi(x)) = \varphi(0) = 0, \tag{1.1.58}$$

starting from relation (1.1.35); it has been assumed that the fundamental function $\varphi(x) \in K$ is such that its support is included in a neighbourhood of the point $x_0 \neq 0$. It follows that for any open set which does not contain the origin, the distribution $\delta(x)$ vanishes; hence, the support of this distribution is a point, namely the origin.

Distributions with a compact support constitute an important class of distributions.

When the support of a distribution f is contained in a set A, the distribution f is said to be *concentrated* on the set A. Thus, we may say that the Dirac distribution $\delta(x)$ is concentrated at the origin.

It can be shown that the distributions $f(x)$ of the type of function with compact support have the same support as the locally integrable functions $f(x)$ which have generated them.

We can compare distributions by using the concept of *positive distribution*. We now give the following

Definition 1.1.7. *A distribution f_1 is said to be at least equal to a distribution f_2 if for any $\varphi \geqslant 0$ belonging to the space K, we have*

$$(f_1, \varphi) \geqslant (f_2, \varphi) \qquad (1.1.59)$$

or

$$(f_1 - f_2, \varphi) \geqslant 0. \qquad (1.1.59')$$

Thus, the distribution $f_1 - f_2$ is positive.

For example, the Dirac distribution is a positive distribution because for any $\varphi(x) \geqslant 0$ belonging to the space K, we have

$$(\delta(x), \varphi(x)) = \varphi(0) \geqslant 0. \qquad (1.1.60)$$

If between the locally integrable functions $f(x)$ and $g(x)$ we have, almost everywhere, the inequality $f(x) \geqslant g(x)$, then the regular distributions too, generated by these functions, will satisfy the same relation, since

$$\int_R f\varphi \, dx \geqslant \int_R g\varphi \, dx, \quad \varphi \geqslant 0. \qquad (1.1.61)$$

We mention also the fact that the distribution $V_p \dfrac{1}{x}$ does not represent a positive distribution.

For $|\alpha| < 1$ we have $\delta(\alpha x) > \delta(x)$ since

$$(\delta(\alpha x), \varphi(x)) = \frac{1}{|\alpha|} \varphi(0) > \varphi(0)$$

and for $|\alpha| > 1$ we have $\delta(\alpha x) < \delta(x)$; for example we can write $\delta\left(\dfrac{1}{2} x\right) > \delta(x)$ and $\delta(2x) < \delta(x)$.

In many applications one can use the following

Theorem 1.1.3. *If $g(x)$ is a locally integrable function and α_i $(i = 1, 2, \ldots, n)$ are constants, the equality*

$$g(x) + \sum_{i=1}^{n} \alpha_i \delta(x - x_i) = 0 \qquad (1.1.62)$$

occurs if, and only if,

$$g(x) = 0, \tag{1.1.62'}$$

$$\alpha_i = 0 \qquad (i = 1, 2, ..., n). \tag{1.1.62''}$$

Indeed, considering the arbitrary fundamental function whose support does not contain the points x_i $(i = 1, 2, ..., n)$, the relation (1.1.62) leads to

$$(g, \varphi) = 0, \tag{1.1.63}$$

whence it follows that the function $g(x)$ vanishes almost everywhere, except eventually at the points x_i.

Let φ be a fundamental function such that its support contains the point x_i only. From (1.1.62) and (1.1.63) we obtain

$$\alpha_i \varphi(x_i) = 0, \tag{1.1.63'}$$

whence there follows $\alpha_i = 0$; proceeding likewise for $i = 1, 2, ..., n$, we obtain conditions (1.1.62''). Replacing these values in relation (1.1.62) we obtain condition (1.1.62') thus completing the proof of the theorem (the sufficiency of the conditions is obvious).

1.1.2.7. Change of variable in distributions

An important problem both from a theoretical and practical point of view is the change of variable in distributions.

We can now give the following

Theorem 1.1.4. *If $f_i(x_1, x_2, ..., x_n)$ $(i = 1, 2, ..., n)$ are functions possessing derivatives of any order, having only the root $x_0(x_1, x_2, ..., x_n)$ and such that the functional determinant* (taken in absolute value)

$$D(x_0) \equiv \left| \frac{\partial(f_1, f_2, ..., f_n)}{\partial(x_1, x_2, ..., x_n)} \right| \neq 0, \tag{1.1.64}$$

then we have the equality

$$\delta(f_1, f_2, ..., f_n) = \frac{1}{D(x_0)} \delta(x_1 - x_1^0, x_2 - x_2^0, ..., x_n - x_n^0). \tag{1.1.64'}$$

Indeed, we can write

$$\int_{R^n} \delta(f_1, f_2, ..., f_n) \varphi(f_1, f_2, ..., f_n) \, df_1 \, df_2 \, ... \, df_n = \varphi(0, 0, ..., 0). \tag{1.1.65}$$

Since

$$df_1\, df_2 \ldots df_n = \left| \frac{\partial(f_1, f_2, \ldots, f_n)}{\partial(x_1, x_2, \ldots, x_n)} \right| dx_1\, dx_2 \ldots dx_n = D(x)\, dx_1\, dx_2 \ldots dx_n, \quad (1.1.66)$$

relation (1.1.66) leads to

$$\int_{R^n} \delta(f_1, f_2, \ldots, f_n) D(x)\varphi(f_1, f_2, \ldots, f_n)\, dx_1\, dx_2 \ldots dx_n = \varphi(0, 0, \ldots, 0). \quad (1.1.65')$$

Introducing the fundamental functions

$$\psi(x_1, x_2, \ldots, x_n) = D(x)\varphi(f_1, f_2, \ldots, f_n), \quad (1.1.67)$$

relation (1.1.65′) becomes

$$\int_{R^n} \delta(f_1, f_2, \ldots, f_n)\psi(x_1, x_2, \ldots, x_n)\, dx_1\, dx_2 \ldots dx_n = \varphi(0, 0, \ldots, 0). \quad (1.1.65'')$$

Taking into account (1.1.67) and the fact that all the functions $f_i (i = 1, 2, \ldots, n)$ admit only the root x_0, we obtain the relation

$$\varphi(0, 0, \ldots, 0) = \frac{1}{D(x_0)} \psi(x_1^0, x_2^0, \ldots, x_n^0). \quad (1.1.68)$$

Relation (1.1.65″) becomes

$$\int_{R^n} \delta(f_1, f_2, \ldots, f_n)\, \psi(x_1, x_2, \ldots, x_n)\, dx_1\, dx_2 \ldots dx_n$$

$$= \frac{1}{D(x_0)} \psi(x_1^0, x_2^0, \ldots, x_n^0), \quad (1.1.65''')$$

whence we obtain relation (1.1.64′); thus the proof of the theorem is completed.

If the functions $f_i (i = 1, 2, \ldots, n)$ admit a certain number (finite or infinite) of simple roots x_j, then relation (1.1.64′) may be generalized in the form

$$\delta(f_1, f_2, \ldots, f_n) = \sum_j \frac{1}{D(x_j)} \delta(x - x_j), \quad x, x_j \in R^n. \quad (1.1.69)$$

For example, in the space R^3 we may write relation

$$\delta(\sin x_1, \sin x_2, \sin x_3) = \sum_{n_1} \sum_{n_2} \sum_{n_3} \delta(x_1 - n_1\pi, x_2 - n_2\pi, x_3 - n_3\pi), \quad (1.1.70)$$

where n_1, n_2, n_3 are integers; in particular, on the real axis R it follows that

$$\delta(\sin x) = \sum_n \delta(x - n\pi).\qquad(1.1.71)$$

The particular case of formula (1.1.69) for the distribution $\delta(f(x))$, where $f(x)$ is a function possessing derivatives of any order, having an arbitrary number of simple roots x_j, is given by the relation

$$\delta(f(x)) = \sum_j \frac{1}{|f'(x_j)|} \delta(x - x_j), \quad x \in R,\qquad(1.1.72)$$

where the summation is performed according to the roots of the equation $f(x)=0$.
For example we may write

$$\delta(x^2 - a^2) = \frac{1}{2|a|}[\delta(x - a) + \delta(x + a)];\qquad(1.1.73)$$

and, similarly, we obtain

$$\delta[(x - a)(x - b)] = \frac{1}{|b - a|}[\delta(x - a) + \delta(x - b)], \quad a \neq b.\qquad(1.1.74)$$

We assume that $\alpha(x)$ is a function of the class C^∞ and which does not vanish, then formula (1.1.72) gives

$$\delta[\alpha(x)(x - a)] = \frac{1}{|\alpha(a)|} \delta(x - a) = \frac{1}{|\alpha(x)|} \delta(x - a),\qquad(1.1.75)$$

where relation (1.1.50′) has been taken into account.
Starting from the general formula (1.1.69) we may write in the space R^2 the relation

$$\delta(x^2 - a^2, y) = \frac{1}{2|a|}[\delta(x - a, y) + \delta(x + a, y)].\qquad(1.1.76)$$

More generally, we may write

$$\delta[(x - a_1)(x - a_2), y - b]$$

$$= \frac{1}{|a_1 - a_2|}[\delta(x - a_1, y - b) + \delta(x - a_2, y - b)]\qquad(1.1.76')$$

and

$$\delta\left[\prod_{i=1}^n (x - a_i), y - b\right] = \sum_{i=1}^n \frac{\delta(x - a_i, y - b)}{\left|\prod_{\substack{j=1 \\ (j \neq i)}}^n (a_i - a_j)\right|}.\qquad(1.1.76'')$$

1.2. Differentiation of distributions.
Representative δ sequences

It is obvious that differential operators play an important part in applications; therefore, in this section we shall define such operators and point out their more important properties.

1.2.1. Differentiation of distributions

It should be noted from the outset that the efficiency of the theory of distributions is due also to the fact that distributions possess derivatives of any order which is not the case with ordinary functions.

The rule of differentiation must be such as to contain the differentiation of ordinary functions as a particular case. For this reason we proceed from distributions defined by differentiable functions.

1.2.1.1. Distributions of a simple variable

Let $f(x)$ be a function of class C^1 and $\varphi(x)$ a fundamental function belonging to the fundamental space K. We denote by $f'(x)$ the derivative of $f(x)$ and we may write

$$(f'(x), \varphi(x)) = \int_{-\infty}^{\infty} f'\varphi \, dx = \int_{-\infty}^{\infty} \varphi \, df = f(x)\varphi(x)\Big|_{-\infty}^{\infty} - \int_{-\infty}^{\infty} f(x)\varphi'(x) \, dx,$$

where we have applied integration by parts; remarking that the function $\varphi(x)$ is fundamental we have $\varphi(-\infty) = \varphi(\infty) = 0$ and we obtain the relation

$$(f', \varphi) = -(f, \varphi'), \tag{1.2.1}$$

which will be considered as the rule for differentiating distributions of a simple variable.

It is easy to see that this functional is linear; from convergence in the space K it follows that this functional is also continuous.

Taking into account the translation of distributions, one may verify that the derivative, considered as a limiting value of a certain ratio, is retained in the theory of distributions; but obviously, the limiting value must be considered in the sense of the theory of distributions.

Thus, if $f(x) \in K'$, we shall write

$$(\Delta f(x))_h = f(x + h) - f(x); \tag{1.2.2}$$

this distribution represents the variation of the distribution $f(x)$ corresponding to the number $h \in R$. Hence, for the *derivative of the distribution* $f(x)$ we may write

$$f'(x) = \lim_{h \to 0} \frac{(\Delta f(x))_h}{h} \, . \tag{1.2.3}$$

Based on the foregoing notion we may introduce also the *concept of the differential* of a distribution $f(x)$. Since the fundamental function $\varphi(x)$ is differentiable (belonging to the class C^∞), we may write the relation

$$\varphi(x - h) - \varphi(x) = -h\varphi'(x) - h\alpha(x, h), \tag{1.2.4}$$

where $\alpha(x, h)$ is also a fundamental function, the expression of which is

$$\alpha(x, h) = \begin{cases} \dfrac{-\varphi(x - h) + \varphi(x)}{h} - \varphi'(x) & \text{for } h \neq 0 \\[2mm] 0 & \text{for } h = 0; \end{cases} \tag{1.2.5}$$

the function $\alpha(x, h)$, thus defined, has the property

$$\lim_{h \to 0} \alpha(x, h) = \alpha(x, 0) = 0, \tag{1.2.6}$$

the convergence occurring in the sense of the space K.

We may write

$$((\Delta f(x))_h, \varphi(x)) = (f(x + h) - f(x), \varphi(x)) = (f(x), \varphi(x - h)) - (f(x), \varphi(x))$$

$$= (f(x), \varphi(x - h) - \varphi(x)) = (f(x), -h\varphi'(x)) + h(f(x), -\alpha(x, h)) ; \tag{1.2.7}$$

we remark that the first term of the right-hand member has the form

$$(f(x) - h\varphi'(x)) = (hf'(x), \varphi(x)), \tag{1.2.8}$$

where $f'(x)$ is the derivative of the distribution $f(x)$.

The second term of the right-hand-side of (1.2.7) defines a new distribution, because $\alpha(x, h)$ is a fundamental function (a function of class C^∞ with a compact support); the new distribution is denoted by $f_h(x)$ and is expressed by

$$(f_h(x), \varphi(x)) = (f(x), -\alpha(x, h)). \tag{1.2.9}$$

Taking into account (1.2.6) we obtain

$$\lim_{h \to 0} (f_h(x), \varphi(x)) = 0, \tag{1.2.10}$$

in the sense of convergence in K', for any $\varphi(x) \in K$; whence we deduce that

$$\lim_{h \to 0} f_h(x) = 0. \tag{1.2.10'}$$

Using relations (1.2.8) and (1.2.9), relation (1.2.7) leads to

$$((\Delta f(x))_h, \varphi(x)) = (hf'(x) + hf_h(x), \varphi(x)), \tag{1.2.7'}$$

whence

$$(\Delta f(x))_h = hf'(x) + hf_h(x), \tag{1.2.7''}$$

which represents a resolution of the variation of the distribution $f(x)$ into two distributions, the first of which is *linear* with respect to the real number h. Taking into account (1.2.10') it follows that the relation (1.2.7'') is similar to the known condition of differentiability of functions; therefore we may state that any distribution is differentiable.

By definition the differential of the distribution $f(x)$ is expressed by

$$df(x) = hf'(x). \tag{1.2.11}$$

In particular, considering the regular distribution $f(x) = x$, we may write

$$dx = h; \tag{1.2.12}$$

by dividing the expressions (1.2.11) and (1.2.12) term by term (which is always possible since h is a number) we obtain

$$f'(x) = \frac{df(x)}{dx}. \tag{1.2.13}$$

Thus, it might be said that the derivative of a distribution is equal to the ratio of the differential of the distribution and the differential of the identical distribution x.

We shall now extend formula (1.2.13) to the case of compound distributions. Let $f(u)$ be an arbitrary distribution and $u(x)$ a function belonging to class C^{∞}; by relation (1.2.7'') we may write

$$\Delta f(u) = \Delta u f'_u(u) + \Delta u f_{\Delta u}(u), \tag{1.2.14}$$

where f'_u is the derivative of the distribution $f(u)$, Δu is an arbitrary increment of u, and $f_{\Delta u}(u)$ is the distribution defined by the formula

$$(f_{\Delta u}(u), \varphi(u)) = (f(u), -\alpha(u, \Delta u)), \tag{1.2.15}$$

having the property

$$\lim_{\Delta u \to 0} f_{\Delta u}(u) = 0. \tag{1.2.16}$$

The function $\alpha(u, \Delta u)$ will be defined by

$$\alpha(u, \Delta u) = \begin{cases} \dfrac{-\varphi(u - \Delta u) + \varphi(u)}{\Delta u} - \varphi'_u(u) & \text{for } \Delta u \neq 0 \\[4mm] 0 & \text{for } \Delta u = 0, \end{cases} \tag{1.2.17}$$

being a fundamental function with respect to the variable u. If we replace this variable by the function $u(x)$ which possesses derivatives of any order, we obtain α as a function of the variable x; this function too, is fundamental with respect to x since $\varphi(x)$ is an arbitrary function which may be taken such that its support contains the values of $u(x)$.

If $h = \Delta x$ is an arbitrary increment of the variable x we may write

$$\lim_{\Delta x \to 0} f_{\Delta u}(u(x)) = 0. \tag{1.2.18}$$

Dividing relation (1.2.14) by the number $h = \Delta x$, which is always possible, we obtain

$$\frac{\Delta f(u(x))}{\Delta x} = \frac{\Delta u(x)}{\Delta x} f'_u + \frac{\Delta u(x)}{\Delta x} f_{\Delta u}(u(x)). \tag{1.2.14'}$$

Passing to the limit in the sense of distributions and taking into account relation (1.2.18), it follows that

$$\frac{df(u(x))}{dx} = f'_u u'_x(x); \tag{1.2.14''}$$

this is the *formula for the differentiation of compound distributions*, which is a generalization of the known formula for the differentiation of compound functions. The formula has a sense since u'_x is a function belonging to class C^∞ and therefore the product in the right-hand-side of the above expression exists.

Applying the differentiation formula (1.2.1) to the Heaviside distribution, $\theta(x)$, introduced in section 1.1.2.4, we obtain

$$\theta'(x) = \delta(x) \tag{1.2.19}$$

or more generally

$$\theta'(x - x_0) = \delta(x - x_0). \tag{1.2.19'}$$

Now let $f(x)$ be a function of class C^∞, which has no root and for which $f'(x) \neq 0$; in that case the function $f(x)$ is either positive or negative on R (see

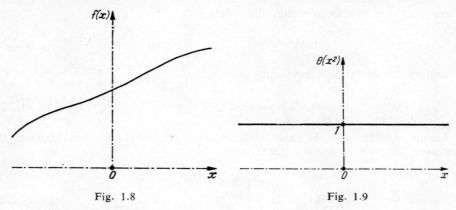

Fig. 1.8   Fig. 1.9

Figure 1.8). Denoting the Heaviside function by $\theta(f(x))$ we may write

$$\theta(f(x)) = \begin{cases} 1 \text{ for } f(x) \geqslant 0, \quad x \in R, \\ \\ 0 \text{ for } f(x) < 0, \quad x \in R. \end{cases} \tag{1.2.20}$$

Differentiating in the sense of the theory of distributions we obtain

$$f'(x)\delta(f(x)) = 0;$$

since $f'(x) \neq 0$ everywhere, we have

$$\delta(f(x)) = 0. \tag{1.2.21}$$

In particular the above conditions are satisfied by the function $f(x) = e^x$; we obtain

$$\delta(e^x) = 0. \tag{1.2.22}$$

Let us consider the case where $f(x) = x^2$; we may write (Figure 1.9)

$$\theta(x^2) = \begin{cases} 1 \text{ for } x \neq 0 \\ \\ 0 \text{ for } x = 0. \end{cases} \tag{1.2.23}$$

By differentiating this relation we obtain

$$x\delta(x^2) = 0; \qquad (1.2.24)$$

since the equation

$$xg(x) = 0, \qquad (1.2.25)$$

where $g(x)$ is a distribution, has the solution *

$$g(x) = c\delta(x), \qquad c = \text{const}, \qquad (1.2.25')$$

we deduce for $\delta(x^2)$ the expression

$$\delta(x^2) = c\delta(x), \qquad c = \text{const.} \qquad (1.2.26)$$

Let us consider also the case of the function $f(x) = x^2 - 1$; we introduce the Heaviside function as follows (see Figure 1.10)

$$\theta(x^2 - 1) = \begin{cases} 1 \text{ for } x^2 - 1 \geqslant 0 \\ \\ 0 \text{ for } x^2 - 1 < 0, \end{cases} \qquad (1.2.27)$$

which may be written in the form

$$\theta(x^2 - 1) = 1 - \theta(x + 1) + \theta(x - 1). \qquad (1.2.27')$$

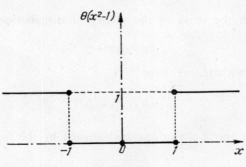

Fig. 1.10

Differentiating relation (1.2.27') we obtain

$$2x\,\delta(x^2 - 1) = \delta(x - 1) - \delta(x + 1); \qquad (1.2.28)$$

---

* See theorem 1.2.2, section 1.2.1.5.

whence

$$\delta(x^2 - 1) = \frac{1}{2x}[\delta(x - 1) - \delta(x + 1)] = \frac{1}{2}[\delta(x - 1) + \delta(x + 1)], \quad (1.2.28')$$

where we have taken relation (1.1.73) into account.

An important property, easily verified, is the following: *If $\psi(x)$ is a function of class C^∞ and $f(x)$ is a distribution we may write*

$$(\psi f)' = \psi' f + \psi f'. \tag{1.2.29}$$

The derivatives and the differentials of higher order are obtained similarly; thus, for *derivatives of the nth* order we apply the formula

$$(f^{(n)}(x), \varphi(x)) = (-1)^n (f(x), \varphi^{(n)}(x)). \tag{1.2.30}$$

In particular, we obtain

$$(\delta^{(n)}(x), \varphi(x)) = (-1)^n (\delta(x), \varphi^{(n)}(x)) = (-1)^n \varphi^{(n)}(0); \tag{1.2.31}$$

we have also

$$(\delta^{(n)}(x - a), \varphi(x)) = (-1)^n \varphi^{(n)}(a), \tag{1.2.31'}$$

which is obtained through a translation.

1.2.1.2. Distributions of several variables

In the case of a distribution of several variables $f(x_1, x_2, \ldots, x_n)$ we may write

$$\left(\frac{\partial}{\partial x_i} f(x_1, x_2, \ldots, x_n), \varphi(x_1, x_2, \ldots, x_n) \right)$$

$$= -\left(f(x_1, x_2, \ldots, x_n), \frac{\partial}{\partial x_i} \varphi(x_1, x_2, \ldots, x_n) \right); \tag{1.2.32}$$

from which we have the property

$$\frac{\partial^2 f}{\partial x_i \, \partial x_j} = \frac{\partial^2 f}{\partial x_j \, \partial x_i} \qquad (i, j = 1, 2, \ldots, n), \tag{1.2.33}$$

which shows that in the case of distributions the derivatives are independent of the sequence in which differentiation is performed.

For *derivatives of higher order* we may write the formula

$$\left(\frac{\partial^m f(x_1, x_2, \ldots, x_n)}{\partial x_1^{k_1} \partial x_2^{k_2} \ldots \partial x_n^{k_n}}, \varphi(x_1, x_2, \ldots, x_n)\right)$$

$$= (-1)^m \left(f(x_1, x_2, \ldots, x_n), \frac{\partial^m \varphi(x_1, x_2, \ldots, x_n)}{\partial x_1^{k_1} \partial x_2^{k_2} \ldots \partial x_n^{k_n}}\right)$$

$$(k_1 + k_2 + \ldots + k_n = m), \qquad (1.2.34)$$

with the property

$$\operatorname{supp} \frac{\partial^m f(x_1, x_2, \ldots, x_n)}{\partial x_1^{k_1} \partial x_2^{k_2} \ldots \partial x_n^{k_n}} \subset \operatorname{supp} f(x_1, x_2, \ldots, x_n)$$

$$(k_1 + k_2 + \ldots + k_n = m). \qquad (1.2.34')$$

The Heaviside function in R^2 leads to

$$\left(\frac{\partial^2 \theta(x, y)}{\partial x \, \partial y}, \varphi(x, y)\right) = \left(\theta(x, y), \frac{\partial^2 \varphi(x, y)}{\partial x \, \partial y}\right) = \int_0^\infty \int_0^\infty \frac{\partial^2 \varphi(x, y)}{\partial x \, \partial y} \, dx \, dy$$

$$= \int_0^\infty dx \int_0^\infty \frac{\partial}{\partial y}\left(\frac{\partial \varphi}{\partial x}\right) dy = \int_0^\infty dx \left[\frac{\partial}{\partial x} \varphi(x, y)\right]_0^\infty = -\int_0^\infty \varphi_x'(x, 0) \, dx$$

$$= -\left[\varphi(x, 0)\right]_0^\infty = \varphi(0, 0) = (\delta(x, y), \varphi(x, y)),$$

whence

$$\frac{\partial^2 \theta(x, y)}{\partial x \, \partial y} = \delta(x, y). \qquad (1.2.35)$$

Similarly, we obtain in R^3

$$\frac{\partial^3 \theta(x, y, z)}{\partial x \, \partial y \, \partial z} = \delta(x, y, z). \qquad (1.2.35')$$

We remark that the formula (1.2.14″) for the differentiation of compound distributions extends to the case of distributions of several variables.

Now let $f(x, y) = xy$, and $\theta(f(x, y))$ the Heaviside function in R^2. We may write

$$\theta(xy) = \begin{cases} 1 \text{ for } xy \geqslant 0 \\ \\ 0 \text{ for } xy < 0; \end{cases} \qquad (1.2.36)$$

we remark that this function may be expressed in the form

$$\theta(xy) = \theta(x, y) + \theta(-x, -y). \qquad (1.2.36')$$

By differentiating this relation we obtain

$$\frac{\partial^2 \theta(xy)}{\partial x \, \partial y} = \delta(x, y) + \delta(-x, -y) = 2\delta(x, y), \tag{1.2.37}$$

where we have taken into account the properties of parity of the Dirac distribution; applying the formula for the differentiation of compound distributions we may write the relation

$$2\delta(x, y) = \delta(xy) + xy\delta'(xy). \tag{1.2.37'}$$

1.2.1.3. Derivatives in the sense of the theory of distributions.
      Derivatives in the usual sense

Let $f(x)$ be a function of class C^1 everywhere except at the point x_0 where the function has a discontinuity of the first species. We shall denote by s_0 the *jump of the function* at the point x_0; the jump is defined by the relation

$$s_0 = f(x_0 + 0) - f(x_0 - 0). \tag{1.2.38}$$

Also, we denote by $f(x)$ the *derivative* of the function $f(x)$ *in the sense of the theory of distributions* and by $\tilde{f}'(x)$ the *derivative* of this function *in the usual sense* wherever the derivative exists. We may write

$$(f'(x), \varphi(x)) = (f(x), -\varphi'(x))$$

$$= -\lim_{\varepsilon \to +0} \left[\int_{-\infty}^{x_0-\varepsilon} f(x) \, d\varphi(x) + \int_{x_0+\varepsilon}^{\infty} f(x) \, d\varphi(x) \right], \quad \varepsilon > 0.$$

By integrating by parts and passing to the limit, we obtain

$$(f'(x), \varphi(x)) = -\lim_{\varepsilon \to +0} \left[f\varphi \Big|_{-\infty}^{x_0-\varepsilon} + f\varphi \Big|_{x_0+\varepsilon}^{\infty} - \int_{-\infty}^{x_0-\varepsilon} \tilde{f}'\varphi \, dx - \int_{x_0+\varepsilon}^{\infty} \tilde{f}'\varphi \, dx \right]$$

$$= s_0\varphi(x_0) + (\tilde{f}', \varphi) = (\tilde{f}'(x) + s_0\delta(x - x_0), \varphi(x)),$$

whence, from the equality of two distributions, it follows that

$$f'(x) = \tilde{f}'(x) + s_0\delta(x - x_0). \tag{1.2.39}$$

If the function $f(x)$ is everywhere of the class C^1 except for the points x_i ($i = 1, 2, \ldots, n$) where it has discontinuities of the first species and denoting by s_i

the jump of the function at the point x_i, we obtain, by a similar procedure, a more general formula, namely

$$f'(x) = \tilde{f}'(x) + \sum_{i=1}^{n} s_i \delta(x - x_i). \tag{1.2.40}$$

It is important to note that if the function $f(x)$ is *continuous at the point* x_0, hence if the jump is null, formula (1.2.35) becomes

$$f'(x) = \tilde{f}'(x) \tag{1.2.41}$$

i.e. the derivative in the sense of the theory of distributions will coincide with the derivative in the usual sense.

For example, let us consider the modulus function (Figure 1.11a)

$$f(x) = |x|. \tag{1.2.42}$$

It is obvious that the function is continuous everywhere; it is also differentiable everywhere, except at the origin where we may write

$$\tilde{f}'(0 - 0) = -1, \quad \tilde{f}'(0 + 0) = 1. \tag{1.2.43}$$

By the formula (1.2.41) we have

$$|x|' = |\tilde{x}|' = \begin{cases} -1 \text{ for } x < 0 \\ \\ 1 \text{ for } x > 0 \end{cases} = \operatorname{sgn} x, \tag{1.2.44}$$

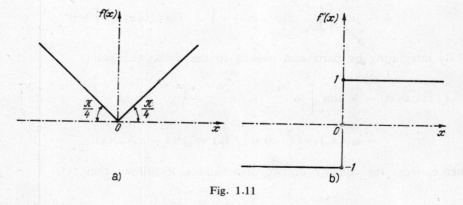

a)   b)

Fig. 1.11

therefore the derivative in the sense of the theory of distributions coincides with the derivative in the usual sense; the graph of the derivative is shown in Figure 1.11b.

We also note that the function $|\tilde{x}|'$ is differentiable, the derivative being zero everywhere except at the origin; on the other hand, the jump of the function $|\tilde{x}|'$ at the origin is $s_0 = 2$. Applying formula (1.2.39) we obtain

$$|x|'' = 2\delta(x). \tag{1.2.45}$$

Applying formula (1.2.30) to the function $f(x) = \theta(x) \cos x$, it follows that

$$[\theta(x) \cos x]' = -\theta(x) \sin x + \delta(x). \tag{1.2.46}$$

This may be obtained also by applying the formula (1.2.31) for the differentiation of a product; indeed we have

$$[\theta(x) \cos x]' = \theta'(x) \cos x - \theta(x) \sin x = \delta(x) \cos x - \theta(x) \sin x$$

and

$$\delta(x) \cos x = \delta(x) \cos 0 = \delta(x).$$

Let us now assume that function $f(x)$ belongs almost everywhere to the class C^n, except the point x_0 where both the function $f(x)$ and its derivatives in the usual sense up to and including the $(n-1)$th order, have discontinuities of the first species.

We denote by $s_k (k = 0, 1, 2, \ldots, n-1)$ the jump of the function $\tilde{f}^{(k)}(x)$ at the point x_0. Differentiating successively relation (1.2.39) we obtain the formula

$$f^{(n)}(x) = \tilde{f}^{(n)}(x) + \sum_{k=0}^{n-1} s_k \delta^{(n-k-1)}(x - x_0). \tag{1.2.47}$$

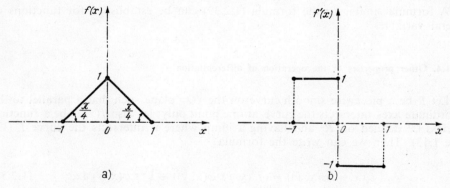

Fig. 1.12

For example, let us consider the function (Figure 1.12a)

$$f(x) = \begin{cases} 0 & \text{for } |x| \geqslant 1 \\ 1 + x & \text{for } -1 \leqslant x \leqslant 0 \\ 1 - x & \text{for } 0 \leqslant x \leqslant 1. \end{cases} \tag{1.2.48}$$

The derivative of the corresponding distribution is (Figure 1.12b)

$$f'(x) = \begin{cases} 0 \text{ for } |x| > 1 \\ 1 \text{ for } -1 < x < 0 \\ -1 \text{ for } 0 < x < 1; \end{cases} \tag{1.2.49}$$

we remark that we can write

$$f'(x) = \theta(x + 1) - 2\theta(x) + \theta(x - 1), \tag{1.2.49'}$$

by introducing the Heaviside distribution.

The derivative of the second order is

$$f''(x) = \delta(x + 1) - 2\delta(x) + \delta(x - 1) \tag{1.2.49''}$$

and the derivative of the third order is

$$f'''(x) = \delta'(x + 1) - 2\delta'(x) + \delta'(x - 1), \tag{1.2.49'''}$$

because $\tilde{f}''(x) = \tilde{f}'''(x) = 0$.

A formula similar to the formula (1.2.39) can be established for functions of several variables.

1.2.1.4. Other properties of the operation of differentiation

Let Γ be a piecewise smooth curve on the xOy plane such that a parallel to the co-ordinate axes intersects the curve at one point only; let also $f(x, y)$ be a function of class C^1 defined on R^2 and having a jump where it intersects the curve Γ (Figure 1.13). Then we can write the formula

$$(f'_y(x, y), \varphi(x, y)) = (\tilde{f}'_y(x, y), \varphi(x, y)) + \int_\Gamma s_y \varphi(x, y)\, \mathrm{d}x, \tag{1.2.50}$$

where s_y represents the jump of the function $f(x, y)$ when it intersects the curve Γ in the positive direction of the Oy axis; the jump is expressed by

$$s_y = f(x_\Gamma, y_\Gamma + 0) - f(x_\Gamma, y_\Gamma - 0), \tag{1.2.50'}$$

where x_Γ and y_Γ are the co-ordinates of a point of the curve Γ.

Indeed we can write

$$(f_y', \varphi) = - (f, \varphi_y') = - \iint_{R^2} f\varphi_y' \, dx \, dy = - \int_{-\infty}^{\infty} dx \left(\int_{-\infty}^{\infty} f\varphi_y' \, dy \right).$$

We remark that

$$\int_{-\infty}^{\infty} f\varphi_y' \, dy = \lim_{\varepsilon \to +0} \left(\int_{-\infty}^{y-\varepsilon} f \, d_y \varphi + \int_{y+\varepsilon}^{\infty} f \, d_y \varphi \right), \quad \varepsilon > 0;$$

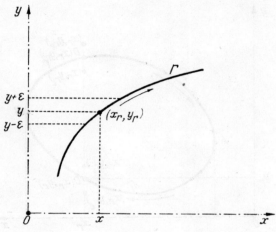

Fig. 1.13

integrating by parts and passing to the limit we obtain

$$\int_{-\infty}^{\infty} f\varphi_y' \, dy = \lim_{\varepsilon \to +0} \left(f\varphi \Big|_{-\infty}^{y-\varepsilon} + f\varphi \Big|_{y+\varepsilon}^{\infty} - \int_{-\infty}^{y-\varepsilon} \varphi \tilde{f}_y' \, dy - \int_{y+\varepsilon}^{\infty} \varphi \tilde{f}_y' \, dy \right)$$

$$= - s_y \varphi(x, y) - \int_{-\infty}^{\infty} \varphi f_y' \, dy.$$

Finally, we obtain

$$(f_y', \varphi) = \int_{-\infty}^{\infty} s_y \varphi(x, y) \, dx + \iint_{R^2} \tilde{f}_y' \, \varphi(x, y) \, dx \, dy;$$

since in the first integral of the right-hand-side, the function $\varphi(x, y)$ is taken on the curve Γ, it follows that this integral is a curvilinear integral and we obtain formula (1.2.50), where $\tilde{f}_y' = \dfrac{\tilde{\partial} f}{\partial y}$ is the partial derivative of the function in the usual sense, and $f_y' = \dfrac{\partial f}{\partial y}$ represents the partial derivative of the same function in the sense of the theory of distributions.

Similarly, we can establish the relation

$$(f_x'(x, y), \varphi(x, y)) = (\tilde{f_x'}(x, y), \varphi(x, y)) + \int_\Gamma s_x \varphi(x, y)\, \mathrm{d}y, \qquad (1.2.51)$$

where the notions used have the same significance.

In establishing formulae (1.2.50) and (1.2.51), it has been assumed that the curve Γ is open; we shall consider now the case where the curve is closed.

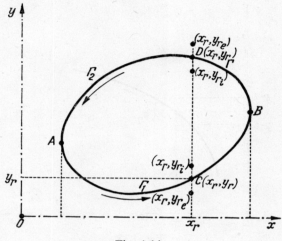

Fig. 1.14

Let $D \in R^2$ be a domain bounded by the piecewise smooth curve Γ (Figure 1.14). We assume that the function $f(x, y)$ defined on R^2 has a jump when it crosses this curve.

We denote by Γ_1 and Γ_2 the two branches of the closed curve $\Gamma = \Gamma_1 \bigcup \Gamma_2$. Applying formula (1.2.50) to the two branches we obtain

$$\left(\frac{\partial f}{\partial y}, \varphi\right) = \left(\frac{\tilde{\partial f}}{\partial y}, \varphi\right) + \int_{\widehat{ACB}} s_y \varphi \, \mathrm{d}x + \int_{\widehat{ADB}} s_y \varphi \, \mathrm{d}x;$$

we remark that Γ_1 is run over clockwise and Γ_2 counter clockwise. Assuming that the branch Γ_2 is also run over clockwise, hence that the whole curve is run over clockwise, we may write

$$(f_y'(x, y), \varphi(x, y)) = (\tilde{f_y'}(x, y), \varphi(x, y)) + \int_{\Gamma_1} s_y \varphi(x, y) \, \mathrm{d}x - \int_{\Gamma_2} s_y \varphi(x, y) \, \mathrm{d}x,$$

hence we have the formula

$$(f_y'(x, y), \varphi(x, y)) = (\tilde{f_y'}(x, y), \varphi(x, y)) + \int_\Gamma (\Delta f(x, y))_y \, \varphi(x, y) \, \mathrm{d}x, \qquad (1.2.52)$$

where $(\Delta f)_y$ represents the jump of the function f when it traverses the curve Γ in the direction of the Oy axis; the jump is expressed by

$$(\Delta f(x, y))_y = f(x_\Gamma, y_{\Gamma_i}) - f(x_\Gamma, y_{\Gamma_e}), \tag{1.2.52'}$$

where the index i corresponds to the interior of the domain whose boundary is the curve Γ, and the index e corresponds to the exterior of that domain.

We can also establish the relationship

$$(f_x'(x, y), \varphi(x, y)) = (\tilde{f}_x'(x, y), \varphi(x, y)) + \int_\Gamma (\Delta f(x, y))_x \, \varphi(x, y) \, dy, \tag{1.2.53}$$

where the notations used have similar significance.

The formulae (1.2.50), ..., (1.2.53), may be extended to the space R^n.

Thus, formula (1.2.50) may be extended into R^3 in the form

$$(f_y'(x, y, z), \varphi(x, y, z)) = (\tilde{f}_y'(x, y, z), \varphi(x, y, z)) + \iint_S s_y \varphi(x, y, z) \, dz \, dx, \tag{1.2.54}$$

where S is a piecewise smooth surface such that a parallel to the co-ordinate axes intersects it in a single point.

Likewise, the analogue of formula (1.2.52) is written in the form

$$(f_y'(x, y, z), \varphi(x, y, z)) = (\tilde{f}_y'(x, y, z), \varphi(x, y, z))$$

$$+ \iint_S (\Delta f(x, y, z))_y \, \varphi(x, y, z) \, dz \, dx \tag{1.2.55}$$

where S is a piecewise smooth surface which is the boundary of a domain in R^3.

We remark that the jumps which figure in the formulae concerning open curves or surfaces are computed in the positive direction of the respective axis; if the direction is changed then the jumps must be taken with opposite signs. In the case of closed curves or surfaces we consider the jumps directed towards the interior of the domain of which they are the boundary.

If $\theta(x, y)$ is the Heaviside function in R^2 we may write

$$\left(\frac{\partial}{\partial y} \theta(x, y), \varphi(x, y) \right) = \int_{y=0} s_y \varphi(x, y) \, dx, \tag{1.2.56}$$

since $\dfrac{\partial \tilde{\theta}}{\partial y} = 0$; the curve of discontinuity is the half-axis Ox so that $s_y = 1$ and relation (1.2.56) becomes

$$\left(\frac{\partial}{\partial y} \theta(x, y), \varphi(x, y) \right) = \int_0^\infty \varphi(x, 0) \, dx. \tag{1.2.56'}$$

Similarly, we obtain

$$\left(\frac{\partial}{\partial x}\,\theta(x, y),\,\varphi(x, y)\right) = \int_0^\infty \varphi(0, y)\,\mathrm{d}y. \qquad (1.2.56'')$$

For example, let us consider the function

$$f(x, y) = \begin{cases} 1 \text{ for } 0 \leqslant y \leqslant 2 \\[2mm] 0 \text{ in the rest,} \end{cases} \qquad (1.2.57)$$

for which the curves of discontinuity are $y = 0$ and $y = 2$.

Applying formula (1.2.50) we obtain

$$\left(\frac{\partial}{\partial y}\,f(x, y),\,\varphi(x, y)\right) = \int_{-\infty}^\infty \left[\varphi(x, 0) - \varphi(x, 2)\right]\mathrm{d}x, \qquad (1.2.58)$$

since, when the curve transverses the straight line $y = 0$ the jump is equal to unity and when it transverses the straight line $y = 2$ the jump is equal to (-1), or since the jump — in the last case — is equal to unity too, if this line is run in the opposite direction; the derivative $\dfrac{\tilde{\partial}f}{\partial y}$ vanishes.

For the derivative with respect to the other variable we may write

$$\left(\frac{\partial}{\partial x}\,f(x, y),\,\varphi(x, y)\right) = 0 \qquad (1.2.59)$$

or

$$\frac{\partial}{\partial x}\,f(x, y) = 0, \qquad (1.2.59')$$

since on the direction of the Ox axis the function has no jump, while $\dfrac{\tilde{\partial}f}{\partial x} = 0$.

In the following we shall use an extension of theorem 1.1.3, namely,

Theorem 1.2.1. *If $A(x)$ is a continuous function and B_j^k ($j = 1, 2, \ldots, m$; $k = 0, 1, 2, \ldots, n$) are constants, the equality*

$$A(x) + \sum_{k=0}^n \sum_{j=1}^m B_j^k \delta^{(k)}(x - x_j) = 0 \qquad (1.2.60)$$

implies

$$A(x) = 0, \qquad (1.2.61)$$

$$B_j^k = 0. \qquad (1.2.61')$$

For the proof let us consider a fundamental function $\varphi(x) \in K$, such that its support contains none of the points x_j; we have

$$(\delta^{(k)}(x - x_j), \varphi(x)) = 0 \qquad (k = 0, 1, 2, ..., n)$$

and, by the equality (1.2.60) we obtain

$$(A(x), \varphi(x)) = 0. \tag{1.2.62}$$

Whence it follows that $A(x) = 0$ almost everywhere, except eventually at the points x_j $(j = 1, 2, ..., m)$. Therefore the continuity of the function $A(x)$ necessarily implies condition (1.2.61).

In that case the equality (1.2.60) takes the form

$$\sum_{k=0}^{n} \sum_{j=1}^{m} B_j^k \delta^{(k)}(x - x_j) = 0. \tag{1.2.60'}$$

We now select a fundamental function $\varphi(x)$ such that its support contains the point x_j only; by relation (1.2.60') we have

$$\sum_{k=0}^{n} (-1)^k B_j^k \varphi^{(k)}(x_j) = 0 \qquad (j = 1, 2, ..., m). \tag{1.2.63}$$

If the fundamental function $\varphi(x)$ is equal to unity in the neighbourhood of the point x_j, we may write

$$\varphi(x_j) = 1, \varphi^{(k)}(x_j) = 0 \qquad (k = 1, 2, ..., n);$$

by relation (1.2.63) we obtain

$$B_j^0 = 0 \qquad (j = 1, 2, ..., m). \tag{1.2.64}$$

and relation (1.2.63) becomes

$$\sum_{k=1}^{n} (-1)^k B_j^k \varphi^{(k)}(x_j) = 0 \qquad (j = 1, 2, ..., m). \tag{1.2.63'}$$

We now select a fundamental function $\varphi(x)$ such that in the neighbourhood of the point x_j we should have $\varphi'(x_j) = 1$; in other words we can write

$$\varphi'(x_j) = 1, \varphi^{(k)}(x_j) = 0 \qquad (k = 2, 3, ..., n)$$

and the relation (1.2.63') leads to

$$B_j^1 = 0 \qquad (j = 1, 2, ..., m). \tag{1.2.64'}$$

Proceeding similarly, we necessarily obtain the conditions (1.2.61') thus proving the theorem (the sufficiency of the conditions is obvious).

A last property which should be mentioned is the following:

If the derivative of a distribution is null then the distribution is a constant.

1.2.1.5. Structure of distributions with a point support

In connection with the structure of distributions with a point support we shall prove the following

Theorem 1.2.2. *The necessary and sufficient condition that a distribution $f(x)$ of a single variable satisfies the equation*

$$P(x)f(x) = 0, \tag{1.2.65}$$

where $P(x)$ is a polynomial, is that $f(x)$ be expressed by

$$f(x) = \sum_{i=1}^{s} c_i \delta(x - x_i) + \sum_{k=1}^{r} \sum_{j=1}^{m_k} c_j^k \delta^{(j-1)}(x - x_k'). \tag{1.2.66}$$

Here x_i $(i = 1, 2, \ldots, s)$ represent the simple roots of the polynomial $P(x)$ and x_k' $(k = 1, 2, \ldots, r)$ the multiple roots whose order of multiplicity is m_k; the magnitudes c_i and c_j^k are constants, and δ is the Dirac distribution.

We shall prove the necessity of relation (1.2.66); its sufficiency is obvious.

The polynomial $P(x)$ may be written in the form

$$P(x) = P_1(x) \, Q(x), \tag{1.2.67}$$

where $Q(x)$ has only complex roots and $P_1(x)$ is expressed by

$$P_1(x) = (x - x_1)(x - x_2) \ldots (x - x_s)(x - x_1')^{m_1}(x - x_2')^{m_2} \ldots (x - x_r')^{m_r}; \tag{1.2.67'}$$

in that case equation (1.2.65) is equivalent to the equation

$$P_1(x)f(x) = 0. \tag{1.2.65'}$$

Denoting a fundamental function by $\varphi(x)$ we may write

$$(P_1 f, \varphi) = (f, P_1 \varphi) = 0. \tag{1.2.68}$$

Introducing the notation

$$\chi(x) = P_1(x)\varphi(x), \tag{1.2.69}$$

it is evident that the new function $\chi(x)$ is a fundamental function too, having the properties

$$\chi(x_i) = 0 \qquad (i = 1, 2, ..., s),$$

$$\chi^{(j-1)}(x'_k) = 0 \qquad (k = 1, 2,..., r; \ j = 1, 2, ..., m_k). \tag{1.2.69'}$$

The general form of the fundamental functions which verify these properties is

$$\psi(x) = \chi(x) + \sum_{k=1}^{r} \sum_{j=1}^{m_k} \lambda_{j-1}^{k}(x - x'_k)^{j-1}\theta_k(x) + \sum_{i=1}^{s} \gamma_i\beta_i(x), \tag{1.2.70}$$

where λ_{j-1}^{k}, γ_i are constants and $\theta_k(x)$, $\beta_i(x)$ are disjoint functions, namely

$$\theta_k(x) = \begin{cases} 1 \ \text{for} \ x \in V(x'_k), \quad V(x'_m) \bigcap V(x'_n) = \emptyset, \quad m \neq n, \\ \\ 0 \ \text{for} \ x \notin V(x'_k) \quad (m, n, k = 1, 2, ..., r), \end{cases} \tag{1.2.71}$$

$$\beta_i(x) = \begin{cases} 1 \ \text{for} \ x \in V(x_i), \quad V(x_i) \bigcap V(x_j) = \emptyset, \quad i \neq j, \\ \\ 0 \ \text{for} \ x \notin V(x_i) \quad (i, j = 1, 2, ..., s). \end{cases} \tag{1.2.71'}$$

We remark that these functions satisfy the relations

$$\theta_k^{(h)}(x'_j) = \begin{cases} 1 \ \text{for} \ k = j \quad (h = 0), \\ \\ 0 \ \text{for} \ k \neq j \quad (h = 1, 2, ...), \end{cases} \tag{1.2.72}$$

$$\beta_i^{(h)}(x_l) = \begin{cases} 1 \ \text{for} \ i = l \quad (h = 0), \\ \\ 0 \ \text{for} \ i \neq l \quad (h = 1, 2, ...). \end{cases} \tag{1.2.72'}$$

Taking into account relations $(1.2.71) - (1.2.72')$ we may write

$$\psi(x_i) = \chi(x_i) + \gamma_i = \gamma_i \qquad (i = 1, 2, ..., s), \tag{1.2.73}$$

$$\psi^{(m)}(x'_k) = m! \, \lambda_m^k \qquad (k = 1, 2, ..., r; \ m = 0, 1, 2, ..., m_k-1). \tag{1.2.73'}$$

It follows that

$$(f, \psi) = (f, \chi) + \sum_{k=1}^{r} \sum_{j=1}^{m_k} \lambda_{j-1}^{k}(f, (x - x'_k)^{j-1}\theta_k) + \sum_{i=1}^{s} \gamma_i(f, \beta_i). \tag{1.2.74}$$

Taking into account relation (1.2.68), (1.2.69) we may write relation (1.2.74) in the form

$$(f, \psi) = \sum_{k=1}^{r} \sum_{j=1}^{m_k} \lambda_{j-1}^{k}(f, (x - x_k')^{j-1}\theta_k) + \sum_{i=1}^{s} \gamma_i(f, \beta_i). \tag{1.2.74'}$$

Since the functions $\theta_k(x)$ and $\beta_i(x)$ are fixed we deduce

$$(f, \beta_i) = c_i, \tag{1.2.75}$$

$$(f, (x - x_k')^{j-1}\theta_k) = b_j^k, \tag{1.2.75'}$$

where c_i and b_j^k are constants.

Applying formula (1.2.31') we may write

$$(\delta^{(j-1)}(x - x_k'), \psi(x)) = (-1)^{j-1}\psi^{(j-1)}(x_k'); \tag{1.2.76}$$

in that case relation (1.2.74') leads to

$$(f, \psi) = \sum_{k=1}^{r} \sum_{i=1}^{m_k} \left(\frac{(-1)^{j-1}b_j^k}{(j-1)!} \delta^{(j-1)}(x - x_k'), \psi \right) + \sum_{i=1}^{s} (c_i\delta(x - x_i), \psi). \tag{1.2.74''}$$

Using the relation

$$c_j^k = (-1)^{j-1} \frac{b_j^k}{(j-1)!}, \tag{1.2.77}$$

we find exactly relation (1.2.66)

We shall now prove

Theorem 1.2.3. *If the support of the distribution $f(x)$ consists of the points* x *($i = 1, 2, \ldots, n$), then $f(x)$ has the form*

$$f(x) = \sum_{i=1}^{n} \sum_{j=0}^{m_i} c_j^i \delta^{(j)}(x - x_i), \tag{1.2.78}$$

where c_j^i and m_i are constants, and δ is the Dirac distribution.

Indeed, for any fundamental function $\psi(x)$ which vanishes at the points x_i ($i = 1, 2, \ldots, n$) together with its derivatives up to and including the mth order, we have

$$(f, \psi) = 0. \tag{1.2.79}$$

Denoting by $\varphi(x)$ an arbitrary fundamental function, we may write

$$\psi(x) = (x - x_1)^{m_1}(x - x_2)^{m_2} \ldots (x - x_n)^{m_n}\varphi(x) = P(x)\varphi(x); \tag{1.2.80}$$

relation (1.2.79) becomes

$$(f, \psi) = (Pf, \varphi) = 0. \tag{1.2.79'}$$

whence we deduce a relation of the form (1.2.65). The previous theorem leads to expression (1.2.78), thus proving theorem 1.2.3.

It should be noted that the last two theorems are equivalent.

In theorem 1.2.2 the simple roots of the polynomial $P(x)$ have been emphasized owing to their importance; it is quite obvious however that their contribution could have been included in the expression corresponding to multiple roots by the consideration of an order of multiplicity equal to unity.

1.2.2. Representative δ sequences

Let us give the following

Definition 1.2.1. *The sequence of functions* $f_n(x_1, x_2, \ldots, x_m)$ *is said to be a representative* δ *sequence, in the sense of the topology in* K', *if we have*

$$\lim_{n \to \infty} f_n(x_1, x_2, \ldots, x_m) = \delta(x_1, x_2, \ldots, x_m); \tag{1.2.81}$$

the condition is obviously equivalent to

$$\lim_{n \to \infty} (f_n(x_1, x_2, \ldots, x_m), \varphi(x_1, x_2, \ldots, x_m)) = \varphi(0, 0, \ldots, 0). \tag{1.2.81'}$$

It may be said that any term of the sequence $f_n(x)$ represents a definite approximation of the δ Dirac distribution; this is a particularly important fact from a practical point of view. Indeed, assuming that we want to obtain numerical values in a problem where the results are expressed in distributions, in order to effect the calculation we can replace a regular distribution by the function which has generated that distribution, while a singular distribution may be replaced by a term of the corresponding representative sequence. Then, the resulting formulae may be programmed on a computer to obtain the desired approximation as a function of the term chosen from the representative sequence.

In the following we shall consider the conditions required by a sequence in order to be a representative δ sequence and examples of such sequences will be given.

1.2.2.1. General results

There are various ways of stating the conditions under which a sequence is a representative δ sequence; for example, by using

Theorem 1.2.4. *If the locally integrable functions* $f_n(x_1, x_2, \ldots, x_m)$ *on* R^m *satisfy the conditions*

(i)
$$\left| \iint \cdots \int_\Omega f_n(x_1, x_2, \ldots, x_m)\, dx_1\, dx_2 \ldots dx_m \right| < M,$$

(ii)
$$\lim_{n \to \infty} \iint \cdots \int_\Omega f_n(x_1, x_2, \ldots, x_m)\, dx_1\, dx_2 \ldots dx_m = \begin{cases} 0 \text{ if } O \notin \Omega \\[2mm] 1 \text{ if } O \in \Omega, \end{cases}$$

where M *does not depend on* n *and* Ω, *while* Ω *is an arbitrary parallelepipedic domain with the sides parallel to the co-ordinate planes and one vertex at the point* $A(x_1, x_2, \ldots, x_m)$ *with* $x_1^0 < 0$, $x_2^0 < 0$, $\ldots, x_m^0 < 0$, *then we may write relation* (1.2.81).

Let there be a function

$$F_n(x_1, x_2, \ldots, x_m) = \int_{x_1^0}^{x_1} \int_{x_2^0}^{x_2} \cdots \int_{x_m^0}^{x_m} f_n(x_1, x_2, \ldots, x_m)\, dx_1\, dx_2 \ldots dx_m, \quad (1.2.82)$$

which leads to

$$\frac{\partial^m F_n(x_1, x_2, \ldots, x_m)}{\partial x_1\, \partial x_2 \ldots \partial x_m} = f_n(x_1, x_2, \ldots, x_m). \tag{1.2.83}$$

We remark that if at least one of the inequalities

$$x_1 < 0, x_2 < 0, \ldots, x_m < 0 \tag{1.2.84}$$

is satisfied, then the domain Ω will not include the origin and we have

$$\lim_{n \to \infty} F_n(x_1, x_2, \ldots, x_m) = 0; \tag{1.2.85}$$

if we have

$$x_1 > 0, x_2 > 0, \ldots, x_m > 0 \tag{1.2.84'}$$

simultaneously, the domain Ω will include the origin and by condition (ii) we may write

$$\lim_{n \to \infty} F_n(x_1, x_2, \ldots, x_m) = 1. \tag{1.2.85'}$$

Since in the passage to the limit the functions F_n are bounded on any interval, it follows by condition (i) that we have

$$\lim_{n \to \infty} F_n(x_1, x_2, \ldots, x_m) = \theta(x_1, x_2, \ldots, x_m), \tag{1.2.85''}$$

where $\theta(x_1, x_2, \ldots, x_m)$ is the Heaviside function in R^m.

From relation (1.2.83) we obtain

$$\lim_{n \to \infty} \frac{\partial^m F_n(x_1, x_2, ..., x_m)}{\partial x_1 \, \partial x_2 \, ... \, \partial x_m} = \frac{\partial^m}{\partial x_1 \, \partial x_2 \, ... \, \partial x_m} \left[\lim_{n \to \infty} F_n(x_1, x_2, ..., x_m) \right]$$

$$= \lim_{n \to \infty} f_n(x_1, x_2, ..., x_m) = \frac{\partial^m \theta(x_1, x_2, ..., x_m)}{\partial x_1 \, \partial x_2 \, ... \, \partial x_m} = \delta(x_1, x_2, ..., x_m), \quad (1.2.83'')$$

which proves the theorem given above.

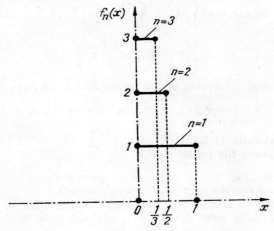

Fig. 1.15

Let us consider for example the function (shown in Figure 1.15 for $m = 1$)

$$f_n(x_1, x_2, ..., x_m) = \begin{cases} n^m \text{ for } 0 \leqslant x_i \leqslant \dfrac{1}{n} & (i = 1, 2, ..., m) \\ 0 \text{ in the rest.} \end{cases} \quad (1.2.86)$$

We note that

$$\iint \cdots \int_\Omega f_n \, \mathrm{d}x = \int_{x_1^0}^{x_1} \int_{x_2^0}^{x_2} \cdots \int_{x_m^0}^{x_m} f_n \, \mathrm{d}x_1 \, \mathrm{d}x_2 \, ... \, \mathrm{d}x_m$$

$$= n^m \int_0^{x_1} \int_0^{x_2} \cdots \int_0^{x_m} \mathrm{d}x_1 \, \mathrm{d}x_2 \, ... \, \mathrm{d}x_m \leqslant n^m \left(\frac{1}{n} \right)^m = 1;$$

hence condition (*i*) is satisfied since the integral is bounded.

Concerning condition (*ii*), we can obviously write

$$\lim_{n \to \infty} \iint \cdots \int_\Omega f_n(x_1, x_2, ..., x_m) \, \mathrm{d}x_1 \, \mathrm{d}x_2 \, ... \, \mathrm{d}x_m = \begin{cases} 1 \text{ for } x_i > 0 & (i = 1, 2, ..., m) \\ 0 \text{ in the rest,} \end{cases}$$

which gives relation (1.2.81).

Let us also consider the integral

$$F_n(x_1, x_2, ..., x_m)$$

$$= \iint ... \int_\Omega \psi(x_1, x_2, ..., x_m) f_n(x_1, x_2, ..., x_m) \, dx_1 \, dx_2 \, ... \, dx_m, \qquad (1.2.87)$$

where $\psi(x)$ is a function of the class C^∞ such that $\psi(0, 0, ..., 0) = 0$, and f_n is the above representative sequence; we have

$$F_n(x_1, x_2, ..., x_m) = n^m \int_0^{x_1} \int_0^{x_2} ... \int_0^{x_m} \psi(x_1, x_2, ..., x_m) \, dx_1 \, dx_2 \, ... \, dx_m$$

$$= n^m \left(\frac{1}{n}\right)^m \psi(\xi_1, \xi_2, ..., \xi_m), \quad 0 \leqslant \xi_i < x_i \leqslant \frac{1}{n} \quad (i = 1, 2, ..., m), \qquad (1.2.87')$$

where we have applied a formula of the mean. Hence we have

$$\lim_{n \to \infty} F_n(x_1, x_2, ..., x_m) = \varphi(0, 0, ..., 0),$$

which leads to relation (1.2.85)

We shall now prove the following

Theorem 1.2.5. *If $f_n(x_1, x_2, ..., x_m)$ is a representative δ sequence and if $\psi(x_1, x_2, ..., x_m)$ is a function of class C^∞ such that $\psi(0, 0, ..., 0) = 0$, then for an arbitrary parallelepipedic domain $\Omega \in R^m$ having the sides parallel to the co-ordinate planes and a vertex at the point $A(x_1^0, x_2^0, ..., x_m^0)$ with $x_1^0 < 0, x_2^0 < 0, ..., x_m^0 < 0$ we have*

$$\lim_{n \to \infty} \iint ... \int_\Omega \psi(x_1, x_2, ..., x_m) f_n(x_1, x_2, ..., x_m) \, dx_1 \, dx_2 \, ... \, dx_m = 0. \qquad (1.2.88)$$

Denoting

$$F_n(x_1, x_2, ..., x_m)$$

$$= \int_{x_1^0}^{x_1} \int_{x_2^0}^{x_2} ... \int_{x_m^0}^{x_m} \psi(x_1, x_2, ..., x_m) f_n(x_1, x_2, ..., x_m) \, dx_1 \, dx_2 \, ... \, dx_m, \qquad (1.2.89)$$

it follows that

$$\frac{\partial^m F_n(x_1, x_2, ..., x_m)}{\partial x_1 \, \partial x_2 \, ... \, \partial x_m} = \psi(x_1, x_2, ..., x_m) f_n(x_1, x_2, ..., x_m). \qquad (1.2.89')$$

Passing to the limit in the sense of the theory of distributions we may write

$$\lim_{n \to \infty} \frac{\partial^m F_n(x_1, x_2, ..., x_m)}{\partial x_1 \, \partial x_2 \, ... \, \partial x_m} = \frac{\partial^m}{\partial x_1 \, \partial x_2 \, ... \, \partial x_m} [\lim_{n \to \infty} F_n(x_1, x_2, ..., x_m)]$$

$$= \lim_{n \to \infty} \psi(x_1, x_2, ..., x_m) f_n(x_1, x_2, ..., x_m) = \psi(x_1, x_2, ..., x_m) \delta(x_1, x_2, ..., x_m)$$

$$= \psi(0, 0, ..., 0) \delta(x_1, x_2, ..., x_m) = 0;$$

hence

$$\lim_{n \to \infty} F_n(x_1, x_2, ..., x_m) = \sum_{i=1}^{m} h_i(x_1, x_2, ..., x_{i-1}, x_{i+1}, ..., x_m), \qquad (1.2.90)$$

where $h_i(x_1, x_2, ..., x_{i-1}, x_{i+1}, ..., x_m)$ are functions of $(m-1)$ variables.

Since relation (1.2.90) holds for any representative δ sequence $f_n(x)$, we may use the previously considered sequences for the determination of the functions h_i; in that case relation (1.2.85) is satisfied. We deduce that $h_i \equiv 0$ which leads to relation (1.2.88), thus proving theorem 1.2.5.

1.2.2.2. Examples of representative δ sequences

We shall give some examples of representative δ sequences. Let there be the function

$$f_n(x) = \frac{2}{\pi} \frac{\varepsilon_n^3}{(x^2 + \varepsilon_n^2)^2}, \quad \varepsilon_n > 0, \quad n \in N, \qquad (1.2.91)$$

where ε_n is a sequence of positive numbers approaching zero; we have

$$\int_a^b f_n(x)\, dx = \frac{1}{\pi} \left(\frac{\varepsilon_n b}{b^2 + \varepsilon_n^2} - \frac{\varepsilon_n a}{a^2 + \varepsilon_n^2} + \arctan \frac{b}{\varepsilon_n} - \arctan \frac{a}{\varepsilon_n} \right). \quad (1.2.92)$$

Since

$$\left| \frac{b\varepsilon_n}{b^2 + \varepsilon_n^2} \right| \leqslant \frac{1}{2}, \quad \left| \frac{a\varepsilon_n}{a^2 + \varepsilon_n^2} \right| \leqslant \frac{1}{2},$$

$$\left| \arctan \frac{b}{\varepsilon_n} - \arctan \frac{a}{\varepsilon_n} \right| \leqslant \pi,$$

we deduce

$$\left| \int_a^b f_n(x)\, dx \right| \leqslant \frac{1}{\pi}(1 + \pi), \qquad (1.2.92')$$

which shows that condition (i) is satisfied. Using relation (1.2.92) we can easily verify condition (ii) too. Hence we obtain

$$\lim_{n \to \infty} f_n(x) = \delta(x), \qquad (1.2.93)$$

the function $f_n(x)$ defining a representative δ sequence; Figure 1.16a gives the graphs corresponding to different values of n.

Similarly, it may be shown that

$$\lim_{n \to \infty} \frac{1}{\pi} \frac{\varepsilon_n}{x^2 + \varepsilon_n^2} = \delta(x); \qquad (1.2.94)$$

for different values of n we obtain the graphs shown in Figure 1.16b.

By differentiating with respect to x we obtain

$$\lim_{n \to \infty} \left[-\frac{2}{\pi} \frac{\varepsilon_n x}{(x^2 + \varepsilon_n^2)^2} \right] = \delta'(x); \qquad (1.2.94')$$

the graphs corresponding to different values of n are given in Figure 1.17.

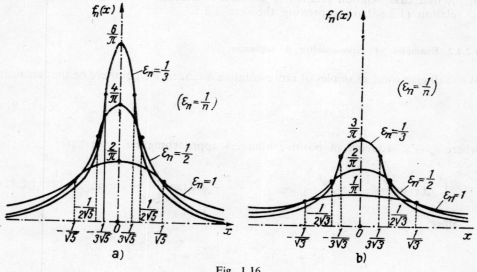

Fig. 1.16

As an application of theorem 1.2.5 we deduce

$$\lim_{n \to \infty} \int_a^b \psi(x) \frac{\sin nx}{x} \, dx = 0, \qquad (1.2.95)$$

where $\psi(x)$ is a function of the class C^∞ and $\psi(0) = 0$.

In particular, taking $\psi(x) = x$ we obtain

$$\lim_{n \to \infty} \int_a^b \sin nx \, dx = 0. \qquad (1.2.96)$$

Integrating by parts, passing to the limit and taking into account relation (1.2.95), we may write

$$\lim_{n \to \infty} \int_a^b \psi(x) \cos nx \, dx = 0. \qquad (1.2.95')$$

In particular, for $\psi(x) = x$ it follows

$$\lim_{n \to \infty} \int_a^b x \cos nx \, dx = 0. \tag{1.2.96'}$$

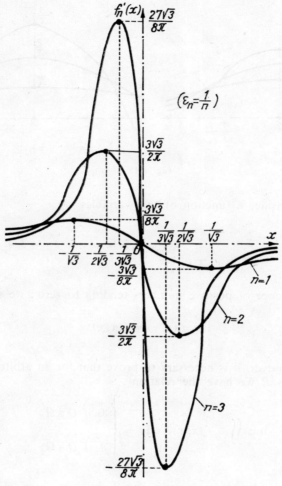

Fig. 1.17

Similarly, we have the relations

$$\lim_{n \to \infty} \sqrt{\frac{n}{2\pi}} \, e^{-\frac{nx^2}{2}} = \delta(x), \tag{1.2.97}$$

$$\lim_{n \to \infty} \frac{1}{\pi} \frac{\sin nx}{x} = \delta(x) \tag{1.2.98}$$

(represented for different values of n by Figures 1.18a and b).

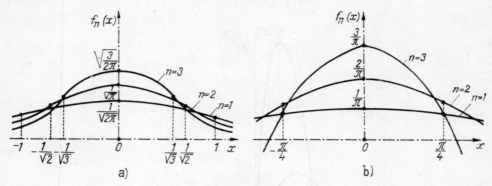

Fig. 1.18

Now let us consider a function of two variables

$$f_n(x, y) = \frac{p-2}{2\pi} \frac{\varepsilon_n^{p-2}}{(x^2 + y^2 + \varepsilon_n^2)^{\frac{p}{2}}}, \quad \varepsilon_n > 0, \quad n \in N \quad (p = 3, 4, 5, \ldots), \quad (1.2.99)$$

where ε_n is a sequence of positive numbers tending to zero; we shall prove that

$$\lim_{n \to \infty} f_n(x, y) = \delta(x, y). \qquad (1.2.100)$$

Before this, however, it is necessary to prove that for an arbitrary circular domain Ω_R of radius R we have the relation

$$\lim_{n \to \infty} \iint_{\Omega_R} f_n(x, y)\, dx\, dy = \begin{cases} 0 \text{ if } O \notin \Omega_R \\[2ex] 1 \text{ if } O \in \Omega_R. \end{cases} \qquad (1.2.101)$$

Indeed, we have

$$\iint_{\Omega_R} f_n(x, y)\, dx\, dy = \frac{p-2}{2\pi} \varepsilon_n^{p-2} \int_{\theta_1}^{\theta_2} d\theta \int_{\rho_1}^{\rho_2} (\rho_2 + \varepsilon_n^2)^{-\frac{p}{2}} \rho\, d\rho$$

$$= -\frac{\varepsilon_n^{p-2}}{2\pi} \int_{\theta_1}^{\theta_2} \left[\frac{1}{(\rho_2^2 + \varepsilon_n^2)^{\frac{p-2}{2}}} - \frac{1}{(\rho_1^2 + \varepsilon_n^2)^{\frac{p-2}{2}}} \right] d\theta,$$

where we have introduced polar co-ordinates (ρ, θ). If $O \in \Omega_R$ then $\rho_1 = 0$ and $\rho_2 > 0$, $\theta_1 = 0$ and $\theta_2 = 2\pi$; hence we have

$$\lim_{n \to \infty} \iint_{\Omega_R} f_n(x, y)\, dx\, dy = -\frac{1}{2\pi} \int_0^{2\pi} \left[\lim_{n \to \infty} \frac{\varepsilon_n^{p-2}}{(\rho_2^2 + \varepsilon_n^2)^{\frac{p-2}{2}}} - 1 \right] d\theta = 1. \quad (1.2.102)$$

In the case $O \notin \Omega_R$ we have $\rho_1, \rho_2 > 0$ and hence

$$\lim_{n \to \infty} \iint_{\Omega_R} f_n(x, y)\, dx\, dy = -\frac{1}{2\pi} \int_{\theta_1}^{\theta_2} \lim_{n \to \infty} \left[\frac{\varepsilon_n^{p-2}}{(\rho_2^2 + \varepsilon_n^2)^{\frac{p-2}{2}}} - \frac{\varepsilon_n^{p-2}}{(\rho_1^2 + \varepsilon_n^2)^{\frac{p-2}{2}}} \right] d\theta = 0,$$

$$(1.2.102')$$

thus proving relation (1.2.101).

We remark further that, for any rectangular domain Ω, there exist two circular domains Ω_r and Ω_R of radii r and R, respectively, such that

$$\Omega_r \subset \Omega \subset \Omega_R; \quad (1.2.103)$$

we may also request that if $O \in \Omega$ then $O \in \Omega_r$ and $O \in \Omega_R$ too, and if $O \notin \Omega$ then $O \notin \Omega_r$ and $O \notin \Omega_R$ too.

Since the function defined by relation (1.2.99) is positive for $\varepsilon_n > 0$, taking into account relation (1.2.103), we may write

$$0 < \iint_{\Omega_r} f_n(x, y)\, dx\, dy < \iint_{\Omega} f_n(x, y)\, dx\, dy < \iint_{\Omega_R} f_n(x, y)\, dx\, dy, \quad (1.2.104)$$

whence we obtain

$$0 < \iint_{\Omega} f_n(x, y)\, dx\, dy < \lim_{R \to \infty} \iint_{\Omega_R} f_n(x, y)\, dx\, dy. \quad (1.2.104')$$

Using the integral computed above in polar co-ordinates and taking the centre of the domain Ω_R at the origin we may write

$$\lim_{R \to \infty} \iint_{\Omega_R} f_n(x, y)\, dx\, dy = -\frac{\varepsilon_n^{p-2}}{2\pi} \int_0^{2\pi} \lim_{R \to \infty} \left[\frac{1}{(R^2 + \varepsilon_n^2)^{\frac{p-2}{2}}} - \frac{1}{\varepsilon_n^{p-2}} \right] d\theta = 1,$$

whence

$$0 < \iint_{\Omega} f_n(x, y)\, dx\, dy < 1, \quad (1.2.105)$$

which shows that condition (*i*) is satisfied.

From the inequalities (1.2.104) we obtain

$$\lim_{n\to\infty} \iint_{\Omega_r} f_n(x, y)\, dx\, dy \leqslant \lim_{n\to\infty} \iint_{\Omega} f_n(x, y)\, dx\, dy \leqslant \lim_{n\to\infty} \iint_{\Omega_R} f_n(x, y)\, dx\, dy.$$

$$(1.2.104'')$$

Taking into account relations (1.2.10') we have

$$\lim_{n\to\infty} \iint_{\Omega_r} f_n(x, y)\, dx\, dy = \lim_{n\to\infty} \iint_{\Omega_R} f_n(x, y)\, dx\, dy = \begin{cases} 0 \text{ if } O \notin \Omega_R \\ 1 \text{ if } O \in \Omega_R, \end{cases} \quad (1.2.106)$$

whence

$$\lim_{n\to\infty} \iint_{\Omega} f_n(x, y)\, dx\, dy = \begin{cases} 0 \text{ if } O \notin \Omega \\ 1 \text{ if } O \in \Omega, \end{cases} \quad (1.2.106')$$

which shows that condition (*ii*) too, is satisfied; this completes the proof of relation (1.2.100).

Differentiating relation (1.2.100) with respect to x and y and taking $p = 3$ we obtain the relations

$$\frac{\partial}{\partial x} \delta(x, y) = \lim_{n\to\infty} \left[\frac{\partial}{\partial x} f_n(x, y) \right] = \lim_{n\to\infty} \left[-\frac{3}{2\pi} \frac{x\varepsilon_n}{(x^2 + y^2 + \varepsilon_n^2)^{\frac{5}{2}}} \right],$$

$$(1.2.107)$$

$$\frac{\partial}{\partial y} \delta(x, y) = \lim_{n\to\infty} \left[\frac{\partial}{\partial y} f_n(x, y) \right] = \lim_{n\to\infty} \left[-\frac{3}{2\pi} \frac{y\varepsilon_n}{(x^2 + y^2 + \varepsilon_n^2)^{\frac{5}{2}}} \right].$$

We shall now show how we can construct very easily representative sequences for the Dirac distribution with one or more variables; such sequences are useful for various applications.

Let us consider the function

$$h(x) = \begin{cases} 1 \text{ for } x \in [-a, a] \\ 0 \text{ in the rest;} \end{cases} \quad (1.2.108)$$

we can show that

$$\lim_{a\to 0} \frac{h(x)}{2a} = \delta(x). \quad (1.2.108')$$

Figure 1.19 shows the graphs of the function $\dfrac{h(x)}{2a}$ for different values of a.

Similarly we may consider in R^2 the function

$$h(x, y) = \begin{cases} 1 \text{ for } (x, y) \in [-a, a] \times [-a, a] \\ \\ 0 \text{ in the rest,} \end{cases} \tag{1.2.109}$$

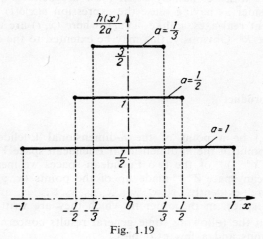

Fig. 1.19

which leads to

$$\lim_{a \to 0} \frac{h(x, y)}{4a^2} = \delta(x, y). \tag{1.2.109'}$$

In R^3 we introduce the function

$$h(x, y, z) = \begin{cases} 1 \text{ for } (x, y, z) \in [-a, a] \times [-a, a] \times [-a, a] \\ \\ 0 \text{ in the rest,} \end{cases} \tag{1.2.110}$$

whence we obtain

$$\lim_{a \to 0} \frac{h(x, y, z)}{8a^3} = \delta(x, y, z). \tag{1.2.110'}$$

1.3. Composition of distributions

The product of two distributions has, in general, no sense; we have seen that the product with a function of the class C^∞ has sense. For this reason we will define products of a special type between distributions *(composition of distributions);* thus we will introduce the *direct* (or *tensor*) *product* and the *convolution product*.

As we shall see the direct product is an extension of the product of two functions; the convolution product is particularly important in the theory of integral transformations (e.g. in the case of the Fourier transform) as well as in solving differential equations.

It should be noted however that sometimes the ordinary product of two distributions may have a sense, namely when these are defined on different spaces. Thus, we may consider as having sense the expression $\delta(x)\delta(t)$ where x and t belong to two different real axes, unlike $\delta(x; t)$, where (x, t) are the co-ordinates of a point in the space R^2. Obviously, this may be extented to the case of several distributions.

1.3.1. The direct product

Let $x(x_1, x_2, \ldots, x_n)$ be a point of the n-dimensional Euclidean space X^n and $y(y_1, y_2, \ldots, y_m)$ a point of the m-dimensional Euclidean space Y^m. By the *direct* or *Cartesian product* $X^n \times Y^m$ of the two Euclidean spaces we mean a new, $(m+n)$-dimensional Euclidean space Z^{m+n} made up of the points $(x, y) = (x_1, x_2, \ldots, x_n, y_1, y_2, \ldots, y_m)$ where, evidently, $x_1, x_2, \ldots, x_n, y_1, y_2, \ldots, y_m$ are the co-ordinates of a point of that space, in the order in which they are written.

We shall give in the following some general results concerning the direct product of two functions and a few examples.

1.3.1.1. General results

Let $f(x_1, x_2, \ldots, x_n)$ be a numerical function defined on X^n and $g(y_1, y_2, \ldots, y_m)$ a numerical function defined on Y^m. *The direct product of the two functions $f(x) \times g(y)$ is the function $h(x, y) = f(x) g(y)$ defined on $X^n \times Y^m$*; therefore, the direct product of two numerical functions coincides with their ordinary product, i.e.

$$f(x) \times g(y) = f(x)g(y), \quad x \in X^n, \quad y \in Y^m. \tag{1.3.1}$$

If the functions $f(x)$ and $g(y)$ are locally integrable functions, their direct product is also a locally integrable function; in this manner we can define the direct product of distributions.

Let $\varphi_1(x)$ be a fundamental function defined on X^n and $\varphi_2(y)$ a fundamental function defined on Y^m. We shall denote by $\varphi(x, y)$ a fundamental function defined on $X^n \times Y^m$; for a fixed x it follows that $\varphi(x, y)$ will be a fundamental function defined on Y^m.

Let us assume that the fundamental function $\varphi(x, y)$ is of the form

$$\varphi(x, y) = \varphi_1(x)\varphi_2(y), \quad x \in X^n, \quad y \in Y^m; \tag{1.3.2}$$

then if $f(x)$, $x \in X^n$ and $g(y)$, $y \in Y^m$ are locally integrable functions, we have

$$(f(x) \times g(y), \varphi(x, y)) = (f(x) \times g(y), \varphi_1(x)\varphi_2(y))$$

$$= \int_{X^n} \int_{Y^m} f(x)g(y)\varphi_1(x)\varphi_2(y) \, dx \, dy = \int_{X^n} f(x)\varphi_1(x) \, dx \int_{Y^m} g(y)\varphi_2(y) \, dy,$$

whence it follows that

$$(f(x) \times g(y), \varphi(x, y)) = (f(x), \varphi_1(x))(g(y), \varphi_2(y)). \tag{1.3.3}$$

If the fundamental function $\varphi(x, y)$ is not of the form (1.3.2) we may write, using Fubini's theorem on the permutation of the order of integration,

$$(f(x) \times g(y), \varphi(x, y)) = \int_{X^n} \int_{Y^m} f(x)g(y)\varphi(x, y) \, dx \, dy$$

$$= \int_{X^n} f(x) \, dx \int_{Y^m} g(y) \, \varphi(x, y) \, dy,$$

whence we obtain

$$(f(x) \times g(y), \varphi(x, y)) = (f(x), (g(y), \varphi(x, y))). \tag{1.3.4}$$

This relation is considered as the definition of the *direct product* $f(x) \times g(y)$ *of two distributions* $f(x)$ *and* $g(y)$ defined on the fundamental spaces K_x, $x \in X^n$, and K_y, $y \in Y^m$, respectively, where $\varphi(x, y)$ is a fundamental function defined on $X^n \times Y^m$. It can be shown that the functional defined by the direct product $f(x) \times g(y)$ (relation (1.3.4)) is a linear and continuous functional; therefore it represents a distribution defined on the fundamental space $K_x \times K_y$, $x \in X^n$, $y \in Y^m$.

Also, it may be shown that the support of the distribution $f(x) \times g(y)$ is equal to the direct product of the supports of the two distributions $f(x)$ and $g(y)$; therefore

$$\text{supp} \, [f(x) \times g(y)] = \text{supp} \, f(x) \times \text{supp} \, g(y), \tag{1.3.5}$$

the support of distribution $f(x) \times g(y)$ consisting of the set of pairs (x, y) where $x \in \text{supp} \, f(x)$ and $y \in \text{supp} \, g(y)$.

It is easily proved that the direct product is *commutative*

$$f(x) \times g(y) = g(y) \times f(x) \tag{1.3.6}$$

and *associative*

$$[f(x) \times g(y)] \times h(z) = f(x) \times [g(y) \times h(z)]. \tag{1.3.6'}$$

Using the first of these properties we can write relation (1.3.4) also in the form

$$(f(x) \times g(y), \varphi(x, y)) = (g(y), (f(x), \varphi(x, y))). \qquad (1.3.4')$$

The second property takes into account the fact that the direct product may be defined for an arbitrary finite number of distributions.

We remark that the direct product of two regular distributions $f(x)$ and $g(y)$ is a regular distribution corresponding to the function $f(x) g(y)$.

Let D_x^p and D_y^q be two differential operators with respect to x and y and of the order p and q, respectively, and let $\varphi(x, y)$ be a fundamental function written as a sum of functions like (1.3.2). For one of the terms of this sum we may write

$$(D_x^p D_y^q [f(x) \times g(y)], \varphi(x, y)) = (-1)^{p+q} (f(x) \times g(y), D_x^p \varphi_1(x) D_y^q \varphi_2(y))$$

$$= (-1)^{p+q} (f(x), (g(y), D_x^p \varphi_1(x) D_y^q \varphi_2(y))) = (-1)^p (f(x), (D_y^q g(y), D_x^p \varphi_1(x) \varphi_2(y)))$$

$$= (-1)^p (f(x), D_x^p \varphi_1(x) (D_y^q g(y), \varphi_2(y))) = (D_x^p f(x), (D_y^q g(y), \varphi_1(x) \varphi_2(y)))$$

$$= (D_x^p f(x) \times D_y^q g(y), \varphi_1(x) \varphi_2(y)) = (D_x^p f(x) \times D_y^q g(y), \varphi(x, y));$$

from the equality of the two distributions it follows that

$$D_x^p D_y^q [f(x) \times g(y)] = D_x^p f(x) \times D_y^q g(y). \qquad (1.3.7)$$

1.3.1.2. Examples

Let $\theta(x)$ and $\theta(y)$ be Heaviside functions of a single variable; their direct product is a Heaviside function of two variables

$$\theta(x) \times \theta(y) = \theta(x)\theta(y) = \theta(x, y). \qquad (1.3.8$$

In general we can obtain a Heaviside function defined on R^n in the form

$$\theta(x_1) \times \theta(x_2) \times \ldots \times \theta(x_n) = \theta(x_1)\theta(x_2) \ldots \theta(x_n) = \theta(x_1, x_2, \ldots, x_n). \qquad (1.3.9)$$

The support of the function $\theta(x)$ is the set $[0, \infty)$; for the support of the direct product we obtain

$$\text{supp} [\theta(x) \times \theta(y)] = \text{supp} \, \theta(x) \times \text{supp} \, \theta(y) = [0, \infty) \times [0, \infty). \qquad (1.3.10)$$

Applying the formula for the differentiation of the direct product of two distributions we may write

$$\frac{\partial^2}{\partial x \partial y} [\theta(x) \times \theta(y)] = \frac{\partial \theta(x)}{\partial x} \times \frac{\partial \theta(y)}{\partial y} = \delta(x) \times \delta(y) = \delta(x, y) \qquad (1.3.11)$$

or, in general,

$$\frac{\partial^n}{\partial x_1 \partial x_2 \dots \partial x_n} [\theta(x_1) \times \theta(x_2) \times \dots \times \theta(x_n)]$$

$$= \delta(x_1) \times \delta(x_2) \times \dots \times \delta(x_n) = \delta(x_1, x_2, \dots, x_n). \qquad (1.3.12)$$

Let there be the function

$$\psi(x, c) = \int_{-c}^{c} \theta(x - t)\, dt = \begin{cases} 0 & \text{for } x < -c \\ x + c & \text{for } -c \leqslant x \leqslant c \\ 2c & \text{for } x > c, \end{cases} \qquad (1.3.13)$$

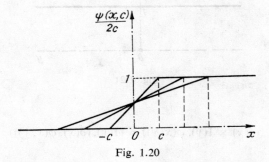

Fig. 1.20

where θ is the Heaviside function; we have (Figure 1.20)

$$\lim_{c \to +0} \frac{\psi(x, c)}{2c} = \theta(x). \qquad (1.3.14)$$

Taking into account relation (1.3.14) we may write

$$\lim_{c \to +0} \left\{ \frac{\partial^2}{\partial x \partial y} \left[\theta(y - y_0) \times \frac{\psi(x, c)}{2c} \right] \right\}$$

$$= \frac{\partial^2}{\partial x \partial y} [\theta(y - y_0) \times \theta(x)] = \delta(y - y_0) \times \delta(x) = \delta(x, y - y_0). \qquad (1.3.15)$$

Similarly, we obtain

$$\lim_{c \to +0} \left\{ \frac{\partial^2}{\partial x \partial y} \left[\frac{\theta(y + c) - \theta(y - c)}{4c^2} \times \psi(x, c) \right] \right\}$$

$$= \frac{\partial^2}{\partial x \partial y} \left[\lim_{c \to +0} \frac{\theta(y + c) - \theta(y - c)}{2c} \times \lim_{c \to +0} \frac{\psi(x, c)}{2c} \right]$$

$$= \frac{\partial^2}{\partial x \partial y} [\delta(y) \times \theta(x)] = \delta_y'(x, y). \qquad (1.3.16)$$

Now let there be

$$\frac{\partial}{\partial x} \left[\theta(x) \times 1(y)\right] = \delta(x) \times 1(y), \tag{1.3.17}$$

where the function $1(y)$ is the function equal to unity for all $y \in R$ (Figure 1.21).

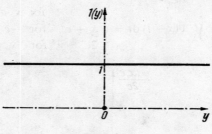

Fig. 1.21

We may write

$$(\delta(x) \times 1(y), \varphi(x, y)) = (1(y), (\delta(x), \varphi(x, y)))$$

$$= (1(y), \varphi(0, y)) = \int_{-\infty}^{\infty} \varphi(0, y) \, dy. \tag{1.3.18}$$

We remark that

$$\theta(x) \times 1(y) = \begin{cases} 1 \text{ for } x \geqslant 0 \\ \\ 0 \text{ for } x < 0. \end{cases}$$

Since the straight line $x = 0$ represents a curve of discontinuity, the jump being $s_x = 1$, and since

$$\frac{\tilde{\partial}}{\partial x} \left[\theta(x) \times 1(y)\right] = 0,$$

the differentiation formula (1.2.51) leads to

$$\left(\frac{\partial}{\partial x} \left[\theta(x) \times 1(y)\right], \varphi(x, y)\right) = \int_{x=0} \varphi(x, y) \, dy, \tag{1.3.19}$$

a relation which is equivalent to relation (1.3.18)

1.3.2. The convolution product

In the following we shall give general results concerning the convolution product of two distributions and examples of such products.

1.3.2.1. General results

Let $f(x)$ and $g(x)$ be locally integrable functions of x; their *convolution product* is the function defined by the formula

$$F(x) = f(x) * g(x) = \int_{-\infty}^{\infty} f(\xi)g(x - \xi)\, \mathrm{d}\xi = (f * g)(x) = (f(\xi), g(x - \xi)). \quad (1.3.20)$$

If the functions $f(x)$ and $g(x)$ are continuous, then their convolution product is continuous too. Obviously the definition is valid for all $x \in R^n$.

In order that the convolution product may exist it is necessary that the functions $f(x)$ and $g(x)$ should satisfy certain conditions. Thus, a sufficiency condition in this respect is that the supports of the two functions $f(x)$ and $g(x)$ be compact.

Denoting by $\varphi(x)$ a fundamental function and by taking into account formula (1.3.20) we may write

$$(F(x), \varphi(x)) = \int_{-\infty}^{\infty} F(x)\varphi(x)\, \mathrm{d}x = \int_{-\infty}^{\infty} \left[\int_{-\infty}^{\infty} f(\xi)g(x - \xi)\, \mathrm{d}\xi \right] \varphi(x)\, \mathrm{d}x$$

$$= \int_{-\infty}^{\infty} \int_{-\infty}^{\infty} f(\xi)g(\eta)\varphi\,(\xi + \eta)\, \mathrm{d}\xi\, \mathrm{d}\eta;$$

since $f(\xi)\, g(\eta)$ represents the direct product of the two locally integrable functions, from the above relation we obtain

$$(F(x), \varphi(x)) = (f(x) \times g(y), \varphi(x + y)), \quad (1.3.21)$$

a formula which may serve as a definition of the convolution product of two distributions, namely

Definition 1.3.1. *If $f(x)$ and $g(x)$ are two distributions on R^n, then their convolution product $f(x) * g(x)$ represents a new distribution on R^n defined by the formula*

$$(f(x) * g(x), \varphi(x)) = (f(x) \times g(y), \varphi(x + y))$$

$$= (f(x), (g(y), \varphi(x + y))) = (g(y), (f(x), \varphi(x + y))). \quad (1.3.22)$$

In order that the convolution product defined above may exist it is necessary that the strip $|x + y| \leqslant a$ should contain the support of the function $\varphi(x + y)$

and have a bounded intersection with the support of the direct product $f(x) \times g(y)$. Thus we may show that the convolution product has a meaning if one of following conditions are satisfied:

(*i*) one of the distributions $f(x)$, $g(x)$ has a compact support;

(*ii*) the distributions $f(x)$ and $g(x)$ have the supports bounded on the same side.

Thus, if $f(x) = 0$ for $x < a$ and $g(x) = 0$ for $x < b$, the supports of the two distributions are bounded to the left.

It is important to remark that in the case of two functions the convolution product has a meaning if one of the functions is integrable on R^n and the other is bounded on R^n; in that case the convolution product is a continuous function bounded on R^n.

If the supports of two functions are bounded to the left and are included in the interval $[0, \infty)$ then the support of the convolution product too is included in the same interval. In general it may be shown that the support of the convolution product of two distributions is included in the closure of the union of the supports of the two distributions, i.e.

$$\operatorname{supp}(f(x) * g(x)) \subset \overline{\operatorname{supp} f(x) \cup \operatorname{supp} g(x)}. \qquad (1.3.23)$$

We remark that the convolution product may be defined for an arbitrary finite number of distributions. Under the conditions required for the existence of the convolution product, one may prove the property of *commutativity*

$$f(x) * g(x) = g(x) * f(x) \qquad (1.3.24)$$

and the property of *associativity*

$$[f(x) * g(x)] * h(x) = f(x) * [g(x) * h(x)]. \qquad (1.3.25)$$

The property of the continuity of a convolution product is particularly important in many applications.

We shall state the following

Theorem 1.3.1. *Let* $f_n(x)$ *be a distribution depending on the parameter n and let* $g(x)$ *be another distribution; if* $f_n(x) \rightarrow f(x)$ *then we have the relation*

$$\lim_{n \to \infty} [f_n(x) * g(x)] = [\lim_{n \to \infty} f_n(x)] * g(x) = f(x) * g(x), \qquad (1.3.26)$$

under one of the following conditions:

(*i*) *the distributions* $f_n(x)$ *are all concentrated on the same bounded set;*

(*ii*) *the distribution* $g(x)$ *is concentrated on a bounded set;*

(*iii*) *the supports of the distributions* $f_n(x)$ *and* $g(x)$ *are bounded on the same side by a constant which does not depend on n.*

Now let D be an arbitrary differential operator. We remark that the product $D\delta(x) * f(x)$, where $f(x)$ is a distribution, has always a meaning since the support of any of the distributions $\delta(x)$, $\delta'(x)$, $\delta''(x)$, ... is a point.

We introduce the notation

$$\overline{D} = (-1)^q D,$$

where q is the order of differentiation. We may write

$$(D[f(x) * g(x)], \varphi(x)) = (D[f(x) \times g(y)], \varphi(x + y))$$

$$= (f(x) \times g(y), \overline{D}\varphi(x + y)) = (g(y), (f(x), \overline{D}\varphi(x + y)))$$

$$= (g(y), (Df(x), \varphi(x + y))) = (Df(x) * g(x), \varphi(x));$$

whence we obtain

$$D[f(x) * g(x)] = Df(x) * g(x) = f(x) * Dg(x), \qquad (1.3.27)$$

where account has been taken of the commutativity of the direct product.

Let us compute the convolution $\delta(x) * f(x)$; we have

$$(\delta(x) * f(x), \varphi(x)) = (f(y), (\delta(x), \varphi(x + y))) = (f(y), \varphi(y)) = (f(x), \varphi(x)),$$

whence

$$\delta(x) * f(x) = f(x). \qquad (1.3.28)$$

In general, we may write

$$D\delta(x) * f(x) = \delta(x) * Df(x) = Df(x). \qquad (1.3.29)$$

We now give the following

Theorem 1.3.2. *If $f_t(x)$ represents a distribution depending on the parameter t and if $\dfrac{\partial f_t(x)}{\partial t}$ exists, then we have*

$$\frac{\partial}{\partial t}(f_t(x) * g(x)) = \frac{\partial f_t(x)}{\partial t} * g(x), \qquad (1.3.30)$$

under one of the following conditions:

(i) the distribution $f_t(x)$ is concentrated on a bounded set;

(ii) the distribution $g(x)$ is concentrated on a bounded set;

(iii) the supports of the distributions $f_t(x)$ and $g(x)$ are bounded on the same side by a constant which is not dependent on t.

1.3.2.2. Examples

We remark that

$$(\delta(x - a) * f(x), \varphi(x)) = (f(y), (\delta(x - a), \varphi(x + y)))$$

$$= (f(y), (\delta(x), \varphi(x + y + a))) = (f(y), \varphi(y + a))$$

$$= (f(x), \varphi(x + a)) = (f(x - a), \varphi(x));$$

thus, relation (1.3.28) is generalized in the form

$$\delta(x - a) * f(x) = f(x - a). \tag{1.3.31}$$

Let $f_n(x)$ be a representative δ sequence and $g(x)$ a distribution. We may write

$$\lim_{n \to \infty} \left[f_n(x) * g(x) \right] = \left[\lim_{n \to \infty} f_n(x) \right] * g(x) = \delta(x) * g(x) = g(x); \tag{1.3.32}$$

the above relation results from the conditions for the definition of representative sequences.

Let us consider the representative δ sequence

$$f_a(x) = \frac{1}{\pi} \frac{a}{x^2 + a^2}, \qquad a > 0. \tag{1.3.33}$$

We remark that the support of this function is the real axis R; the function is at the same time bounded since

$$|f_a(x)| < \frac{1}{\pi a}. \tag{1.3.34}$$

The convolution product $f_a(x) * f_b(x)$ has a meaning, since $f_a(x)$ and $f_b(x)$ are functions, one of which is integrable on the axis R, while the other is bounded on R. Effecting the product we obtain

$$f_a(x) * f_b(x) = f_{a+b}(x). \tag{1.3.35}$$

1.4. Distributions concentrated on curves and surfaces. Homogeneous distributions

We shall give in the following section some results concerning the distributions concentrated on curves and surfaces, as well as the homogeneous distributions. These types of distributions have many applications.

1.4.1. Distributions concentrated on curves and surfaces

Let f be an arbitrary distribution and A a set. If the support of the distribution f is included in the set A ($\operatorname{supp} f \subset A$) the distribution f is said to be concentrated on the set A.

Thus the Dirac distribution $\delta(x - x_0)$ is concentrated in the point x_0 since its support is the point x_0.

Such distributions are particularly useful in applications because they intervene in the expression of derivatives of distributions of several variables.

1.4.1.1. General results

Let $f(x, y, z)$ be a locally integrable function, $\varphi(x, y, z)$ a fundamental function and S a piecewise smooth surface; the formula

$$(f(x, y, z), \varphi(x, y, z)) = \iint_S f(x, y, z)\, \varphi(x, y, z)\, \mathrm{d}S \qquad (1.4.1)$$

where $\mathrm{d}S$ is an area element, defines a *distribution concentrated on that surface*.

Similarly, if C is a skew, piecewise-smooth curve, the formula

$$(f(x, y, z), \varphi(x, y, z)) = \int_C f(x, y, z)\, \varphi(x, y, z)\, \mathrm{d}s \qquad (1.4.2)$$

where $\mathrm{d}s$ is an arc element, defines a *distribution concentrated on that curve*.

For the definition of distributions concentrated on varieties, the differential forms attached to those varieties are of particular importance.

Thus, let be an $(n - 1)$-dimensional surface in the n-dimensional space and let

$$P(x_1, x_2, \ldots, x_n) = 0 \qquad (1.4.3)$$

be the equation of that surface. We assume that the surface is regular, hence that it has no singular points and that the condition

$$\operatorname{grad} P \neq 0 \qquad (1.4.4)$$

is satisfied.

The differential form ω_P of J. Leray attached to the $(n - 1)$-dimensional surface S whose equation is (1.4.3), is defined by the expression

$$\omega_P\, \mathrm{d}P = \mathrm{d}V, \qquad (1.4.5)$$

where $\mathrm{d}P$ is the differential of the function $P(x_1, x_2, \ldots, x_n)$ and

$$\mathrm{d}V = \mathrm{d}x_1\, \mathrm{d}x_2 \ldots \mathrm{d}x_n \qquad (1.4.6)$$

is the volume element in the n-dimensional space.

It may be shown that the differential form ω_P of the $(n-1)$-order, attached to the $(n-1)$-dimensional surface S whose equation is (1.4.3), exists and is determined by the relation

$$\omega_P = (-1)^{j-1} \frac{\mathrm{d}x_1\, \mathrm{d}x_2 \ldots \mathrm{d}x_{j-1}\, \mathrm{d}x_{j+1} \ldots \mathrm{d}x_n}{\dfrac{\partial P}{\partial x_j}} \qquad (1.4.7)$$

or by the relation

$$\omega_P = (-1)^{j-1} \mathrm{D}\left(\frac{x}{u}\right) \mathrm{d}u_1\, \mathrm{d}u_2 \ldots \mathrm{d}u_{j-1}\, \mathrm{d}u_{j+1} \ldots \mathrm{d}u_n, \qquad (1.4.7')$$

where $\mathrm{D}\left(\dfrac{x}{u}\right)$ is the positive Jacobian of the change of variable

$$u_1 = x_1, \quad u_2 = x_2, \ldots, u_{j-1} = x_{j-1}, \quad u_j = P,$$

$$(1.4.8)$$

$$u_{j+1} = x_{j+1}, \ldots, u_n = x_n,$$

namely

$$\mathrm{D}\left(\frac{x}{u}\right) = \frac{1}{\mathrm{D}\left(\dfrac{u}{x}\right)} = \frac{1}{\dfrac{\partial P}{\partial x_j}}. \qquad (1.4.8')$$

It may be shown that the differential form ω_P which verifies relation (1.4.5) is independent of the index j, hence of the manner in which the co-ordinates u_1, $u_2, \ldots, u_n$ are chosen; however, it is obvious that this form does depend on the function P used for defining the surface S.

Thus, the Dirac distribution $\delta(P)$ concentrated on the surface S whose equation is (1.4.3), is defined by the formula

$$(\delta(P), \varphi(x)) = \int_{P=0} \varphi(x)\omega_P = \int_{R^n} \delta(P)\varphi(x)\, \mathrm{d}x. \qquad (1.4.9)$$

Let $\theta(P)$ be the *characteristic function* for the domain $P \geqslant 0$, defined by the relationship

$$\theta(P) = \begin{cases} 1 \text{ for } P \geqslant 0 \\ \\ 0 \text{ for } P < 0; \end{cases} \qquad (1.4.10)$$

it operates according to the formula

$$(\theta(P), \varphi(x)) = \int_{P \geqslant 0} \varphi(x)\, dx. \tag{1.4.10'}$$

We can write

$$\theta'(P) = \delta(P), \tag{1.4.11}$$

a relation which must be taken in the sense that

$$\frac{\partial}{\partial x_i}\theta(P) = P_i'\delta(P) \qquad (i = 1, 2, ..., n); \tag{1.4.11'}$$

we can write relation (1.4.11') also in the form

$$\text{grad}\,\theta(P) = \delta(P)\,\text{grad}\,P. \tag{1.4.11''}$$

Also, we have

$$\frac{\partial}{\partial x_i}\delta^{(k)}(P) = P_i'\delta^{(k+1)}(P) \qquad (k = 0, 1, 2, ...). \tag{1.4.12}$$

If the surfaces $P = 0$ and $Q = 0$ do not intersect then

$$\delta(PQ) = \frac{1}{Q}\,\delta(P) + \frac{1}{P}\,\delta(Q); \tag{1.4.13}$$

in particular, for a function $\alpha(x) \neq 0$, we may write

$$\delta(\alpha P) = \frac{1}{\alpha}\,\delta(P). \tag{1.4.13'}$$

Let us compute for example the distribution $\delta(ax + by + cz + d)$. We have

$$P(x, y, z) \equiv ax + by + cz + d = 0 \tag{1.4.14}$$

and

$$\omega_P = (-1)^{3-1}\frac{dx\,dy}{\dfrac{\partial P}{\partial z}} = \frac{1}{c}\,dx\,dy, \tag{1.4.14'}$$

whence we obtain

$$(\delta(ax + by + cz + d), \varphi(x, y, z)) = \int_{R^3}\delta(P)\varphi(x, y, z)\,dx\,dy\,dz$$

$$= \frac{1}{c}\int_{P=0}\varphi(x, y, z)\,dx\,dy = \frac{1}{c}\int_{R^2}\varphi\left[x, y, -\frac{1}{c}(ax + by + d)\right]dx\,dy, \tag{1.4.15}$$

which defines the *Dirac distribution concentrated on the plane* $P = 0$; the relation may be written also in the form

$$(\delta(P), \varphi(x, y, z)) = \frac{1}{c} \int_{P=0} \varphi(x, y, z) \frac{dS}{\cos(n, z)}, \qquad (1.4.15')$$

where $\cos(n, z)$ is the direction cosine of the normal to the plane corresponding to the Oz axis.

In particular, for

$$P \equiv ax = 0 \qquad (1.4.16)$$

we have

$$\omega_P = (-1)^{1-1} \frac{dy\,dz}{a} = \frac{1}{a} dy\,dz, \qquad (1.4.16')$$

which leads to

$$(\delta(ax), \varphi(x, y, z)) = \frac{1}{a} \int_{P=0} \varphi(x, y, z)\,dy\,dz = \frac{1}{a} \int_{R^2} \varphi(0, y, z)\,dy\,dz. \quad (1.4.17)$$

1.4.1.2. Applications of the Dirac distribution $\delta(P)$

Using the Dirac distribution $\delta(P)$ we may express the partial derivatives of a function with discontinuities of the first species.

Let $f(x, y)$ be a function of the class C^1, defined on R^2, having a jump at the intersection with the curve Γ, piecewise smooth and such that a parallel to the co-ordinate axes intersects it at one point only (Figure 1.22); the formula (1.2.50) which links the derivative f'_y in the sense of the theory of distributions and the derivative $\widetilde{f'_y}$ in the ordinary sense has been established in section 1.2.1.4, and the jump s_y which occurs is expressed by formula (1.2. 50'). We also mention that the jump s_y is considered in the positive direction of the Oy axis when it intersects the curve Γ.

As the equation of the curve Γ is

$$P(x, y) = 0 \qquad (1.4.18)$$

we may write the differential form

$$\omega_P = (-1)^{2-1} \frac{dx}{\dfrac{\partial P}{\partial y}} = -\frac{dx}{P'_y}. \qquad (1.4.19)$$

We have

$$dx = ds \cos(\mathbf{u}, x) = ds \cos(\mathbf{n}_1, y),$$

$$dy = ds \cos(\mathbf{u}, y) = - ds \cos(\mathbf{n}_1, x),$$

(1.4.20)

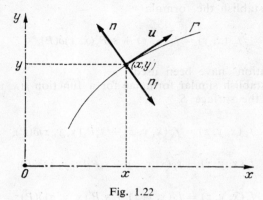

Fig. 1.22

where ds is the arc element on the curve Γ, $\mathbf{n}_1$ is the unit vector of the principal normal to the curve Γ, and $\mathbf{u}$ is the unit vector of the tangent to the same curve; the differential form (1.4.19) becomes

$$\omega_P = - \frac{dx}{P'_y} = - \frac{1}{P'_y} ds \cos(\mathbf{n}_1, y) = - \frac{ds}{\sqrt{P_x'^2 + P_y'^2}}.$$

(1.4.19′)

We may write

$$(P'_y(x, y)s_y\delta(P), \varphi(x, y)) = (\delta(P), P'(x, y)s_y\varphi(x, y))$$

$$= \int_\Gamma P'(x, y)s_y\varphi(x, y)\omega_P = - \int_\Gamma s_y\varphi(x, y)\,dx = - \int_\Gamma s_y\varphi(x, y) \cos(\mathbf{n}_1, y)\,ds;$$

formula (1.2.50) takes the form

$$(f'_y(x, y), \varphi(x, y)) = (\tilde{f}'_y(x, y), \varphi(x, y)) - (P'_y(x, y)s_y\delta(P), \varphi(x, y)),$$

(1.4.21)

whence

$$f'_y(x, y) = \tilde{f}'_y(x, y) - s_y P'_y(x, y)\delta(P),$$

(1.4.22)

where the principal normal $\mathbf{n}_1$ has been considered.

If we use the normal **n** directed in the positive direction of transversion of the curve Γ (Figure 1.22), formula (1.4.21) becomes

$$f'_y(x, y) = \tilde{f}'_y(x, y) + s_y P'_y(x, y)\delta(P);$$

(1.4.23)

similarly we can establish the formula

$$f_x'(x, y) = \tilde{f}_x'(x, y) + s_x P'_x(x, y)\delta(P),$$

(1.4.23')

where similar notations have been used.

Also, we can establish similar formulae for a function $f(x, y, z)$ having a jump when transversing the surface S

$$f_x'(x, y, z) = \tilde{f}_x'(x, y, z) + s_x P'_x(x, y, z)\delta(P),$$

$$f_y'(x, y, z) = \tilde{f}_y'(x, y, z) + s_y P'_y(x, y, z)\delta(P),$$

(1.4.24)

$$f_z'(x, y, z) = \tilde{f}_z'(x, y, z) + s_z P'_z(x, y, z)\delta(P);$$

here too, the normal to the surface is directed in the positive direction of transversing the surface.

In the case where there are several surfaces of discontinuity S_k ($k = 1, 2, \ldots, h$), the equations of which are

$$P_k(x_1, x_2, \ldots, x_n) = 0 \qquad (k = 1, 2, \ldots, h),$$

(1.4.25)

we may write the formulae

$$\frac{\partial}{\partial x_i} f(x_1, x_2, \ldots, x_n) = \frac{\tilde{\partial}}{\partial x_i} f(x_1, x_2, \ldots, x_n)$$

$$+ \sum_{k=1}^{h} (\Delta f)_i^k \frac{\partial}{\partial x_i} P_k(x_1, x_2, \ldots, x_n)\delta(P_k),$$

(1.4.26)

where $(\Delta f)_i^k$ represents the jump of the function $f(x_1, x_2, \ldots, x_n)$ in the positive direction of the Ox_i axis, corresponding to the surface of discontinuity S_k.

Let us consider the function (1.2.57) for which the straight lines $y = 0$ and $y = 2$ are curves of discontinuity. The corresponding jumps in the positive direction of the Oy axis are 1 and (-1), respectively; we have

$$\frac{\partial f(x, y)}{\partial y} = \delta(y) - \delta(y - 2),$$

(1.4.27)

since $\tilde{f}'_y = 0$ and $P'_y = 1$ for both the curves $P \equiv y = 0$ and $P \equiv y - 2 = 0$. Formula (1.4.27) is equivalent to formula (1.2.58).

Similarly, let us consider the function (Figure 1.23)

$$f(x, y) = \begin{cases} 1 \text{ for } |x| \leqslant y, \, y > 0, \\[2mm] 0 \text{ for } |x| > y, \, y > 0; \end{cases} \tag{1.4.28}$$

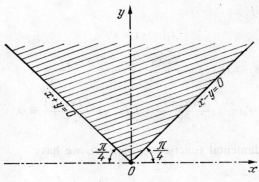

Fig. 1.23

therefore the function is equal to unity in the zone delimited by the lines $x - y = 0$ and $x + y = 0$ and is equal to zero outside that zone.

The lines mentioned are curves of discontinuity for the function $f(x, y)$ and the jumps which occur are equal to unity; we denote

$$P_1 \equiv x + y = 0, \tag{1.4.29}$$
$$P_2 \equiv x - y = 0.$$

Since $\tilde{f}'_x = \tilde{f}'_y = 0$ we may write

$$\frac{\partial f(x, y)}{\partial x} = \frac{\partial P_1}{\partial x} \delta(x + y) - \frac{\partial P_2}{\partial x} \delta(x - y) = \delta(x + y) - \delta(x - y),$$

$$\tag{1.4.30}$$

$$\frac{\partial f(x, y)}{\partial y} = \frac{\partial P_1}{\partial y} \delta(x + y) + \frac{\partial P_2}{\partial y} \delta(x - y) = \delta(x + y) + \delta(x - y),$$

where we have taken account of the fact that the normals to the curves of discontinuity must be taken in the positive direction of transversing corresponding to the respective axis. This result may be obtained also by applying the definition of the partial derivative.

We shall now prove the following

Theorem 1.4.1. *If $f_1(x, y)$ is a continuous function, $f_i(x, y)$ ($i = 2, 3, 4$) are functions possessing derivatives of all orders and $\delta(P)$ is the Dirac distribution concentrated on the plane curve $P(x, y) = 0$, P possessing derivatives of all orders, then the equality*

$$f_1(x, y) + f_2(x, y)\delta(P) + \frac{\partial}{\partial x}[f_3(x, y)\delta(P)] + \frac{\partial}{\partial y}[f_4(x, y)\delta(P)] = 0 \quad (1.4.31)$$

occurs if and only if

$$f_1(x, y) = 0, \quad (1.4.32)$$

$$f_2(x, y) + \frac{\partial f_3(x, y)}{\partial x} + \frac{\partial f_4(x, y)}{\partial y} = 0,$$

$$\quad (1.4.32')$$

$$f_3(x, y)\frac{\partial P(x, y)}{\partial x} + f_4(x, y)\frac{\partial P(x, y)}{\partial y} = 0.$$

Denoting a fundamental function by $\varphi(x, y)$, we have

$$(f_1, \varphi) + (f_2\delta(P), \varphi) + \left(\frac{\partial}{\partial x}[f_3\delta(P)], \varphi\right) + \left(\frac{\partial}{\partial y}[f_4\delta(P)], \varphi\right) = 0. \quad (1.4.31')$$

Remarking that

$$(\delta(P), \varphi(x, y)) = \int_{P=0} \varphi(x, y) \; \frac{ds}{\sqrt{P_x'^2 + P_y'^2}} \quad (1.4.33)$$

and choosing the function $\varphi(x, y)$ such that its support contains no point belonging to the curve $P = 0$, we obtain

$$(f_2\delta(P), \varphi) = \left(\frac{\partial}{\partial x}[f_3\delta(P)], \varphi\right) = \left(\frac{\partial}{\partial y}[f_4\delta(P)], \varphi\right) = 0;$$

thus relation (1.4.31') leads to

$$(f_1, \varphi) = 0. \quad (1.4.34)$$

This shows that condition (1.4.32) is satisfied almost everywhere, except eventually at the points belonging to the curve $P = 0$; from the continuity of the function $f_1(x, y)$ it may be stated that condition (1.4.32) is satisfied everywhere.
Relation (1.4.31) becomes

$$\left(f_2 + \frac{\partial f_3}{\partial x} + \frac{\partial f_4}{\partial y}\right)\delta(P) + f_3\frac{\partial}{\partial x}\delta(P) + f_4\frac{\partial}{\partial y}\delta(P) = 0 \quad (1.4.31'')$$

and may be written also as

$$\left(f_2 + \frac{\partial f_3}{\partial x} + \frac{\partial f_4}{\partial y}\right)\delta(P) + \left(f_3 \frac{\partial P}{\partial x} + f_4 \frac{\partial P}{\partial y}\right)\delta'(P) = 0, \qquad (1.4.31''')$$

whence the conditions (1.4.32′) follow necessarily, which proves the theorem (the sufficiency of the conditions is obvious).

In particular, if f_2, f_3, f_4 are constants we have

$$f_1 = f_2 = f_3 = f_4 = 0. \qquad (1.4.35)$$

In a similar way we can prove the following

Theorem 1.4.2. *If $f(x_1, x_2, \ldots, x_n)$ is a continuous function in R^n, $g_k(x_1, x_2, \ldots, x_n)$ and $P_k(x_1, x_2, \ldots, x_n)$ are functions possessing derivatives of all orders and $\delta(P_k)$ is a Dirac distribution concentrated on the $(n-1)$-dimensional surface $P_k = 0$ ($k = 1, 2, \ldots, m$), then the equality*

$$f(x_1, x_2, \ldots, x_n) + \sum_{k=1}^{m} g_k(x_1, x_2, \ldots, x_n)\delta(P_k) = 0 \qquad (1.4.36)$$

occurs if and only if

$$f(x_1, x_2, \ldots, x_n) = 0, \qquad (1.4.37)$$

$$g_k(x_1, x_2, \ldots, x_n) = 0 \qquad (k = 1, 2, \ldots, m). \qquad (1.4.37')$$

This may be considered as an extension of theorem 1.1.3.

Let us consider a fundamental function $\varphi(x_1, x_2, \ldots, x_n)$ such that its support contains no point belonging to the surfaces $P_k = 0$ ($k = 1, 2, \ldots, m$). Taking into account the relation which defines the distribution $\delta(P_k)$ we obtain

$$(\delta(P_k), \varphi(x_1, x_2, \ldots, x_n))$$

$$= \iint\ldots\int_{P_k=0} \frac{\varphi(x_1, x_2, \ldots, x_n)}{\sqrt{\left(\frac{\partial P_k}{\partial x_1}\right)^2 + \left(\frac{\partial P_k}{\partial x_2}\right)^2 + \ldots + \left(\frac{\partial P_k}{\partial x_n}\right)^2}} \, d\sigma_k = 0 \quad (k = 1, 2, \ldots, m),$$

where $d\sigma_k$ is the $(n-1)$-dimensional area element.

From relation (1.4.36) it follows that

$$(f(x_1, x_2, \ldots, x_n), \varphi(x_1, x_2, \ldots, x_n)) = 0,$$

from which we deduce that $f(x_1, x_2, \ldots, x_n) = 0$ almost everywhere except eventually on the surfaces $P_k = 0$ ($k = 1, 2, \ldots, m$); since the function $f(x_1, x_2, \ldots, x_n)$ is a continuous function, we deduce that condition (1.4.37) is satisfied everywhere.

Thus, relation (1.4.36) becomes

$$\sum_{k=1}^{m} g_k(x_1, x_2, ..., x_n)\delta(P_k) = 0. \tag{1.4.36'}$$

Now choosing the fundamental function $\varphi(x_1, x_2, ..., x_n)$ so as to contain only points belonging to the surface $P_k = 0$ we have

$$(g_k(x_1, x_2, ..., x_n)\,\delta(P_k),\, \varphi(x_1, x_2, ..., x_n)) = 0,$$

in other words

$$\iint ... \int_{P_k=0} \frac{g_k(x_1, x_2, ..., x_n)\varphi(x_1, x_2, ..., x_n)}{\sqrt{\left(\dfrac{\partial P_k}{\partial x_1}\right)^2 + \left(\dfrac{\partial P_k}{\partial x_2}\right)^2 + ... + \left(\dfrac{\partial P_k}{\partial x_n}\right)^2}}\, d\sigma_k = 0 \;\; (k = 1, 2, ..., m),$$

a relation which holds only if the conditions (1.4.37') are satisfied; the sufficiency of these conditions is obvious.

1.4.2. Homogeneous distributions

A function $f(x, y, z)$ defined on R^3 is said to be *homogeneous of degree* λ if we have the relation

$$f(kx, ky, kz) = k^\lambda f(x, y, z), \tag{1.4.38}$$

where k is a real positive number.

Since in the case of distributions homothety has been introduced as a linear transformation, there follows that relation (1.4.38) may be extended also to distributions; hence a distribution $f(x, y, z)$ is homogeneous of degree n if it satisfies relation (1.4.38).

In the following discussion, general results and examples are given concerning homogeneous distributions.

1.4.2.1. General results

It can be shown that the *necessary and sufficient condition* that the distribution $f(x_1, x_2, ..., x_n)$ be homogeneous of degree λ, is to satisfy *Euler's relation*

$$\sum_{i=1}^{n} x_i \frac{\partial}{\partial x_i} f(x_1, x_2, ..., x_n) = \lambda f(x_1, x_2, ..., x_n). \tag{1.4.39}$$

Also, we remark that formula (1.1.42) may be written

$$(f(kx_1, kx_2, ..., kx_n), \varphi(x_1, x_2, ..., x_n))$$

$$= k^{-n}\left(f(x_1, x_2, ..., x_n), \varphi\left(\frac{x_1}{k}, \frac{x_2}{k}, ..., \frac{x_n}{k} \right) \right); \qquad (1.4.40)$$

assuming that function $f(x_1, x_2, ..., x_n)$ is homogeneous of degree λ we obtain relation

$$k^{\lambda}(f(x_1, x_2, ..., x_n), \varphi(x_1, x_2, ..., x_n))$$

$$= k^{-n}\left(f(x_1, x_2, ..., x_n), \varphi\left(\frac{x_1}{k}, \frac{x_2}{k}, ..., \frac{x_n}{k} \right) \right), \qquad (1.4.41)$$

whence

$$k^{n+\lambda}(f(x_1, x_2, ..., x_n), \varphi(x_1, x_2, ..., x_n))$$

$$= \left(f(x_1, x_2, ..., x_n), \varphi\left(\frac{x_1}{k}, \frac{x_2}{k}, ..., \frac{x_n}{k} \right) \right). \qquad (1.4.41')$$

For $\delta(x)$ we may write

$$\delta(kx) = k^{-1}\delta(x); \qquad (1.4.42)$$

hence the distribution $\delta(x)$ of a single variable is homogeneous of degree (-1).
For $\delta(x, y)$ we have

$$\delta(kx, ky) = k^{-2}\delta(x, y); \qquad (1.4.42')$$

hence $\delta(x, y)$ is a homogeneous distribution of degree (-2).
In general, we have

$$\delta(kx_1, kx_2, ..., kx_n) = k^{-n}\delta(x_1, x_2, ...x_n); \qquad (1.4.43)$$

hence $\delta(x_1, x_2, ..., x_n)$ is a homogeneous distribution of degree $(-n)$.
The sum of two homogeneous distributions of degree λ is a homogeneous distribution of degree λ.

Homogeneous distributions of different degree are linearly independent.

The product of a homogeneous distribution of degree λ and a function $\alpha(x)$ of degree μ and possessing derivatives of all orders is a homogeneous distribution of degree $(\lambda + \mu)$.

The function

$$f(x_1, x_2, ..., x_n) = r_n = \sqrt{x_1^2 + x_2^2 + ... + x_n^2} \qquad (1.4.44)$$

is homogeneous of degree (-1); it has no singular point. The derivatives of this function are again homogeneous functions of degree zero but they have a singular point at the origin. In general, the derivative of a homogeneous function of degree λ is again a homogeneous function of degree $(\lambda - 1)$. This property holds also in the case of a distribution f of degree λ; indeed

$$
\left(\frac{\partial}{\partial x_i} f(x_1, x_2, ..., x_n), \ \varphi\left(\frac{x_1}{k}, \frac{x_2}{k}, ..., \frac{x_n}{k} \right) \right)
$$

$$
= -\left(f(x_1, x_2, ..., x_n), \frac{\partial}{\partial x_i} \varphi\left(\frac{x_1}{k}, \frac{x_2}{k}, \ \cdots \ , \frac{x_n}{k} \right) \right)
$$

$$
= -\frac{1}{k} k^{n+\lambda} \left(f(x_1, x_2, ..., x_n), \frac{\partial}{\partial x_i} \varphi(x_1, x_2, ..., x_n) \right)
$$

$$
= k^{n+\lambda-1} \left(\frac{\partial}{\partial x_i} f(x_1, x_2, ..., x_n), \varphi(x_1, x_2, ..., x_n) \right). \tag{1.4.45}
$$

We mention also the function

$$
f(x_1, x_2, ..., x_n) = \frac{1}{r_n} = \frac{1}{\sqrt{x_1^2 + x_2^2 + ... + x_n^2}}, \tag{1.4.46}
$$

which is homogeneous, of degree (-1) and has a singular point at the origin. The function is not locally integrable in the neighbourhood of the origin; for this reason we attach to the function a definite functional which coincides with it everywhere except at the origin.

It may be shown that for homogeneous functions $f(x_1, x_2, \ldots, x_n)$ of degree $(1-n)$, locally integrable, the derivative is obtained by applying the formula

$$
\frac{\partial}{\partial x_i} f(x_1, x_2, ..., x_n) = \frac{\tilde{\partial}}{\partial x_i} f(x_1, x_2, ..., x_n)
$$

$$
+ (-1)^{i-1} \delta(x_1, x_2, ..., x_n) \int_{\Gamma} f(x_1, x_2, ..., x_n) \, dx_1 \, dx_2 ... dx_{i-1} \, dx_{i+1} ... dx_n, \tag{1.4.47}
$$

where Γ is an arbitrary, $(n - 1)$-dimensional, closed surface while the integral of the right-hand-side is independent of that surface.

1.4.2.2. Examples

The function

$$
\frac{1}{R} = \frac{1}{\sqrt{x^2 + y^2 + z^2}} \tag{1.4.48}
$$

which is homogeneous of degree $-1 \neq 1 - 3$ has the derivatives

$$\frac{\partial}{\partial x}\left(\frac{1}{R}\right) = -\frac{x}{R^3}, \frac{\partial}{\partial y}\left(\frac{1}{R}\right) = -\frac{y}{R^3}, \frac{\partial}{\partial z}\left(\frac{1}{R}\right) = -\frac{z}{R^3}; \qquad (1.4.49)$$

thus we obtain homogeneous functions of degree $1 - 3 = -2$.

We may apply the differentiation formula (1.4.47) and we obtain

$$\frac{\partial^2}{\partial x^2}\left(\frac{1}{R}\right) = \frac{3x^2 - R^2}{R^5} - (-1)^{1-1}\delta(x, y, z)\int_\Gamma \frac{x\,dy\,dz}{R^3},$$

$$\frac{\partial^2}{\partial y^2}\left(\frac{1}{R}\right) = \frac{3y^2 - R^2}{R^5} - (-1)^{2-1}\delta(x, y, z)\int_\Gamma \frac{y\,dx\,dz}{R^3}, \qquad (1.4.50)$$

$$\frac{\partial^2}{\partial z^2}\left(\frac{1}{R}\right) = \frac{3z^2 - R^2}{R^5} - (-1)^{3-1}\delta(x, y, z)\int_\Gamma \frac{z\,dx\,dy}{R^3};$$

adding these three relations, we may write

$$\Delta\left(\frac{1}{R}\right) = -\delta(x, y, z)\int_\Gamma \frac{x\,dy\,dz - y\,dx\,dz + z\,dx\,dy}{R^3}, \qquad (1.4.51)$$

where Δ is the Laplace operator of three variables.

If we take for Γ a sphere with unit radius and with the centre at the origin, and if we remark that $dx\,dz = -dz\,dx$, we obtain

$$\Delta\left(\frac{1}{R}\right) = -4\pi\delta(x, y, z), \quad R = \sqrt{x^2 + y^2 + z^2}. \qquad (1.4.52)$$

Similarly, in the real two-dimensional space we find

$$\Delta[\log(x^2 + y^2)] = 4\pi\delta(x, y) \qquad (1.4.53)$$

or

$$\Delta\log\frac{1}{r} = -2\pi\delta(x, y), r = \sqrt{x^2 + y^2}, \qquad (1.4.53')$$

where now Δ is the Laplace operator corresponding to the two-dimensional space.

Returning to the real three-dimensional space, we consider the function

$$\frac{x^2}{R^3} = \frac{1}{R} - \frac{\partial}{\partial x}\left(\frac{x}{R}\right) ; \qquad (1.4.54)$$

remarking that

$$\Delta\left(\frac{x}{R}\right) = 2\frac{\partial}{\partial x}\left(\frac{1}{R}\right) \qquad (1.4.55)$$

and applying the Laplace operator to the function (1.4.54) we obtain

$$\Delta\left(\frac{x^2}{R^3}\right) = \Delta\left(\frac{1}{R}\right) - \frac{\partial}{\partial x}\left[\Delta\left(\frac{x}{R}\right)\right] = \Delta\left(\frac{1}{R}\right) - 2\frac{\partial^2}{\partial x^2}\left(\frac{1}{R}\right). \qquad (1.4.56)$$

Taking into account formulae (1.4.50) and integrating over a sphere of unit radius and with the centre at the origin, we have

$$\frac{\partial^2}{\partial x^2}\left(\frac{1}{R}\right) = \frac{1}{R^5}(3x^2 - R^2) - \frac{4\pi}{3}\delta(x, y, z); \qquad (1.4.50')$$

relation (1.4.56) takes the form

$$\Delta\left(\frac{x^2}{R^3}\right) = \frac{2}{R^5}(R^2 - 3x^2) - \frac{4\pi}{3}\delta(x, y, z), \qquad (1.4.56')$$

where formula (1.4.52) has also been used.

Let us consider the function

$$\frac{2x}{r^2} = \frac{2x}{x^2 + y^2} = \frac{\partial}{\partial x}\log(x^2 + y^2) \qquad (1.4.57)$$

in the real two-dimensional space; we can write

$$\Delta\left(\frac{2x}{x^2 + y^2}\right) = \Delta\left[\frac{\partial}{\partial x}\log(x^2 + y^2)\right] = \frac{\partial}{\partial x}\{\Delta[\log(x^2 - y^2)]\}, \qquad (1.4.58)$$

whence

$$\Delta\left(\frac{x}{r^2}\right) = 2\pi\frac{\partial\delta(x, y)}{\partial x}, \qquad (1.4.58')$$

where account has been taken of relation (1.4.53).

We remark that homogeneous distributions of degree $(1 - n)$ occur frequently in the theory of elasticity*. Thus, let there be the homogeneous functions of degree $1 - 2 = -1$

$$f_1(x, y) = - \frac{1}{4\pi(1 - v)} \frac{x}{x^2 + y^2} \left(1 - 2v + \frac{2x^2}{x^2 + y^2}\right),$$

$$\tag{1.4.59}$$

$$f_2(x, y) = - \frac{1}{4\pi(1 - v)} \frac{y}{x^2 + y^2} \left(1 - 2v + \frac{2x^2}{x^2 + y^2}\right),$$

where v is a constant; applying formula (1.4.47) we may write

$$\frac{\partial f_1}{\partial x} + \frac{\partial f_2}{\partial y} = \frac{\tilde{\partial} f_1}{\partial x} + \frac{\tilde{\partial} f_2}{\partial y} + \delta(x, y) \int_\Gamma f_1 \, dy - f_2 \, dx. \tag{1.4.60}$$

Remarking that the sum of the derivatives in the ordinary sense is equal to zero

$$\frac{\tilde{\partial} f_1}{\partial x} + \frac{\tilde{\partial} f_2}{\partial y} = 0 \tag{1.4.61}$$

and taking for Γ a circle of unit radius and with the centre at the origin, we obtain

$$\frac{\partial f_1}{\partial x} + \frac{\partial f_2}{\partial y} = - \delta(x, y). \tag{1.4.60'}$$

---

* See e.g. section 9.1.1.

2

Integral transformations
in distributions

2.1. Fourier transform

Among integral transformations an important place is taken by the Fourier transform which is a powerful computation tool in the study of the equations of mathematical physics.

In the theory of elasticity, for example, the Fourier transform is particularly useful for the integration of the differential equations of the problems encountered; in many cases, with its help we can determine effectively the fundamental solution of ordinary or partial differential equations, the general solution being then expressed with the help of the convolution product.

2.1.1. General results

In the following we shall first give results concerning the Fourier transform of a distribution of a simple variable; the results will be then extended to the Fourier transform of a distribution of several variables.

2.1.1.1. Fourier transform of a distribution of a single variable

If $f(x)$ is a real or complex function of the real variable $x \in R$, which satisfies the *Dirichlet conditions* (it is bounded, piecewise monotonic and has at most a finite number of points of discontinuity of the first species) and is *absolutely integrable* $\left(\text{i.e. } \int_{-\infty}^{\infty} |f(x)| \, \mathrm{d}x \text{ exists}\right)$, then the function

$$F(u) = \int_{-\infty}^{\infty} f(x) \mathrm{e}^{iux} \, \mathrm{d}x \tag{2.1.1}$$

exists and is called the *Fourier transform* of the function $f(x)$; we shall write

$$F[f(x)] = F(u).   \tag{2.1.1'}$$

We remark that the variable u is real; in general the *image function* $F(u)$ is complex, although the function $f(x)$ may be a real function.

Assuming that the function $F(u)$ is given, equality (2.1.1) may be considered as an integral equation with respect to the unknown function $f(x)$ under the symbol " $\int$ " ; the solution of this integral equation is written in the form

$$f(x) = \frac{1}{2\pi} \int_{-\infty}^{\infty} F(u)\, e^{-iux}\, du.   \tag{2.1.2}$$

The function $f(x)$ is called the *inverse Fourier transform* of the function $F(u)$; we have

$$f(x) = F^{-1}[F[f(x)]] = F^{-1}[F(u)].   \tag{2.1.2'}$$

The formulae (2.1.1) and (2.1.2) show that the Fourier transform represents a *linear operator;* indeed, the relations

$$F[Af_1(x) + Bf_2(x)] = AF[f_1(x)] + BF[f_2(x)],   \tag{2.1.3}$$

$$F^{-1}[A_1F_1(u) + B_1F_2(u)] = A_1F^{-1}[F_1(u)] + B_1F^{-1}[F_2(u)],   \tag{2.1.3'}$$

where A, B and A_1, B_1 are constants, are verified.

From the definition, formula (2.1.1), it will be seen that the use of the Fourier transform is limited since it cannot be applied to a great number of usual functions such as $\sin x$, $\operatorname{sh} x$, $\operatorname{ch} x$, $1(x)$, the Heaviside function $\theta(x)$, etc; the necessity of extending the range of applicability of the Fourier transform to wider classes of functions has initiated the *introduction of distributions.* It may be said that the theory of distributions creates the natural (and at the same time the general) application range of the Fourier transform, giving a sense to certain Fourier transforms which otherwise had no meaning at all.

In general, *the Fourier transform of a distribution may be defined on an arbitrary fundamental space;* thus, if the distributions are defined on a fundamental space Φ, the Fourier transform of these distributions are linear and continuous functionals on the space $F(\Phi)$, the elements of which are the Fourier transforms of the corresponding fundamental functions. In many cases the spaces K and S are particularly useful; in the following discussion we shall consider only the Fourier transforms of distributions defined on these fundamental spaces.

We remark that instead of the fundamental space K with real values we shall use the fundamental space K with complex values, as well as the space K' corresponding to complex distributions.

Thus, to a locally integrable function $f(x)$ with complex values there may correspond a distribution

$$(f(x), \varphi(x)) = (\overline{f}(x), \varphi(x)), \tag{2.1.4}$$

where $\overline{f}$ represents the conjugate of the function f. Also, multiplication by a complex function $\alpha(x)$ possessing derivatives of all orders is defined by the relation

$$(\alpha(x)f(x), \varphi(x)) = (f(x), \overline{\alpha}(x)\varphi(x)); \tag{2.1.5}$$

in particular if $\alpha(x) = a$ is a complex number, it follows that

$$(af(x), \varphi(x)) = \overline{a}(f(x), \varphi(x)) = (f(x), \overline{a}\varphi(x)). \tag{2.1.5'}$$

Thus, to a distribution $f(x)$ we may associate a complex conjugate distribution $\overline{f}(x)$ through the agency of the relation

$$(\overline{f}(x), \varphi(x)) = \overline{(f(x), \overline{\varphi}(x))}. \tag{2.1.6}$$

Let $\varphi(x)$ be a fundamental complex function of a real variable x; therefore $\varphi(x)$ is a function possessing derivatives of all orders and a compact support (e.g. $|x| \leqslant a$). By formula (2.1.1) the Fourier transform of this fundamental function is

$$F[\varphi(x)] = \psi(u) = \int_{-\infty}^{\infty} \varphi(x)\, e^{iux}\, dx. \tag{2.1.7}$$

The function $\psi(u)$ can be defined also for complex values $s = u + iv$ namely

$$\psi(s) = \int_{-\infty}^{\infty} \varphi(x)\, e^{isx}\, dx = \int_{-\infty}^{\infty} \varphi(x)\, e^{-vx}\, e^{iux}\, dx. \tag{2.1.7'}$$

Relation (2.1.7') defines a holomorphic function over all the complex plane (except the point at infinity) such that it can be differentiated with respect to the complex variable s; we obtain in general

$$F[\varphi^{(m)}(x)] = (-is)^m F[\varphi(x)] \qquad (m = 1, 2, \ldots). \tag{2.1.8}$$

In the case of a differential polynomial $P = P\left(\dfrac{d}{dx}\right)$, relation (2.1.8) takes the more general form

$$F\left[P\left(\frac{d}{dx}\right)\varphi(x)\right] = P(-is)F[\varphi(x)]. \tag{2.1.8'}$$

The functions $\psi(s)$ satisfy the inequality

$$|s^p \psi(s)| \leqslant C_p \, e^{a|v|},$$ (2.1.9)

where p is a non-negative integer and C_p a constant depending on p (eventually also on $\psi(s)$); it is assumed that the support of the fundamental function $\varphi(x)$ is the segment $[-a, a]$.

The set of functions $\psi(s) = F[\varphi(x)]$, where the support of the fundamental functions is included in the segment $[-a, a]$, forms *the vector space* $Z(a)$ whose elements satisfy the inequality (2.1.9). We shall denote by

$$Z = \bigcup_a Z(a) \ \ (K = \bigcup_a K(a))$$ (2.1.10)

the new *complex linear space*.

Convergence in the space Z is given by

Definition 2.1.1. *The sequence* $\psi_n(s)$ *converges to* $\psi(s)$ *in* Z *if the terms of the sequence verify the inequality*

$$|s^p \psi_n(s)| \leqslant C_p \, e^{a|v|},$$ (2.1.11)

where C_p *and* a *do not depend on* n, p *is a non-negative integer and all the functions* $\psi_n(s)$ *tend uniformly to* $\psi(s)$ *on any interval of the real axis.*

We shall denote by Z' the set of linear and continuous functionals defined on Z; the functional $F(s) \in Z'$ represents a distribution defined on $Z^{\star}$. Between the spaces K and Z there exists a one-to-one correspondence. Moreover we can write

$$F[K] = Z, \qquad F^{-1}[Z] = K.$$ (2.1.10')

The space Z permits us to define the Fourier transforms of a distribution so as to include the classical Fourier transform as a particular case. We shall introduce the following

Definition 2.1.2. *If* $F(s)$ *is a distribution defined on* Z *and* $f(x)$ *is a distribution defined on* K, *the functional* $F(s) \in Z'$ *specified by the equality of the Parseval type*

$$(F(s), \psi(s)) = 2\pi(f(x), \varphi(x))$$ (2.1.12)

is termed the Fourier transform of the distribution $f(x)$ *and will be denoted by*

$$F(s) = F[f(x)].$$ (2.1.13)

---

[*] Some authors denote a distribution defined on the fundamental space Z by the term of *ultra-distribution*.

We can also write

$$(F[f(x)], F[\varphi(x)]) = 2\pi(f(x), \varphi(x)). \tag{2.1.12'}$$

The *classical properties of the Fourier transform* are maintained in the form

$$P\left(\frac{d}{ds}\right) F(s) = P\left(\frac{d}{ds}\right) F[f(x)] = F[P(ix)f(x)], \tag{2.1.14}$$

$$F\left[P\left(\frac{d}{dx}\right) f(x)\right] = P(-is)F[f(x)] = P(-is)F(s), \tag{2.1.15}$$

$$F^{-1}[F[f(x)]] = f(x), \tag{2.1.16}$$

$$F[F[\varphi(x)]] = 2\pi\varphi(-x), \tag{2.1.16'}$$

$$F[F[f(x)]] = 2\pi f(-x), \tag{2.1.16''}$$

where F^{-1} is the inverse operator, defined on Z'.

For the *direct product* of two distributions we can establish the relation

$$F[f(x) \times g(y)] = F[f(x)] \times F[g(y)]. \tag{2.1.17}$$

Also, in connection with the *convolution product* one can prove the relations

$$F[f(x) * g(x)] = F[f(x)]F[g(x)], \tag{2.1.18}$$

$$F[f(x)g(x)] = F[f(x)] * F[g(x)]. \tag{2.1.18'}$$

The first relation is valid if $f(x) \in S'$ and $g(x)$ is a distribution with a bounded support; the second relation is valid if $f(x) \in S'$ and the function $g(x) \in C^\infty$ is such that $f(x)g(x) \in S'$ and the support of its Fourier transform is bounded.

An important method of introducing the Fourier transform of a distribution consists in considering such a transform as the *limit of a sequence*. Thus, if the sequence $f_n(x)$ determines the distribution $f(x) = \lim_{n\to\infty} f_n(x)$ and if we denote by $F_n(s)$ the Fourier transform of $f_n(x)(F_n(s) = F[f_n(x)])$, then we have

$$F(s) = F[f(x)] = \lim_{n\to\infty} F_n(s) = \lim_{n\to\infty} F[f_n(x)]; \tag{2.1.19}$$

the limit is obviously considered in the sense of convergence in the space Z'.

The Fourier transform of a *concentrated distribution on a bounded domain* is a regular distribution and may be computed by the formula

$$F[f(x)] = (\bar{f}(x), e^{iux}) = (\overline{f(x), e^{-iux}}) = F(u), \tag{2.1.20}$$

where e^{iux} is a fundamental function of the space K, which coincides with the function e^{iux} on the support of $f(x)$.

We often have to deal with *temperate distributions* $f(x)$ whose Fourier transforms verify the equality

$$(F[f(x)], \varphi(u)) = (f(u), F[\varphi(x)]), \tag{2.1.21}$$

where $F[\varphi(x)] = \psi(u)$ and $F[f(x)] = F(u)$; these distributions depend only on the variable u.

2.1.1.2. Fourier transform of a distribution of several variables

In a similar way we introduce the Fourier transform of a distribution of several variables. Let us consider the fundamental functions of several variables $\varphi(x_1, x_2, \ldots, x_n)$ and their Fourier transforms $\psi(s_1, s_2, \ldots, s_n)$; we shall now give the following

Definition 2.1.3. *The Fourier transform of the function* $f(x_1, x_2, \ldots, x_n)$ *is defined by the equality of the Parseval type*

$$(F(s_1, s_2, \ldots, s_n), \psi(s_1, s_2, \ldots, s_n)) = (2\pi)^n (f(x_1, x_2, \ldots, x_n), \varphi(x_1, x_2, \ldots, x_n)) \tag{2.1.22}$$

and we can write

$$F(s_1, s_2, \ldots, s_n) = F[f(x_1, x_2, \ldots, x_n)], \tag{2.1.23}$$

$$f(x_1, x_2, \ldots, x_n) = F^{-1}[F(s_1, s_2, \ldots, s_n)]. \tag{2.1.23'}$$

We can establish easily the relations

$$F\left[P\left(\frac{\partial}{\partial x_1}, \frac{\partial}{\partial x_2}, \ldots, \frac{\partial}{\partial x_n} \right) f(x_1, x_2, \ldots, x_n) \right]$$

$$= P(-is_1, -is_2, \ldots, -is_n) F(s_1, s_2, \ldots, s_n), \tag{2.1.24}$$

$$P\left(\frac{\partial}{\partial s_1}, \frac{\partial}{\partial s_2}, \ldots, \frac{\partial}{\partial s_n} \right) F(s_1, s_2, \ldots, s_n)$$

$$= F[P(ix_1, ix_2, \ldots, ix_n) f(x_1, x_2, \ldots, x_n)]. \tag{2.1.24'}$$

Also, the considerations concerning the Fourier transform of a distribution with a single variable may be extended to the present case.

2.1.2. Examples

In the following we give a few examples of computation of Fourier transforms for some important distributions.

2.1.2.1. Dirac distribution

Let us consider the Dirac distribution, which is concentrated at the origin; using formula (2.1.20) we may write

$$F[\delta(x)] = \overline{(\delta(x), e^{-iux})} = e^0 = 1. \tag{2.1.25}$$

In general, we have

$$F[\delta(x_1, x_2, ..., x_n)] = 1, \tag{2.1.26}$$

whence

$$F[\delta'(x)] = -isF[\delta(x)] = -is = -v + iu, \tag{2.1.27}$$

$$F[\delta^{(m)}(x)] = (-is)^m. \tag{2.1.27'}$$

Also we can write

$$F[\delta(x - a)] = \overline{(\delta(x - a), e^{-iux})} = e^{iua}. \tag{2.1.28}$$

If $f(x)$ is a distribution we have in general

$$F[f(x - a)] = e^{iua}F[f(x)]. \tag{2.1.28'}$$

Introducing the function $1(x)$ and applying the formula (2.1.12') we obtain

$$(F[1(x)], \psi(s)) = 2\pi(1(x), \varphi(x)) = 2\pi \int_{-\infty}^{\infty} \varphi(x) \, dx = 2\pi\psi(0) = 2\pi(\delta(s), \psi(s));$$

hence

$$F[1(x)] = 2\pi\delta(s). \tag{2.1.29}$$

On the other hand the function $f(x) = 1(x)$ is bounded and defines a temperate distribution such that the Fourier transform of that distribution will depend on the variable u only; in that case we can write by formula (2.1.21)

$$(F[1(x)], \varphi(u)) = (1(u), F[\varphi(x)]) = \int_{-\infty}^{\infty} \psi(u) \, du$$

and taking into account (2.1.7) and (2.1.9′), it follows that

$$F[1(x)] = \int_{-\infty}^{\infty} e^{iux} \, dx = \lim_{n \to \infty} \int_{-n}^{n} e^{iux} \, dx = 2\pi\delta(u). \qquad (2.1.29')$$

Taking into account Euler's formula

$$e^{iux} = \cos ux + i \sin ux$$

and integrating we obtain

$$2\pi\delta(u) = \lim_{n \to \infty} \frac{\sin ux}{u} \Big|_{-n}^{n} - i \lim_{n \to \infty} \frac{\cos ux}{u} \Big|_{-n}^{n} = 2 \lim_{n \to \infty} \frac{\sin nu}{u},$$

i.e.

$$\delta(u) = \lim_{n \to \infty} \frac{\sin nu}{\pi u}; \qquad (2.1.30)$$

this is an important limit and

$$f_n(u) = \frac{\sin nu}{\pi u} \qquad (2.1.30')$$

is a representative δ sequence which corresponds to that given by formula (1.2.96). In the case of a distribution with several variables we have

$$F[1(x_1, x_2, ..., x_n)] = (2\pi)^n \delta(s_1, s_2, ..., s_n), \qquad (2.1.31)$$

a transform which may be connected to the transform (2.1.26).

2.1.2.2. Heaviside distribution

To compute the Fourier transform of the Heaviside distribution $\theta(x)$ we write

$$F\left[\frac{d}{dx} \theta(x)\right] = F[\delta(x)] = 1,$$

whence

$$- \text{is } F[\theta(x)] = 1;$$

finally we obtain

$$F[\theta(x)] = -\frac{1}{\text{i}s} = \frac{\text{i}}{s} = \frac{v}{u^2 + v^2} + \text{i}\frac{u}{u^2 + v^2}.$$

Since the Heaviside function $\theta(x)$ is bounded, it defines a temperate distribution and the Fourier transform of that distribution will depend on the variable u only (this is due to the fact that we may extend a regular distribution from the fundamental space K to the fundamental space S); noting that formula (1.2.94) leads to

$$\lim_{v \to +0} \frac{1}{\pi} \frac{v}{u^2 + v^2} = \delta(u); \qquad (2.1.32)$$

it follows that

$$F[\theta(x)] = \pi\delta(u) = \frac{\text{i}}{u}. \qquad (2.1.33)$$

Since

$$\theta(x) + \theta(-x) = 1(x), \qquad (2.1.34)$$

we have also

$$F[\theta(-x)] = F[1(x)] - F[\theta(x)] = \pi\delta(u) - \frac{\text{i}}{u}. \qquad (2.1.35)$$

Taking into account (1.1.56) we may introduce the distribution $\frac{1}{x}$ and compute

$$\lim_{\varepsilon \to +0} \int_{-\infty}^{\infty} \frac{\varphi(x)}{x + \text{i}\varepsilon} \, dx = -\text{i}\pi\varphi(0) + Vp\int_{-\infty}^{\infty} \frac{\varphi(x)}{x} \, dx,$$

where we have also used relation (2.1.32). This allows to write Sohocki's relations

$$\frac{1}{x + \text{i}0} = \lim_{\varepsilon \to +0} \frac{1}{x + \text{i}\varepsilon} = \frac{1}{x} - \text{i}\pi\delta(x), \qquad (2.1.36)$$

$$\frac{1}{x - \text{i}0} = \lim_{\varepsilon \to +0} \frac{1}{x - \text{i}\varepsilon} = \frac{1}{x} + \text{i}\pi\delta(x); \qquad (2.1.36')$$

hence we can write

$$F[\theta(x)] = \frac{i}{u + i0}, \tag{2.1.33'}$$

$$F[\theta(-x)] = -\frac{i}{u - i0}. \tag{2.1.35'}$$

2.1.2.3. Heisenberg's distributions

W. Heisenberg has introduced in quantum mechanics the distributions $\delta_+(x)$ and $\delta_-(x)$ defined by the relations

$$\delta_+(x) = \frac{1}{2}\delta(x) - \frac{1}{2\pi i x} = \frac{1}{2}\delta(x) + \frac{i}{2\pi x} = \frac{1}{2\pi}\frac{i}{x + i0}, \tag{2.1.37}$$

$$\delta_-(x) = \frac{1}{2}\delta(x) + \frac{1}{2\pi i x} = \frac{1}{2}\delta(x) - \frac{i}{2\pi x} = -\frac{1}{2\pi}\frac{i}{x - i0} = \delta(x) - \delta_+(x). \tag{2.1.37'}$$

Taking into account relations (2.1.33) and (2.1.35') we obtain

$$F[\theta(x)] = 2\pi\delta_+(u), \tag{2.1.38}$$

$$F[\theta(-x)] = 2\pi\delta_-(u). \tag{2.1.38'}$$

2.1.2.4. Distributions x_+ and x_-

We introduce now the distributions corresponding to the functions x_+ and x_- defined as follows (Figure 2.1a, b)

$$x_+ = x\theta(x) = \begin{cases} 0 \text{ for } x \leqslant 0 \\ \\ x \text{ for } x > 0, \end{cases} \tag{2.1.39}$$

$$x_- = -x\theta(-x) = \begin{cases} -x \text{ for } x < 0 \\ \\ 0 \text{ for } x \geqslant 0; \end{cases} \tag{2.1.39'}$$

We remark that these distributions are temperate distributions (defined on the fundamental space S).

For the computation of the Fourier transform of the distribution x_+ we note that

$$\frac{\mathrm{d}}{\mathrm{d}x}\left[x\theta(x)\right] = \theta(x) + x\delta(x) = \theta(x),$$

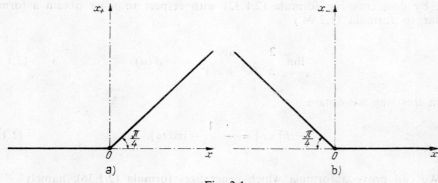

Fig. 2.1

whence

$$F\left[\frac{\mathrm{d}}{\mathrm{d}x}(x_+)\right] = F[\theta(x)] = \frac{\mathrm{i}}{s}$$

i.e.

$$- \mathrm{i}sF[x_+] = \frac{\mathrm{i}}{s}.$$

Hence, we obtain over the space Z

$$F[x_+] = -\frac{1}{s^2}.$$

But the Fourier transform $F[x_+]$ depends on u only, since x_+ is a temperate distribution which may be extended over the space S; we may write

$$F[x_+] = -\lim_{v \to +0} \frac{1}{s^2} = -\lim_{v \to +0} \frac{1}{(u + \mathrm{i}v)^2}$$

$$= -\lim_{v \to +0} \frac{u^2 - v^2}{(u^2 + v^2)^2} + 2\mathrm{i} \lim_{v \to +0} \frac{uv}{(u^2 + v^2)^2}.$$

8—c. 767

We note that

$$\lim_{v \to +0} \frac{u^2 - v^2}{(u^2 + v^2)^2} = \frac{1}{u^2};$$

also, by differentiating formula (2.1.32) with respect to u we obtain a formula similar to formula (1.2.94')

$$\lim_{v \to +0} \frac{2}{\pi} \frac{uv}{(u^2 + v^2)^2} = -\delta'(u). \qquad (2.1.32')$$

In this way we obtain

$$F[x_+] = -\frac{1}{u^2} - i\pi\delta'(u). \qquad (2.1.40)$$

We can prove a formula which generalizes formula (2.1.36), namely

$$\frac{1}{(x + i0)^m} = \frac{1}{x^m} - (-1)^{m-1} \frac{i\pi}{(m-1)!} \delta^{(m-1)}(x) \qquad (m = 1, 2, \ldots); \quad (2.1.41)$$

in that case the Fourier transform (2.1.40) may be written also in the form

$$F[x_+] = -\frac{1}{(u + i0)^2}. \qquad (2.1.40')$$

Using the relation

$$\frac{d}{dx}(x_+^n) = nx_+^{n-1}$$

and applying the method of complete induction it may be proved that

$$F[x_+^n] = n! \left(\frac{i}{u}\right)^{n+1} + (-i)^n \pi \delta^{(n)}(u); \qquad (2.1.42)$$

using formula (2.1.41) we may write this Fourier transform also in the form

$$F[x_+^n] = \frac{i^{n+1} n!}{(u + i0)^{n+1}}. \qquad (2.1.42')$$

With regard to the distribution x_-, we may write

$$\frac{\mathrm{d}}{\mathrm{d}x}\left[-x\theta(-x)\right] = -\theta(-x),$$

whence

$$F\left[\frac{\mathrm{d}}{\mathrm{d}x}(x_-)\right] = F[-\theta(-x)] = -F[\theta(-x)] = \frac{\mathrm{i}}{s},$$

that is

$$-\mathrm{i}sF[x_-] = \frac{\mathrm{i}}{s};$$

as in the preceding case, we may write

$$F[x_-] = -\lim_{v \to -0}\frac{1}{s^2} = -\lim_{v \to -0}\frac{1}{(u+\mathrm{i}v)^2} = -\lim_{v \to +0}\frac{1}{(u-\mathrm{i}v)^2}$$

$$= -\lim_{v \to +0}\frac{u^2 - v^2}{(u^2 + v^2)^2} - 2\mathrm{i}\lim_{v \to +0}\frac{uv}{(u^2 + v^2)^2},$$

whence, taking into account relation (2.1.32′), it follows that

$$F[x_-] = -\frac{1}{u^2} + \mathrm{i}\pi\delta'(u). \qquad (2.1.40'')$$

Formula (2.1.36′) may be generalized in the form

$$\frac{1}{(x-\mathrm{i}0)^m} = \frac{1}{x^m} + (-1)^{m-1}\frac{\mathrm{i}\pi}{(m-1)!}\,\delta^{(m-1)}(x)\ (m = 1, 2, \ldots); \quad (2.1.41')$$

in that case the Fourier transform (2.1.40″) may be written also

$$F[x_-] = -\frac{1}{(u-\mathrm{i}0)^2}. \qquad (2.1.40''')$$

Remarking that

$$x = x_+ - x_-, \qquad (2.1.43)$$

we may also write

$$F[x] = F[x_+] - F[x_-] = -2i\pi\delta'(u).$$ (2.1.44)

2.1.2.5. Other Fourier transforms

Using the property (2.1.14) we may write

$$F[P(x)] = P\left(-i\frac{d}{ds}\right)F[1(x)] = 2\pi P\left(-i\frac{d}{ds}\right)\delta(s);$$ (2.1.45)

in particular, by extension from the space K over the space S, it follows that

$$F[x^n] = 2\pi(-1)^n\delta^{(n)}(u) \qquad (n = 0, 1, 2, \ldots).$$ (2.1.46)

Starting from the function $\theta(x)e^{-ax}$, $a > 0$, we obtain

$$F[\theta(x)e^{-ax}] = \int_{-\infty}^{\infty}\theta(x)e^{-ax}\,e^{iux}\,dx = \int_0^\infty e^{-ax}\cos ux\,dx + i\int_0^\infty e^{-ax}\sin ux\,dx;$$

but we have

$$\int_0^\infty e^{-ax}\cos ux\,dx = \frac{a}{a^2+u^2},$$

$$\int_0^\infty e^{-ax}\sin ux\,dx = \frac{u}{a^2+u^2};$$ (2.1.47)

hence we may write

$$F[\theta(x)\,e^{-ax}] = \frac{a}{a^2+u^2} + i\frac{u}{a^2+u^2}$$

$$= \frac{1}{a-iu} = F[\theta(x)\,e^{-a|x|}], \quad a > 0.$$ (2.1.48)

In a similar way we may write

$$F[\theta(-x)\,e^{-a|x|}] = \int_{-\infty}^0 e^{-a|x|}\,e^{iux}\,dx = -\int_\infty^0 e^{-a|y|}\,e^{-iuy}\,dy$$

$$= \int_0^\infty e^{-a|x|}\,e^{-iux}\,dx = \int_0^\infty e^{-a|x|}\cos ux\,dx - i\int_0^\infty e^{-a|x|}\sin ux\,dx,$$

whence

$$F[\theta(-x)\,e^{-a|x|}] = \frac{a}{a^2 + u^2} - i\,\frac{u}{a^2 + u^2} = \frac{1}{a + iu}, \quad a > 0. \qquad (2.1.48')$$

Starting from relations (2.1.48) and (2.1.48′) and taking into account relation (2.1.36) we may also write

$$F[e^{-a|x|}] = F[\theta(x)\,e^{-a|x|} + \theta(-x)\,e^{-a|x|}]$$

$$= \frac{1}{a - iu} + \frac{1}{a + iu} = \frac{2a}{a^2 + u^2}, \quad a > 0. \qquad (2.1.49)$$

We shall now compute the Fourier transform of the distribution $\mathrm{Vp}\,\dfrac{1}{x}$ (which is denoted by $\dfrac{1}{x}$); we start from the Heaviside distribution and apply formula (2.1.16″). It follows that

$$F[F[\theta(u)]] = \pi F[\delta(x)] + iF\left[\frac{1}{x}\right] = \pi + iF\left[\frac{1}{x}\right] = 2\pi\theta(-u),$$

whence

$$iF\left[\frac{1}{x}\right] = 2\pi\theta(-u) - \pi = \begin{cases} \pi & \text{for } u < 0, \\[2mm] -\pi & \text{for } u > 0; \end{cases}$$

hence we may write

$$F\left[\frac{1}{x}\right] = i\pi\,\mathrm{sgn}\,u. \qquad (2.1.50)$$

2.1.2.6. Fourier transform with respect to several variables

In the case of the Fourier transform with respect to two variables of the distribution $\delta(y)$, we may write

$$(\delta(y), \varphi(x, y)) = \int_{-\infty}^{\infty} \varphi(x, 0)\,dx. \qquad (2.1.51)$$

The defining relation (2.1.22) becomes

$$(F[\delta(y)], \psi(\alpha, \beta)) = 4\pi^2 \int_{-\infty}^{\infty} \varphi(x, 0)\,dx. \qquad (2.1.52)$$

Taking into account relation

$$\varphi(x, y) = \frac{1}{4\pi^2} \int_{-\infty}^{\infty} \int_{-\infty}^{\infty} \psi(\alpha, \beta) \, e^{-i(\alpha x + \beta y)} \, d\alpha \, d\beta \qquad (2.1.53)$$

and relation (2.1.29'), we obtain

$$(F[\delta(y)], \psi(\alpha, \beta)) = \int_{-\infty}^{\infty} \int_{-\infty}^{\infty} \psi(\alpha, \beta) \, d\alpha \, d\beta \int_{-\infty}^{\infty} e^{-i\alpha x} \, dx$$

$$= 2\pi \int_{-\infty}^{\infty} \int_{-\infty}^{\infty} \psi(\alpha, \beta) \, d\alpha \, d\beta = (2\pi \delta(\alpha), \psi(\alpha, \beta)).$$

Hence

$$F[\delta(y)] = 2\pi \delta(\alpha). \qquad (2.1.54)$$

We remark that $\delta(y)$ is a distribution of two variables concentrated on the line $y = 0$ and should not be mistaken for $\delta(y)$ which may represent a distribution concentrated at the origin on the Oy line.

Proceeding similarly we obtain

$$F[\delta(x)] = 2\pi \delta(\beta). \qquad (2.1.54')$$

Using formulae (2.1.54) and (2.1.54') we may also write the relationships

$$F\left[\frac{d}{dx} \delta(x) \right] = - i\alpha F[\delta(x)] = - 2i\pi\alpha\delta(\beta), \qquad (2.1.55)$$

$$F\left[\frac{d}{dy} \delta(y) \right] = - i\beta F[\delta(y)] = - 2i\pi\beta\delta(\alpha). \qquad (2.1.55')$$

In the case of the Fourier transform with respect to n variables the formulae (2.1.54) and (2.1.54') become

$$F[\delta(x_2)] = (2\pi)^{n-1} \delta(\alpha_1, \alpha_3, \alpha_4, ..., \alpha_n), \qquad (2.1.56)$$

$$F[\delta(x_1)] = (2\pi)^{n-1} \delta(\alpha_2, \alpha_3, ..., \alpha_n). \qquad (2.1.56')$$

We remark that formula (2.1.56) for example, could also have been obtained starting from the direct product

$$\delta(x_2) = \delta(x_2) \times 1(x_1, x_3, x_4, ..., x_n), \qquad (2.1.57)$$

which leads to

$$F[\delta(x_2)] = F[\delta(x_2)] \times F[1(x_1, x_3, x_4, ..., x_n)], \qquad (2.1.58)$$

where the first factor of the product represents a Fourier transform with respect to a single variable.

Similarly, starting from

$$\theta(x) = \theta(x) \times 1(y), \qquad (2.1.59)$$

we find

$$F[\theta(x)] = F[\theta(x)] \times F[1(y)]$$

$$= \left[\pi\delta(\alpha) + \frac{i}{\alpha} \right] \times [2\pi\delta(\beta)] = 4\pi^2 \delta_+(\alpha) \times \delta(\beta), \qquad (2.1.60)$$

where the foregoing remark applies again.

Taking into account the distributivity of the direct product with respect to addition and the relation

$$\delta(\alpha) \times \delta(\beta) = \delta(\alpha, \beta), \qquad (2.1.61)$$

we may write relation (2.1.60) in the form

$$F[\theta(x)] = 2\pi^2 \delta(\alpha, \beta) + 2i\pi \frac{1}{\alpha} \times \delta(\beta). \qquad (2.1.60')$$

2.2. Laplace transform

Among the integral transformations closely connected to the Fourier transform, an important part is played by the Laplace transform.

In the following discussion we shall deal first with the Laplace transform of a distribution of a single variable and then with the Laplace transform of a distribution of several variables; also, the connection existing between the Laplace and the Fourier transforms will be pointed out.

2.2.1. Laplace transform of a distribution of a single variable

In the following discussion we shall treat of some general results and then we shall deal with the Laplace transform of a derivative of a distribution.

2.2.1.1. General results

We shall introduce

Definition 2.2.1. *If $f(x)$ is a complex function of a real variable which satisfies the conditions*

(i) $f(x) = 0$ *for* $x < 0$,

(ii) $f(x)$ *is piecewise differentiable*,

(iii) $|f(x)| \leqslant Me^{ax}$,

where M is a positive constant and the non-negative constant a represents the incremental ratio of the function, then the function $L(p)$ of the complex variable $p = u + iv$, defined by the expression

$$L(p) = \int_0^\infty f(x)\,e^{-px}\,dx, \tag{2.2.1}$$

is called the Laplace transform of the function $f(x)$ and is denoted by

$$L[f(x)] = L(p). \tag{2.2.1'}$$

The function $f(x)$ is also called the *original function* and $L(p)$ the *image function*. The functions $\theta(x), x_+^n$, $\theta(x)\,P(x)\sin \omega x$ are original functions, the function $\theta(x)e^{x^2}$, for example, is not an original function.

It may be shown easily that the sum and the product of two original functions are also original functions. In particular if the original function $f(x)$ is a bounded function then its incremental ratio is $a = 0$; for example the original function $\theta(x)e^{-x^2}$ has this property.

It may be shown that when conditions (i), (ii), (iii) are satisfied, the Laplace transform (2.2.1) exists in the half-plane $u > a$ and is holomorphic in that half-plane; also, the integral (2.2.1) is absolutely convergent. Consequently the function $L(p)$ possesses derivatives of all orders in the half-plane mentioned and the following relation *(the theorem of the differentiation of the image function)* is valid

$$L[(-x)^n f(x)] = L^{(n)}(p) = \frac{d^n}{dp^n}\,L(p). \tag{2.2.2}$$

The incremental ratio of the original function $(-x)^n f(x)$ is the same as that of the original function $f(x)$.

Also we remark that if $a = 0$ the Laplace transform exists in the half-plane $u > 0$.

To the Laplace transform defined by formula (2.2.1) there corresponds an inverse transform. Thus, if $f(x)$ is the original function with the incremental ratio *a* and $L(p)$ is its image, then the *inverse Laplace transform* is given by the relation

$$L^{-1}[L(p)] = \frac{1}{2\pi i} \int_{u-i\infty}^{u+i\infty} L(p)\, e^{px}\, dp, \ u > a. \tag{2.2.3}$$

In general, a holomorphic function $L(p)$ may represent the Laplace transform of a function $f(x)$ if, and only if, it exists in the half-plane $u > a$ and its modulus can be majorized in that half-plane by a polynomial in $|p|$.

We remark that the defining relation (2.2.1) may be extended to distributions whose support is on the half-line $x \geqslant 0$. This leads to the following

Definition 2.2.2. *If $f(x)$ is a distribution having its support on the half-line $x \geqslant 0$ and is such that the distribution $f(x)e^{-px}$ is a temperate distribution, then*

$$L[f(x)] = (f(x), e^{-px}) \tag{2.2.4}$$

represents the Laplace transform of that distribution.

It is obvious that relation (2.2.4) generalizes relation (2.2.1). We also note that relation (2.2.2) is maintained in the case of distributions, owing to the differentiation formula of the latter.

2.2.1.2. Examples

It is well known that any distribution with a bounded support may be extended from the fundamental space K to the fundamental space S. Since the support of the distribution $\delta(x)$ is the origin of the Ox axis, the distribution $\delta(x)e^{-px}$ will be a temperate distribution.

Applying formula (2.2.4) we obtain

$$L[\delta(x)] = (\delta(x), e^{-px}) = e^0 = 1 \tag{2.2.5}$$

and also

$$L[\delta'(x)] = (\delta'(x), e^{-px}) = -(\delta(x), -pe^{-px}) = p; \tag{2.2.6}$$

in general we may write

$$L[\delta^{(n)}(x)] = p^n \qquad (n = 0, 1, 2, ...). \tag{2.2.7}$$

For $\delta(x - a)$ there follows

$$L[\delta(x - a)] = (\delta(x - a), e^{-px}) = (\delta(x), e^{-p(x+a)}) = e^{-pa} \tag{2.2.8}$$

and in general we have

$$L[f(x - a)] = (f(x - a), e^{-px}) = (f(x), e^{-p(x+a)}) = e^{-pa}L[f(x)]; \quad (2.2.9)$$

this constitutes the *delay theorem* which is valid in the case of distributions, too. For the distribution $\delta(kx)$ we may write

$$L[\delta(kx)] = (\delta(kx), e^{-px}) = \frac{1}{k}(\delta(x), e^{-\frac{px}{k}}) = \frac{1}{k}, \quad k > 0. \quad (2.2.10)$$

More generally, we may write

$$L[f(kx)] = (f(kx), e^{-px}) = \frac{1}{k}(f(x), e^{-\frac{px}{k}}), \quad k > 0$$

or

$$L[f(kx)] = \frac{1}{k}L\left(\frac{p}{k}\right), \quad k > 0, \quad (2.2.11)$$

where we have used the notation (2.2.1'); this constitutes the *theorem of similitude*. Since the Laplace transform is a linear transform, we may write

$$L[\delta(a^2x^2 - b^2)] = \frac{1}{a^2}L\left[\delta\left(x^2 - \frac{b^2}{a^2}\right)\right]$$

$$= \frac{1}{2|ab|}\left\{L\left[\delta\left(x + \frac{b}{a}\right)\right] + L\left[\delta\left(x - \frac{b}{a}\right)\right]\right\},$$

where account has been taken of (1.1.73); there follows

$$L[\delta(a^2x^2 - b^2)] = \frac{1}{2|ab|}(e^{p\frac{b}{a}} + e^{-p\frac{b}{a}}) = \frac{1}{|ab|}\cosh p\,\frac{b}{a}. \quad (2.2.12)$$

Similarly, taking into account (1.1.71), we obtain

$$L[\delta(\sin x)] = L[\sum_n \delta(x - n\pi)] = \sum_n e^{-n\pi p}. \quad (2.2.13)$$

We remark that we may write

$$L[\theta(x)] = (\theta(x), e^{-px}) = \int_0^\infty e^{-px}\,dx = \frac{1}{p}. \quad (2.2.14)$$

Using relation (2.2.2) we obtain

$$L[(-x)^n\theta(x)] = \left(\frac{1}{p}\right)^{(n)} = (-1)^n \frac{n!}{p^{n+1}}, \qquad (2.2.15)$$

whence

$$L[x_+^n] = (-1)^n L[(-x)^n\theta(x)] = \frac{n!}{p^{n+1}} = (-1)^n \left(\frac{1}{p}\right)^{(n)}. \qquad (2.2.16)$$

It may be also shown that the *theorem of translation (damping theorem)* is valid in the theory of distributions. Thus we have

$$L[f(x)e^{qx}] = (f(x)e^{qx}, e^{-px}) = (f(x), e^{-(p-q)x}),$$

whence

$$L[f(x)e^{qx}] = L(p - q), \qquad (2.2.17)$$

with the notation (2.2.1'). For example we may write

$$L[\theta(x)e^{\lambda x}] = L(p - \lambda); \qquad (2.2.18)$$

using the transform (2.2.14) we obtain

$$L[\theta(x)e^{\lambda x}] = \frac{1}{p-\lambda}. \qquad (2.2.18')$$

2.2.1.3. Laplace transform of the derivative of a distribution

For the image of the original distribution $f'(x)$ we obtain

$$L[f'(x)] = (f'(x), e^{-px}) = -(f(x), -pe^{-px}),$$

whence

$$L[f'(x)] = pL[f(x)]. \qquad (2.2.19)$$

It should be noted that the above derivative is considered in the sense of the theory of distributions. In particular, assuming that $f(x)$ is a distribution of the function type, the origin is in general a discontinuity point of the first species; assuming that the function $f(x)$ has in addition other points of discontinuity of the first species corresponding to the abscissae $c_1, c_2, \ldots, c_n$, we may write

$$f'(x) = \tilde{f}'(x) + f(0 + 0)\delta(x) + \sum_{j=1}^n \Delta f(c_j)\delta(x - c_j), \qquad (2.2.20)$$

where $\Delta f(c_j)$ is the jump of the function $f(x)$ at the point of abscissa c_j ($j = 1, 2, \ldots, n$), $f'(x)$ is the derivative in the ordinary sense and where we have used the formula (1.2.40).

Replacing in relation (2.2.19) we obtain

$$L[\tilde{f}'(x)] = pL[f(x)] - f(0 + 0) - \sum_{j=1}^{n} \Delta f(c_j)e^{-pc_j}. \qquad (2.2.21)$$

If the function $f(x)$ has no discontinuity point other than, eventually, the origin, the relation (2.2.21) becomes

$$L[\tilde{f}'(x)] = pL[f(x)] - f(0 + 0). \qquad (2.2.22)$$

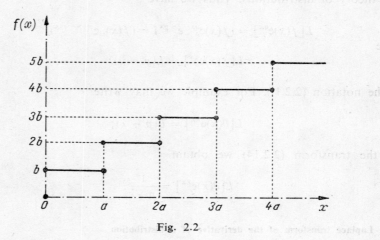

Fig. 2.2

For the *step function* shown in Figure 2.2., we have

$$\tilde{f}'(x) = 0, \quad f(0 + 0) = b, \quad c_j = ja, \quad \Delta f(c_j) = b, \qquad (2.2.23)$$

so that formula (2.2.21) yields

$$L[\tilde{f}'(x)] = pL[f(x)] - b - \sum_{j=1}^{\infty} be^{-jap},$$

whence

$$L[f(x)] = \frac{b}{p} \sum_{j=0}^{\infty} e^{-jap}. \qquad (2.2.24)$$

We remark that the step function may be expressed in the form

$$f(x) = b \sum_{j=0}^{\infty} \theta(x - ja); \qquad (2.2.25)$$

taking into account the delay theorem (2.2.9) and (2.2.14) we find again the expression (2.2.24), which may be written also in the form

$$L[f(x)] = \frac{b}{p} \frac{1}{1 - e^{-pa}} = \frac{b}{2p} \frac{e^{\frac{pa}{2}}}{\sinh \frac{pa}{2}}, \tag{2.2.24'}$$

as the sum of a series whose terms vary in geometrical progression.

2.2.2. Laplace transform of a distribution of several variables

We shall first give some general results and then we shall show how to compute the Laplace transform of the derivative of a distribution of several variables as well as the Laplace transform of a convolution product.

2.2.2.1. General results

We shall extend the definition (2.2.1) to the case of a function of several variables. Thus we give the following

Definition 2.2.3. *If the function* $f(x_1, x_2, \ldots, x_n)$ *of n real variables satisfies the conditions*

(i) $f(x_1, x_2, \ldots, x_n) = 0$ *for* $x_1 < 0$ *or* $x_2 < 0 \ldots$ *or* $x_n < 0$,
(ii) $f(x_1 x_2, \ldots, x_n)$ *has partial derivatives of the first order*,
(iii) $f(x_1, x_2, \ldots, x_n) \leqslant M e^{a_1 x_1 + a_2 x_2 + \ldots + a_n x_n}$,

where M is a positive constant and a_j $(j = 1, 2, \ldots, n)$ *are also non-negative constants, then the function* $L(p_1, p_2, \ldots, p_n)$ *of the complex variables* $p_j = u_j + iv_j$ $(j = 1, 2, \ldots, n)$, *defined by the expression*

$$L(p_1, p_2, \ldots, p_n) = \int_0^\infty \int_0^\infty \ldots \int_0^\infty f(x_1, x_2, \ldots, x_n) e^{-(p_1 x_1 + p_2 x_2 + \ldots + p_n x_n)} \, dx_1 \, dx_2 \ldots dx_n, \tag{2.2.26}$$

is called the Laplace transform of the function $f(x_1, x_2, \ldots, x_n)$ *and is denoted by*

$$L[f(x_1, x_2, \ldots, x_n)] = L(p_1, p_2, \ldots, p_n). \tag{2.2.26'}$$

It may be easily shown that under the conditions (*i*), (*ii*), (*iii*), the transform (2.2.26) exists in the half-space Re $p_j > a_j$ $(j = 1, 2, \ldots, n)$.

Proceeding as we did in the case of functions of a single variable, we consider the partial derivatives with respect to the complex variables p_j, and obtain the *theorem of the differentiation of the image function*

$$\frac{\partial^{k_1 + k_2 + \dots + k_n}}{\partial p_1^{k_1} \partial p_2^{k_2} \dots \partial p_n^{k_n}} L(p_1, p_2, \dots, p_n)$$

$$= L[(-x_1)^{k_1}(-x_2)^{k_2} \dots (-x_n)^{k_n} f(x_1, x_2, \dots, x_n)]. \tag{2.2.27}$$

The *inverse transform* is given by the relation

$$f(x_1, x_2, \dots, x_n)$$

$$= \frac{1}{(2\pi i)^n} \int_{u_1 - i\infty}^{u_1 + i\infty} \int_{u_2 - i\infty}^{u_2 + i\infty} \dots \int_{u_n - i\infty}^{u_n + i\infty} L(p_1, p_2, \dots, p_n) e^{p_1 x_1 + p_2 x_2 + \dots + p_n x_n} \, dp_1 \, dp_2 \dots dp_n, \tag{2.2.28}$$

for $u_j > a_j \, (j = 1, 2, \dots, n)$; we shall also use the notation

$$L^{-1}[L(p_1, p_2, \dots, p_n)] = f(x_1, x_2, \dots, x_n). \tag{2.2.28'}$$

For a function of the form

$$f(x_1, x_2, \dots, x_n) = f(x_1) f_2(x_2) \dots f_n(x_n), \tag{2.2.29}$$

the defining relation (2.2.26) leads to

$$L[f(x_1, x_2, \dots, x_n)] = L[f_1(x_1)] L[f_2(x_2)] \dots L[f_n(x_n)]. \tag{2.2.30}$$

For example if $\theta(x_1, x_2, \dots, x_n)$ is the Heaviside function in R^n and if we take into account (1.3.9) we may write

$$L[\theta(x_1, x_2, \dots, x_n)] = L[\theta(x_1)] L[\theta(x_2)] \dots L[\theta(x_n)] = \frac{1}{p_1 p_2 \dots p_n}. \tag{2.2.31}$$

It should be noted that the defining relation (2.2.26) may be extended to distributions of several variables. Thus, we give

Definition 2.2.4. *If $f(x_1, x_2, \dots, x_n)$ is a distribution with the support on the domain $x_1 > 0$, $x_2 > 0$, $\dots, x_n > 0$ so that the following distribution $f(x_1, x_2, \dots, x_n) \, e^{-(p_1 x_1 + p_2 x_2 + \dots + p_n x_n)}$ is a temperate distribution, then*

$$L[f(x_1, x_2, \dots, x_n)] = (f(x_1, x_2, \dots, x_n), e^{-(p_1 x_1 + p_2 x_2 + \dots + p_n x_n)}) \tag{2.2.32}$$

represents the Laplace transform of that distribution.

We remark that relation (2.2.27) *(theorem of the diferentiation of the image function)* remains valid in the case of distributions, owing to the rule of differentiation of the latter.

2.2.2.2. Examples

For the distribution $\delta(x_1, x_2, ..., x_n)$, whose support is the origin, we may write

$$L[\delta(x_1, x_2, ..., x_n)] = (\delta(x_1, x_2, ..., x_n), e^{-(p_1x_1 + p_2x_2 + ... + p_nx_n)}) = e^0 = 1. \quad (2.2.33)$$

In general we have

$$L[\delta(k_1x_1, k_2x_2, ..., k_nx_n)] = \frac{1}{k_1k_2 ... k_n}, \quad k_1 > 0, k_2 > 0, ..., k_n > 0. \quad (2.2.34)$$

We can also write

$$L[\delta(x_1 - a_1, x_2 - a_2, ..., x_n - a_n)] = e^{-(p_1a_1 + p_2a_2 + ... + p_na_n)} \quad (2.2.35)$$

or, more generally,

$$L[f(x_1 - a_1, x_2 - a_2, ..., x_n - a_n)]$$
$$= e^{-(p_1a_1 + p_2a_2 + ... + p_na_n)}L[f(x_1, x_2, ..., x_n)]. \quad (2.2.36)$$

The Laplace transform of the distribution $\delta(x_1)$ in R^n is given by

$$L[\delta(x_1)] = (\delta(x_1), e^{-(p_1x_1 + p_2x_2 + ... + p_nx_n)})$$

$$= \int_0^\infty \int_0^\infty ... \int_0^\infty e^{-(p_2x_2 + p_3x_3 + ... + p_nx_n)} \, dx_2 \, dx_3 ... dx_n = \frac{1}{p_2p_3 ... p_n}. \quad (2.2.37)$$

In general, we may write

$$L[\delta(x_1, x_2, ..., x_m)] = \frac{1}{p_{m+1} \, p_{m+2} ... p_n}. \quad (2.2.38)$$

We note that the Laplace transform (2.2.31) is valid also in distributions.

2.2.2.3. Iterated Laplace transform

In addition to the general Laplace transform of n complex variables $p_1, p_2, ..., p_n$, defined by formula (2.2.26), we shall introduce a particular case by considering $p_1 = p_2 = ... = p_n = p$. This will be termed the *iterated Laplace transform* and is defined by the relation

$$\bar{L}[f(x_1, x_2, ..., x_n)] = \bar{L}(p)$$

$$= \int_0^\infty \int_0^\infty ... \int_0^\infty f(x_1, x_2, ..., x_n)e^{-p(x_1 + x_2 + ... + x_n)} \, dx_1 \, dx_2 ... dx_n, \quad (2.2.39)$$

under the same conditions as the general Laplace transform.

Thus we have

$$\bar{L}[\delta(x_1, x_2, ..., x_n)] = 1 \qquad (2.2.40)$$

as well as

$$\bar{L}[\theta(x_1, x_2, ..., x_n)] = \frac{1}{p^n}. \qquad (2.2.41)$$

Obviously, other general Laplace transforms may be particularized in a similar way.

2.2.2.4 Laplace transform of the derivative of a distribution of several variables

For the Laplace transform of the derivative of an arbitrary order of a distribution of several variables, we obtain easily

$$L\left[\frac{\partial^{k_1 + k_2 + ... + k_n}}{\partial x_1^{k_1} \partial x_2^{k_2} ... \partial x_n^{k_n}} f(x_1, x_2, ..., x_n)\right] = p_1^{k_1} p_2^{k_2} ... p_n^{k_n} L[f(x_1, x_2, ..., x_n)]. \qquad (2.2.42)$$

In particular, we have

$$L\left[\frac{\partial^{k_1 + k_2 + ... + k_n}}{\partial x_1^{k_1} \partial x_2^{k_2} ... \partial x_n^{k_n}} \delta(x_1, x_2, ..., x_n)\right] = p_1^{k_1} p_2^{k_2} ... p_n^{k_n}, \qquad (2.2.43)$$

$$L\left[\frac{\partial}{\partial x_1} \theta(x_1, x_2, ..., x_n)\right] = p_1 L[\theta(x_1, x_2, ..., x_n)] = \frac{1}{p_2 p_3 ... p_n}. \qquad (2.2.44)$$

Let us now consider a function $f(x_1, x_2)$ which vanishes for $x_1 < 0$, $x_2 < 0$ and has discontinuities of the first species when it crosses the curve $P(x_1, x_2) = 0$. In accordance with the formulae (1.4.22), (1.4.22') for the differentiation of distributions, we may write

$$\frac{\partial}{\partial x_1} f(x_1, x_2) = \frac{\tilde{\partial}}{\partial x_1} f(x_1, x_2) + (\Delta f)_{x_1} P'_{x_1}(x_1, x_2)\delta(P) + f(0, x_2)\delta(x_1), \qquad (2.2.45)$$

$$\frac{\partial}{\partial x_2} f(x_1, x_2) = \frac{\tilde{\partial}}{\partial x_2} f(x_1, x_2) + (\Delta f)_{x_2} P'_{x_2}(x_1, x_2)\delta(P) + f(x_1, 0)\delta(x_2), \qquad (2.2.45')$$

where $(\Delta f)_{x_j}$ $(j = 1, 2)$ represents the jump of the function $f(x, y)$ as it crosses the curve $P(x, y) = 0$ in the direction of the Ox_j axis, while $\dfrac{\tilde{\partial}}{\partial x_1}f(x_1, x_2)$, $\dfrac{\tilde{\partial}}{\partial x_2}f(x_1, x_2)$ are the derivatives in the ordinary sense of the function $f(x_1, x_2)$; we have taken into account that besides the curve $P(x_1, x_2) = 0$ the function $f(x_1, x_2)$ has also the curves of discontinuity $x_1 = 0$ and $x_2 = 0$.

Starting from formula (2.2.42) we obtain

$$L\left[\frac{\tilde{\partial}}{\partial x_1} f(x_1, x_2)\right] = p_1 L[f(x_1, x_2)]$$

$$- L[f(0, x_2)\delta(x_1)] - L[(\Delta f)_{x_1} P'_{x_1}(x_1, x_2)\delta(P)], \qquad (2.2.46)$$

$$L\left[\frac{\tilde{\partial}}{\partial x_2} f(x_1, x_2)\right] = p_2 L[f(x_1, x_2)]$$

$$- L[f(x_1, 0)\delta(x_2)] - L[(\Delta f)_{x_2} P'_{x_2}(x_1, x_2)\delta(P)]. \qquad (2.2.46')$$

In particular, if the function $f(x_1, x_2)$ has only the curves of discontinuity $x_1 = 0$ and $x_2 = 0$, the above formulae become simplified

$$L\left[\frac{\tilde{\partial}}{\partial x_1} f(x_1, x_2)\right] = p_1 L[f(x_1, x_2)] - \frac{1}{p_2} L[f(0, x_2)], \qquad (2.2.47)$$

$$L\left[\frac{\tilde{\partial}}{\partial x_2} f(x_1, x_2)\right] = p_2 L[f(x_1, x_2)] - \frac{1}{p_1} L[f(x_1, 0)]. \qquad (2.2.47')$$

We note that

$$\frac{\partial}{\partial x_1} \theta(x_1, x_2) = \frac{\tilde{\partial}}{\partial x_1} \theta(x_1, x_2) + \delta(x_1) = \delta(x_1), \qquad (2.2.48)$$

where we have taken account of relations (1.2.56') and (1.2.56''); there follows

$$L\left[\frac{\partial}{\partial x_1} \theta(x_1, x_2)\right] = L[\delta(x_1)] = \frac{1}{p_2}, \qquad (2.2.49)$$

which could have been obtained also from (2.2.24).

2.2.2.5. Laplace transform of a convolution product

Let $f(x)$, $g(x)$ be two distributions with the support in $x \geqslant 0$; we assume that their Laplace transforms exist for $\operatorname{Re} p > a_1$ and $\operatorname{Re} p > a_2$, respectively. We have

$$L[f(x) * g(x)] = (f(x) * g(x), e^{-px}) = (f(x) \times g(y), e^{-p(x+y)})$$

$$= (f(x), (g(y), e^{-p(x+y)})) = (f(x), (g(y), e^{-py})e^{-px});$$

hence

$$L[f(x) * g(x)] = L[f(x)]L[g(x)], \qquad (2.2.50)$$

i.e. *the image of a convolution product is equal to the product of the images,* in the case where such a product has a meaning.

In particular we have

$$L[f(x) * \delta(x)] = L[f(x)]L[\delta(x)] = L[f(x)] \qquad (2.2.51)$$

or more generally

$$L[f(x) * \delta^{(n)}(x)] = L[f^{(n)}(x) * \delta(x)] = L[f^{(n)}(x)]. \qquad (2.2.52)$$

Also, we can write

$$L[\theta(x)e^{\alpha x} * \theta(x)e^{\beta x}]$$

$$= L[\theta(x)e^{\alpha x}]L[\theta(x)e^{\beta x}] = \frac{1}{(p-\alpha)(p-\beta)} \qquad (2.2.53)$$

or, in particular,

$$L[\theta(x) * \theta(x)e^{\alpha x}] = \frac{1}{p(p-\alpha)} \qquad (2.2.54)$$

and also

$$L[\theta(x) * \theta(x)] = \frac{1}{p^2} = L[x_+]; \qquad (2.2.55)$$

it follows that

$$\theta(x) * \theta(x) = x_+. \qquad (2.2.56)$$

Similarly we have

$$L[\delta(x-a) * \delta(x-b)] = e^{-ap}e^{-bp} = e^{-(a+b)p} = L[\delta(x-a-b)], \qquad (2.2.57)$$

whence there follows

$$\delta(x-a) * \delta(x-b) = \delta(x-a-b). \qquad (2.2.58)$$

More generally, we have

$$L[\delta(x-a) * f(x)] = e^{-ap}L[f(x)] = L[f(x-a)], \qquad (2.2.59)$$

which corresponds to the formula (1.3.31).

Taking into account (2.2.16) we may write

$$L[x_+^{n-1}] = \frac{(n-1)!}{p^n} = \frac{p}{n}\frac{n!}{p^{n+1}} = \frac{p}{n}L[x_+^n] \qquad (2.2.60)$$

and

$$L[\theta(x) * x_+^{n-1}] = L[\theta(x)]L[x_+^{n-1}] = \frac{1}{p}\frac{p}{n}L[x_+^n] = \frac{1}{n}L[x_+^n], \qquad (2.2.61)$$

whence there follows

$$\theta(x) * x_+^{n-1} = \frac{1}{n} x_+^n. \tag{2.2.62}$$

In particular

$$\theta(x) * x_+ = \frac{1}{2} x_+^2. \tag{2.2.63}$$

The above results have been given, for the sake of simplicity of notation, for distributions of a single variable only; they can be easily extended to distributions of several variables.

2.2.3. Connection between the Fourier and the Laplace transforms

In the following discussion we shall deal with the connection between the Fourier and the Laplace transform in the case of distributions both of a single and of several variables.

2.2.3.1 Case of a distribution of a single variable

Let $f(x)$ be a distribution with a bounded support contained in the half-line $x \geqslant 0$. The Laplace transform is defined by the relation

$$L(p) = (f(x), e^{-px}), \quad \text{Re } p > a, \tag{2.2.64}$$

where $p = u + iv$; the Fourier transform of this distribution is

$$F(s) = (f(x), e^{isx}), \tag{2.2.65}$$

where $s = u + iv$.

From relation (2.2.64) we obtain

$$L(-ip) = L(v - iu) = (f(x), e^{ipx}) = (f(x)e^{-vx}, e^{iux}) \tag{2.2.64'}$$

and (2.2.65) leads to

$$F(u + iv) = (f(x)e^{-vx}, e^{iux}). \tag{2.2.65'}$$

Comparing relations (2.2.64') and (2.2.65') we obtain

$$L(-is) = F(s), \quad s = u + iv, \tag{2.2.66}$$

where

$$L(s) = L[f(x)], \quad F(s) = F[f(x)]. \tag{2.2.66'}$$

Since the Fourier transforms of distributions with bounded support are complex functions of the real variable u, by extension from the space K to the space S we deduce from relation (2.2.66)

$$F(u) = \lim_{v \to +0} F(s) = \lim_{v \to +0} L(-is). \qquad (2.2.67)$$

Also, if $f(x)$ is a distribution of the function type, we deduce that the Fourier transform of the distribution $f(x)$ is given by

$$F(u) = \lim_{v \to +0} F[f(x)e^{-vx}] = \lim_{v \to +0} L(-is), \; v > 0, \qquad (2.2.68)$$

where account has been taken of the notation (2.2.66′).

Thus, for $f(x) = \theta(x)$ we have

$$F[\theta(x)e^{-vx}] = -\frac{1}{is} = \frac{i}{s} = \frac{1}{v - iu} = \frac{v + iu}{u^2 + v^2}, \quad v > 0, \qquad (2.2.69)$$

so that

$$F[\theta(x)] = \lim_{v \to +0} L(-is) = \lim_{v \to +0} \frac{1}{v - iu} = \pi\delta(u) + \frac{i}{u} \qquad (2.2.70)$$

and we obtain the same result as that supplied by formula (2.1.33).

Similarly, since

$$L[\theta(x) \sin x] = \frac{1}{1 + s^2} \qquad (2.2.71)$$

we obtain

$$F[\theta(x) \sin x \, e^{-vx}] = \frac{1}{1 + (-is)^2} = \frac{1}{1 - (u + iv)^2}$$

$$= \frac{1}{(1 - u - iv)(1 + u + iv)}, \quad v > 0; \qquad (2.2.72)$$

this leads to

$$F[\theta(x) \sin x] = \lim_{v \to +0} \frac{1}{1 - (u + iv)^2}$$

$$= \frac{1}{1 - u^2} + i\frac{\pi}{2}[\delta(u - 1) - \delta(u + 1)] = \frac{1}{1 - u^2} + i\pi\delta(u^2 - 1), \qquad (2.2.73)$$

if we remark that

$$\frac{1}{(1 - u - iv)(1 + u + iv)} = \frac{(1 - u + iv)(1 + u - iv)}{[(1 + u)^2 + v^2][(1 - u)^2 + v^2]}$$

$$= \frac{1 - (u - iv)^2}{(1 - u^2 + v^2)^2 + 4u^2v^2} = \frac{1 - u^2 + v^2}{(1 - u^2 + v^2)^2 + 4u^2v^2}$$

$$+ \frac{i}{2}\left[\frac{v}{(u - 1)^2 + v^2} - \frac{v}{(u + 1)^2 + v^2}\right]$$

and if we take account into (2.1.32) and (1.2.28′).

2.2.3.2. Case of a distribution of several variables

In the case of a distribution $f(x_1, x_2, \ldots, x_n)$ of n variables, the Laplace transform is defined by the relation

$$L(p_1, p_2, \ldots, p_n) = (f(x_1, x_2, \ldots, x_n), e^{-(p_1x_1 + p_2x_2 + \ldots + p_nx_n)}), \quad \text{Re } p_j > a_j, \quad (2.2.74)$$

where $p_j = u_j + iv_j$ $(j = 1, 2, \ldots, n)$; the Fourier transform of this distribution is

$$F(s_1, s_2, \ldots, s_n) = (f(x_1, x_2, \ldots, x_n), e^{i(s_1x_1 + s_2x_2 + \ldots + s_nx_n)}), \quad (2.2.75)$$

where $s_j = u_j + iv_j$ $(j = 1, 2, \ldots, n)$.

We remark that

$$L(-is_1, -is_2, \ldots, -is_n) = F(s_1, s_2, \ldots, s_n), \quad (2.2.76)$$

where

$$L(s_1, s_2, \ldots, s_n) = L[f(x_1, x_2, \ldots, x_n)], \quad (2.2.77)$$

$$F(s_1, s_2, \ldots, s_n) = F[f(x_1, x_2, \ldots, x_n)]. \quad (2.2.77')$$

Also, for distributions of several variables with bounded support, we may write

$$F(u_1, u_2, \ldots, u_n) = \lim_{\substack{v_j \to +0 \\ (j=1,2,\ldots,n)}} F(s_1, s_2, \ldots, s_n)$$

$$= \lim_{\substack{v_j \to +0 \\ (j=1,2,\ldots,n)}} L(-is_1, -is_2, \ldots, -is_n), \quad (2.2.78)$$

where we have used the notation (2.2.77).

For example, taking into account (2.2.31) we obtain

$$F[\theta(x_1, x_2)e^{-(v_1 x_1 + v_2 x_2)}] = \frac{1}{(-is_1)(-is_2)} = -\frac{1}{(u_1 + iv_1)(u_2 + iv_2)}$$

$$= \frac{1}{(v_1 - iu_1)(v_2 - iu_2)}, \quad v_1 > 0, \ v_2 > 0, \tag{2.2.79}$$

whence we may obtain $F[\theta(x_1, x_2)]$.

3

Variational calculus and differential equations in distributions

3.1. Variational calculus in distributions

Variational calculus methods are particularly efficient for the study of many physical phenomena, particularly in mechanics; with their help we may thoroughly investigate the overall behaviour of a mechanical system. Taking into account the results of variational calculus, the principles of mechanics may be expressed in a general form which is often convenient in calculations.

Since the conditions under which some of the results of the variational calculus have been established affect directly their range of applicability to different physical phenomena, we shall generalize some of these results, particularly the concept of variation of a functional as well as the Euler-Poisson and Euler-Ostrogradski equations; we shall assume that the functions to be integrated are continuous functions of all the arguments and that the admissible lines are distributions or curves of class C^n everywhere, except a finite number of points where the derivatives of the nth order may have discontinuities of the first order.

We shall consider first the case of a distribution of a single variable and then the case of distributions of several variables.

3.1.1. Distributions of a single variable

In the following we shall first introduce the variation of the first order of a functional and the corresponding equations of the Euler-Poisson type; then we shall show also how to introduce the variation of the second order.

3.1.1.1. Variation of the first order

Let

$$I(y_1, y_2, ..., y_m) = I(y_i) = \int_a^b F(x; y_i, y_i', y_i'', ..., y_i^{(n)}) \, dx$$

$$= \int_a^b F(x; y_1, y_2, ..., y_m; y_1', y_2', ..., y_m'; y_1'', y_2'', ..., y_m''; ...; y_1^{(n)}, y_2^{(n)}, ..., y_m^{(n)}) \, dx \quad (3.1.1)$$

be a functional where F is a function of class C^0 with respect to all its arguments.

We shall consider as *admissible lines* the curves expressed by distributions in the case where the operations indicated by F have a meaning, or, if they have not, the curves of class C^n for which the derivatives of order n may have discontinuities of the first species at a finite number of points x_j ($j = 1, 2, ..., p$).

For example the functional

$$I(y) = \int_{-1}^{1} x^3 yy' \, dx \qquad (3.1.2)$$

has a meaning if we take as admissible line $y = \delta(x)$. In that case we obtain

$$F(x, y, y') = x^3 yy' = 0, \qquad (3.1.3)$$

regardless of the order of the operations $(x^3 y')y$ or $(x^3 y)y'$; we remark that the operation $(yy')x^3$ is meaningless for $y = \delta(x)$.

We shall denote by $\mathcal{K}$ a subset of $K (\mathcal{K} \subset K)$ which includes the functions $\varphi(x)$ possessing derivatives of all orders, with the support inside the open interval (a, b). Hence we have

$$\varphi^{(k)}(a) = \varphi^{(k)}(b) = 0 \qquad (k = 0, 1, 2, ...). \qquad (3.1.4)$$

If $y_i(x)$ is an admissible line expressed by a distribution in order that F has a meaning, then by definition the neighbouring admissible lines are expressed by

$$\bar{y}_i(x) = y_i(x) + \varphi_i(x), \qquad (3.1.5)$$

with

$$\varphi_i(x) \in \mathcal{K} \subset K. \qquad (3.1.6)$$

Obviously, in the case where $y_i(x)$ is a distribution the neighbouring admissible line will be expressed by a distribution too.

In this way, the notion of neighbourhood is retained; the curve $\bar{y}_i(x)$ will belong to the neighbourhood of order r of the curve $y_i(x)$ if the following conditions are satisfied

$$\left| \bar{y}_i^{(k)}(x) - y_i^{(k)}(x) \right| < \varepsilon \qquad (k = 0, 1, 2, ..., r), \qquad (3.1.7)$$

which are equivalent to

$$\left| \varphi_i^{(k)}(x) \right| < \varepsilon, \qquad (3.1.7')$$

where $\varepsilon > 0$ is arbitrary.

Taking into account the results concerning the support of a distribution presented in section 1.1.2.6, we may state the following

Theorem 3.1.1. *The necessary and sufficient condition that the distribution* $f(x)$ *be zero over the interval* (a, b) *is that*

$$(f(x), \varphi(x)) = 0 \tag{3.1.8}$$

for any $\varphi(x) \in \mathscr{K} \subset K$.

It is important to note that theorem 3.1.1 includes as particular cases the fundamental lemmas of the variational calculus: the lemmas of Lagrange and of Du Bois-Reymond.

Indeed, assuming that $f(x)$ is of class C^0, the equality (3.1.8) may be written as

$$(f(x), \varphi(x)) = \int_a^b f(x)\varphi(x)\,\mathrm{d}x = 0 \tag{3.1.9}$$

and takes place only if $f(x) = 0$; this is the *lemma of Lagrange*.

Considering, instead of the function $\varphi(x)$, the function $\varphi'(x) \in \mathscr{K}$ we may write

$$(f(x), \varphi'(x)) = -(f'(x), \varphi(x)) = -\int_a^b f'(x)\varphi(x)\,\mathrm{d}x = 0, \tag{3.1.10}$$

where we assume that $f'(x)$ is a function of the class C^0; this results in $f'(x) = 0$ and hence $f = \mathrm{const}$, which constitutes the *lemma of Du Bois-Reymond*.

Consequently, theorem 3.1.1 represents the fundamental lemma in the case where the admissible lines are distributions.

Let us introduce the following

Definition 3.1.1. *The distribution* $\delta I = \delta I(y_i)$ *defined in the form*

$$\left(\delta I, \sum_{i=1}^m \varphi_i\right) = \sum_{i=1}^m \sum_{k=0}^n (F_{y_i^{(k)}}, \varphi_i^{(k)}), \tag{3.1.11}$$

where $\varphi_i(x) \in \mathscr{K} \subset K$ *and*

$$F_{y_i^{(k)}} = \frac{\partial F}{\partial y_i^{(k)}} \tag{3.1.12}$$

represents derivatives in the sense of the theory of distributions, is called a variation of the 1st order of the functional (3.1.1).

It is obvious that expression (3.1.11) of the first variation of the functional (3.1.1) includes, as a particular case, the expression of the variation δI known in the classical variational calculus.

We remark that

$$(F_{y_i^{(k)}}, \varphi_i^{(k)}) = (-1)^k \left(\frac{\mathrm{d}^k}{\mathrm{d}x^k} F_{y_i^{(k)}}, \varphi_i\right), \tag{3.1.13}$$

and expression (3.1.11) becomes

$$\left(\delta I, \sum_{i=1}^{m} \varphi_i\right) = \sum_{i=1}^{m} \sum_{k=0}^{n} (-1)^k \left(\frac{d^k}{dx^k} F_{y_i^{(k)}}, \varphi_i\right) \qquad (3.1.11')$$

where, as has been mentioned before, the differentiations are performed in the sense of the theory of distributions.

Expanding relation (3.1.11') we may write

$$\left(\delta I, \sum_{i=1}^{m} \varphi_i\right) = \sum_{i=1}^{m} \left(F_{y_i} - \frac{d}{dx} F_{y_i'} + \frac{d^2}{dx^2} F_{y_i''} - \ldots + (-1)^n \frac{d^n}{dx^n} F_{y_i^{(n)}}^{(n)}, \varphi_i\right). \qquad (3.1.11'')$$

3.1.1.2. Equations of the Euler-Poisson type

The extremals $y_i(x)$ $(i = 1, 2, \ldots, m)$ of the functional (3.1.1) are obtained by imposing the condition

$$\delta I = 0. \qquad (3.1.14)$$

Using for the distribution δI the expression (3.1.11'') and taking into account theorem 3.1.1 we find the following necessary equations

$$F_{y_i} - \frac{d}{dx} F_{y_i'} + \frac{d^2}{dx^2} F_{y_i''} - \ldots + (-1)^n \frac{d^n}{dx^n} F_{y_i^{(n)}} = 0 \quad (i = 1, 2, \ldots, m). \qquad (3.1.15)$$

We remark that these equations are of the same form as the classical Euler-Poisson equations; however, they differ substantially from the latter since the function F has been assumed to be of class C^0 with respect to all its arguments, and the differentiation is performed in the sense of the theory of distributions. For these reasons we shall say that such equations are of the Euler-Poisson type.

Let us assume that the *extremal curves* $y_i(x)$ $(i = 1, 2, \ldots, m)$ are of class C^n everywhere, except at the points $x_j (j = 1, 2, \ldots, p)$ where the derivatives of the nth order may have discontinuities of the first species.

If we mark with a tilde the ordinary derivative we have

$$\frac{d^h}{dx^h} y_i(x) = \frac{\tilde{d}^h}{dx^h} y_i(x) \qquad (i = 1, 2, \ldots, m; \; h = 1, 2, \ldots, n), \qquad (3.1.16)$$

so that the operations indicated by F have a meaning; thus the function F, considered as a function of the variable x on the integral curves, may have discontinuities of the first species at the above mentioned points only.

Let $f(x)$ be a function of class C^n except at the points $x_j (j = 1, 2, \ldots, p)$, where it has discontinuities of the first species together with its derivatives in the usual sense of the order up to and including $(n - 1)$; we denote by $(\Delta \tilde{f}^{(h)})_j$ the jump of the function $\dfrac{\tilde{d}^h}{dx^h} f(x) = \tilde{f}^{(h)}(x)$ at the point x_j, that is

$$(\Delta \tilde{f}^{(h)})_j = \tilde{f}^{(h)}(x_j + 0) - \tilde{f}^{(h)}(x_j - 0) \qquad (h = 0, 1, 2, \ldots, n - 1). \quad (3.1.17)$$

Applying formula (1.2.47) for the differentiation of a distribution we may write

$$\frac{d^k}{dx^k} f(x) = \frac{\tilde{d}^k}{dx^k} f(x) + \sum_{j=1}^{p} (\Delta f)_j \delta^{(k-1)}(x - x_j)$$

$$+ \sum_{j=1}^{p} (\Delta \tilde{f}')_j \delta^{(k-2)}(x - x_j) + \ldots + \sum_{j=1}^{p} (\Delta \tilde{f}^{(k-1)})_j \delta(x - x_j), \qquad (3.1.18)$$

for $i = 1, 2, \ldots, m$.

Applying formula (3.1.18) to the functions $F_{y_i'}, F_{y_i''}, \ldots, F_{y_i^{(n)}}$ which possess the properties stated above and using notations similar to the notations (3.1.17) we obtain the relations

$$\frac{d}{dx} F_{y_i'} = \frac{\tilde{d}}{dx} F_{y_i'} + \sum_{j=1}^{p} (\Delta F_{y_i'})_j \delta(x - x_j),$$

$$\frac{d^2}{dx^2} F_{y_i''} = \frac{\tilde{d}^2}{dx^2} F_{y_i''} + \sum_{j=1}^{p} \left(\Delta \frac{\tilde{d}}{dx} F_{y_i''} \right) \delta(x - x_j) + \sum_{j=1}^{p} (\Delta F_{y_i''})_j \delta'(x - x_j),$$

$$\cdots\cdots\cdots\cdots\cdots\cdots\cdots\cdots\cdots\cdots\cdots\cdots\cdots\cdots\cdots\cdots$$

$$(3.1.19)$$

$$\frac{d^n}{dx^n} F_{y_i^{(n)}} = \frac{\tilde{d}^n}{dx^n} F_{y_i^{(n)}} + \sum_{j=1}^{p} \left(\Delta \frac{\tilde{d}^{n-1}}{dx^{n-1}} F_{y_i^{(n)}} \right)_j \delta(x - x_j)$$

$$+ \sum_{j=1}^{p} \left(\Delta \frac{\tilde{d}^{n-2}}{dx^{n-2}} F_{y_i^{(n)}} \right)_j \delta'(x - x_j) + \ldots + \sum_{j=1}^{p} (\Delta F_{y_i^{(n)}})_j \delta^{(n-1)}(x - x_j);$$

replacing the expressions (3.1.19) in equations (3.1.15) we obtain the equations

$$A_i + \sum_{k=0}^{n-1} \sum_{j=1}^{p} (-1)^{k+1} B_{i,j}^k \delta^{(k)}(x - x_j) = 0 \qquad (i = 1, 2, \ldots, m), \qquad (3.1.20)$$

where we have introduced the notations

$$A_i = F_{y_i} - \frac{\tilde{\mathrm{d}}}{\mathrm{d}x} F_{y_i'} + \frac{\tilde{\mathrm{d}}^2}{\mathrm{d}x^2} F_{y_i''} - \dots + (-1)^n \frac{\tilde{\mathrm{d}}^n}{\mathrm{d}x^n} F_{y_i^{(n)}}, \tag{3.1.21}$$

$$B_{i,j}^k = (\Delta F_{y_i^{(k+1)}})_j - \left(\Delta \frac{\tilde{\mathrm{d}}}{\mathrm{d}x} F_{y_i^{(k+2)}} \right)_j + \dots$$

$$+ (-1)^{n-(k+1)} \left(\Delta \frac{\tilde{\mathrm{d}}^{n-(k+1)}}{\mathrm{d}x^{n-(k+1)}} F_{y_i^{(n)}} \right)_j. \tag{3.1.21'}$$

Taking into account theorem 1.2.1 (see section 1.2.1.4) the equations (3.1.20) are satisfied if and only if,

$$A_i = 0 \qquad (i = 1, 2, \dots, m), \tag{3.1.22}$$

$$B_{i,j}^k = 0 \qquad (i = 1, 2, \dots, m; \ j = 1, 2, \dots, p; \ k = 0, 1, 2, \dots, n-1). \tag{3.1.22'}$$

Hence, the extremum equations for the functional (3.1.1) may be written in the form

$$F_{y_i} - \frac{\tilde{\mathrm{d}}}{\mathrm{d}x} F_{y_i'} + \frac{\tilde{\mathrm{d}}^2}{\mathrm{d}x^2} F_{y_i''} - \dots + (-1)^n \frac{\tilde{\mathrm{d}}^n}{\mathrm{d}x^n} F_{y_i^{(n)}} = 0, \tag{3.1.23}$$

where $i = 1, 2, \dots, m$; also, in the case where the extremal curves $y_i (i = 1, 2, \dots, m)$ are of class C^n everywhere except at the points x_j where the derivatives of the nth order may have discontinuities of the first species, the conditions at the points of discontinuity x_j $(j = 1, 2, \dots, p)$ will be

$$(\Delta F_{y_i^{(k+1)}})_j - \left(\Delta \frac{\tilde{\mathrm{d}}}{\mathrm{d}x} F_{y_i^{(k+2)}} \right)_j + \dots$$

$$+ (-1)^{n-(k+1)} \left(\Delta \frac{\tilde{\mathrm{d}}^{n-(k+1)}}{\mathrm{d}x^{n-(k+1)}} F_{y_i^{(n)}} \right)_j = 0, \tag{3.1.24}$$

where $i = 1, 2, \dots, m, \ j = 1, 2, \dots, p, \ k = 0, 1, 2, \dots, n-1$.
In the particular case of the functional

$$I(y) = \int_a^b F(x; y, y', y'', \dots, y^{(n)}) \, \mathrm{d}x \tag{3.1.25}$$

the necessary condition of extremum is

$$F_y - \frac{\tilde{d}}{dx} F_{y'} + \frac{\tilde{d}^2}{dx^2} F_{y''} - \ldots + (-1)^n \frac{\tilde{d}^n}{dx^n} F_{y^{(n)}} = 0 \qquad (3.1.26)$$

and the conditions at the points of discontinuity are written in the form

$$(\Delta F_{y'})_j - \left(\Delta \frac{\tilde{d}}{dx} F_{y''} \right)_j + \ldots + (-1)^{n-1} \left(\Delta \frac{\tilde{d}^{n-1}}{dx^{n-1}} F_{y^{(n)}} \right)_j = 0,$$

$$(\Delta F_{y''})_j - \left(\Delta \frac{\tilde{d}}{dx} F_{y'''} \right)_j + \ldots + (-1)^{n-2} \left(\Delta \frac{\tilde{d}^{n-2}}{dx^{n-2}} F_{y^{(n)}} \right)_j = 0,$$

$$\cdots\cdots\cdots\cdots\cdots\cdots\cdots\cdots\cdots\cdots\cdots\cdots\cdots\cdots\cdots\cdots\cdots\cdots \qquad (3.1.27)$$

$$(\Delta F_{y^{(n-1)}})_j - \left(\Delta \frac{\tilde{d}}{dx} F_{y^{(n)}} \right) = 0,$$

$$(\Delta F_{y^{(n)}})_j = 0 \qquad (j = 1, 2, \ldots, p).$$

The conditions (3.1.27) represent the natural limiting conditions which correspond to the functional (3.1.25) at points of discontinuity x_j.

Thus, in the case of the functional

$$I(y) = \int_a^b F(x; y, y', y'') \, dx \qquad (3.1.28)$$

the necessary equation for an extremum is

$$F_y - \frac{\tilde{d}}{dx} F_{y'} + \frac{\tilde{d}^2}{dx^2} F_{y''} = 0 \qquad (3.1.29)$$

and the natural limiting conditions may be written in the form

$$(\Delta F_{y'})_j - \left(\Delta \frac{\tilde{d}}{dx} F_{y''} \right)_j = 0,$$

$$(\Delta F_{y''})_j = 0 \qquad (j = 1, 2, \ldots, p). \qquad (3.1.30)$$

Let us consider again the particular functional (3.1.2) and let us determine the function which realizes the maximum of that functional such that

$$y(-1) = y(1) = 0. \tag{3.1.31}$$

We remark that the functional may be written in the form

$$I(y) = \frac{1}{2} \int_{-1}^{1} x^3 \, dy^2, \tag{3.1.2'}$$

whence, integrating by parts and taking into account conditions (3.1.31), we obtain

$$I(y) = -\frac{3}{2} \int_{-1}^{1} x^2 y^2 \, dx. \tag{3.1.2''}$$

Obviously, the mxaimum of the functional occurs when the integral vanishes; in that case we obtain

$$y = 0. \tag{3.1.32}$$

However, by writing the equation of the Euler-Poisson type in distributions, we have

$$3x^2 y = 0 \tag{3.1.33}$$

whence, besides the solution (3.1.32), we obtain also the solutions

$$y = \delta(x), \ y = \delta'(x). \tag{3.1.32'}$$

Indeed, we have

$$I(\delta(x)) = 0, \ I(\delta'(x)) = 0, \tag{3.1.34}$$

whether the operations in the function F are performed in the order $(x^3 y')y$ or $(x^3 y)y'$.

It should be noted that $\delta(x)$ and $\delta'(x)$ constitute extremals of the functional (3.1.2) since they vanish in the neighbourhood of the points (-1) and 1 as specified in conditions (3.1.31).

The existence of three solutions is due to the fact that $y = 0$ belongs to the class of admissible lines of the class C^1 (classical case), whereas $y = \delta(x)$ and $y = \delta'(x)$ are distributions and hence admissible lines considered as distributions. This emphasizes the need of extending in distributions the concept of the functional and its first variation.

Similarly, let us consider the functional

$$I(y) = \int_{-1}^{1} x^2 y'^2 \, dx; \tag{3.1.35}$$

K. Weierstrass has attempted to determine a function $y(x)$ of class C^1 on the interval $[-1, 1]$, which verifies the bilocal conditions

$$y(-1) = -1, \; y(1) = 1 \tag{3.1.36}$$

and which minimizes the functional (3.1.35). Indeed, this functional has a minimum since $I(y) \geqslant 0$, but this minimum cannot be realized by any of the curves (γ) of class C^1 which pass through the points $A(-1, -1)$ and $B(1, 1)$ (Figure 3.1).

Writing the necessary condition of extremum (3.1.29) we obtain the equation

$$2xy' + x^2 y'' = 0 \tag{3.1.37}$$

which admits the solution

$$y(x) = 2\theta(x) - 1 = \text{sgn } x , \tag{3.1.38}$$

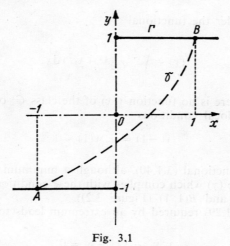

Fig. 3.1

where we have introduced the Heaviside distribution $\theta(x)$; indeed

$$y'(x) = 2\delta(x) \tag{3.1.39}$$

and we may write

$$2xy' + x^2 y'' = 4x\delta(x) + 2x^2 \delta'(x) = 0.$$

The limiting conditions (3.1.36) are also verified, as well as the natural limiting conditions (3.1.30), since by (3.1.39) we may write

$$2x^2 y' = 0.$$

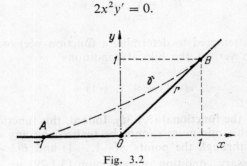

Fig. 3.2

In this way the minimum of the functional (3.1.35) is realized by the curve (Γ) which has a discontinuity of the first species at the origin; this minimum is equal to zero.

In order to obtain this result it has been necessary to use the methods of the theory of distributions; the derivatives which occur are computed in the sense of that theory.

Let us also consider the functional

$$I(y) = \int_{-1}^{1} y^2(1 - y'^2)\, dx; \tag{3.1.40}$$

we find again that there is no function $y(x)$ of the class C^1 on the interval $[-1, 1]$ which verifies the bilocal conditions

$$y(-1) = 0, \quad y(1) = 1 \tag{3.1.41}$$

and minimizes the functional (3.1.40), although a minimum exists (hence we cannot determine a curve (γ) which complies with these conditions and passes through the points $A(-1, 0)$ and $B(1, 1)$ (Figure 3.2)).

The condition (3.1.29) required by an extremum leads to the equation

$$y^2 y'' + yy' - y = 0, \tag{3.1.42}$$

which admits the solution

$$y(x) = x_+ = x\theta(x); \tag{3.1.43}$$

this solution may be easily verified by remarking that

$$y'(x) = \theta(x), \quad y''(x) = \delta(x). \tag{3.1.44}$$

The boundary conditions (3.1.41) and the natural limiting conditions (3.1.30) are also verified. Hence, the minimum of the functional (3.1.40) is realized by the curve (Γ) which is continuous at the origin and has a discontinuous derivative at that point; the minimum is equal to zero. This is an example of a different type which emphasizes the usefulness of the distribution theory.

3.1.1.3. Variation of the second order

For the simplicity of notation we shall consider the particular functional (3.1.25). By analogy with the expression of the variation of the first order of this functional we shall state

Definition 3.1.2. *The distribution* $\delta^2 I = \delta^2 I(y)$ *defined in the form*

$$(\delta^2 I, \varphi) = \frac{1}{2}\left[(\mathscr{F}_y, \varphi) + (\mathscr{F}_{y'}, \varphi') + \ldots + (\mathscr{F}_{y^{(n)}}, \varphi^{(n)}) \right], \tag{3.1.45}$$

where

$$\mathscr{F} = F_y\varphi + F_{y'}\varphi' + \ldots + F_{y^{(n)}}\varphi^{(n)} \tag{3.1.46}$$

and where it is assumed that F is a function of class C^1 *with respect to all its arguments, is termed a variation of the IInd order of the functional* (3.1.25).

In the particular case of the functional

$$I(y) = \int_a^b F(x; y, y')\,\mathrm{d}x \tag{3.1.47}$$

the expression (3.1.45) takes the form

$$(\delta^2 I, \varphi) = \frac{1}{2}\left[(\mathscr{F}_y, \varphi) + (\mathscr{F}_{y'}, \varphi') \right], \tag{3.1.48}$$

where we have introduced the notation

$$\mathscr{F} = F_y\varphi + F_{y'}\varphi'. \tag{3.1.49}$$

Replacing (3.1.49) in (3.1.48) we obtain

$$(\delta^2 I, \varphi) = \frac{1}{2}\left[(F_{yy}\varphi, \varphi) + (F_{yy'}\varphi', \varphi) + (F_{yy'}\varphi, \varphi') + (F_{y'y'}\varphi', \varphi') \right] \tag{3.1.48'}$$

or

$$(\delta^2 I, \varphi) = \frac{1}{2}\left[(F_{yy}, \varphi^2) + 2(F_{yy'}, \varphi\varphi') + (F_{y'y'}, \varphi'^2) \right]; \tag{3.1.48''}$$

noting that

$$\varphi\varphi' = \frac{1}{2}(\varphi^2)',$$

we have

$$(F_{yy'}, \varphi\varphi') = -\frac{1}{2}\left(\frac{d}{dx}F_{yy'}, \varphi^2\right),$$

so that

$$(\delta^2 I, \varphi) = \frac{1}{2}\left[\left(F_{yy} - \frac{d}{dx}F_{yy'}, \varphi^2\right) + (F_{y'y'}, \varphi'^2)\right]. \qquad (3.1.48''')$$

In a similar way we may introduce variations of any order. Starting from relation (3.1.48''') let us introduce the notations

$$P(x) = F_{yy} - \frac{d}{dx}F_{yy'}, \quad R(x) = F_{y'y'}; \qquad (3.1.50)$$

then we may write

$$2(\delta^2 I, \varphi) = (P(x), \varphi^2(x)) + (R(x), \varphi'^2(x)). \qquad (3.1.51)$$

We remark that the study of the second variation $\delta^2 I$, which represents a distribution defined by relation (3.1.51) involves the study of the structure of the operator A expressed by

$$A\varphi = P\varphi - \frac{d}{dx}\left(R\frac{d\varphi}{dx}\right), \qquad (3.1.52)$$

where $P(x)$ and $R(x)$ are distributions.

Assuming that $\varphi \in \mathcal{H}^m(a, b)$, where $\mathcal{H}^m(a, b)$ is a subset of the fundamental space K^m consisting of the functions of class C^m with a compact support and included in the interval $[a, b]$, it follows that the operator A has a meaning on the set of these fundamental functions φ, if $P(x)$ and $R(x)$ are distributions of the $(m-1)$th order.

If $u(x)$ and $v(x)$ are fundamental functions of the space $\mathcal{H}^m(a, b)$ then Au will be a distribution and an application of it in $\mathcal{H}^m(a, b)$ will be written in the form (Au, v).

We now introduce the following

Definition 3.1.3. *An operator* B *defined on* $\mathcal{H}^m(a, b)$ *is said to be positive if for any* $u \in \mathcal{H}^m(a, b)$ *we have*

$$(Bu, u) \geqslant 0, \qquad (3.1.53)$$

where B*u* *is a distribution* (this is also the condition that the distribution B*u* be positive).

Under these conditions the problem of the positivity of the quadratic functional defined by the relation (3.1.51) reduces to the determination of the conditions which must be satisfied by the distributions $P(x)$ and $R(x)$ in order that the operator A be positive. Thus we may proove the following

Theorem 3.1.2. *The necessary condition that the operator A with*

$$P(x) = \overline{P}(x) + \sum_{j=1}^{m} \sum_{i=1}^{n} a_i^j \delta^{(j-1)}(x - x_i), \qquad (3.1.54)$$

$$R(x) = \overline{R}(x) + \sum_{j=1}^{m} \sum_{i=1}^{n} b_i^j \delta^{(j-1)}(x - x_i) \qquad (3.1.54')$$

is positive, is that the distribution $\overline{R}(x)$ be positive on the interval $[a, b]$ *;* $P(x)$ *and* $\overline{R}(x)$ *are regular distributions of the order* $(m - 1)$ *defined on* $\mathscr{H}^m(a, b)$*, the Dirac distribution and its derivatives are concentrated at the points* $x_i \in (a, b)$ $(i = 1, 2, \ldots, n)$ *and a_i^j, b_i^j are constants.*

The above theorem is a generalization of the theorem of A. M. Legendre concerning the condition which must be satisfied in order that a curve realizes the minimum of a functional. This theorem is very useful for the study of many problems of mathematical physics in general and of mechanics in particular.

Referring now to the functional (3.1.47) we find the necessary condition that $I(y)$ should have a minimum which satisfies $\delta^2 I \geqslant 0$; using the notation (3.1.50), this will be

$$F_{y'y'} \geqslant 0 \qquad (3.1.55)$$

along the extremal curve.

We remark that condition (3.1.55) is satisfied by all the examples considered in the preceding subsection.

3.1.2. Distributions of several variables

In the following we shall introduce the variation of the Ist order of a functional, assuming that the admissible lines may be also distributions of several variables; in this way we shall find the corresponding equation of the Euler-Ostrogradski type.

3.1.2.1. Variation of the first order

Let us consider a functional in the form of a multiple integral in an n-dimensional space

$$I(u^1, u^2, ..., u^m) = \iint ... \int_{\Omega_n} F\left(x_1, x_2, ..., x_n; u^1, u^2, ..., u^m; \right.$$

$$\left. \frac{\partial u^1}{\partial x_1}, \frac{\partial u^2}{dx_1}, ..., \frac{\partial u^m}{\partial x_1}; ...; \frac{\partial u^1}{\partial x_n}, \frac{\partial u^2}{\partial x_n}, ..., \frac{\partial u^m}{\partial x_n} \right) dx_1 \, dx_2 \, ... \, dx_n; \qquad (3.1.56)$$

in a more compact form we may write

$$I(u^j) = \iint ... \int_{\Omega_n} F\left(x_i; u^j, \frac{\partial u^j}{\partial x_i} \right) dx_1 \, dx_2 \, ... \, dx_n$$

$$(i = 1, 2, ..., n; \ j = 1, 2, ..., m). \qquad (3.1.56')$$

We assume that the function F is of class C^0 with respect to all its arguments. We shall consider the n-dimensional admissible hypersurfaces $u^j(x_i) = u^j(x_1, x_2, ..., x_n)$, which are expressible either by distributions in the case where the operations indicated in F have a meaning, or by functions of class C^1 everywhere except on the $(n-1)$-dimensional hypersurfaces $P_k(x_1, x_2, ..., x_n)$ $(k = 1, 2, ..., h)$, indefinitely differentiable, where they have discontinuities of the first species.

We shall denote by $\mathscr{K}$ the subset of $K(\mathscr{K} \subset K)$ which includes the functions $\varphi(x_1, x_2, ..., x_n)$ whose supports belong to the domain Ω_n; we have

$$\varphi|_{\Gamma_n} = 0, \quad \varphi|_{\overline{\Omega}_n} = 0, \qquad (3.1.57)$$

where Γ_n is the boundary of the domain Ω_n, and $\overline{\Omega}_n$ the complementary domain of the latter.

Assuming that $u^j(x_i)$ is an admissible hypersurface which corresponds to the functional (3.1.56'), then, by definition, the neighbouring admissible hypersurfaces are given by the expression

$$\bar{u}^j(x_i) = u^j(x_i) + \varphi^j(x_i), \qquad (3.1.58)$$

with

$$\varphi^j(x_i) \in \mathscr{K} \subset K. \qquad (3.1.59)$$

Obviously, if an admissible hypersurface is expressed by a distribution all the neighbouring admissible hypersurfaces will be also expressed by distributions.

We can now state the following

Definition 3.1.4. *The distribution* $\delta I = \delta I(u^j)$ *defined by*

$$\left(\delta I, \sum_{j=1}^{m} \varphi^j\right) = \sum_{i=0}^{n} \sum_{j=1}^{m} (F_{u_i^j}, \varphi_i^j), \qquad (3.1.60)$$

where we have introduced the notations

$$\varphi_i^j = \frac{\partial \varphi^j}{\partial x_i}, \quad \varphi_0^j = \varphi^j$$

$$(i = 1, 2, ..., n; \; j = 1, 2, ..., m), \qquad (3.1.61)$$

$$u_i^j = \frac{\partial u^j}{\partial x_i}, \quad u_0^j = u^j$$

is called variation of the 1st order of the functional (3.1.56').

It may be verified easily that the expression (3.1.60) of the variation of the first order is a generalization of the known expression of the classical variational calculus.

By taking into account the relation

$$(F_{u_i^j}, \varphi_i^j) = -\left(\frac{\partial F_{u_i^j}}{\partial x_i}, \varphi^j\right), \qquad (3.1.62)$$

the expression (3.1.60) of the variation of the first order becomes

$$\left(\delta I, \sum_{j=1}^{m} \varphi^j\right) = \sum_{i=1}^{n} \sum_{j=1}^{m} \left[(F_{t^j}, \varphi^j) - \left(\frac{\partial}{\partial x_i} F_{u_i^j}, \varphi^j\right)\right]; \qquad (3.1.60')$$

it may be also expanded in the form

$$\left(\delta I, \sum_{j=1}^{m} \varphi^j\right) = \sum_{j=1}^{m} \left(F_{u^j} - \frac{\partial}{\partial x_1} F_{u_1^j} - \frac{\partial}{\partial x_2} F_{u_2^j} - ... - \frac{\partial}{\partial x_n} F_{u_n^j}, \varphi^j\right). \qquad (3.1.60'')$$

3.1.2.2. Equations of the Euler-Ostrogradski type

In order that $u^j(x_i)$ may represent the extremals of the functional (3.1.56') the condition

$$\delta I(u^j) = 0 \qquad (3.1.63)$$

must be satisfied.

Using expression (3.1.60″) for δI and taking into account the theorem 3.1.1, we obtain the equations

$$F_{u^j} - \frac{\partial}{\partial x_1} F_{v_1^j} - \frac{\partial}{\partial x_2} F_{u_2^j} - \dots - \frac{\partial}{\partial x_n} F_{u_n^j} = 0 \qquad (j = 1, 2, \dots, m); \quad (3.1.64)$$

these equations are formally identical to the Euler-Ostrogradski equations. However, they differ from the latter since the differentiations are performed in the sense of the theory of distributions; the extremals $u^j(x_i)$ are distributions or at least functions of class C^1, except the hypersurfaces $P_k(x_i) = 0$ $(k = 1, 2, \dots, h)$ where the derivatives of the first order may have discontinuities of the first species. For this reason they will be termed equations of the Euler-Ostrogradski *type*.

In the case where the extremals $u^j(x_i)$ are functions of class C^1 everywhere except the surfaces $P_k(x_i) = 0$ where they have discontinuities of the first species, the function F has a meaning and has discontinuities of the first species only.

Using formula (1.4.27) we may write

$$\frac{\partial}{\partial x_i} F_{u_i^j} = \frac{\tilde{\partial}}{\partial x_i} F_{u_i^j} + \sum_{k=1}^{h} (\Delta F_{u_i^j})_k \frac{\partial P_k}{\partial x_i} \delta(P_k), \qquad (3.1.65)$$

where $i = 1, 2, \dots, n$ and $j = 1, 2, \dots, m$; $\delta(P_k)$ is the Dirac distribution concentrated on the surface $P_k = 0$, and $(\Delta F_{u_i^j})_k$ represents the jump of the function $F_{u_i^j}$ as it transverses the surface $P_k = 0$, in the direction of the x_i axis.

Introducing the expressions (3.1.65) in equation (3.1.64) we obtain

$$A_j + \sum_{k=1}^{h} B_k^j \delta(P_k) = 0 \qquad (j = 1, 2, \dots, m), \qquad (3.1.66)$$

where we have introduced the notations

$$A_j = F_{u^j} - \sum_{i=1}^{n} \frac{\tilde{\partial}}{\partial x_i} F_{u_i^j}, \qquad (3.1.67)$$

$$B_k^j = \sum_{i=1}^{n} (\Delta F_{u_i^j})_k \frac{\partial P_k}{\partial x_i}. \qquad (3.1.67')$$

Using theorem 1.4.2 of section 1.4.1.2, the equations (3.1.66) are satisfied if and only if

$$A_j = 0 \qquad (j = 1, 2, \dots, m), \qquad (3.1.68)$$

$$B_k^j = 0 \quad (j = 1, 2, \dots, m \; ; \; k = 1, 2, \dots, h). \qquad (3.1.68')$$

Hence, the equations required for an extremum will be written in the form

$$F_{u^j} - \sum_{i=1}^{n} \frac{\tilde{\partial}}{\partial x_i} F_{u_i^j} = 0 \qquad (j = 1, 2, \ldots, m), \tag{3.1.69}$$

and the conditions on the surfaces of discontinuity $P_k = 0$ are given by

$$\sum_{i=1}^{n} (\Delta F_{u_i^j})_k \frac{\partial P_k}{\partial x_i} = 0 \qquad (j = 1, 2, \ldots, m; \; k = 1, 2, \ldots, h). \tag{3.1.70}$$

It should be mentioned that the surfaces $P_k = 0$ are assumed to be indefinitely differentiable.

In the particular case of the functional

$$I(u) = \iint_{\Omega} F(x, y; u, u_x, u_y) \, dx \, dy \tag{3.1.71}$$

the necessary equation for an extremum is

$$F_u - \frac{\tilde{\partial}}{\partial x} F_{u_x} - \frac{\tilde{\partial}}{\partial y} F_{u_y} = 0; \tag{3.1.72}$$

the condition along the curve of discontinuity $P(x, y) = 0$ will be of the form

$$(\Delta F_{u_x})_x \frac{\partial P}{\partial x} + (\Delta F_{u_y})_y \frac{\partial P}{\partial y} = 0 \tag{3.1.73}$$

or of the form

$$(\Delta F_{u_x})_x \, dy - (\Delta F_{u_y})_y \, dx = 0. \tag{3.1.73'}$$

3.1.2.3. Case where derivatives of higher order of the functions $u^j(x_i)$ occur.

In the foregoing we have considered the case where the functional $I(u^j)$ depends only on the derivatives of the first order of the functions $u^j(x_i)$ which are functions of class C^1; in a similar way we may also consider functionals depending on derivatives of a higher order of functions $u^j(x_i)$ which obviously must belong to a class of a higher order.

In the following we shall consider only a particular case, namely the functional

$$I(u) = \iint_{\Omega} F\left(x, y; u, \frac{\partial u}{\partial x}, \frac{\partial u}{\partial y}, \frac{\partial^2 u}{\partial x^2}, \frac{\partial^2 u}{\partial x \partial y}, \frac{\partial^2 u}{\partial y^2}\right) dx \, dy, \tag{3.1.74}$$

where it is assumed that F is of class C^0 with respect to all its arguments, and $u(x, y)$ is everywhere of the class C^2 except at the points of the curve $P(x, y) = 0$ where the derivatives of the second order may have points of discontinuity of the first species.

The variation of the first order of this functional is

$$(\delta I(u), \varphi) = (F_u, \varphi) + (F_{u_x}, \varphi_x') + (F_{u_y}, \varphi_y')$$

$$+ (F_{u_{xx}}, \varphi_{xx}'') + (F_{u_{xy}}, \varphi_{xy}'') + (F_{u_{yy}}, \varphi_{yy}''), \qquad (3.1.75)$$

which may be written in the form

$$(\delta I(u), \varphi) = \left(F_u - \frac{\partial}{\partial x} F_{u_x} - \frac{\partial}{\partial y} F_{u_y} + \frac{\partial^2}{\partial x^2} F_{u_{xx}} \right.$$

$$\left. + \frac{\partial^2}{\partial x \partial y} F_{u_{xy}} + \frac{\partial^2}{\partial y^2} F_{u_{yy}}, \varphi \right). \qquad (3.1.75')$$

Hence the condition necessary for the extremal is

$$F_u - \frac{\partial}{\partial x} F_{u_x} - \frac{\partial}{\partial y} F_{u_y} + \frac{\partial^2}{\partial x^2} F_{u_{xx}} + \frac{\partial^2}{\partial x \partial y} F_{u_{xy}} + \frac{\partial^2}{\partial y^2} F_{u_{yy}} = 0. \qquad (3.1.76)$$

Using formula (1.4.26) we may express the derivatives which occur in equation (3.1.76) in the form

$$\frac{\partial}{\partial x} F_{u_x} = \frac{\tilde{\partial}}{\partial x} F_{u_x} + (\Delta F_{u_x})_x \frac{\partial P}{\partial x} \delta(P),$$

$$\qquad (3.1.77)$$

$$\frac{\partial}{\partial y} F_{u_y} = \frac{\tilde{\partial}}{\partial y} F_{u_y} + (\Delta F_{u_y})_y \frac{\partial P}{\partial y} \delta(P);$$

$$\frac{\partial^2}{\partial x^2} F_{u_{xx}} = \frac{\tilde{\partial}^2}{\partial x^2} F_{u_{xx}} + \left(\Delta \frac{\tilde{\partial}}{\partial x} F_{u_{xx}} \right)_x \frac{\partial P}{\partial x} \delta(P) + \frac{\partial}{\partial x} \left[(\Delta F_{u_{xx}})_x \frac{\partial P}{\partial x} \delta(P) \right],$$

$$\frac{\partial^2}{\partial x \partial y} F_{u_{xy}} = \frac{\tilde{\partial}^2}{\partial x \partial y} F_{u_{xy}} + \left(\Delta \frac{\tilde{\partial}}{\partial y} F_{u_{xy}} \right)_y \frac{\partial P}{\partial y} \delta(P) + \frac{\partial}{\partial y} \left[(\Delta F_{u_{xy}})_x \frac{\partial P}{\partial x} \delta(P) \right],$$

$$\qquad (3.1.77')$$

$$\frac{\partial^2}{\partial y^2} F_{u_{yy}} = \frac{\tilde{\partial}^2}{\partial y^2} F_{u_{yy}} + \left(\Delta \frac{\tilde{\partial}}{\partial y} F_{u_{yy}} \right)_y \frac{\partial P}{\partial x} \delta(P) + \frac{\partial}{\partial y} \left[(\Delta F_{u_{yy}})_y \frac{\partial P}{\partial y} \delta(P) \right],$$

where $(\Delta f)_x$, $(\Delta f)_y$ represent the jump of the function f as it crosses the curve $P=0$ in the direction of the Ox and Oy axes, respectively. By placing the expressions (3.1.77) and (3.1.77') in equation (3.1.76) we may write the latter in the form

$$A_1 - A_2\delta(P) - \frac{\partial}{\partial x}\left[A_3\delta(P)\right] - \frac{\partial}{\partial y}\left[A_4\delta(P)\right] = 0, \qquad (3.1.76')$$

where the functions A_i $(i = 1, 2, 3, 4)$ are given by

$$A_1 = F_u - \frac{\tilde{\partial}}{\partial x}F_{u_x} - \frac{\tilde{\partial}}{\partial y}F_{u_y} + \frac{\tilde{\partial}^2}{\partial x^2}F_{u_{xx}} + \frac{\tilde{\partial}^2}{\partial x\partial y}F_{u_{xy}} + \frac{\tilde{\partial}^2}{\partial y^2}F_{u_{yy}},$$

$$A_2 = \left[\Delta\left(F_{u_x} - \frac{\tilde{\partial}}{\partial x}F_{u_{xx}}\right)\right]_x \frac{\partial P}{\partial x} + \left[\Delta\left(F_{u_y} - \frac{\tilde{\partial}}{\partial x}F_{u_{xy}} - \frac{\tilde{\partial}}{\partial y}F_{u_{yy}}\right)\right]_y \frac{\partial P}{\partial y},$$

$$(3.1.78)$$

$$A_3 = -(\Delta F_{u_{xx}})_x \frac{\partial P}{\partial x},$$

$$A_4 = -(\Delta F_{u_{yy}})_y \frac{\partial P}{\partial y} - (\Delta F_{u_{xy}})_x \frac{\partial P}{\partial x}.$$

Taking into account theorem 1.4.1 of section 1.4.1.2, the equation (3.1.76') is satisfied if and only if

$$A_1 = 0, \quad A_2 + \frac{\partial A_3}{\partial x} + \frac{\partial A_4}{\partial y} = 0, \quad A_3\frac{\partial P}{\partial x} + A_4\frac{\partial P}{\partial y} = 0. \qquad (3.1.79)$$

Hence, the equation necessary for an extremum is

$$F_u - \frac{\tilde{\partial}}{\partial x}F_{u_x} - \frac{\tilde{\partial}}{\partial y}F_{u_y} + \frac{\tilde{\partial}^2}{\partial x^2}F_{u_{xx}} + \frac{\tilde{\partial}^2}{\partial x\partial y}F_{u_{xy}} + \frac{\tilde{\partial}^2}{\partial y^2}F_{u_{yy}} = 0; \qquad (3.1.80)$$

the conditions along the curves of discontinuity $P(x, y) = 0$ are

$$\left[\Delta\left(F_{u_x} - \frac{\tilde{\partial}}{\partial x}F_{u_{xx}}\right)\right]_x \frac{\partial P}{\partial x} + \left[\Delta\left(F_{u_y} - \frac{\tilde{\partial}}{\partial y}F_{u_{yy}} - \frac{\tilde{\partial}}{\partial x}F_{u_{xy}}\right)\right]_y \frac{\partial P}{\partial y}$$

$$- \frac{\partial}{\partial x}\left[(\Delta F_{u_{xx}})_x\frac{\partial P}{\partial x}\right] - \frac{\partial}{\partial y}\left[(\Delta F_{u_{yy}})_y\frac{\partial P}{\partial y} + (\Delta F_{u_{xy}})_x\frac{\partial P}{\partial x}\right] = 0,$$

$$(3.1.81)$$

$$(\Delta F_{u_{xx}})_x\left(\frac{\partial P}{\partial x}\right)^2 + \left[(\Delta F_{u_{yy}})_y\frac{\partial P}{\partial y} + (\Delta F_{u_{xy}})_x\frac{\partial P}{\partial x}\right]\frac{\partial P}{\partial y} = 0.$$

3.1.2.4. Variation of the IInd order. Examples

Let us consider the functional defined by the integral

$$I(u) = \iint \ldots \int_{\Omega_n} F\left((x_i; u, \frac{\partial u}{\partial x_i} \right) dx_1 \, dx_2 \ldots dx_n, \tag{3.1.82}$$

where we assume that F is a function of class C^1 with respect to all its arguments. In accordance with the definition given in section 3.1.2.1 the variation of the Ist order of the functional (3.1.82) is expressed by the relation

$$(\delta I, \varphi) = \sum_{i=0}^{n} (F_{u_i}, \varphi_i), \tag{3.1.83}$$

where

$$F_{u_i} = \frac{\partial F}{\partial u_i}, \; u_i = \frac{\partial u}{\partial x_i}, \; \varphi_i = \frac{\partial \varphi}{\partial x_i} \qquad (i = 1, 2, \ldots, n), \tag{3.1.84}$$

$$u_0 = u(x_i), \; \varphi_0 = \varphi(x_i) \in \mathscr{K}^m(\Omega_n), \tag{3.1.84'}$$

where $\mathscr{K}^m(\Omega_n)$ is the subspace of K^m which includes the functions $\varphi \in C^m$ with a compact support included in Ω_n. Concerning the variation of the IInd order, we give the following

Definition 3.1.5. *The distribution* $\delta^2 I = \delta^2 I(y)$ *expressed in the form*

$$(\delta^2 I, \varphi) = \sum_{k=0}^{n} (\mathscr{F}_{u_k}, \varphi_k), \tag{3.1.85}$$

where the notation

$$\mathscr{F} = \sum_{i=0}^{n} F_{u_i} \varphi_i \tag{3.1.86}$$

has been introduced, is termed the variation of the second order of the functional (3.1.82).

Introducing the notations

$$F''_{u_i u_k} = F''_{u_k u_i} = \frac{\partial^2 F}{\partial u_i \partial u_k} = P_{ik} = P_{ki} \qquad (i, k = 0, 1, 2, \ldots, n) \tag{3.1.87}$$

and

$$P_0 = \sum_{k=0}^{n} \frac{\partial P_{0k}}{\partial x_k} \tag{3.1.87'}$$

and taking into account notation (3.1.86), we may write relation (3.1.85) in the form

$$(\delta^2 I, \varphi) = (P_0, \varphi^2) + \sum_{i=1}^{n} \sum_{k=1}^{n} (P_{ik}, \varphi_i \varphi_k). \tag{3.1.88}$$

As in the case considered in the section 3.1.1.3, the study of the second variation $\delta^2 I$ may be related to the study of the structure of the operator A defined by the expression

$$A\varphi = P_0 \varphi - \sum_{i=1}^{n} \sum_{k=1}^{n} \frac{\partial}{\partial x_k} \left(P_{ik} \frac{\partial \varphi}{\partial x_i} \right), \tag{3.1.89}$$

where P_0 and P_{ik} are distributions of the $(m-1)$th order, and $\varphi \in \mathscr{K}^m(\Omega_n)$; we remark that under these conditions the operator (3.1.89) has a meaning from the point of view of the theory of distributions.

We now introduce the following

Definition 3.1.6. *The operator* B *defined on* $\mathscr{K}^m(\Omega_n)$ *is said to be positive on* Ω_n *if for any* $u(x_i) \in \mathscr{K}^m(\Omega_n)$ *we have*

$$(Bu, u) \geqslant 0. \tag{3.1.90}$$

We admit coefficients P_0 and P_{ik} of the form

$$P_0 = \bar{P}_0 + a_0 \delta(x_1 - x_1^0, x_2 - x_2^0, ..., x_n - x_n^0), \tag{3.1.91}$$

$$P_{ik} = \bar{P}_{ik} + a_{ik} \delta(x_1 - x_1^0, x_2 - x_2^0, ..., x_n - x_n^0) \qquad (i, k = 1, 2, ..., n), \tag{3.1.91'}$$

where a_0, $a_{ik} = a_{ki}$ and $\bar{P}_{ik} = \bar{P}_{ki}$ $(k \neq i; i, k = 1, 2, ..., n)$ are constants, $\bar{P}_0$ and $\bar{P}_{ii}$ $(i = 1, 2, ..., n)$ are regular distributions and $(x_1^0, x_2^0, ..., x_n^0) \in \Omega_n$; then we may state the following

Theorem 3.1.3. *The necessary condition that the operator* A *defined by expression* (3.1.89) *under the conditions* (3.1.91), (3.1.91') *be positive, is that we have*

$$\bar{P}_{ii} \geqslant 0 \qquad (i = 1, 2, ..., n). \tag{3.1.92}$$

Since $\bar{P}_0$ and $\bar{P}_{ii}$ are measures it is sufficient to admit that $\varphi \in \mathscr{K}^1(\Omega_n)$.

Returning to the functional (3.1.82) we remark that the conditions (3.1.92) are equivalent to

$$\frac{\partial^2 F}{\partial u_{x_i}^2} \geqslant 0 \qquad (i = 1, 2, ..., n) \tag{3.1.93}$$

and we can state

Theorem 3.1.4. *The necessary condition that the distribution $u(x_1, x_2, \ldots, x_n)$ with the support in Ω_n may realize the minimum of the functional (3.1.82) is that along the extremal $u(x_1, x_2, \ldots, x_n)$ the conditions (3.1.93) be satisfied and the derivatives (3.1.87), (3.1.87') be of the form (3.1.91) and (3.1.91'), respectively.*

In that case we have $\delta^2 I \geqslant 0$.

Let us now consider the Dirichlet integral of the function $u(x_1, x_2, \ldots, x_n)$, extended to the domain $\Omega_n \subset R^n$,

$$D(u) = \iint \ldots \int_{\Omega_n} \sum_{i=1}^{n} \left(\frac{\partial u}{\partial x_i} \right)^2 dx_1 \, dx_2 \ldots dx_n. \tag{3.1.94}$$

The corresponding Euler-Ostrogradski equations will be

$$\Delta u = \sum_{i=1}^{n} \frac{\partial^2 u}{\partial x_i^2} = 0; \tag{3.1.95}$$

we remark that

$$\frac{\partial^2 F}{\partial u_{x_i} \partial u_{x_k}} = 0, \ i \neq k, \quad \frac{\partial^2 F}{\partial u_{x_i}^2} = 2 > 0 \qquad (i, k = 1, 2, \ldots, n);$$

hence the conditions required by a minimum are satisfied.

Let us also consider the functional

$$I(u) = \iint \ldots \int_{\Omega_n} \sum_{i=1}^{n} x_i^2 \left(\frac{\partial u}{\partial x_i} \right)^2 dx_1 \, dx_2 \ldots dx_n,$$

$$\Omega_n = [0, 1] \times [0, 1] \times \ldots \times [0, 1] \subset R^n, \tag{3.1.96}$$

which generalizes the functional (3.1.35) considered by Weierstrass. From the structure of the functional it follows that $I(u) \geqslant 0$; the minimum value is zero.

It has been found that in the class of functions $C^1(\Omega_n)$ there is no function which realizes the minimum of the functional (3.1.96) and satisfies the conditions

$$u(x_1, x_2, \ldots, x_n)$$

$$= \begin{cases} -1 \text{ for } x_n = -1, (x_1, x_2, \ldots, x_{n-1}) \in [0, 1] \times [0, 1] \times \ldots \times [0, 1] \\ \\ 1 \text{ for } x_n = 1, (x_1, x_2, \ldots, x_{n-1}) \in [0, 1] \times [0, 1] \times \ldots \times [0, 1]; \end{cases} \tag{3.1.97}$$

hence we must seek a solution assuming that the extremals are distributions.

The equation of the Euler-Ostrogradski type where the operations are considered in the sense of the theory of distributions is written in the form

$$\sum_{i=1}^{n} x_i^2 \frac{\partial^2 u}{\partial x_i^2} + 2 \sum_{i=1}^{n} x_i \frac{\partial u}{\partial x_i} = 0. \tag{3.1.98}$$

If $\theta(x_1, x_2, \ldots, x_n)$ represents the Heaviside function

$$\theta(x_1, x_2, \ldots, x_n) = \begin{cases} 1 \text{ for } x_i \geqslant 0 \quad (i = 1, 2, \ldots, n) \\ \\ 0 \text{ in the rest,} \end{cases} \tag{3.1.99}$$

the solution of the problems is realized by the regular distribution

$$u(x_1, x_2, \ldots, x_n) = 2\theta(x_1, x_2, \ldots, x_n) - 1; \tag{3.1.100}$$

indeed, both equation (3.1.98) and the conditions (3.1.97) are satisfied.

On the other hand the conditions required for a minimum of the functional

$$\frac{\partial^2 F}{\partial u_{x_i} \partial u_{x_k}} = 0, \; i \neq k, \quad \frac{\partial^2 F}{\partial u_{x_i}^2} = 2x_i^2 \geqslant 0 \qquad (i, k = 1, 2, \ldots, n) \tag{3.1.101}$$

are satisfied too.

This shows the need of a complete study of the extremals of a functional within the frame of the theory of distributions in the case of distributions of several variables too.

3.2. Differential equations in distributions

It is particularly important for practical applications to establish in the range of the theory of distributions the differential equations which describe various physical phenomena and to find their solution in distributions. The solution of differential equations in distributions supplies new solutions that could not be obtained by classical methods. In many cases the solution of a differential equation may be expressed in a unitary and general form with the help of the convolution product in distributions; an important part is played in this sense by the fundamental solutions.

It should be noted that some equations of mathematical physics cannot always be deduced directly in the space of distributions owing to the difficulties encountered in modelling physical phenomena.

In general, the equations which describe such phenomena are obtained first by classical methods. Next, an extension is effected where the unknown functions take zero values, so that these are defined on the whole space; the derivatives, considered in the ordinary sense, are replaced by other expressions given by relations which connect derivatives in the sense of the theory of distributions to derivatives in the ordinary sense of a function continuous almost everywhere, and having a finite number of discontinuities of the first species. In this way the unknowns of the problem will be regular distributions; then it will be assumed that these unknowns may be arbitrary distributions. Another possibility which is frequently used is to assume from the beginning that the unknowns of the problem are arbitrary distributions, assuming the same form in distributions, for the differential equations obtained by classical methods (obviously these are no longer valid for the whole space). However, there is no general method for passing to differential equations in distributions.

In the following we shall treat first of ordinary differential equations and then of partial differential equations.

3.2.1. Ordinary differential equations

We shall give first some general results and then we shall deal with the problem of finding the fundamental solution of a differential equation and point out its usefulness.

3.2.1.1. General results

We shall first state

Theorem 3.2.1. *The differential equation in distributions.*

$$\frac{dy(x)}{dx} = 0 \tag{3.2.1}$$

has no solution other than the classical solution which is

$$y = C = \text{const.} \tag{3.2.2}$$

Indeed, denoting a fundamental function by $\varphi(x)$, equation (3.2.1) is equivalent to

$$(y'(x), \varphi(x)) = 0, \tag{3.2.3}$$

whence

$$(y(x), \varphi'(x)) = 0. \tag{3.2.3'}$$

The last relation shows that the distribution $y(x)$ is defined on the set of the fundamental functions $\chi(x) \in K$ of the form

$$\chi(x) = \varphi'(x), \quad \varphi(x) \in K. \tag{3.2.4}$$

Hence, the characteristic property of the fundamental functions $\chi(x)$ consists in the relation

$$\int_{-\infty}^{\infty} \chi(x)\,dx = \int_{-\infty}^{\infty} \varphi'(x)\,dx = \varphi(x)\Big|_{-\infty}^{\infty},$$

which leads to

$$\int_{-\infty}^{\infty} \chi(x)\,dx = 0. \tag{3.2.5}$$

Let us consider a fundamental function $\varphi_1(x)$ which satisfies the condition

$$\int_{-\infty}^{\infty} \varphi_1(x)\,dx = 1. \tag{3.2.6}$$

Thus, any fundamental function $\varphi(x)$ may be written in the form

$$\varphi(x) = \varphi_1(x)\int_{-\infty}^{\infty} \varphi(x)\,dx + \chi(x); \tag{3.2.7}$$

the function $\varphi_1(x)$ is uniquely specified.

Considering relations (3.2.3') and (3.2.8), we may write

$$(y(x), \varphi(x)) = (y(x), \varphi_1(x))\int_{-\infty}^{\infty} \varphi(x)\,dx + (y(x), \chi(x)); \tag{3.2.8}$$

remarking that

$$(y(x), \varphi_1(x)) = C = \text{const}, \tag{3.2.9}$$

it follows that

$$(y(x), \varphi(x)) = (C, \varphi(x)), \tag{3.2.10}$$

which leads to relation (3.2.2), thus proving the theorem.

We shall give now the following

Definition 3.2.1. *The differential equation in distributions*

$$\frac{dy(x)}{dx} = f(x) \tag{3.2.11}$$

being given where $f(x)$ is a distribution, the distribution $y(x)$ is said to be the primitive of the distribution $f(x)$ and is denoted by

$$y(x) = \int f(x)\,\mathrm{d}x. \tag{3.2.12}$$

Equation (3.2.11) is equivalent to

$$(y'(x), \varphi(x)) = (f(x), \varphi(x)), \tag{3.2.13}$$

whence

$$(y(x), \varphi'(x)) = - (f(x), \varphi(x)). \tag{3.2.13'}$$

It follows that the distribution $y(x)$ is defined on the set of the fundamental functions $\mathcal{X}(x)$, of the form (3.2.4). Hence, in order to determine a solution of equation (3.2.11), i.e. a distribution $y(x)$ which verifies relation (3.2.13'), it is sufficient to take its value in $\mathcal{X}(x)$. Therefore, we can state the following

Theorem 3.2.2. *The differential equation* (3.2.11) *considered in the sense of the theory of distributions, where $f(x)$ is an arbitrary distribution, always admits a solution.*
Let $y_1(x)$ and $y_2(x)$ be two solutions of the equation (3.2.11); we may write

$$\frac{\mathrm{d}y_1(x)}{\mathrm{d}x} = f(x), \quad \frac{\mathrm{d}y_2(x)}{\mathrm{d}x} = f(x), \tag{3.2.14}$$

whence, by substraction

$$\frac{\mathrm{d}}{\mathrm{d}x}\left[y_1(x) - y_2(x)\right] = 0; \tag{3.2.14'}$$

applying the theorem (3.2.1), there follows

$$y_1(x) = y_2(x) + C, \ C = \text{const}, \tag{3.2.15}$$

and we may state the following

Theorem 3.2.2'. *Two primitives of the differential equation* (3.2.11) *differ by a constant.*
In the case where $f(x)$ is a distribution of the function type we note that we always have a primitive which is also a distribution of the same class and, by theorem 3.2.2', all its primitives are distributions of the function type; we may thus state the following

Theorem 3.2.2''. *The equation* (3.2.11) *considered in the sense of the theory of distributions, where $f(x)$ is a function, has no solution other than the classical solution.*
We remark that it is possible — $f(x)$ being a singular distribution — to obtain a primitive which is a distribution of the function type; this occurs for example for $f(x) = \delta(x)$, where the primitive is the Heaviside distribution $\theta(x)$.

Also, if $f(x)$ is a distribution of the function type, the primitive may be a function. For example the equation

$$\frac{dy(x)}{dx} = \theta(x),$$ (3.2.16)

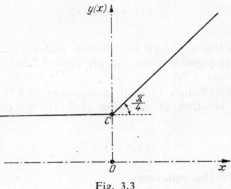

Fig. 3.3

where $\theta(x)$ is the Heaviside distribution, obviously has the solution

$$y(x) = \begin{cases} C \text{ for } x \leqslant 0 \ \ C = \text{const,} \\ \\ x + C \text{ for } x > 0, \end{cases}$$ (3.2.17)

the graph of which is shown in Figure 3.3; the solution (3.2.17) is a continuous function which may be written also in the form

$$y(x) = x\theta(x) + C = x_+ + C.$$ (3.2.17′)

Let us now consider the linear and homogeneous differential equation of the first order, in the sense of the theory of distributions,

$$y'(x) + P(x)y(x) = 0,$$ (3.2.18)

where $P(x)$ is a function of class C^∞; we state the following

Theorem 3.2.3. *The differential equation (3.2.18) considered in the sense of the theory of distributions, has no solution except the classical solution which is given by*

$$u(x) = C e^{-\int P(x)\,dx},$$ (3.2.19)

where C is an arbitrary constant.

We remark that the function $u(x)$ possesses derivatives of all orders and does not vanish at any point $(u(x) \neq 0$ for all $x \in R)$.

Let us now assume that equation (3.2.18) in distributions admits the distribution $y(x)$ as a solution; in order to determine this distribution we perform the substitution

$$y(x) = u(x)z(x), \qquad (3.2.20)$$

where $z(x)$ represents the unknown distribution and $C = 1$.

This substitution is possible since the product (3.2.20) has a meaning as the function $u(x)$ is of class C^∞.

Introducing the distribution (3.2.20) in equation (3.2.18) and taking into account that $u(x)$ is a solution of equation (3.2.18), we obtain the equation in distributions

$$u(x)z'(x) = 0, \qquad (3.2.21)$$

which is equivalent to the equation

$$z'(x) = 0 \qquad (3.2.21')$$

since, as has been shown above, $u(x)$ does not vanish at any point; by theorem 3.2.1 the equation (3.2.21') admits only the solution

$$z = C = \text{const}, \qquad (3.2.22)$$

which, taking into account substitution (3.2.20), proves the theorem.

If $P(x)$ is of class C^0 the solution of the equation is of class C^1.

Let us now consider the homogeneous equation of order n

$$\mathrm{D}y(x) \equiv y^{(n)}(x) + a_1(x)y^{(n-1)}(x) + \ldots + a_n(x)y(x) = 0, \qquad (3.2.23)$$

where $a_i(x)$ $(i = 1, 2, \ldots, n)$ are functions of class C^0; we can state the following

Theorem 3.2.4. *The differential equation* (3.2.23) *has always a solution of class* C^n *and different from zero for all* $x \in R$.

Indeed, in that case it exists a fundamental system of solutions $y_i(x)(i=1, 2, \ldots, n)$ of class C^n, for $x \in R$; we shall show that we can always choose the constants $C_i(i = 1, 2, \ldots, n)$ in order to have

$$y(x) = \sum_{i=1}^{n} C_i y_i(x) \neq 0, \qquad x \in R. \qquad (3.2.24)$$

With that end in view we remark that we can determine a solution which is piecewise of class C^n and different from zero (Figure 3.4). Let

$$y(x) = f_1(x), \qquad x \in [x_1, x_2],$$

$$(3.2.25)$$

$$y(x) = f_2(x), \qquad x \in [x_2, x_3]$$

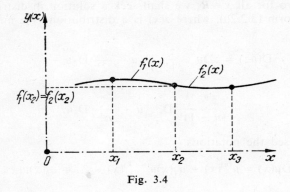

Fig. 3.4

be two solutions of equation (3.2.23) corresponding to the intervals $[x_1, x_2]$ and $[x_2, x_3]$; if we choose for the solution $y = f_2(x)$, $x \in [x_2, x_3]$ the initial conditions

$$f_2(x_2) = f_1(x_2),$$

$$f_2'(x_2 + 0) = f_1'(x_2 - 0),$$

$$(3.2.25')$$

$$\dots\dots\dots\dots\dots\dots\dots\dots\dots\dots\dots\dots$$

$$f_2^{(n-1)}(x_2 + 0) = f_1^{(n-1)}(x_2 - 0),$$

we can determine the constants C_i. Taking into account that both solutions (3.2.25) satisfy equation (3.2.23) we deduce that

$$f_2^{(n)}(x_2 + 0) = f_1^{(n)}(x_2 - 0). \qquad (3.2.25'')$$

Hence the function

$$y(x) = \begin{cases} f_1(x) \text{ for } x \in [x_1, x_2] \\ \\ f_2(x) \text{ for } x \in [x_2, x_3] \end{cases} \qquad (3.2.24')$$

is a solution of class C^n of the equation (3.2.23) on the interval $[x_1, x_3]$, a solution which does not vanish in that interval. We continue to construct the solution by this method for all $x \in R$, thus proving the theorem 3.2.4.

Based on that theorem we can prove the following

Theorem 3.2.5. *The differential equation* (3.2.23) *considered in the sense of the theory of distributions has no solution except the classical solution.*

We remark that equation (3.2.23) has always a solution $u(x)$ of class C^n and different from zero for all $x \in R$; we shall seek a solution in distributions of that equation in the form (3.2.20), where $z(x)$ is a distribution. We may write

$$D(uz) = zDu + \frac{z'}{1!} D_1u + \frac{z''}{2!} D_2u + \ldots$$

$$+ \frac{z^{(n-1)}}{(n-1)!} D_{n-1}u + \frac{z^{(n)}}{n!} D_nu, \tag{3.2.26}$$

where we have used the notations

$$Du(x) = u^{(n)}(x) + a_1(x)u^{(n-1)}(x) + \ldots + a_n(x)u(x),$$

$$D_1u(x) = nu^{(n-1)}(x) + (n-1)a_1(x)u^{(n-2)}(x) + \ldots + 2a_{n-2}(x)u'(x) + a_{n-1}(x)u(x),$$

$$D_2u(x) = n(n-1)u^{(n-2)}(x) + (n-1)(n-2)a_1(x)u^{(n-3)}(x) + \ldots 2.1.a_{n-2}(x)u(x),$$

$$\tag{3.2.26'}$$

$\ldots\ldots\ldots\ldots\ldots\ldots\ldots\ldots\ldots\ldots\ldots\ldots\ldots\ldots\ldots\ldots\ldots\ldots\ldots\ldots$

$$D_{n-1}u(x) = n(n-1)\ldots 2u'(x) + (n-1)(n-2)\ldots 2.1.a_1(x)u(x),$$

$$D_nu(x) = n(n-1)\ldots 2.1.u(x).$$

Taking into account that $u(x)$ is a solution of the equation (3.2.23), it follows that, after performing the substitution (3.2.21), the equation becomes

$$\frac{z'(x)}{1!}D_1u(x) + \frac{z''(x)}{2!}D_2u(x) + \ldots$$

$$+ \frac{z^{(n-1)}(x)}{(n-1)!} D_{n-1}u(x) + \frac{z^{(n)}(x)}{n!}D_nu(x) = 0. \tag{3.2.27}$$

Performing the substitution

$$z'(x) = v(x), \tag{3.2.27'}$$

equation (3.2.27) becomes a homogeneous differential equation of the $(n - 1)$th order where all the coefficients are functions of class C^0, namely

$$v^{(n-1)}(x) + \frac{1}{(n-1)!} \frac{D_{n-1}u(x)}{u(x)} v^{(n-2)}(x) + \cdots$$

$$+ \frac{1}{2!} \frac{D_2 u(x)}{u(x)} v'(x) + \frac{1}{1!} \frac{D_1 u(x)}{u(x)} v(x) = 0 \qquad (3.2.27'')$$

Therefore, any homogeneous ordinary differential equation of the nth order of the form (3.2.23) may be reduced in this way to a homogeneous differential equation of the $(n - 1)$th order where the coefficient of the derivative of the $(n-1)$th order is equal to unity and the other coefficients are functions of class C^0. Since the theorem has been proved for the case $n = 1$ (theorem 3.2.3), the method of complete induction and the considerations stated above ensure its validity for an arbitrary n.

By introducing the Wronskian corresponding to the fundamental system of solutions considered above

$$W(y_1, y_2, \ldots, y_n) \equiv \begin{vmatrix} y_1 & y_2 & \cdots & y_n \\ y_1' & y_2' & \cdots & y_n' \\ \cdots\cdots\cdots\cdots\cdots\cdots\cdots\cdots \\ y_1^{(n-1)} & y_2^{(n-1)} & \cdots & y_n^{(n-1)} \end{vmatrix}, \qquad (3.2.23')$$

equation (3.2.23) is equivalent to

$$W(y_1, y_2, \ldots, y_n; y) \equiv \begin{vmatrix} y_1 & y_2 & \cdots & y_n & y \\ y_1' & y_2' & \cdots & y_n' & y' \\ \cdots\cdots\cdots\cdots\cdots\cdots\cdots\cdots\cdots\cdots \\ y_1^{(n-1)} & y_2^{(n-1)} & \cdots & y_n^{(n-1)} & y^{(n-1)} \\ y_1^{(n)} & y_2^{(n)} & \cdots & y_n^{(n)} & y^{(n)} \end{vmatrix} = 0. \qquad (3.2.23'')$$

Since $W(y_1, y_2, \ldots, y_n) \neq 0$ for all $x \in R$ we may express the coefficients $a_i(x)$ $(i = 1, 2, \ldots, n)$ in the form

$$a_i(x) = \frac{(-1)^i}{W(y_1, y_2, \ldots, y_n)} \begin{vmatrix} y_1 & y_2 & \cdots & y_n \\ y_1' & y_2' & \cdots & y_n' \\ \cdots\cdots\cdots\cdots\cdots\cdots\cdots\cdots \\ y_1^{(n-i-1)} & y_2^{(n-i-1)} & \cdots & y_n^{(n-i-1)} \\ y_1^{(n-i+1)} & y_2^{(n-i+1)} & \cdots & y_n^{(n-i+1)} \\ \cdots\cdots\cdots\cdots\cdots\cdots\cdots\cdots \\ y_1^{(n)} & y_2^{(n)} & \cdots & y_n^{(n)} \end{vmatrix}. \qquad (3.2.23''')$$

We shall now assume that $a_i(x)$ $(i = 1, 2, \ldots, n)$ are functions of class C^∞, and that at least one of these coefficients is not identically zero*. In that case it follows that $W(y_1, y_2, \ldots, y_n)$ has derivatives of the first order; relation (3.2.23''') shows that the fundamental system of solutions has derivatives of the order $(n+1)$. Returning to the expression (3.2.23') of the Wronskian it follows that the latter admits now derivatives of the second order a.s.o. We conclude that the Wronskian possesses derivatives of all orders and that the fundamental system of solutions is of class C^∞ for all $x \in R$.

With reference to theorem 3.2.4 we now remark that the *differential equation* (3.2.23) *for which the* $a_i(x)$ *are of class* C^∞ *has always a solution which is indefinitely differentiable almost everywhere and is different from zero for all* $x \in R$.

3.2.1.2. Fundamental solution

We shall first consider the non-homogeneous differential equation of the first order

$$y'(x) + P(x)y(x) = Q(x), \tag{3.2.28}$$

where $P(x)$ is a function of class C^∞ and $Q(x)$ is a distribution. By theorem 3.2.3 the homogeneous equation corresponding to the equation (3.2.28) has no solution except the classical solution, namely

$$u(x) = e^{-\int P(x)dx}. \tag{3.2.29}$$

Since $u(x)$ is of class C^∞ and does not vanish at any point $(u(x) \neq 0)$ we shall seek a solution of equation (3.2.28) in the form

$$y(x) = u(x)v(x), \tag{3.2.30}$$

where $v(x)$ is an arbitrary distribution; performing the differentiations in the sense of the theory of distributions, equation (3.2.28) becomes

$$u(x)\frac{dv(x)}{dx} = Q(x). \tag{3.2.31}$$

Since $u(x) \neq 0$ for all $x \in R$, we have

$$\frac{dv(x)}{dx} = \frac{Q(x)}{u(x)}; \tag{3.2.31'}$$

---

* Indeed, for the equation $y^{(n)}(x) = 0$ it exists always a fundamental system of solutions of class C^∞ for all $x \in R$.

the corresponding primitive is

$$v(x) = \int \frac{Q(x)}{u(x)} \, \mathrm{d}x. \tag{3.2.32}$$

In particular if $Q(x) = \delta(x)$, equation (3.2.31′) becomes

$$\frac{\mathrm{d}v(x)}{\mathrm{d}x} = \frac{1}{u(x)} \delta(x) = \frac{1}{u(0)} \delta(x), \tag{3.2.33}$$

whence

$$v(x) = \frac{1}{u(0)} \, \theta(x) + C, \tag{3.2.34}$$

where $\theta(x)$ is the Heaviside distribution.
Therefore, the general solution of the equation

$$y'(x) + P(x)y(x) = \delta(x) \tag{3.2.35}$$

is

$$y(x) = \frac{1}{u(0)} u(x)\theta(x) + Cu(x) \tag{3.2.36}$$

and is termed the *fundamental solution* of the differential equation (3.2.28).
For $Q(x) = \delta'(x)$, equation (3.2.31′) becomes

$$v'(x) = \frac{1}{u(x)} \delta'(x). \tag{3.2.37}$$

From the relation

$$\frac{1}{u(x)} \, \delta(x) = \frac{1}{u(0)} \, \delta(x)$$

we obtain by differentiation

$$\frac{1}{u(x)} \delta'(x) = \frac{1}{u(0)} \delta'(x) + \frac{u'(x)}{u^2(x)} \delta(x) = \frac{1}{u(0)} \delta'(x) + \frac{u'(0)}{u^2(0)} \delta(x), \tag{3.2.38}$$

whence

$$v(x) = \frac{1}{u(0)} \delta(x) + \frac{u'(0)}{u^2(0)} \theta(x) + C. \tag{3.2.39}$$

Therefore the general solution of the equation

$$y'(x) + P(x)y(x) = \delta'(x) \tag{3.2.40}$$

is

$$y(x) = \delta(x) + \frac{u'(0)}{u^2(0)} u(x)\theta(x) + Cu(x). \tag{3.2.41}$$

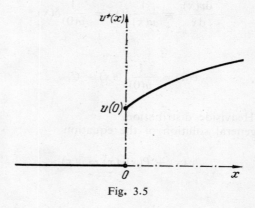

Fig. 3.5

Introducing the function

$$u^+(x) = \theta(x)u(x) = \begin{cases} 0 & \text{for } x < 0, \\ u(x) & \text{for } x \geqslant 0, \end{cases} \tag{3.2.42}$$

the graph of which is shown in Figure 3.5 and which is termed the *positive part of the function* $u(x)$, we may write the general solutions (3.2.36) and (3.2.41) in the form

$$y(x) = \frac{1}{u(0)} u^+(x) + Cu(x), \tag{3.2.36'}$$

$$y(x) = \delta(x) + \frac{u'(0)}{u^2(0)} u^+(x) + Cu(x). \tag{3.2.41'}$$

In particular, the positive part of the function $u(x) = x$ is x_+.

Let us now consider the non-homogeneous linear ordinary differential equation of order n whose coefficients are functions possessing derivatives of all orders

$$Dy(x) \equiv y^{(n)}(x) + a_1(x)y^{(n-1)}(x) + \ldots + a_n(x)y(x) = f(x), \tag{3.2.43,}$$

where $f(x)$ is a distribution.

Since by theorem 3.2.5 the homogeneous differential equation

$$y^{(n)}(x) + a_1(x)y^{(n-1)}(x) + \ldots + a_n(x)y(x) = 0, \qquad (3.2.44)$$

corresponding to equation (3.2.43), considered in the sense of the theory of distributions, has no solution except the classical one, we shall denote by $u_i(x)$ $(i = 1, 2, \ldots, n)$ a fundamental system of solutions and we shall write the general solution of that equation denoted by $Y(x)$, in the form

$$Y(x) = C_1u_1(x) + C_2u_2(x) + \ldots + C_nu_n(x). \qquad (3.2.45)$$

For the determination of the general solution in distributions of the non-homogeneous equation (3.2.43), we apply the method of the variation of constants; this leads to the system of equations

$$C_1'(x)u_1(x) + C_2'(x)u_2(x) + \ldots + C_n'(x)u_n(x) = 0,$$

$$C_1'(x)u_1'(x) + C_2'(x)u_2'(x) + \ldots + C_n'(x)u_n'(x) = 0,$$

$$\ldots\ldots\ldots\ldots\ldots\ldots\ldots\ldots\ldots\ldots\ldots\ldots\ldots\ldots\ldots\ldots\ldots\ldots\ldots\ldots\ldots\ldots \qquad (3.2.46)$$

$$C_1'(x)u_1^{(n-2)}(x) + C_2'(x)u_2^{(n-2)}(x) + \ldots + C_n'(x)u_n^{(n-2)}(x) = 0,$$

$$C_1'(x)u_1^{(n-1)}(x) + C_2'(x)u_2^{(n-1)}(x) + \ldots + C_n'(x)u_n^{(n-1)}(x) = f(x).$$

Since the determinant of the coefficients of the system of equations in C_i' $(i = 1, 2, \ldots, n)$ is the Wronskian corresponding to the fundamental system of solutions $u_i(x)$, it follows that it will be different from zero and indefinitely differentiable; solving by Cramer's rule, we obtain

$$C_i'(x) = \frac{A_i(x)}{W(x)}f(x) \qquad (i = 1, 2, \ldots, n), \qquad (3.2.47)$$

where $A_i(x)$ $(i = 1, 2, \ldots, n)$ represent the algebraic complements of the Wronskian $W(x)$ corresponding to the last row and column i.

Integrating we obtain

$$C_i(x) = \int \frac{A_i(x)}{W(x)}f(x)\,\mathrm{d}x, \qquad (3.2.48)$$

so that the general solution of the non-homogeneous equation (3.2.43) may be written in the form

$$y(x) = \sum_{i=1}^{n} u_i(x)\int \frac{A_i(x)}{W(x)}f(x)\,\mathrm{d}x. \qquad (3.2.49)$$

Let us now consider the case $f(x) = \delta(x)$; from relation (3.2.47) it follows that

$$C_i'(x) = \frac{A_i(x)}{W(x)}\delta(x) = \frac{A_i(0)}{W(0)}\delta(x) \qquad (i = 1, 2, ..., n), \qquad (3.2.50)$$

whence

$$C_i(x) = \frac{A_i(0)}{W(0)}\theta(x) \qquad (i = 1, 2, ..., n). \qquad (3.2.50')$$

Hence, the general solution of the equation

$$y^{(n)}(x) + a_1(x)y^{(n-1)}(x) + ... + a_n(x)y(x) = \delta(x) \qquad (3.2.51)$$

is given by

$$E(x) = Y(x) + \sum_{i=1}^{n} \frac{A_i(0)}{W(0)}\theta(x)u_i(x) \qquad (3.2.52)$$

and is termed the *fundamental solution* of the equation (3.2.43).

We introduce the positive part of the function $u_i(x)$ with the help of a formula of the form (3.2.42); the expression

$$E^+(x) = \sum_{i=1}^{n} \frac{A_i(0)}{W(0)}u_i^+(x) \qquad (3.2.53)$$

will be called the *particular fundamental solution* of equation (3.2.43).

Thus, the general solution of the non-homogeneous equation (3.2.51) is given by

$$y(x) = Y(x) + E^+(x). \qquad (3.2.52')$$

3.2.1.3. Ordinary linear differential equations with constant coefficients

The notions of fundamental solution and particular fundamental solution play an important part in the case of ordinary linear differential equations with constant coefficients. Thus, let us consider

$$Dy(x) \equiv y^{(n)}(x) + a_1 y^{(n-1)}(x) + ... + a_n y(x) = f(x) \qquad (3.2.54)$$

an ordinary linear differential equation with constant coefficients, where $f(x)$ is a distribution.

Definition 3.2.2. *The distribution $E(x)$ which satisfies the equation*

$$DE(x) = \delta(x) \qquad (3.2.55)$$

is called the fundamental solution of equation (3.2.54).

This fundamental solution will be of the form

$$E(x) = Y(x) + E^+(x),$$ (3.2.56)

where $E^+(x)$ is the particular fundamental solution.

Taking into account formula (1.3.27) we may write

$$D[E(x) * f(x)] = DE(x) * f(x) = \delta(x) * f(x) = f(x);$$

hence, the general solution of equation (3.2.54) may be written, using the product of convolution, in the form

$$y(x) = E(x) * f(x).$$ (3.2.57)

This result, valid for linear differential equations with constant coefficients, justifies the terms of fundamental and particular fundamental solution introduced above; by extension, we used the same terms in the case of differential equations with variable coefficients, but this property is verified only for certain classes of such equations.

In case the fundamental system of solutions $u_i(x)$ $(i = 1, 2, \ldots, n)$ of the homogeneous equation $Dy(x) = 0$ is normal, then $W(0) = 1$, $A_i(0) = 0$ $(i = 1, 2, \ldots, n-1)$, $A_n(0) = 1$, and the fundamental solution $E(x)$ is expressed by

$$E(x) = Y(x) + u_n^+(x)$$ (3.2.58)

and the particular fundamental solution by

$$E^+(x) = u_n^+(x) = \theta(x)u_n(x).$$ (3.2.58')

Therefore the general solution of the non-homogeneous equation with constant coefficients (3.2.54) is

$$y(x) = [Y(x) + u_n^+(x)] * f(x),$$ (3.2.59)

where the general solution of the corresponding homogeneous equation is

$$Y(x) = y_0 u_1(x) + y_1 u_2(x) + \ldots + y_{n-1} u_n(x),$$ (3.2.60)

with the initial conditions

$$C_1 = y(x)|_{x=x_0} = y(x_0) = y_0,$$

$$C_2 = y'(x)|_{x=x_0} = y'(x_0) = y_1,$$

$$\cdots\cdots\cdots\cdots\cdots\cdots\cdots\cdots\cdots\cdots\cdots\cdots$$ (3.2.61)

$$C_n = y^{(n-1)}(x)|_{x=x_0} = y^{(n-1)}(x_0) = y_{n-1}.$$

In particular, for $y_0 = y_1 = y_2 = \ldots = y_{n-2} = 0$, $y_{n-1} = 1$ the solution $Y(x)$ of the homogeneous equation, under the above conditions, may be written

$$Y(x) = u_n(x); \tag{3.2.62}$$

therefore, the particular fundamental solution $E^+(x)$ is obtained by starting from (3.2.62) and is given by

$$E^+(X) = \theta(x)Y(x). \tag{3.2.63}$$

Thus we have obtained a simple method for determining the particular fundamental solution $E^+(x)$: we determine first the solution $Y(x)$ of the homogeneous equation which satisfies the initial conditions

$$Y(0) = 0,\ Y'(0) = 0, \ldots,\ Y^{(n-2)}(0) = 0,\ Y^{(n-1)}(0) = 1 \tag{3.2.64}$$

and then we compute the particular fundamental solution using formula (3.2.63).

Let us now consider equation (3.2.54) for which a solution is sought for all $x \geqslant 0$, $f(x)$ being an arbitrary function with the support in $[0, \infty)$; the initial conditions are assumed to be of the Cauchy type

$$y^{(k)}(0) = y_k \qquad (k = 0, 1, 2, \ldots, n-1). \tag{3.2.65}$$

We shall consider the positive part of the function $y(x)$ which is given by

$$\theta(x)y(x) = \begin{cases} 0 & \text{for } x < 0 \\[2mm] y(x) & \text{for } x \geqslant 0 \end{cases} \tag{3.2.66}$$

and which presents a discontinuity of the first species for $x = 0$. Taking into account the differentiation formula (1.2.47) we obtain the relation

$$D[\theta(x)y(x)] = \theta(x)\tilde{D}y(x) + \sum_{k=0}^{n-1} h_k \delta^{(k)}(x), \tag{3.2.67}$$

where the jumps h_k are expressed by

$$h_k = y_{n-k-1} + a_1 y_{n-k-2} + \ldots + a_{n-k-1}y_0, \tag{3.2.68}$$

for $k = 0, 1, 2, \ldots, n-1$.

The particular fundamental solution may be obtained by the method shown previously and satisfies the relation

$$DE^+(x) = \delta(x). \tag{3.2.69}$$

Therefore, the solution of the differential equation (3.2.54) with the initial conditions (3.2.65) may be written in the form

$$\theta(x)y(x) = E^+(x) * \left[\theta(x)f(x) + \sum_{k=0}^{n-1} h_k \delta^{(k)}(x) \right]. \tag{3.2.70}$$

Indeed we have

$$D[\theta(x)y(x)] = DE^+(x) * \left[\theta(x)f(x) + \sum_{k=0}^{n-1} h_k \delta^{(k)}(x) \right] = \theta(x)f(x) + \sum_{k=0}^{n-1} h_k \delta^{(k)}(x),$$

whence, comparing with relation (3.2.67), it follows that equation $\tilde{D}y(x) = f(x)$ is satisfied.

Writing explicitly the expression (3.2.70), the solution takes the form

$$y(x) = E^+(x) * \theta(x)f(x) + \sum_{k=0}^{n-1} h_k \frac{d^k}{dx^k} E^+(x) \tag{3.2.70'}$$

or the form

$$y(x) = \int_0^x E^+(x-t)f(t)\, dt + \sum_{k=0}^{n-1} h_k \frac{\tilde{d}^k}{dx^k} E^+(x) \tag{3.2.70''}$$

for all $x \geqslant 0$; account has been taken of the fact that

$$\frac{d^k}{dx^k} E^+(x) = \frac{\tilde{d}^k}{dx^k} E^+(x) \qquad (k = 0, 1, 2, ..., n-1), \tag{3.2.71}$$

where the tilde corresponds to the derivative in the ordinary sense.

In particular, the solution of the homogeneous equation

$$Dy(x) = 0 \tag{3.2.72}$$

for $x \geqslant 0$, with the initial conditions of the Cauchy type (3.2.65), is given by

$$y(x) = \sum_{k=0}^{n-1} h_k \frac{\tilde{d}^k}{dx^k} E^+(x). \tag{3.2.73}$$

The notion of fundamental solution may be extended in the same way also to systems of ordinary differential equations.

3.2.1.4. Particular case

In the following we shall consider the particular case of equation

$$\frac{d^2 f(x)}{dx^2} = 0; \tag{3.2.74}$$

thus we shall show, by means of an example, a method for solving a differential equation with different initial or boundary conditions.

Let us first consider *initial conditions of the Cauchy type*

$$x = x_0 : f(x) = f(x_0), \quad \frac{df(x)}{dx} = f'(x_0); \tag{3.2.75}$$

we seek a solution for all $x \geqslant x_0$.

With that end in view we introduce the function $\bar{f}(x)$ defined by

$$\bar{f}(x) = \begin{cases} 0 & \text{for } x < x_0 \\ f(x_0) & \text{for } x = x_0 \\ f(x) & \text{for } x > x_0, \end{cases} \tag{3.2.76}$$

and the corresponding regular distribution; we may write

$$\frac{\widetilde{d\bar{f}(x)}}{dx} = \begin{cases} 0 & \text{for } x < x_0 \\ f'(x_0) & \text{for } x = x_0 \\ \dfrac{df(x)}{dx} & \text{for } x > x_0 \end{cases} \tag{3.2.77}$$

and

$$\frac{\widetilde{d^2 \bar{f}(x)}}{dx^2} = \begin{cases} 0 & \text{for } x < x_0 \\ \dfrac{d^2 f}{dx^2} & \text{for } x \geqslant x_0. \end{cases} \tag{3.2.77'}$$

Taking into account the jumps of the functions $\bar{f}(x)$ and $\dfrac{d\bar{f}(x)}{dx}$ at the point $x = x_0$ it follows that

$$\frac{d\bar{f}(x)}{dx} = \frac{\widetilde{d\bar{f}(x)}}{dx} + f(x_0)\delta(x - x_0), \tag{3.2.78}$$

$$\frac{d^2 \bar{f}(x)}{dx^2} = \frac{\widetilde{d^2 \bar{f}(x)}}{dx^2} + f'(x_0)\delta(x - x_0) + f(x_0)\delta'(x - x_0). \tag{3.2.78'}$$

Returning to equation (3.2.74), the derivative (3.2.77′) may be written

$$\frac{\tilde{\mathrm{d}}^2 \overline{f(x)}}{\mathrm{d}x^2} = 0, \qquad x \in R, \tag{3.2.79}$$

so that using relation (3.2.78′) we may write the equation in distributions, namely

$$\frac{\mathrm{d}^2 \overline{f(x)}}{\mathrm{d}x^2} = f'(x_0)\delta(x - x_0) + f(x_0)\delta'(x - x_0). \tag{3.2.80}$$

Integrating we obtain

$$\frac{\mathrm{d}\overline{f(x)}}{\mathrm{d}x} = f'(x_0)\theta(x - x_0) + f(x_0)\delta(x - x_0), \tag{3.2.81}$$

where the Heaviside distribution occurs, while the constant which appears must vanish so that relation (3.2.78) is satisfied by (3.2.77) for $x < x_0$. Integrating again, we may write

$$\overline{f}(x) = f'(x_0)(x - x_0)\theta(x - x_0) + f(x_0)\theta(x - x_0), \tag{3.2.82}$$

where the constant which occurs must vanish too, so as to satisfy the relation (3.2.76) for $x < x_0$.
The solution sought is given by

$$\overline{f}(x) = f(x)\theta(x - x_0), \qquad x \in R; \tag{3.2.76'}$$

remarking that

$$f(x) = \overline{f}(x) \quad \text{for} \quad x \geqslant x_0 \tag{3.2.83}$$

and introducing the positive part of the function $(x - x_0)$ in the form

$$(x - x_0)_+ = (x - x_0)\theta(x - x_0), \tag{3.2.84}$$

we may write

$$\overline{f}(x) = f'(x_0)(x - x_0)_+ + f(x_0)\theta(x - x_0), \qquad x \in R \tag{3.2.85}$$

or

$$f(x) = f'(x_0)(x - x_0) + f(x_0), \qquad x \geqslant x_0. \tag{3.2.85'}$$

In particular, for $x_0 = 0$ there follows

$$\overline{f}(x) = f'(0)x_+ + f(0)\theta(x), \qquad x \in R. \tag{3.2.86}$$

We shall now apply the method indicated in the previous section and the formula (3.2.73); to this end let us compute the particular fundamental solution in the form

$$E^+(x) = x_+ \qquad\qquad (3.2.87)$$

so that

$$\frac{d^2}{dx^2} E^+(x) = \delta(x). \qquad\qquad (3.2.87')$$

Remarking that $a_1 = a_2 = 0$, $h_0 = y_1$, $h_1 = y_0$, we obtain

$$\bar{f}(x) = y_1 x_+ + y_0 \theta(x), \qquad x \in R, \qquad\qquad (3.2.86')$$

a solution which coincides with (3.2.86) since $y_0 = f(0)$ and $y_1 = f'(0)$.

Let us now consider a *bilocal* problem for which the boundary conditions are of the form

$$x = x_1 : f(x) = f(x_1), \ x = x_2 : f(x) = f(x_2), \qquad x_1 < x_2, \qquad (3.2.88)$$

and the equation must be solved for $x \in [x_1, x_2]$. This is in fact the one-dimensional case of a problem of the Dirichlet type for the harmonic equation.

We introduce the function $f(x)$ defined by

$$\bar{f}(x) = \begin{cases} 0 & \text{for } x < x_1 \\ f(x_1) & \text{for } x = x_1 \\ f(x) & \text{for } x_1 < x < x_2 \\ f(x_2) & \text{for } x = x_2 \\ 0 & \text{for } x > x_2 \end{cases} \qquad\qquad (3.2.89)$$

and the corresponding regular distributions; there follows

$$\frac{\widetilde{d\bar{f}(x)}}{dx} = \begin{cases} 0 & \text{for } x < x_1 \\ f'(x_1) & \text{for } x = x_1 \\ \dfrac{df(x)}{dx} & \text{for } x_1 < x < x_2 \\ f'(x_2) & \text{for } x = x_2 \\ 0 & \text{for } x > x_2] \end{cases} \qquad\qquad (3.2.90)$$

and

$$\frac{\tilde{\mathrm{d}}^2\overline{f}(x)}{\mathrm{d}x^2} = \begin{cases} 0 & \text{for } x < x_1 \\ \dfrac{\mathrm{d}^2 f(x)}{\mathrm{d}x^2} & \text{for } x_1 \leqslant x \leqslant x_2 \\ 0 & \text{for } x > x_2, \end{cases} \tag{3.2.90'}$$

where $f'(x_1)$, $f'(x_2)$ are for the present unknown magnitudes.

Taking into account the jumps of the functions $\overline{f}(x)$ and $\dfrac{\overline{\mathrm{d}f}(x)}{\mathrm{d}x}$ at the points $x = x_1$ and $x = x_2$ as well as the signs of these jumps, it follows that

$$\frac{\overline{\mathrm{d}f}(x)}{\mathrm{d}x} = \frac{\tilde{\overline{\mathrm{d}f}}(x)}{\mathrm{d}x} + f(x_1)\delta(x - x_1) - f(x_2)\delta(x - x_2), \tag{3.2.91}$$

$$\frac{\mathrm{d}^2\overline{f}(x)}{\mathrm{d}x^2} = \frac{\tilde{\mathrm{d}}^2\overline{f}(x)}{\mathrm{d}x^2} + f'(x_1)\delta(x - x_1) - f'(x_2)\delta(x - x_2)$$

$$+ f(x_1)\delta'(x - x_1) - f(x_2)\delta'(x - x_2). \tag{3.2.92}$$

Returning to equation (3.2.74), the derivative (3.2.90') takes the form (3.2.79), so that the equation may be written in distributions as follows

$$\frac{\mathrm{d}^2\overline{f}(x)}{\mathrm{d}x^2} = f'(x_1)\delta(x - x_1) - f'(x_2)\delta(x - x_2)$$

$$+ f(x_1)\delta'(x - x_1) - f(x_2)\delta'(x - x_2). \tag{3.2.93}$$

By integrating, we obtain

$$\frac{\overline{\mathrm{d}f}(x)}{\mathrm{d}x} = f'(x_1)\theta(x - x_1) - f'(x_2)\theta(x - x_2)$$

$$+ f(x_1)\delta(x - x_1) - f(x_2)\delta(x - x_2), \tag{3.2.94}$$

where the constant which occurs must be taken equal to zero so that the relation (3.2.91) with (3.2.90) is satisfied. Putting the condition that the same relation is verified for $x > x_2$ it follows

$$f'(x_1) = f'(x_2);$$
(3.2.95)

hence the two unknown magnitudes are equal.

By a new integration we may write

$$\bar{f}(x) = f'(x_1)[(x - x_1)_+ - (x - x_2)_+] + f(x_1)\theta(x - x_1) - f(x_2)\theta(x - x_2),$$
(3.2.96)

where account has been taken of condition (3.2.95) and where we have introduced the positive parts of the functions $(x - x_1)$ and $(x - x_2)$ in the form

$$(x - x_1)_+ = (x - x_1)\theta(x - x_1), \quad (x - x_2)_+ = (x - x_2)\theta(x - x_2);$$
(3.2.84')

the constant of integration must also be zero so that relation (3.2.89) be satisfied for $x < x_1$. By putting the condition that the relation must be satisfied also for $x > x_2$ it follows that

$$f'(x_1) = \frac{f(x_2) - f(x_1)}{x_2 - x_1},$$
(3.2.95')

so that we may write

$$\bar{f}(x) = \frac{f(x_2) - f(x_1)}{x_2 - x_1} [(x - x_1)_+ - (x - x_2)_+]$$

$$+ f(x_1)\theta(x - x_1) - f(x_2)\theta(x - x_2).$$
(3.2.97)

The solution is given by

$$\bar{f}(x) = f(x)[\theta(x - x_1) - \theta(x - x_2)], \qquad x \in R;$$
(3.2.89')

remarking that

$$\theta(x - x_1)\theta(x - x_2) = \theta(x - x_2), \qquad x_1 < x_2,$$
(3.2.98)

it follows that

$$\bar{f}(x) = \frac{f(x_2) - f(x_1)}{x_2 - x_1} (x - x_1)_+[1 - \theta(x - x_2)]$$

$$+ f(x_1)[\theta(x - x_1) - \theta(x - x_2)], \qquad x \in R,$$
(3.2.99)

or

$$f(x) = \frac{f(x_2) - f(x_1)}{x_2 - x_1} (x - x_1) + f(x_1), \qquad x \in [x_1, x_2]. \qquad (3.2.99')$$

Bilocal problems of this kind are often encountered in mechanics in addition to those with initial conditions of the Cauchy type; thus, there are problems of ballistics where we know the position at the initial and final moments and we want to determine the trajectory or the angle of launching, problems of oscillations where we know the initial position and the position at a given moment and we want to determine the frequency, etc. In all these cases, the method expounded above may be applied.

3.2.2. Partial differential equations

In the following discussion we shall consider problems of the type dealt with previously in the case of ordinary differential equations, but concerning the more general case of partial differential equations. We shall give first some general results and then we shall treat an important particular class of such equations.

3.2.2.1. General results

Let us consider

$$P\left(\frac{\partial}{\partial x_1}, \frac{\partial}{\partial x_2}, ..., \frac{\partial}{\partial x_m}; \frac{\partial}{\partial t} \right) u(x_1, x_2, ..., x_m; t) = 0 \qquad (3.2.100)$$

a homogeneous linear partial differential equation, of the order n with respect to the variable t, with constant coefficients. The problem of Cauchy for the equation consists of determining the distribution $u(x_1, x_2, ..., x_m; t)$ which satisfies equation (3.2.100) and the initial conditions

$$u(x_1, x_2, ..., x_m; t_0) = u_0(x_1, x_2, ..., x_m),$$

$$\frac{\partial}{\partial t} u(x_1, x_2, ..., x_m; t_0) = u_1(x_1, x_2, ..., x_m),$$

$$\cdots\cdots\cdots\cdots\cdots\cdots\cdots\cdots\cdots\cdots\cdots\cdots\cdots\cdots\cdots\cdots\cdots\cdots \qquad (3.2.101)$$

$$\frac{\partial^{n-1}}{\partial t^{n-1}} u(x_1, x_2, ..., x_m; t_0) = u_{n-1}(x_1, x_2, ..., x_m).$$

Let us introduce the following

Definition 3.2.3. *The distribution* $E(x_1, x_2, ..., x_m; t)$ *which satisfies equation* (3.2.100) *and the initial conditions*

$$E(x_1, x_2, ..., x_m; t_0) = 0,$$

$$\frac{\partial}{t\partial} E(x_1, x_2, ..., x_m; t_0) = 0,$$

$$\dots\dots\dots\dots\dots\dots\dots\dots\dots\dots\dots\dots\dots\dots\dots\dots\dots \quad (3.2.102)$$

$$\frac{\partial^{n-2}}{\partial t^{n-2}} E(x_1, x_2, ..., x_m; t_0) = 0,$$

$$\frac{\partial^{n-1}}{\partial t^{n-1}} E(x_1, x_2, ..., x_m; t_0) = \delta(x_1, x_2, ..., x_m)$$

is said to be the fundamental solution of the Cauchy problem corresponding to the equation (3.2.100) *and the initial conditions* (3.2.101).

The distribution given by the convolution

$$u(x_1, x_2, ..., x_m; t) = E(x_1, x_2, ..., x_m; t) * u_{n-1}(x_1, x_2, ..., x_m) \quad (3.2.103)$$

represents the solution of Cauchy's problem for the initial conditions

$$u(x_1, x_2, ..., x_m; t_0) = 0,$$

$$\frac{\partial}{\partial t} u(x_1, x_2, ..., x_m; t_0) = 0,$$

$$\dots\dots\dots\dots\dots\dots\dots\dots\dots\dots\dots\dots\dots\dots\dots\dots\dots \quad (3.2.103')$$

$$\frac{\partial^{n-2}}{\partial t^{n-2}} u(x_1, x_2, ..., x_m; t_0) = 0,$$

$$\frac{\partial^{n-1}}{\partial t^{n-1}} u(x_1, x_2, ..., x_m; t_0) = u_{n-1}(x_1, x_2, ..., x_m).$$

Indeed, we have

$$P\left(\frac{\partial}{\partial x_1}, \frac{\partial}{\partial x_2}, ..., \frac{\partial}{\partial x_m}; \frac{\partial}{\partial t}\right)[E(x_1, x_2, ..., x_m; t) * u_{n-1}(x_1, x_2, ..., x_m)]$$

$$= P\left(\frac{\partial}{\partial x_1}, \frac{\partial}{\partial x_2}, ..., \frac{\partial}{\partial x_m}; \frac{\partial}{\partial t}\right)E(x_1, x_2, ..., x_m; t) * u_{n-1}(x_1, x_2, ..., x_m) = 0,$$

and

$$\frac{\partial^k}{\partial t^k} u(x_1, x_2, ..., x_m; t_0) = \frac{\partial^k}{\partial t^k} E(x_1, x_2, ..., x_m; t_0) * u_{n-1}(x_1, x_2, ..., x_m) = 0$$

$$(k = 0, 1, 2, ..., n - 2),$$

$$\frac{\partial^{n-1}}{\partial t^{n-1}} u(x_1, x_2,..., x_m; t_0) = \frac{\partial^{n-1}}{\partial t^{n-1}} E(x_1, x_2, ..., x_m; t_0) * u_{n-1}(x_1, x_2, ..., x_m)$$

$$= \delta(x_1, x_2, ..., x_m) * u_{n-1}(x_1, x_2, ..., x_m) = u_{n-1}(x_1, x_2, ..., x_m).$$

It may be easily shown that the solution of Cauchy's general problem (the initial conditions (3.2.101)) reduces to the repeated application of the above procedure.

To solve the problem mentioned above let us consider the function

$$\bar{u}(x_1, x_2, ..., x_m; t) = \begin{cases} 0 & \text{for } t < t_0 \\ \\ u(x_1, x_2, ..., x_m; t) & \text{for } t \geq t_0 \end{cases} \tag{3.2.104}$$

and the corresponding regular distribution; taking into account the formula which connects the derivative in the sense of the theory of distributions to the derivative in the usual sense, and using the initial conditions (3.2.101) we obtain

$$\frac{\partial}{\partial t} \bar{u}(x_1, x_2, ..., x_m; t) = \frac{\tilde{\partial}}{\partial t} \bar{u}(x_1, x_2, ..., x_m; t) + u_0(x_1, x_2, ..., x_m)\delta(t - t_0),$$

$$\frac{\partial^2}{\partial t^2} \bar{u}(x_1, x_2, ..., x_m; t) = \frac{\tilde{\partial}^2}{\partial t^2} \bar{u}(x_1, x_2, ..., x_m; t) + u_1(x_1, x_2, ..., x_m)\delta(t - t_0)$$

$$+ u_0(x_1, x_2, ..., x_m)\delta'(t - t_0), \tag{3.2.101'}$$

$$\dots\dots\dots\dots\dots\dots\dots\dots\dots\dots\dots\dots\dots\dots\dots\dots\dots\dots$$

$$\frac{\partial^n}{\partial t^n} \bar{u}(x_1, x_2, ..., x_m; t) = \frac{\tilde{\partial}^n}{\partial t^n} \bar{u}(x_1, x_2, ..., x_m) + u_{n-1}(x_1, x_2, ..., x_m)\delta(t - t_0)$$

$$+ u_{n-2}(x_1, x_2, ..., x_m)\delta'(t - t_0) + ... + u_0(x_1, x_2, ..., x_m)\delta^{(n-1)}(t - t_0).$$

We remark that the derivatives in the ordinary sense of the new function $\bar{u}(x_1, x_2, ..., x_m; t)$ with regard to the variable t are equal to the corresponding derivatives of the function $u(x_1, x_2, \ldots, x_n; t)$ for $t > t_0$; it follows that equation (3.2.100) takes in distributions the form

$$P\left(\frac{\partial}{\partial x_1}, \frac{\partial}{\partial x_2}, \ldots, \frac{\partial}{\partial x_m}; \frac{\partial}{\partial t}\right) \bar{u}(x_1, x_2, \ldots, x_m; t) = f(x_1, x_2, \ldots, x_m; t), \quad (3.2.100')$$

where $f(x_1, x_2, \ldots, x_m; t)$ is a given distribution which includes the initial conditions considered above.

In that case we give the following

Definition 3.2.3′. *The distribution $E(x_1, x_2, \ldots, x_m; t)$ which satisfies the equation*

$$P\left(\frac{\partial}{\partial x_1}, \frac{\partial}{\partial x_2}, \ldots, \frac{\partial}{\partial x_m}; \frac{\partial}{\partial t}\right) E(x_1, x_2, \ldots, x_m; t) = \delta(x_1, x_2, \ldots, x_m; t) \quad (3.2.105)$$

is termed the fundamental solution of equation (3.2.100′).

The solution of Cauchy's problem, as stated above, is expressed by

$$\bar{u}(x_1, x_2, \ldots, x_m; t) = u(x_1, x_2, \ldots, x_m; t)\theta(t - t_0), \quad (3.2.104')$$

where $\theta(t - t_0)$ is the Heaviside function and

$$\bar{u}(x_1, x_2, \ldots, x_m; t) = E(x_1, x_2, \ldots, x_m; t) * f(x_1, x_2, \ldots, x_m; t) \quad (3.2.106)$$

is the convolution product with reference to all $(n + 1)$ variables.

This procedure of introducing the fundamental solution is equivalent to that used previously.

In the case of systems of partial differential equations the problems are treated in a similar way. General solutions are sought which are expressed with the help of convolution products with respect to some unknown functions; the determination of the latter is obtained from the limiting (initial and boundary) conditions of the problem. The method will be applied to problems pertaining to the mechanics of deformable solids. Also, use will be made of the fundamental solution introduced by definition (3.2.3′).

3.2.2.2. An important class of partial differential equations

The general method of determining the fundamental solutions of partial differential equations is based on the application of the Fourier and Laplace transforms; this method will be illustrated throughout the book for various particular problems of mechanics.

In the following discussion we shall present a simple method of a particular character which may be applied to an important class of equations.

Let us consider the differential equation with constant coefficients

$$Lu(x, y) \equiv a_{11} \frac{\partial^2 u(x, y)}{\partial x^2} + 2a_{12} \frac{\partial^2 u(x, y)}{\partial x \partial y} + a_{22} \frac{\partial^2 u(x, y)}{\partial y^2}$$

$$+ 2a_{13} \frac{\partial u(x, y)}{\partial x} + 2a_{23} \frac{\partial u(x, y)}{\partial y} + a_{33}u(x, y) = 0; \qquad (3.2.107)$$

we shall introduce the notation

$$L_0u(x, y) \equiv a_{11} \frac{\partial^2 u(x, y)}{\partial x^2} + 2a_{12} \frac{\partial^2 u(x, y)}{\partial x \partial y} + a_{22} \frac{\partial^2 u(x, y)}{\partial y^2} . \qquad (3.2.108)$$

The fundamental solution of Cauchy's problem is the distribution $E(x, y)$ which satisfies equation (3.2.107) and the limiting conditions

$$E(x, 0) = 0, \frac{\partial}{\partial y} E(x, 0) = \delta(x). \qquad (3.2.109)$$

We shall show that for a class of differential equations which will be completely determined, the fundamental solution of Cauchy's problem may be written in the form

$$E(x, y) = A(x, y)\theta(\varphi(x, y)) + B(x, y)\theta(\psi(x, y)), \qquad (3.2.110)$$

where θ is the Heaviside function, $\varphi(x, y) = C_1$ and $\psi(x, y) = C_2$, $C_1, C_2 = $ const, are the equations of the two families of characteristic curves and $A(x, y)$ and $B(x,y)$ are functions of class C^2 which are univocally determined by certain conditions.

Differentiating expression (3.2.11) in the sense of the theory of distributions, we obtain

$$\frac{\partial E(x, y)}{\partial x} = A'_x\theta(\varphi) + B'_x\theta(\psi) + A\varphi'_x\delta(\varphi) + B\psi'_x\delta(\psi),$$

$$\frac{\partial E(x, y)}{\partial y} = A'_y\theta(\varphi) + B'_y\theta(\psi) + A\varphi'_y\delta(\varphi) + B\psi'_y\delta(\psi),$$

$$\frac{\partial^2 E(x, y)}{\partial x^2} = A''_{xx}\theta(\varphi) + B''_{xx}\theta(\psi) + (A\varphi''_{xx} + 2A'_x\varphi'_x)\delta(\varphi)$$

$$+ (B\psi''_{xx} + 2B'_x\psi'_x)\delta(\psi) + A\varphi'^2_x\delta'(\varphi) + B\psi'^2_x\delta'(\psi),$$

$$\frac{\partial^2 E(x, y)}{\partial x \partial y} = A_{xy}'' \theta(\varphi) + B_{xy}'' \theta(\psi) + (A\varphi_{xy}'' + A_x' \varphi_y' + A_y' \varphi_x')\delta(\varphi)$$

$$+ (B\psi_{xy}'' + B_x' \psi_y' + B_y' \psi_x')\delta(\psi) + A\varphi_x' \varphi_y' \delta'(\varphi) + B\psi_x' \psi_y' \delta'(\psi),$$

$$\frac{\partial^2 E(x, y)}{\partial y^2} = A_{yy}'' \theta(\varphi) + B_{yy}'' \theta(\psi) + (A\varphi_{yy}'' + 2A_y' \varphi_y')\delta(\varphi)$$

$$+ (B\psi_{yy}'' + 2B_y' \psi_y')\delta(\psi) + A\varphi_y'^2 \delta'(\varphi) + B\psi_y'^2 \delta'(\psi);$$

introducing this in equation (3.2.107) we have

$$\theta(\varphi)LA + \theta(\psi)LB + \delta(\varphi)\{AL_0\varphi + 2A(a_{13}\varphi_x' + a_{23}\varphi_y')$$

$$+ 2[a_{11}A_x' \varphi_x' + a_{12}(A_x' \varphi_y' + A_y' \varphi_x') + a_{22}A_y' \varphi_y']\} + \delta(\psi)\{BL_0\psi$$

$$+ 2B(a_{13}\psi_x' + a_{23}\psi_y') + 2[a_{11}B_x' \psi_x' + a_{12}(B_x' \psi_y' + B_y' \psi_x')$$

$$+ a_{22}B_{yy}' \psi_{yy}']\} + \delta'(\varphi)A(a_{11}\varphi_x'^2 + 2a_{12}\varphi_x' \varphi_y' + a_{22}\varphi_y'^2)$$

$$+ \delta'(\psi)B(a_{11}\psi_x'^2 + 2a_{12}\psi_x' \psi_y' + a_{22}\psi_y'^2) = 0. \qquad (3.2.107')$$

The characteristic curves are straight lines which satisfy the equations

$$a_{11}\varphi_x'^2 + 2a_{12}\varphi_x' \varphi_y' + a_{22}\varphi_y'^2 = 0,$$

$$\qquad\qquad (3.2.111)$$

$$a_{11}\psi_x'^2 + 2a_{12}\psi_x' \psi_y' + a_{22}\psi_y'^2 = 0.$$

Hence the equation (3.2.107') becomes

$$\theta(\varphi)LA + \theta(\psi)LB + 2\delta(\varphi)[A(a_{13}\varphi_x' + a_{23}\varphi_y') + a_{11}A_x' \varphi_x'$$

$$+ a_{12}(A_x' \varphi_y' + A_y' \varphi_x') + a_{22}A_y' \varphi_y'] + 2\delta(\psi)[B(a_{13}\psi_x' + a_{23}\psi_y') + a_{11}B_x' \psi_x'$$

$$+ a_{12}(B_x' \psi_y' + B_y' \psi_x') + a_{22}B_y' \psi_y'] = 0. \qquad (3.2.107'')$$

In order that this equation is satisfied it is sufficient to set the conditions that the coefficients of θ and δ are zero; by introducing the notations

$$\alpha = a_{11}\varphi'_x + a_{12}\varphi'_y, \quad \beta = a_{12}\varphi'_x + a_{22}\varphi'_y, \quad \gamma = a_{13}\varphi'_x + a_{23}\varphi'_y, \tag{3.2.112}$$

$$\alpha' = a_{11}\psi'_x + a_{12}\psi'_y, \quad \beta' = a_{12}\psi'_x + a_{22}\psi'_y, \quad \gamma' = a_{13}\psi'_x + a_{23}\psi'_y, \tag{3.2.112'}$$

where $\alpha, \beta, \ldots, \gamma'$ are numbers, we obtain the systems

$$LA = 0,$$
$$\alpha A'_x + \beta A'_y + \gamma A = 0, \tag{3.2.113}$$

$$LB = 0,$$
$$\alpha' B'_x + \beta' B'_y + \gamma' B = 0, \tag{3.2.113'}$$

which determine the functions $A(x, y)$ and $B(x, y)$.

An interesting generalization of this result may be obtained by putting the condition that the coefficient of δ is a function of the variable φ, hence of the form $f(\varphi)$, with $f(0) = 0$.

Let us give the following

Theorem 3.2.6. *The necessary and sufficient condition that the first equation* (3.1.113) *is equivalent to the second equation of the same system is that relation*

$$a_{22}\gamma^2 - 2a_{23}\beta\gamma + a_{33}\beta^2 = 0 \tag{3.2.114}$$

is satisfied.

Indeed, the second equation of (3.2.113) is equivalent to the system

$$\frac{dx}{\alpha} = \frac{dy}{\beta} = -\frac{dA}{\gamma A}, \tag{3.2.115}$$

whence the first integrals

$$\frac{x}{\alpha} - \frac{y}{\beta} = C_1, \quad A = C_2 e^{-\frac{\gamma}{\beta}y}, \quad C_1, C_2 = \text{const.} \tag{3.2.115'}$$

The general integral of the system (3.2.115) is

$$C_2 = \Phi(C_1), \tag{3.2.116}$$

where Φ is an arbitrary function; therefore the general integral of the second equation (3.2.113) is

$$A(x, y) = e^{-\frac{\beta}{\gamma}y} \, \Phi\left(\frac{x}{\alpha} - \frac{y}{\beta}\right). \tag{3.2.116'}$$

Since this function must satisfy also the first equation (3.2.113) we obtain for the function Φ the equation

$$\left(\frac{a_{11}}{\alpha^2} - 2\frac{a_{12}}{\alpha\beta} + \frac{a_{22}}{\beta^2}\right)\Phi'' - 2\left(\frac{\gamma}{\alpha\beta}a_{12} - \frac{\gamma}{\beta^2}a_{22} - \frac{a_{13}}{\alpha} + \frac{a_{23}}{\beta}\right)\Phi'$$

$$+ \left(\frac{\gamma^2}{\beta^2}a_{22} - 2\frac{\gamma}{\beta}a_{23} + a_{33}\right)\Phi = 0. \tag{3.2.117}$$

Taking into account the notations (3.2.112), the first of the equations (3.2.111) takes the form

$$\alpha\varphi'_x + \beta\varphi'_y = 0. \tag{3.2.118}$$

By eliminating φ'_x and φ'_y from relations (3.2.112) and (3.2.118) we obtain

$$\frac{a_{11}}{\alpha^2} - 2\frac{a_{12}}{\alpha\beta} + \frac{a_{22}}{\beta^2} = 0,$$

$$\frac{\gamma}{\alpha\beta}a_{12} - \frac{\gamma}{\beta^2}a_{22} - \frac{a_{13}}{\alpha} + \frac{a_{23}}{\beta} = 0,$$

so that the coefficients of Φ' and Φ'' in equation (3.2.117) are zero; in order that the function Φ may satisfy equation (3.2.117), the coefficient of Φ in that equation must be zero.

Thus, the condition (3.2.114) stated in the theorem has been proved to be necessary; the sufficiency of the condition is obvious.

The system (3.2.113') may be studied in a similar way.

3.2.2.3. A particular case

In particular, if equation (3.2.107) reduces to the homogeneous part of the second degree,

$$L_0 u(x, y) = 0 \tag{3.2.119}$$

we shall have $\gamma = 0$, $a_{33} = 0$, and the condition (3.2.114) is identically satisfied.

We remark that, in that case, the functions $A(x, y)$ and $B(x, y)$ may be considered as constants. We shall show that these constants are determined uniquely so that the fundamental solution of Cauchy's problem is of the form (3.2.110); equation (3.2.119) is evidently satisfied.

The constants A and B are obtained from the conditions

$$E(x, 0)' = A\theta(\varphi(x, 0)) + B\theta(\psi(x, 0)) = 0,$$

$$\frac{\partial E(x, 0)}{\partial y} = A\varphi_y'(x, 0)\delta(\varphi(x, 0)) + B\psi_y'(x, 0)\delta(\psi(x, 0)) = \delta(x). \tag{3.2.120}$$

Since $\varphi(x, y)$ and $\psi(x, y)$ are linear functions with respect to x and y, we may choose them so that the coefficients of x in both equations are positive; then we may write

$$\theta(\varphi(x, 0)) = \theta(\psi(x, 0)),$$

$$\delta(\varphi(x, 0)) = \frac{1}{a}\,\delta(x), \quad \delta(\psi(x, 0)) = \frac{1}{b}\,\delta(x), \tag{3.2.121}$$

where a and b are the positive coefficients of x in φ and ψ. Thus the conditions (3.2.120) lead to

$$A + B = 0,$$

$$\frac{A}{a}\,\varphi_y' + \frac{B}{b}\,\psi_y' = 1, \tag{3.2.120'}$$

whence

$$A = -B = \frac{ab}{b\varphi_y' - a\psi_y'}. \tag{3.2.122}$$

The fundamental solution of Cauchy's problem for the equation (3.2.119) is thus given by

$$E(x, y) = \frac{ab}{b\varphi_y' - a\psi_y'}\,[\theta(\varphi) - \theta(\psi)]. \tag{3.2.123}$$

The above considerations may be now extended to the general case.

Thus, we remark that in order that the two equations (3.2.113') are equivalent it is necessary and sufficient to satisfy the condition

$$a_{22}\gamma'^2 - 2a_{23}\gamma'\beta' + a_{33}\beta'^2 = 0; \tag{3.2.114'}$$

the function $A(x, y)$ is given by relation (3.2.116′) and the function $B(x, y)$ by

$$B(x, y) = e^{-\frac{\gamma'}{\beta'}y} \, \bar{\Phi}\left(\frac{x}{\alpha'} - \frac{y}{\beta'}\right), \tag{3.2.116″}$$

where $\bar{\Phi}$ is also an arbitrary function.

These functions are obtained from the conditions (3.2.120) which take the form

$$E(x, 0) = \Phi\left(\frac{x}{\alpha}\right)\theta(\varphi(x, 0)) + \bar{\Phi}\left(\frac{x}{\alpha'}\right)\theta(\psi(x, 0)) = 0,$$

$$\tag{3.2.120″}$$

$$\frac{\partial E(x, 0)}{\partial y} = -\frac{1}{\beta}\left[\gamma\Phi\left(\frac{x}{\alpha}\right) + \bar{\Phi}'\left(\frac{x}{\alpha}\right)\right]\theta(\varphi(x, 0)) + \Phi\left(\frac{x}{\alpha}\right)\varphi'_y(x, 0)\delta(\varphi(x, 0))$$

$$-\frac{1}{\beta'}\left[\gamma'\bar{\Phi}\left(\frac{x}{\alpha}\right) + \bar{\Phi}'\left(\frac{x}{\alpha'}\right)\right]\theta(\psi(x, 0)) + \bar{\Phi}\left(\frac{x}{\alpha'}\right)\psi'_y(x, 0)\delta(\psi(x, 0)) = \delta(x).$$

On the basis of considerations similar to those of the particular case considered, we may choose φ and ψ such that conditions (3.2.121) are satisfied; in this way conditions (3.2.120″) lead to

$$\Phi\left(\frac{x}{\alpha}\right) + \bar{\Phi}\left(\frac{x}{\alpha'}\right) = 0 \text{ for } x \geqslant 0, \tag{3.2.124}$$

$$\frac{1}{\beta}\left[\gamma\Phi\left(\frac{x}{\alpha}\right) + \Phi'\left(\frac{x}{\alpha}\right)\right] + \frac{1}{\beta'}\left[\gamma'\bar{\Phi}\left(\frac{x}{\alpha'}\right) + \bar{\Phi}'\left(\frac{x}{\alpha'}\right)\right] = 0 \text{ for } x \geqslant 0, \tag{3.2.124′}$$

$$\frac{1}{a}\varphi'_y(x, 0)\Phi(0) + \frac{1}{b}\psi'_y(x, 0)\bar{\Phi}(0) = 1. \tag{3.2.124″}$$

Eliminating $\bar{\Phi}$ from (3.2.124) and (3.2.124′) and integrating we obtain

$$\Phi\left(\frac{x}{\alpha}\right) = Ce^{-\varepsilon x}, \tag{3.2.125}$$

where

$$\varepsilon = -\frac{\gamma\beta' - \gamma'\beta}{\alpha\beta' - \alpha'\beta} \tag{3.2.126}$$

and C is an arbitrary constant; similarly

$$\bar{\Phi}\left(\frac{x}{\alpha'}\right) = -Ce^{-\varepsilon x}. \tag{3.2.125'}$$

From (3.2.124) we obtain

$$\Phi(0) + \bar{\Phi}(0) = 0, \tag{3.2.124'''}$$

so that relation (3.2.124'') leads to

$$\Phi(0) = -\bar{\Phi}(0) = \frac{ab}{b\varphi_y'(x,0) - a\psi_y'(x,0)}; \tag{3.2.127}$$

finally, the formula (3.2.125) determines the constant C in the form

$$C = \Phi(0). \tag{3.2.127'}$$

Therefore

$$\Phi\left(\frac{x}{\alpha} - \frac{y}{\beta}\right) = Ce^{-\varepsilon\left(x - \frac{\alpha}{\beta}y\right)}, \tag{3.2.128}$$

$$\bar{\Phi}\left(\frac{x}{\alpha'} - \frac{y}{\beta'}\right) = -Ce^{-\varepsilon\left(x - \frac{\alpha'}{\beta'}y\right)}, \tag{3.2.128'}$$

and the functions $A(x, y)$ and $B(x, y)$ given by relations (3.2.116') and (3.2.116'') are thus completely specified.

3.2.2.4. Examples

Let us consider equation

$$\frac{\partial^2 u(x,y)}{\partial x^2} + 2\frac{\partial^2 u(x,y)}{\partial x \partial y} - 3\frac{\partial^2 u(x,y)}{\partial y^2} = 0 \tag{3.2.129}$$

with the limiting conditions

$$u(x,0) = 0, \quad \frac{\partial u(x,0)}{\partial y} = u_1(x). \tag{3.2.130}$$

The corresponding characteristic equation is

$$dy^2 - 2\,dx\,dy - 3\,dx^2 = 0 \tag{3.2.131}$$

with the general integrals

$$x + y = C_1, \ 3x - y = C_2; \tag{3.2.131'}$$

hence

$$\varphi(x, y) \equiv x + y, \ \psi(x, y) \equiv 3x - y. \tag{3.2.131''}$$

Being in the particular case considered above we shall use formula (3.2.123), whence we obtain

$$E(x, y) = \frac{3}{4} [\theta(x + y) - \theta(3x - y)]. \tag{3.2.132}$$

Hence the solution of the boundary-value problem is

$$u(x, y) = E(x, y) * u_1(x). \tag{3.2.133}$$

Let us now consider, in the general case, the equation

$$-\frac{\partial^2 u(x, y)}{\partial x^2} + \frac{\partial^2 u(x, y)}{\partial y^2} + 2\frac{\partial u(x, y)}{\partial y} + u(x, y) = 0. \tag{3.2.134}$$

The characteristic curves are given by

$$\varphi(x, y) \equiv x + y = C_1, \ \psi(x, y) \equiv x - y = C_2; \tag{3.2.135}$$

we remark that the conditions (3.2.114) and (3.2.114') are satisfied since $\alpha = -1$, $\beta = \gamma = 1$ and $\alpha' = \beta' = \gamma' = -1$.

Applying formula (3.2.127), we have $\Phi(0) = \frac{1}{2}$; also $\varepsilon = 0$. Thus, the fundamental solution may be written in the form

$$E(x, y) = \frac{1}{2} e^{-y} [\theta(x + y) - \theta(x - y)]. \tag{3.2.136}$$

Similarly, let us consider the equation

$$\frac{\partial^2 u(x, y)}{\partial x^2} + 2\frac{\partial^2 u(x, y)}{\partial x \partial y} - 3\frac{\partial^2 u(x, y)}{\partial y^2} + 2\frac{\partial u(x, y)}{\partial x} + 6\frac{\partial u(x, y)}{\partial y} = 0, \tag{3.2.137}$$

with the characteristic curves

$$\varphi(x, y) \equiv x + y = C_1, \ \psi(x, y) \equiv 3x - y = C_2. \tag{3.2.138}$$

We have $\alpha = 2$, $\beta = -2$, $\gamma = 4$, $\alpha' = 2$, $\beta' = 6$, $\gamma' = 0$ so that conditions (3.2.114), (3.2.114′) are satisfied; it also follows that $\varepsilon = \dfrac{3}{2}$ and $C = \dfrac{3}{4}$. The fundamental solution may be written in the form

$$E(x, y) = \frac{3}{4} e^{-\frac{1}{2}(3x - y)} [\theta(x + y) - \theta(3x - y)]. \tag{3.2.139}$$

3.3. Green's functions. Green's distributions

The solution of many problems of mathematical physics is related to the construction of the so-called Green's functions with the help of which the solutions of boundary-value problems may be determined explicitly and presented in an integral form. The very important part played by these functions in writing the solution of boundary-value problems consists in their close connection to the Dirac distribution δ and, more exactly, to the concept of fundamental solution introduced for differential equations. Thus, the determination of Green's function corresponding to a given problem amounts to the determination of the fundamental solution of the boundary-value problem of the equation considered. In case the fundamental solution is a distribution, we have to deal with Green's distributions.

In the following discussion we shall first treat the ordinary and then the partial differential equations.

3.3.1. Ordinary differential equations

The notion of Green's function for ordinary differential equations, the procedure by which it is obtained and its usefulness will be emphasized by the study of a few particular differential equations; the conclusions may be extended to other types of ordinary differential equations.

3.3.1.1. The equation $y^{(n)} = f(x)$. Cauchy's formula

Let us consider the equation

$$y^{(n)}(x) = f(x). \tag{3.3.1}$$

We seek a solution satisfying the initial conditions

$$y^{(k)}(0) = 0 \qquad (k = 0, 1, 2, ..., n - 1) \tag{3.3.2}$$

for $x = 0$; it is assumed that $f(x)$ is a function of class C^0 on $[0, \infty)$.

It is well known that the solution of the problem with the initial conditions (3.3.2) is given by *Cauchy's formula*

$$y(x) = \frac{1}{(n-1)!} \int_0^x (x - \xi)^{n-1} f(\xi)\, d\xi, \qquad x \geqslant 0. \tag{3.3.3}$$

Green's function for the equation (3.3.1) will be

$$G(x; \xi) = \frac{1}{(n-1)!}(x - \xi)^{n-1} \text{ for } x \geqslant \xi; \tag{3.3.4}$$

this function represents the solution of the equation

$$y^{(n)}(x) = \delta(x - \xi), \tag{3.3.5}$$

for $x \geqslant \xi$, where ξ is a parameter.

We remark that by determining the particular fundamental solution $E^+(x)$ of the equation

$$y^{(n)}(x) = \delta(x), \tag{3.3.5'}$$

for $x \in R$, Green's distribution, hence the solution of equation (3.3.5), may be written

$$G(x; \xi) = E^+(x - \xi). \tag{3.3.6}$$

Indeed, from

$$\frac{d^n}{dx^n} E^+(x) = \delta(x) \tag{3.3.7}$$

there follows

$$\frac{d^n}{dx^n} E^+(x - \xi) = \delta(x - \xi). \tag{3.3.7'}$$

A normal fundamental system of solutions of the homogeneous equation corresponding to equation (3.3.5′) is

$$1, x, \frac{x^2}{2!}, \frac{x^3}{3!}, \ldots, \frac{x^{n-1}}{(n-1)!}. \tag{3.3.8}$$

In that case the particular fundamental solution is

$$E^+(x) = \theta(x) \frac{x^{n-1}}{(n-1)!},$$ (3.3.9)

where $\theta(x)$ is the Heaviside distribution; as may be seen, equation (3.3.7) is easily verified.

Taking into account relation (3.3.6), the corresponding Green's distribution is given by

$$G(x;\xi) = \frac{1}{(n-1)!} \theta(x-\xi)(x-\xi)^{n-1},$$ (3.3.4')

a formula equivalent to (3.3.4).

The solution of equation (3.3.1) with the initial conditions (3.3.2) is

$$y(x) = E^+(x) * f(x),$$ (3.3.10)

whence

$$y(x) = \int_0^x E^+(x-\xi) f(\xi)\, d\xi = \int_0^x G(x;\xi) f(\xi)\, d\xi,$$ (3.3.10')

a formula equivalent to Cauchy's formula (3.3.3).

The general solution of equation (3.3.1) with the initial conditions

$$y(0) = y_0, y'(0) = y_1, y''(0) = y_2, \ldots, y^{(n-1)}(0) = y_{n-1}$$ (3.3.11)

is given by

$$y(x) = y_0 + y_1 x + y_2 \frac{x^2}{2!} + \ldots + y_{n-1} \frac{x^{n-1}}{(n-1)!}$$

$$+ \frac{1}{(n-1)!} \int_0^x (x-\xi)^{n-1} f(\xi)\, d\xi, \qquad x \geqslant 0.$$ (3.3.12)

3.3.1.2. Linear equations with constant coefficients

Let us consider the linear differential equation with constant coefficients (3.2.54), where $f(x)$ is an integrable function over $[0, \infty)$. We have seen in section 3.2.1.3 how to construct the particular fundamental solution $E^+(x)$ which, by a translation at the point ξ, satisfies the equation

$$DE^+(x-\xi) = \delta(x-\xi).$$ (3.3.13)

In this way Green's distribution for equation (3.2.54) will be given by a relation of the form (3.3.6); when the Green's distribution is determined, the solution of the equation (3.2.54) is given by a formula of the form (3.3.10) or (3.3.10′).

We remark that we have thus obtained a particular solution of the non-homogeneous equation (3.2.54); if we add to it the general solution of the corresponding homogeneous equation, we obtain the general solution of equation (3.2.54). A similar procedure has been used in the preceding subsection.

The Green's distribution used above has been introduced with the help of the particular fundamental solution which satisfies equation (3.3.13). However, we may introduce Green's distribution $G_1(x; \xi)$ for equation (3.2.54), noting that it satisfies equation

$$DG_1(x; \xi) = \delta(x - \xi). \tag{3.3.13′}$$

Green's distribution thus defined includes also Green's distribution $G(x; \xi)$ defined by equation (3.3.13); moreover, Green's distribution $G(x; \xi)$ represents the *singular part* of the Green's distribution $G_1(x; \xi)$. The second part of Green's distribution $G_1(x; \xi)$ represents the non-singular part which is nothing else than the general solution of the homogeneous equation $Dy(x) = 0$; denoting by $Y(x)$ the general solution of that homogeneous equation, Green's distribution $G_1(x; \xi)$ is expressed by

$$G_1(x; \xi) = (x; \xi) + Y(x). \tag{3.3.14}$$

Obviously, the singular part $G(x; \xi)$ which we use is the essential part of Green's distribution $G_1(x; \xi)$.

3.3.2. Partial differential equations

In the following discussion we shall first consider partial differential equations of the parabolic type such as the equation of heat conduction. We shall study next Poisson's equation and a generalized form of the latter as well as another equation of the elliptic type the homogeneous part of which is the biharmonic equation; finally, we shall also introduce equations of the hyperbolic type the homogeneous part of which is a simple or double wave equation. In all the cases, assuming that we have to deal with one, two, or three spatial variables, we shall show how to determine the fundamental solution or the corresponding Green's distribution.

3.3.2.1. Equation of heat conduction

Let us consider the equation of parabolic type

$$\frac{\partial u(x; t)}{\partial t} = a \frac{\partial^2 u(x; t)}{\partial x^2}, \qquad x \in R, t > 0; a > 0, \tag{3.3.15}$$

with the initial condition

$$u(x; 0) = \delta(x - \xi).$$ (3.3.16)

If $u(x; t)$ is the temperature variation, then the coefficient a will represent the *thermal diffusivity* and will be expressed by

$$a = \frac{\lambda}{c\gamma},$$ (3.3.17)

where c is the *specific heat*, γ the *specific weight* and λ the *thermal conductivity;* the equation is in that case Fourier's equation of heat conduction in solids in the absence of sources of heat.

The solution $E(x, \xi; t)$ which satisfies equation (3.3.15) and the initial condition (3.3.16) represents Green's distribution corresponding to the phenomenon of thermal conduction.

To show this, it is sufficient to determine the distribution $E(x; t)$ which satisfies equation (3.3.15) and the initial condition

$$E(x; 0) = \delta(x).$$ (3.3.16′)

By translating the distribution at the point ξ, we obtain the Green's distribution in the form

$$(x, \xi; t) = E(x - \xi; t).$$ (3.3.18)

We remark that $E(x; t)$ represents the fundamental solution of Cauchy's problem: the solution of equation (3.3.15) with the initial condition

$$u(x; 0) = \delta(x).$$ (3.3.16″)

To solve the problem we shall replace the function $u(x; t)$, defined for $t \geqslant 0$, by the function

$$\bar{u}(x; t) = \begin{cases} 0 & \text{for } t < 0 \\ \\ u(x; t) & \text{for } t \geqslant 0, \end{cases}$$ (3.3.19)

defined for all t; then we introduce the corresponding regular distribution.

In this way equation (3.3.15) is replaced by the equation in distributions

$$\frac{\partial \bar{u}(x; t)}{\partial t} = a^2 \frac{\partial^2 \bar{u}(x; t)}{\partial x^2} + u(x; 0)\delta(t),$$ (3.3.20)

which, taking into account the initial condition (3.3.16''), becomes

$$\frac{\partial \overline{u}(x;t)}{\partial t} = a\frac{\partial^2 \overline{u}(x;t)}{\partial x^2} + \delta(x)\delta(t), \qquad (3.3.20')$$

which includes also that initial condition.

By applying successively the Fourier and the Laplace transforms in distributions we obtain

$$pL[F[\overline{u}(x;t)]] = -a\alpha^2 L[F[\overline{u}(x;t)]] + 1, \qquad (3.3.21)$$

whence

$$L[F[\overline{u}(x;t)]] = \frac{1}{p + a\alpha^2}; \qquad (3.3.21')$$

here α is the complex variable due to the Fourier transform with respect to the variable x, and p is the complex variable due to the Laplace transform with respect to the variable t.

Effecting the inverse Laplace transform we obtain

$$F[\overline{u}(x;t)] = L^{-1}\left[\frac{1}{p + a\alpha^2}\right] = \theta(t)e^{-a\alpha^2 t}; \qquad (3.3.22)$$

applying now also the inverse Fourier transform, we may write

$$\overline{u}(x;t) = F^{-1}[\theta(t)e^{-a\alpha^2 t}] = \frac{1}{2\sqrt{\pi a t}}\,\theta(t)e^{-\frac{x^2}{4at}}. \qquad (3.3.22')$$

Whence there follows

$$E(x;t) = \theta(t)\overline{u}(x;t) = \frac{1}{2\sqrt{\pi a t}}\,\theta(t)e^{-\frac{x^2}{4at}}, \qquad t\in(-\infty,\infty) \qquad (3.3.23)$$

or

$$E(x;t) = \frac{1}{2\sqrt{\pi a t}}\,e^{-\frac{x^2}{4at}}, \qquad t \geqslant 0. \qquad (3.3.23')$$

In order to verify that the function $E(x;t)$ given by (3.3.23') satisfies equation (3.3.15), it is important to remark that the differentiation with respect to t is performed only for $t > 0$, a case where the derivative in the sense of the theory of distributions coincides with the derivative in the usual sense; indeed, the equation of thermal conduction (3.3.15) is considered only for $t > 0$. If the differentiation of the solution $E(x; t)$ is performed for all $t \in R$, then the discontinuity of the first species which might exist for $t = 0$ could cause the equation to be no longer verified.

It is easily seen that

$$\lim_{t \to +0} E(x; t) = \delta(x), \tag{3.3.24}$$

which shows that $E(x; t)$ constitues a representative δ sequence, where the time t is a parameter.

But Green's distribution for the equation of thermal conduction represents the fundamental solution of Cauchy's problem for that equation, translated at the point ξ. By relation (3.3.18) Green's distribution will be

$$G(x, \xi; t) = \frac{1}{2\sqrt{\pi a t}} \, \theta(t) e^{-\frac{(x - \xi)^2}{4at}} ; \tag{3.3.25}$$

obviously $G(x, \xi, t) \geqslant 0$.

With the help of Green's function the solution of equation (3.3.15) with the initial condition

$$u(x; 0) = f(x), \tag{3.3.16'''}$$

where $f(x)$ is an integrable function, may be written

$$\bar{u}(x; t) = E(x; t) * f(x), \quad t \in (-\infty, \infty), \tag{3.3.26}$$

whence we obtain

$$u(x; t) = \int_{-\infty}^{\infty} E(x - \xi; t) f(\xi) \, d\xi = \int_{-\infty}^{\infty} G(x, \xi; t) f(\xi) \, d\xi, \quad t > 0. \tag{3.3.26'}$$

We remark that expression (3.3.26) represents also the solution of the problem, under the assumption that $f(x)$ is a distribution and equation (3.3.15) is considered in the sense of the theory of distributions as, in fact, it has been considered above.

Green's distribution may be found also when seeking a solution of the differential equation

$$\frac{\partial u(x; t)}{\partial t} - a \frac{\partial^2 u(x; t)}{\partial x^2} = \delta(x; t), \tag{3.3.27}$$

where both variables x and t vary from $(-\infty)$ to ∞.

Applying the Fourier transform with respect to both variables and denoting the complex variables by α and s, we obtain

$$- isF(\alpha, s) - a(- i\alpha)^2 F(\alpha, s) = 1, \qquad (3.3.28)$$

whence

$$F(\alpha, s) = \frac{1}{a\alpha^2 - is}. \qquad (3.3.28')$$

Effecting the inverse Fourier transform with respect to the variable s (corresponding to the time t), we have

$$F(\alpha, t) = F^{-1}\left[\frac{1}{a\alpha^2 - is}\right] = \theta(t)e^{-a\alpha^2 t}; \qquad (3.3.29)$$

a new inverse Fourier transform leads to

$$u(x; t) = F^{-1}[\theta(t)e^{-a\alpha^2 t}]$$

$$= \frac{1}{2\sqrt{\pi at}}\, \theta(t)e^{-\frac{x^2}{4at}} = E(x; t), \qquad t \in (-\infty, \infty), \qquad (3.3.29')$$

so that Green's distribution is given by

$$G(x, \xi; t) = u(x - \xi; t) \qquad (3.3.30)$$

and obviously, coincides with that expressed by relation (3.3.25).

Thus, the solution (3.3.29') represents the fundamental solution for the equation

$$\frac{\partial u(x; t)}{\partial t} - a\frac{\partial^2 u(x; t)}{\partial x^2} = f(x; t), \qquad (3.3.27')$$

where $f(x; t)$ is a distribution; we remark that this fundamental solution coincides with the one defined previously.

If $a < 0$ we obtain the inverse equation of thermal conduction while, if a is a purely imaginary number, we obtain *Schrödinger's equation* of quantum mechanics for a particle in a field of conservative forces.

In the case of two spatial variables the equation considered above takes the form

$$\frac{\partial}{\partial t} u(x, y; t) = a\Delta u(x, y; t), \ (x, y) \in R^2, \ t > 0; \ a > 0, \qquad (3.3.31)$$

where the Laplace operator is

$$\Delta = \frac{\partial^2}{\partial x^2} + \frac{\partial^2}{\partial y^2};$$ (3.3.32)

we put the initial condition

$$u(x, y; 0) = \delta(x, y).$$ (3.3.33)

The corresponding fundamental solution may be written in the form

$$E(x, y; t) = \frac{1}{4\pi at} \theta(t) e^{-\frac{r^2}{4at}}, \qquad t \in (-\infty, \infty)$$ (3.3.34)

or in the form

$$E(x, y; t) = \frac{1}{4\pi at} e^{-\frac{r^2}{4at}}, \qquad t \geqslant 0,$$ (3.3.34')

with the radius vector

$$r = \sqrt{x^2 + y^2}.$$ (3.3.35)

For three spatial variables the equation becomes

$$\frac{\partial}{\partial t} u(x, y, z; t) = a\Delta u(x, y, z; t), \ (x, y, z) \in R^3, \ t > 0; \ a > 0.$$ (3.3.31')

the Laplace operator being

$$\Delta = \frac{\partial^2}{\partial x^2} + \frac{\partial^2}{\partial y^2} + \frac{\partial^2}{\partial z^2}$$ (3.3.32')

and the initial condition

$$u(x, y, z; 0) = \delta(x, y, z).$$ (3.3.33')

The fundamental solution is

$$E(x, y, z; t) = \frac{1}{8\pi at \sqrt{\pi at}} \theta(t) e^{-\frac{R^2}{4at}}, \qquad t \in (-\infty, \infty)$$ (3.3.36)

or

$$E(x, y, z; t) = \frac{1}{8\pi at \sqrt{\pi at}} e^{-\frac{R^2}{4at}}, \qquad t \geqslant 0,$$ (3.3.36')

with the radius vector

$$R = \sqrt{x^2 + y^2 + z^2}. \qquad (3.3.35')$$

Also, we remark that solution (3.3.34) is the fundamental solution of equation

$$\frac{\partial}{\partial t} u(x, y; t) - a\Delta u(x, y; t) = f(x, y; t), \qquad (3.3.37)$$

where $f(x, y; t)$ is a distribution, and solution (3.3.36) is a fundamental solution of equation

$$\frac{\partial}{\partial t} u(x, y, z; t) - a\Delta u(x, y, z; t) = f(x, y, z; t), \qquad (3.3.37')$$

where $f(x, y, z; t)$ is a distribution too.

This shows that the two definitions (3.2.3) and (3.2.3') of the fundamental solution are equivalent.

3.3.2.2. Poisson's equation

We shall now consider Poisson's equation of three spatial variables.

$$\Delta u(x, y, z) = f(x, y, z), \qquad (3.3.38)$$

where $f(x, y, z)$ is a distribution; this equation has interesting applications in Newton's theory of gravitation.

With a view to obviate certain difficulties which occur in that theory, Neumann and Seelinger have introduced a complementary term $(-k^2 u(x, y, z))$; this makes us consider also the generalized equation of Poisson

$$\Delta u(x, y, z) - k^2 u(x, y, z) = f(x, y, z), \qquad k = \text{const.} \qquad (3.3.39)$$

By definition, Green's function which corresponds to equation (3.3.39) represents the solution of equation

$$\Delta u(x, y, z) - k^2 u(x, y, z) = \delta(x - \xi, v - \eta, z - \zeta), \qquad (3.3.40)$$

where ξ, η, ζ are parameters. It is sufficient to determine the fundamental solution $E(x, y, z)$ which satisfies equation

$$\Delta E(x, y, z) - k^2 E(x, y, z) = \delta(x, y, z), \qquad (3.3.40')$$

since in that case Green's function is given by

$$(x, y, z; \xi, \eta, \zeta) = E(x - \xi, y - \eta, z - \zeta); \qquad (3.3.41)$$

hence, it can be obtained by a translation of the function $E(x, y, z)$ at the point (ξ, η, ζ).

Taking into account equation (3.3.40′) and formula (3.3.41) we obviously have

$$\Delta \ (x, y, z; \xi, \eta, \zeta) - k^2 \ (x, y, z; \xi, \eta, \zeta) = \delta(x - \xi, y - \eta, z - \zeta). \qquad (3.3.40'')$$

Applying to equation (3.3.40′) the Fourier transform in R^3, where α, β, γ are complex variables, we obtain

$$- (\alpha^2 + \beta^2 + \gamma^2)F[E(x, y, z)] - k^2 F[E(x, y, z)] = 1, \qquad (3.3.42)$$

whence

$$F(\alpha, \beta, \gamma) = F[E(x, y, z)] = - \frac{1}{k^2 + \alpha^2 + \beta^2 + \gamma^2}. \qquad (3.3.42')$$

We denote by $F_x[\ ]$, $F_y[\ ]$, $F_z[\ ]$ the Fourier transforms in R with regard to the variables x, y, and z respectively; we have

$$F_x\left[\frac{1}{R} e^{-kR} \right] = 2K_0(\sqrt{\alpha^2 + k^2}\sqrt{y^2 + z^2}), \qquad (3.3.43)$$

where $K_0(y, z)$ is Bessel's function of the zeroth order and R is the radius vector given by (3.3.35′).

Then we may write

$$F_y[K_0(\sqrt{\alpha^2 + k^2}\sqrt{y^2 + z^2})] = \pi \frac{e^{-|z|\sqrt{\alpha^2 + \beta^2 + k^2}}}{\sqrt{\alpha^2 + \beta^2 + k^2}} \qquad (3.3.44)$$

and

$$F_z\left[\frac{e^{-|z|\sqrt{\alpha^2 + \beta^2 + k^2}}}{\sqrt{\alpha^2 + \beta^2 + k^2}} \right] = \frac{2}{\alpha^2 + \beta^2 + \gamma^2 + k^2}. \qquad (3.3.45)$$

If $F[\ ]$ represents the Fourier transform in R^3 with respect to all the three variables, we have

$$F\left[\frac{1}{R} e^{-kR} \right] = \frac{4\pi}{\alpha^2 + \beta^2 + \gamma^2 + k^2}, \qquad (3.3.46)$$

whence

$$F^{-1}\left[\frac{1}{\alpha^2 + \beta^2 + \gamma^2 + k^2}\right] = \frac{1}{4\pi R} e^{-kR}. \tag{3.3.46'}$$

Taking into account relation (3.3.42'), it follows that

$$E(x, y, z) = F^{-1}[F(\alpha, \beta, \gamma)] = -\frac{1}{4\pi R} e^{-kR}, \tag{3.3.47}$$

which constitutes the fundamental solution of equation (3.3.39); by relation (3.3.41) Green's function is expressed by

$$G(x, y, z; \xi, \eta, \zeta) = -\frac{1}{4\pi\rho} e^{-k\rho}, \tag{3.3.48}$$

where we have used the notation

$$\rho = \sqrt{(x - \xi)^2 + (y - \eta)^2 + (z - \zeta)^2}. \tag{3.3.49}$$

In particular, for $k = 0$ we obtain Green's function corresponding to Poisson's equation (3.3.38) in the form

$$G(x, y, z; \xi, \eta, \zeta) = -\frac{1}{4\pi\rho}. \tag{3.3.48'}$$

We remark that the function

$$E(x, y, z) = -\frac{1}{4\pi R} \tag{3.3.50}$$

represents the particular fundamental solution of Poisson's equation (3.3.38) and corresponds also to formula (1.4.52); in order to have the fundamental solution of that equation we must add the general solution of Laplace's equation

$$\Delta u(x, y, z) = 0, \tag{3.3.51}$$

which is the homogeneous equation corresponding to Poisson's equation considered above.

Using Green's function determined above, the solution of the generalized equation of Poisson including the solution of Poisson's equation is given by the expression

$$u(x, y, z) = E(x, y, z) * f(x, y, z). \tag{3.3.52}$$

Assuming that $f(x, y, z)$ is an integrable function in R^3 we may write

$$u(x, y, z) = \int_{-\infty}^{\infty} \int_{-\infty}^{\infty} \int_{-\infty}^{\infty} E(x - \xi, y - \eta, z - \zeta) f(\xi, \eta, \zeta)\, d\xi\, d\eta\, d\zeta$$

$$= \int_{-\infty}^{\infty} \int_{-\infty}^{\infty} \int_{-\infty}^{\infty} G(x, y, z; \xi, \eta, \zeta) f(\xi, \eta, \zeta)\, d\xi\, d\eta\, d\zeta. \qquad (3.3.52')$$

If equation (3.3.39) is considered in the sense of the theory of distributions and $f(x, y, z)$ is a distribution with a bounded support, then the solution of this equation is still (3.3.52), where $E(x, y, z)$ is the distribution defined by (3.3.47) and the convolution product is also taken in the sense of the theory of distributions.

In order to obtain the general solution of equation (3.3.39), we have to add to the particular solution (3.3.52) the general solution of the homogeneous equation

$$\Delta u(x, y, z) - k^2 u(x, y, z) = 0. \qquad (3.3.53)$$

Replacing k by $(-k)$, we obtain for equation (3.3.39) also the conjugate fundamental solution

$$\overline{E}(x, y, z) = -\frac{1}{4\pi R} e^{kR}. \qquad (3.3.47')$$

Taking into account the relation

$$\sinh kR = \frac{1}{2}(e^{kR} + e^{-kR}), \qquad (3.3.54)$$

we obtain also the fundamental solution

$$\mathscr{E}(x, y, z) = -\frac{1}{4\pi R} \sinh kR; \qquad (3.3.47'')$$

using relation

$$\sinh kR = \frac{1}{2}(e^{kR} - e^{-kR}), \qquad (3.3.54')$$

we obtain also an integral of the homogeneous equation (3.3.53) in the form

$$u(x, y, z) = \frac{1}{R} \sinh kR. \qquad (3.3.55)$$

Replacing k by $(\pm\, ik)$ where i is the square root of (-1) (the unit imaginary number), we obtain the complex conjugate fundamental solutions

$$E(x, y, z) = -\frac{1}{4\pi R}\,\mathrm{e}^{-ikR}, \tag{3.3.56}$$

$$\overline{E}(x, y, z) = -\frac{1}{4\pi R}\,\mathrm{e}^{ikR} \tag{3.3.56'}$$

for the equation

$$\Delta u(x, y, z) + k^2 u(x, y, z) = f(x, y, z), \qquad k = \mathrm{const}, \tag{3.3.57}$$

where $f(x, y, z)$ is a given distribution.

Using relation

$$\cos kR = \frac{1}{2}(\mathrm{e}^{ikR} + \mathrm{e}^{-ikR}), \tag{3.3.58}$$

we obtain also the fundamental solution

$$\mathscr{E}(x, y, z) = -\frac{1}{4\pi R}\cos kR \tag{3.3.56''}$$

and with the relation

$$\sin kR = \frac{1}{2i}(\mathrm{e}^{ikR} - \mathrm{e}^{-ikR}) \tag{3.3.58'}$$

we obtain an integral of the homogeneous equation of Helmholtz

$$\Delta u(x, y, z) + k^2 u(x, y, z) = 0 \tag{3.3.59}$$

in the form

$$u(x, y, z) = \frac{1}{R}\sin kR. \tag{3.3.60}$$

We shall consider also equation,

$$\Delta\Delta u(x, y, z) = f(x, y, z), \tag{3.3.61}$$

where $f(x, y, z)$ is a given distribution.

To find the fundamental solution, we shall use the relation

$$\Delta(hg) = h\Delta g + g\Delta h + 2\left(\frac{\partial g}{\partial x}\frac{\partial h}{\partial x} + \frac{\partial g}{\partial y}\frac{\partial h}{\partial y} + \frac{\partial g}{\partial z}\frac{\partial h}{\partial z}\right), \tag{3.3.62}$$

where $h = h(x, y, z)$ is a function indefinitely differentiable and $g = g(x, y, z)$ is a distribution; in that case we may write

$$\Delta\left(\frac{1}{2}R^2 g\right) = \frac{1}{2}g\Delta(R^2) + \frac{1}{2}R^2\Delta g + \frac{\partial(R^2)}{\partial x}\frac{\partial g}{\partial x} + \frac{\partial(R^2)}{\partial y}\frac{\partial g}{\partial y} + \frac{\partial(R^2)}{\partial z}\frac{\partial g}{\partial z},$$

where

$$g = -\frac{1}{4\pi R}, \quad \Delta g = \delta(x, y, z)$$

is the fundamental solution of Poisson's equation. Remarking that we have

$$\frac{\partial(R^2)}{\partial x} = 2x, \quad \Delta(R^2) = 6, \quad \frac{\partial}{\partial x}\left(\frac{1}{R}\right) = -\frac{x}{R^3} \tag{3.3.63}$$

and other similar relations, and taking into account that

$$R^2\delta(x, y, z) = 0,$$

there follows

$$\Delta\left(\frac{1}{2}R^2 g\right) = -\frac{1}{4\pi R},$$

so that

$$\Delta\Delta\left(\frac{1}{2}R^2 g\right) = \delta(x, y, z);$$

we obtain thus a fundamental solution of equation (3.3.61) in the form

$$E(x, y, z) = -\frac{1}{8\pi}R. \tag{3.3.64}$$

In the case of only two spatial variables, formula (1.4.53′) permits us to write a fundamental solution of Poisson's equation

$$\Delta u(x, y) = f(x, y), \tag{3.3.65}$$

where $f(x, y)$ is a given distribution in the form

$$E(x, y) = -\frac{1}{2\pi}\log\frac{1}{r} \tag{3.3.66}$$

and the radius vector is given by (3.3.35).

Using the formula

$$\Delta(hg) = h\Delta g + g\Delta h + 2\left(\frac{\partial g}{\partial x}\frac{\partial h}{\partial x} + \frac{\partial g}{\partial y}\frac{\partial h}{\partial y}\right), \qquad (3.3.62')$$

where $h = h(x, y)$ is a function possessing derivatives of all orders and $g = g(x, y)$ is a distribution, we may write

$$\Delta\left(\frac{1}{4}r^2 g\right) = \frac{1}{4}g\Delta(r^2) + \frac{1}{4}r^2\Delta g + \frac{1}{2}\left[\frac{\partial(r^2)}{\partial x}\frac{\partial g}{\partial x} + \frac{\partial(r^2)}{\partial y}\frac{\partial g}{\partial y}\right],$$

with

$$g = -\frac{1}{2\pi}\log\frac{1}{r}, \quad \Delta g = \delta(x, y);$$

noting that we have

$$\frac{\partial(r^2)}{\partial x} = 2x, \quad \Delta(r^2) = 4, \quad \frac{\partial}{\partial x}\log\frac{1}{r} = -\frac{x}{r^2} \qquad (3.3.63')$$

and other similar relations, and taking into account that

$$r^2\delta(x, y) = 0,$$

we obtain

$$\Delta\left(\frac{1}{4}r^2 g\right) = \frac{1}{2\pi}\left(1 - \log\frac{1}{r}\right),$$

whence

$$\Delta\Delta\left(\frac{1}{4}r^2 g\right) = \delta(x, y).$$

Thus we obtain the fundamental solution

$$E(x, y) = -\frac{1}{8\pi}r^2\log\frac{1}{r} \qquad (3.3.67)$$

for the equation

$$\Delta\Delta u(x, y) = f(x, y), \qquad (3.3.68)$$

where $f(x, y)$ is a given distribution.

3.3.2.3. Equations of the hyperbolic type

Let us consider the equations

$$\Box_i u(x, y, z; t) = f(x, y, z; t) \qquad (i = 1, 2), \tag{3.3.69}$$

where the d'Alembert hyperbolic operator is given by

$$\Box_i = \Delta - \frac{1}{c_i^2} \frac{\partial^2}{\partial t^2} \tag{3.3.70}$$

and c_i is the propagation velocity of longitudinal $(i = 1)$ or transverse $(i = 2)$ waves, while $f(x, y, z)$ is a given distribution. We assume that the equation is valid over the whole space and the initial conditions are homogeneous (equal to zero for $t = 0$); using the Fourier transform with regard to the spatial variable and the Laplace transform with regard to time, we may write

$$-\left(\alpha^2 + \beta^2 + \gamma^2 + \frac{p^2}{c_i^2}\right) F[L[u]] = F[L[f]],$$

where α, β, γ and p are the new complex variables corresponding to the spatial and time variables, respectively. Since we have

$$F\left[\frac{1}{R} e^{-p\frac{R}{c_i}}\right] = \frac{4\pi}{\alpha^2 + \beta^2 + \gamma^2 + \frac{p^2}{c_i^2}} \qquad (i = 1, 2), \tag{3.3.71}$$

we obtain

$$u(x, y, z; t) = -\frac{1}{4\pi} L^{-1}\left[\frac{1}{R} e^{-p\frac{R}{c_i}} * L[f]\right] \qquad (i = 1, 2), \tag{3.3.72}$$

where we have considered the product of convolution with respect to the spatial variables.

If

$$f(x, y, z; t) = -4\pi \varkappa(t)\delta(x, y, z), \tag{3.3.73}$$

where $\varkappa(t)$ is a distribution, then it follows that

$$u(x, y, z; t) = L^{-1}\left[\frac{1}{R} e^{-p\frac{R}{c_i}}\right] \underset{(t)}{*} \varkappa(t) \qquad (i = 1, 2), \tag{3.3.74}$$

where convolution occurs with respect to the time variable. Thus, the solution of equation

$$\square_i u(x, y, z; t) + 4\pi \varkappa(t)\delta(x, y, z) = 0 \qquad (i = 1, 2) \tag{3.3.75}$$

is given by

$$u(x, y, z; t) = \frac{1}{R} \varkappa\left(t - \frac{R}{c_i}\right) \qquad (i = 1, 2), \tag{3.3.76}$$

where account has been taken of

$$L\left[\delta\left(t - \frac{R}{c_i}\right)\right] = e^{-p\frac{R}{c_i}} \qquad (i = 1, 2). \tag{3.3.77}$$

In particular, a fundamental solution of the equations (3.3.69) will be written in the form

$$E(x, y, z; t) = -\frac{1}{4\pi R} \delta\left(t - \frac{R}{c_i}\right) \qquad (i = 1, 2). \tag{3.3.78}$$

Let us now consider the equation

$$\square_1 \square_2 u(x, y, z; t) = f(x, y, z; t), \tag{3.3.79}$$

where $f(x, y, z; t)$ is a given distribution. Under the same conditions as in the preceding case, we shall use the integral transformations and we shall write

$$\left(\alpha^2 + \beta^2 + \gamma^2 + \frac{p^2}{c_1^2}\right)\left(\alpha^2 + \beta^2 + \gamma^2 + \frac{p^2}{c_2^2}\right)F[L[u]] = F[L[f]],$$

whence it follows that

$$F[L[u(x, y, z; t)]] = \frac{c_1^2 c_2^2}{4\pi p^2(c_1^2 - c_2^2)} F\left[\frac{1}{R}(e^{-p\frac{R}{c_1}} - e^{-p\frac{R}{c_2}})\right]F[L[f(x, y, z; t)]],$$

$$\tag{3.3.80}$$

where account has been taken of the break down into simple fractions

$$\frac{1}{\left(\alpha^2 + \beta^2 + \gamma^2 + \dfrac{p^2}{c_1^2}\right)\left(\alpha^2 + \beta^2 + \gamma^2 + \dfrac{p^2}{c_2^2}\right)}$$

$$= \frac{c_1^2 c_2^2}{p^2(c_1^2 - c_2^2)}\left(\frac{1}{\alpha^2 + \beta^2 + \gamma^2 + \dfrac{p^2}{c_1^2}} - \frac{1}{\alpha^2 + \beta^2 + \gamma^2 + \dfrac{p^2}{c_2^2}}\right)$$

and of the formula (3.3.71). In that case the result is

$$u(x, y, z; t) = \frac{c_1^2 c_2^2}{4\pi(c_1^2 - c_2^2)} L^{-1}\left[\frac{1}{p^2 R}\left(e^{-p\frac{R}{c_1}} - e^{-p\frac{R}{c_2}}\right) * L[f]\right], \qquad (3.3.81)$$

where convolution with regard to the spatial variables has been considered.
If the distribution $f(x, y, z; t)$ is of the form (3.3.73), there follows

$$u(x, y, z; t) = -\frac{c_1^2 c_2^2}{c_1^2 - c_2^2} L^{-1}\left[\frac{1}{p^2 R}\left(e^{-p\frac{R}{c_1}} - e^{-p\frac{R}{c_2}}\right)\right]_{(t)} * \varkappa(t), \qquad (3.3.82)$$

where convolution occurs with respect to the time variable. Thus the solution of equation

$$\square_1\square_2 u(x, y, z; t) + 4\pi\varkappa(t)\delta(x, y, z) = 0 \qquad (3.3.83)$$

is given by

$$u(x, y, z; t) = -\frac{c_1^2 c_2^2}{c_1^2 - c_2^2}\left[\left(t - \frac{R}{c_1}\right)_+ - \left(t - \frac{R}{c_2}\right)_+\right]_{(t)} * \varkappa(t), \qquad (3.3.84)$$

where account has been taken of

$$L\left[\left(t - \frac{R}{c_i}\right)_+\right] = \frac{1}{p^2} e^{-p\frac{R}{c_i}} \qquad (i = 1, 2), \qquad (3.3.85)$$

and where the positive part of the above functions has been introduced in the form

$$\left(t - \frac{R}{c_i}\right)_+ = \left(t - \frac{R}{c_i}\right)\theta\left(t - \frac{R}{c_i}\right) \qquad (i = 1, 2). \qquad (3.3.86)$$

In particular, a fundamental solution of equation (3.3.79) will be of the form

$$E(x, y, z; t) = \frac{c_1^2 c_2^2}{4\pi(c_1^2 - c_2^2)}\left[\left(t - \frac{R}{c_1}\right)_+ - \left(t - \frac{R}{c_2}\right)_+\right]. \qquad (3.3.87)$$

In the case of only two spatial variables, we shall consider the equations

$$\Box_i u(x, y; t) = f(x, y; t) \qquad (i = 1, 2),$$ (3.3.88)

$$\Box_1 \Box_2 u(x, y; t) = f(x, y; t),$$ (3.3.89)

where $f(x, y, z; t)$ is a given distribution. Proceeding as before and considering the Fourier transform

$$F\left[K_0\left(p\frac{r}{c_i}\right)\right] = \frac{2\pi}{\alpha^2 + \beta^2 + \dfrac{p^2}{c_i^2}}, \quad \operatorname{Re} p > 0 \qquad (i = 1, 2),$$ (3.3.90)

where K_0 represents the modified Bessel function of order zero, we obtain for the equations (3.3.88) the integral

$$u(x, y; t) = -\frac{1}{2\pi} L^{-1}\left[K_0\left(p\frac{r}{c_i}\right) * L[f]\right] \qquad (i = 1, 2),$$ (3.3.91)

where the convolution refers to the spatial variables.

We introduce the distribution defined by the function

$$f_0\left(t; \frac{r}{c_i}\right) = L^{-1}\left[K_0\left(p\frac{r}{c_i}\right)\right] = \begin{cases} 0 & \text{for } 0 < t < \dfrac{r}{c_i} \\ \dfrac{1}{\sqrt{t^2 - \dfrac{r^2}{c_i^2}}} & \text{for } t > \dfrac{r}{c_i} \end{cases} \qquad (i = 1, 2); \quad (3.3.92)$$

in that case, to the equation (3.3.88) there corresponds the fundamental solution

$$E(x, y; t) = -\frac{1}{2\pi} f_0\left(t; \frac{r}{c_i}\right) \qquad (i = 1, 2).$$ (3.3.93)

Similarly, we may write for equation (3.3.89)

$$u(x, y; t) = \frac{c_1^2 c_2^2}{2\pi(c_1^2 - c_2^2)} L^{-1}\left[\frac{1}{p^2}\left[K_0\left(p\frac{r}{c_1}\right) - K_0\left(p\frac{r}{c_2}\right)\right] * L[f]\right],$$ (3.3.94)

where convolution refers to the spatial variables.

If we introduce the distribution defined by the function

$$f_{-2}\left(t; \frac{r}{c_i}\right) = L^{-1}\left[\frac{1}{p^2} K_0\left(p\frac{r}{c_i}\right)\right]$$

$$= \begin{cases} 0 & \text{for } 0 < t < \dfrac{r}{c_i} \\[2mm] t\log\left(t + \sqrt{t^2 - \dfrac{r^2}{c_i^2}}\right) - \sqrt{t^2 - \dfrac{r^2}{c_i^2}} & \text{for } t > \dfrac{r}{c_i} \end{cases} \qquad (i = 1, 2), \qquad (3.3.95)$$

we obtain the fundamental solution

$$E(x, y; t) = \frac{c_1^2 c_2^2}{2\pi(c_1^2 - c_2^2)}\left[f_{-2}\left(t; \frac{r}{c_1}\right) - f_{-2}\left(t; \frac{r}{c_2}\right)\right], \qquad (3.3.96)$$

corresponding to equation (3.3.89).

Part 2

Applications of the theory of distributions in general mechanics

4

Mathematical model
of Newtonian mechanics

4.1. General concepts and results

The mathematical model of Newtonian mechanics is sufficient in most cases to deal with a large class of natural phenomena where mechanical motion plays an essential part. The model has been constructed by Isaac Newton and is based on the introduction of the concepts of space, time and mass; the magnitudes corresponding to these concepts are independent from one another. Essentially, there are three principles on which mechanics is based; they were stated by Isaac Newton and are the result of many previous investigations.

An important part is played by the *second principle of mechanics* which may be expressed in the form

$$F = \frac{d}{dt}(m\dot{r}),\qquad (4.1.1)$$

where F is the force acting on a particle of mass m, and $r = r(t)$ is the radius vector of that point with respect to a fixed reference point. If the mass of the particle is constant in time, as will be assumed, then the law (4.1.1) may be written in the form

$$F = m\ddot{r};\qquad (4.1.1')$$

here the dot represents the derivative with respect to the time variable t.

In general, it is assumed that the radius vector $r(t)$ has derivatives of the first and second order (its components are functions of the class C^2); also, it is assumed that the force $F = F(r, \dot{r}; t)$ is continuous with respect to the radius vector, velocity and time. In that case, the equations of the problem can be integrated when certain initial conditions are given.

However, in a great many mechanical phenomena the continuity conditions mentioned above are not complied with; such phenomena do not fit easily into the classical schemes based on the use of the usual functions. This justifies the need to extend the Newtonian mechanics and to complete its mathematical model by adjoining the results of the theory of distributions.

In the following discussion we shall treat of the form taken by various classical mechanical magnitudes and we shall give a modified form of the fundamental equation of Newtonian mechanics.

4.1.1. Mechanical magnitudes

The mechanical magnitudes which will be defined within the frame of the theory of distributions are the following: velocity, acceleration, force, percussion, linear and angular momentum, mechanical work and kinetic energy; as we shall see, the discontinuities that may occur in connection with these magnitudes are easily expressed in the theory of distributions.

4.1.1.1. Preliminary considerations

Among the mechanical phenomena which necessitate the introduction of the theory of distributions we mention those where the radius vector is a continuous function but the velocity and acceleration have discontinuities of the first species for $t = t_0$, i.e.

$$\mathbf{v}(t_0 - 0) \neq \mathbf{v}(t_0 + 0), \quad \dot{\mathbf{v}}(t_0 - 0) \neq \dot{\mathbf{v}}(t_0 + 0); \tag{4.1.2}$$

under these conditions the functions $\mathbf{v}(t)$ and $\mathbf{a}(t) = \dot{\mathbf{v}}(t)$ can be integrated.

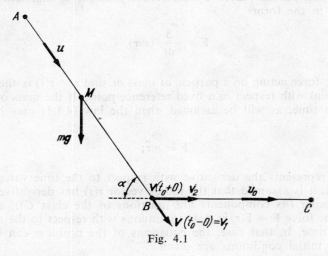

Fig. 4.1

An example of such a mechanical phenomenon in the motion of a heavy particle M on a trajectory ABC (a broken line, Figure 4.1) where, for simplicity, we assume that there is no friction; in fact, taking friction into account, the character of the mechanical phenomenon does not change.

As may be easily seen, the trajectory of the particle M is a continuous function which is made up of the segments AB and BC. We remark that the trajectory has no derivatives of the first and second order at the point B so that we cannot determine at that point the velocity, the acceleration or the constraint force. We

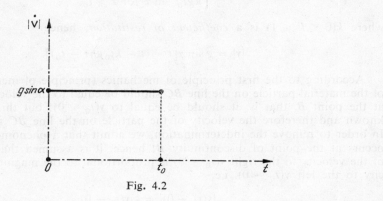

Fig. 4.2

shall assume that at the moment $t = 0$ the particle is at A and at the moment $t = t_0$ at B; in that case the magnitude of the acceleration may be written in the form

$$|\dot{\mathbf{v}}| = \begin{cases} g \sin \alpha \ \text{ for } 0 \leqslant t < t_0 \\ \\ 0 \qquad \text{ for } t > t_0 \end{cases} \qquad (4.1.3)$$

or

$$|\dot{\mathbf{v}}| = g \sin \alpha [1 - \theta(t - t_0)], \qquad t \geqslant 0, \qquad (4.1.3')$$

where g is the acceleration of gravity (Figure 4.2) and θ is the Heaviside function. In this way a discontinuity of the first species is pointed out, since at the moment $t = t_0$ we may write

$$\lim_{t \to t_0 - 0} |\dot{\mathbf{v}}| = g \sin \alpha, \quad \lim_{t \to t_0 + 0} |\dot{\mathbf{v}}| = 0. \qquad (4.1.4)$$

We remark that besides the discontinuity of the module, the acceleration has also a discontinuity of direction; thus, the acceleration vector has a discontinuity of the first species at the moment $t = t_0$ expressed by

$$\lim_{t \to t_0 - 0} \dot{\mathbf{v}}(t) = \mathbf{a}_1 = \mathbf{u} g \sin \alpha, \quad \lim_{t \to t_0 + 0} \dot{\mathbf{v}}(t) = 0, \qquad (4.1.4')$$

where $\mathbf{u}$ is the unit vector of the line AB, directed in the sense of motion of the particle.

The magnitude of the velocity is

$$|\mathbf{v}| = \begin{cases} gt \sin \alpha & \text{for } 0 \leqslant t < t_0 \\ \\ kgt_0 \sin \alpha & \text{for } t > t_0, \end{cases} \tag{4.1.5}$$

where $k(0 \leqslant k \leqslant 1)$ is a *coefficient of restitution*; hence

$$|\mathbf{v}| = g \sin \alpha [t - (t - kt_0)\theta(t - t_0)]. \tag{4.1.5'}$$

According to the first principle of mechanics (principle of inertia) the velocity of the material particle on the line BC should be equal to the velocity on the right at the point B, that is, it should be equal to $\mathbf{v}(t_0 + 0)$; but this velocity is unknown and therefore the velocity of the particle on the line BC is indeterminate. In order to remove the indetermination, we admit that a phenomenon of *collision* occurs at the point of discontinuity B; hence, it is assumed that the magnitude of the velocity to the right $\mathbf{v}(t_0 + 0)$ is proportional to the magnitude of the velocity to the left $\mathbf{v}(t_0 - 0)$, i.e.

$$|\mathbf{v}(t_0 + 0)| = k|\mathbf{v}(t_0 - 0)|. \tag{4.1.5''}$$

If the coefficient of restitution is equal to unity ($k = 1$), the collision in B is *perfectly elastic* and the magnitude of the velocity is a continuous function. In practice this is the usually adopted case, which, however, is not the only possibility.

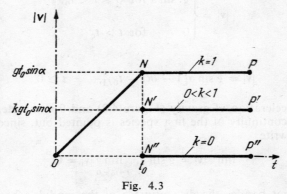

Fig. 4.3

Whatever the magnitude of the velocity on BC, the velocity vector has a discontinuity of the first species at the point B because

$$\lim_{t \to t_0 - 0} \mathbf{v}(t) = \mathbf{v}(t_0 - 0) = \mathbf{u}v(t_0 - 0) = \mathbf{v}_1,$$

$$\lim_{t \to t_0 + 0} \mathbf{v}(t) = \mathbf{v}(t_0 + 0) = \mathbf{u}v(t_0 + 0) = \mathbf{v}_2, \tag{4.1.6}$$

where $\mathbf{u}_0$ is the unit vector of BC in the direction of motion of the particle M; $\mathbf{v}_1$ is the velocity of the point M on AB, and $\mathbf{v}_2$ is the velocity of the same point on BC. Figure 4.3 gives the magnitude of the velocity as a function of time (the segments ON and NP for $k = 1$, the segments ON and $N'P'$ for $0 < k < 1$ and the segments ON and $N''P''$ for $k = 0$ (case of *plastic collision*).

The study of problems of the kind considered above involves the introduction of various mechanical magnitudes in the frame of the theory of distributions.

4.1.1.2. Trajectory. Velocity. Acceleration

Let us denote by $\mathbf{r}(t)$ the radius vector of a particle of mass m; the trajectory of the particle is expressed by

$$\mathbf{r} = \mathbf{r}(t) \tag{4.1.7}$$

and, from a physical point of view, it cannot be admitted that it has points of discontinuity. Therefore, $\mathbf{r}(t)$ represents a continuous function of the variable t.

We assume that *velocity*

$$\mathbf{v}(t) = \frac{\mathrm{d}}{\mathrm{d}t}\mathbf{r}(t) \tag{4.1.8}$$

and *acceleration*

$$\mathbf{a}(t) = \frac{\mathrm{d}}{\mathrm{d}t}\mathbf{v}(t) = \frac{\mathrm{d}^2}{\mathrm{d}t^2}\mathbf{r}(t) \tag{4.1.9}$$

are continuous functions in the interval $[t', t'']$ considered except at a finite number of values of t (the values $t = t_i$, $i = 1, 2, ..., n$) to which there correspond discontinuities of the first species. Under these conditions we can state that the integrals

$$\int_{t'}^{t''} \mathbf{v}(t)\,\mathrm{d}t, \quad \int_{t'}^{t''} \dot{\mathbf{v}}(t)\,\mathrm{d}t, \quad t' < t_i < t'' \qquad (i = 1, 2, ..., n) \tag{4.1.10}$$

exist.

Taking into account formula (1.2.40) we may write

$$\frac{\mathrm{d}^2\mathbf{r}(t)}{\mathrm{d}t^2} = \frac{\tilde{\mathrm{d}}^2\mathbf{r}(t)}{\mathrm{d}t^2} + \sum_{i=1}^{n} \mathbf{V}_i\delta(t - t_i), \tag{4.1.11}$$

where

$$\mathbf{V}_i = \mathbf{v}(t_i + 0) - \mathbf{v}(t_i - 0) \qquad (i = 1, 2, ..., n) \tag{4.1.12}$$

represents the velocity jump corresponding to the moment of discontinuity t_i, and the tilde corresponds to the derivative in the ordinary sense; thus we can state the following

Theorem 4.1.1. *The acceleration of a particle in the sense of the theory of distributions is equal to the distribution defined by the acceleration of that particle in the ordinary sense, where the latter exists, to which is added the sum of the products of the jumps of the velocity of the particle and the Dirac distribution.*

We introduce the notations

$$\mathbf{a}(t) = \frac{d^2\mathbf{r}(t)}{dt^2}, \ \tilde{\mathbf{a}}(t) = \frac{\tilde{d^2\mathbf{r}}(t)}{dt^2}, \ \mathbf{a}_c(t) = \sum_{i=1}^{n} \mathbf{V}_i\delta(t - t_i), \tag{4.1.13}$$

where $\mathbf{a}(t)$ is the *acceleration in the sense of the theory of distributions*, $\tilde{\mathbf{a}}(t)$ is the *acceleration in the ordinary sense* and $\tilde{\mathbf{a}}_c(t)$ is the *complementary acceleration* due to the discontinuities.

With these notations, relation (4.1.11) becomes

$$\mathbf{a}(t) = \tilde{\mathbf{a}}(t) + \mathbf{a}_c(t). \tag{4.1.14}$$

The acceleration in the sense of the theory of distributions will be termed also *generalized acceleration.*

4.1.1.3. Momentum. Force. Principles of mechanics

The product of the mass and the velocity of a particle is a vector which is termed the *linear momentum* or the *momentum of the particle* and is given by

$$\mathbf{H}(t) = m\mathbf{v}(t). \tag{4.1.15}$$

The momentum of a particle is a continuous function in the time interval considered except the moments $t_i \ (i = 1, 2, \ldots, n)$, where we have discontinuities of the first species.

In the case of a system of particles, the resultant momentum is the sum of the momenta of each separate particle. In the case of a continuous body the total momentum is obtained by integrating, over the whole domain occupied by the body, the momentum of a mass element; in this connection, an important part is played by the Stieltjes integral as we shall see in the next paragraph.

Such considerations are valid for all the magnitudes which will be introduced in the following discussion; however, we shall restrict ourselves to considerations concerning a single particle.

We introduce the notations

$$\mathbf{F}(t) = m\mathbf{a}(t), \ \tilde{\mathbf{F}}(t) = m(\tilde{\mathbf{a}}t), \ \mathbf{F}_c(t) = m\mathbf{a}_c(t), \tag{4.1.16}$$

where $\mathbf{F}(t)$ is the *generalized force* (in the sense of the theory of distributions), $\tilde{\mathbf{F}}(t)$ is the *force in the ordinary sense* and $\mathbf{F}_c(t)$ is the *complementary force* due to discontinuities. In that case relation (4.1.14) becomes

$$\mathbf{F}(t) = \tilde{\mathbf{F}}(t) + \mathbf{F}_c(t).$$ (4.1.17)

Hence we may state the following

Theorem 4.1.2. *The generalized force (in the sense of the theory of distributions) which acts on a particle is equal to the sum of the force in the ordinary sense and the complementary force (due to the discontinuities) acting on the same particle.*

From the foregoing considerations it follows that if we attach the complementary forces corresponding to the moments of discontinuity, the *second principle of mechanics* stated in equation (4.1.1) may be applied provided the conditions of the problem are complied with (velocity and acceleration have discontinuities of the first species)*.

Let us now consider the *first principle of mechanics (the inertia principle)* which represents a criterion by which we know when a force is acting on a particle. By the inertia principle the absence of acceleration of a particle means that no force is acting on it; in that case the particle moves uniformly along a straight line. We shall assume that for the generalized force $\mathbf{F} = 0$ too, we have $\mathbf{a} = 0$, i.e.:

$$\tilde{\mathbf{a}} + \sum_{i=1}^{n} \mathbf{V}_i \delta(t - t_i) = 0;$$ (4.1.18)

taking into account theorem 1.1.3 (see subsection 1.1.2.6) it follows that $\mathbf{V}_i = 0$ $(i = 1, 2, \ldots, n)$ and $\tilde{\mathbf{a}}(t) = 0$, hence the particle moves uniformly on a straight line. We conclude that *the inertia principle applies also in the case of generalized forces.*

The *third principle of mechanics (the principle of action and reaction)* applies as in the case of ordinary forces.

Differentiating relation (4.1.15) in the sense of the theory of distributions we may write

$$\frac{\mathrm{d}}{\mathrm{d}t}\mathbf{H}(t) = \frac{\tilde{\mathrm{d}}}{\mathrm{d}t}\mathbf{H}(t) + \sum_{i=1}^{n} (\Delta\mathbf{H})_i \delta(t - t_i),$$ (4.1.19)

where the jump of the momentum is given by

$$(\Delta\mathbf{H})_i = m\mathbf{V}_i = m[\mathbf{v}(t_i + 0) - \mathbf{v}(t_i - 0)].$$ (4.1.20)

Thus, we have introduced the velocity jump (4.1.12), corresponding to the moment of discontinuity t_i. Moreover, we remark that formula (4.1.19) corresponds to formula (4.1.11) for the differentiation of velocity.

---

* Detailed considerations concerning the second principle are presented in section 4.1.2.

In the frame of the theory of distributions the second principle of mechanics takes the form

$$\mathbf{F}(t) = \frac{\mathrm{d}}{\mathrm{d}t}\big[m\mathbf{v}(t)\big]; \tag{4.1.21}$$

in the case where there are no discontinuities, it becomes

$$\tilde{\mathbf{F}}(t) = \frac{\tilde{\mathrm{d}}}{\mathrm{d}t}\big[m\mathbf{v}(t)\big]. \tag{4.1.21'}$$

Taking into account the relation (4.1.19) and the relation which supplies the complementary force

$$\mathbf{F}_c(t) = \sum_{i=1}^{n} (\Delta\mathbf{H})_i \delta(t - t_i), \tag{4.1.22}$$

we obtain the relation

$$\frac{\mathrm{d}}{\mathrm{d}t}\mathbf{H}(t) = \mathbf{F}(t) = \tilde{\mathbf{F}}(t) + \mathbf{F}_c(t), \tag{4.1.23}$$

which allows us to state the following

Theorem 4.1.3. *The derivative with respect to time, in the sense of the theory of distributions, of the momentum of a particle is equal to the generalized force which acts on that particle.* This constitutes the *theorem of momentum.*

We shall introduce the following magnitudes

$$\int_{t'}^{t''} \mathbf{F}(t)\,\mathrm{d}t, \quad \int_{t'}^{t''} \tilde{\mathbf{F}}(t)\,\mathrm{d}t, \quad \int_{t'}^{t''} \mathbf{F}_c(t)\,\mathrm{d}t, \tag{4.1.24}$$

which represent the *linear impulse or the impulse of the generalized force* corresponding to the time interval $[t', t'']$, respectively the *impulse of the force in the ordinary sense* and the *impulse of the complementary force* in the same time interval; we have adopted, for the sake of uniformity, the classical notations for the first and third of these magnitudes, although the respective integrals have no sense from the point of wiev of the theory of distributions.

Remarking that

$$\int_{t'}^{t''} \mathbf{F}_c(t)\,\mathrm{d}t = \sum_{i=1}^{n} \int_{t'}^{t''} m\mathbf{V}_i\delta(t - t_i)\,\mathrm{d}t = \sum_{i=1}^{n} m\mathbf{V}_i, \quad t' < t_i < t'', \tag{4.1.25}$$

and taking into account (4.1.17), we may write

$$\int_{t'}^{t''} \mathbf{F}(t)\,\mathrm{d}t = \int_{t'}^{t''} \tilde{\mathbf{F}}(t)\,\mathrm{d}t + \sum_{i=1}^{n} (\Delta\mathbf{H})_i, \tag{4.1.26}$$

from which the following theorem results

Theorem 4.1.3′. *The impulse of a generalized force which acts on a particle in a given time interval is equal to the impulse of the ordinary force acting on that particle in the same time interval to which is added the sum of the jumps of the particle corresponding to the moments of discontinuity.*

The above considerations are valid in the case of a free particle; in the case of a particle subject to constraints we must apply the axiom of replacement of constraints, which leads to the introduction of the constraint force $\mathbf{R} = \mathbf{R}(t)$.

Using considerations applied previously we may write

$$\mathbf{R}(t) = \widetilde{\mathbf{R}}(t) + \mathbf{R}_c(t), \tag{4.1.27}$$

where $\mathbf{R}(t)$ is the *generalized constraint force* (in the sense of the theory of distributions), $\widetilde{\mathbf{R}}(t)$ is the *constraint force in the usual sense*, and $\mathbf{R}_c(t)$ is the *complementary constraint force* due to discontinuities. Therefore the theorem of momentum expressed by relation (4.1.23) takes the form

$$\frac{\mathrm{d}}{\mathrm{d}t}\mathbf{H}(t) = \mathbf{F}(t) + \mathbf{R}(t). \tag{4.1.28}$$

The complementary constraint force may be written

$$\mathbf{R}_c(t) = \sum_{i=1}^{n} \mathbf{R}_i(t), \tag{4.1.29}$$

where the constraint force $\mathbf{R}_i(t)$ is of the form

$$\mathbf{R}_i(t) = m\mathbf{V}_i\delta(t - t_i) \qquad (i = 1, 2, \ldots, n) \tag{4.1.30}$$

corresponding to the velocity jump at the moment of discontinuity t_i. In this way we may assume that the normal to the constraint force $\mathbf{R}_i(t)$, corresponding to the moment of discontinuity t_i, defines the direction of the tangent to the trajectory at that moment.

Let us return now to the problem considered in section 4.1.1.1. We have in this case ($t' = 0$, $t'' = t_0$)

$$\int_0^{t_0} \mathbf{F}(t)\,\mathrm{d}t = mgt_0 \sin\alpha\,\mathbf{u} + m(\mathbf{v}_2 - \mathbf{v}_1); \tag{4.1.31}$$

the jump of the constraint force corresponding to the point B may be assumed to be

$$\mathbf{R} = m\mathbf{V}\delta(t - t_0), \tag{4.1.32}$$

where

$$\mathbf{V} = \mathbf{v}_2 - \mathbf{v}_1. \tag{4.1.33}$$

If $k = 1$ *(perfectly elastic collision)*, we have $|\mathbf{v}_1| = |\mathbf{v}_2|$; the jump of the constraint force will have the direction of the bisector of the angle ABC and the external bisector of that angle will be tangent to the trajectory.

If $k = 0$ *(plastic collision)*, we have $\mathbf{v}_2 = 0$ which constitutes the second limiting case.

4.1.1.4. Moment of momentum

We shall now introduce the notion of *moment of momentum* (or *angular momentum*) *of a particle* by the relation

$$\mathbf{K}_0(t) = \mathbf{r}(t) \times \mathbf{H}(t) = \mathbf{r}(t) \times [m\mathbf{v}(t)]; \qquad (4.1.34)$$

hence the angular momentum of a particle is the moment of the linear momentum of that particle about the origin of the considered frame; here "$\times$" is the symbol of the cross product.

Let us consider first the case where velocity and acceleration are continuous functions (in the ordinary sense); we may write the theorem of moment of momentum in the form

$$\frac{\mathrm{d}}{\mathrm{d}t}\mathbf{K}_0(t) = \frac{\tilde{\mathrm{d}}}{\mathrm{d}t}\{\mathbf{r}(t) \times [m\mathbf{v}(t)]\} = \mathbf{r}(t) \times \tilde{\mathbf{F}}(t) = \mathbf{M}_0\tilde{\mathbf{F}}(t). \qquad (4.1.35)$$

If the velocity and the acceleration have discontinuities of the first species, we note that the moment of momentum $\mathbf{K}(t)$ and its derivative in the sense of the theory of distributions have the same moments of discontinuity; we may write

$$\frac{\mathrm{d}}{\mathrm{d}t}\mathbf{K}_0(t) = \frac{\tilde{\mathrm{d}}}{\mathrm{d}t}\mathbf{K}_0(t) + \sum_{i=1}^{n} (\Delta\mathbf{K})_i\delta(t - t_i), \qquad (4.1.36)$$

where the jump of the moment of momentum is given by

$$(\Delta\mathbf{K})_i = m\mathbf{r}(t_i) \times \mathbf{v}(t_i + 0) - m\mathbf{r}(t_i) \times \mathbf{v}(t_i - 0)$$

$$= m\mathbf{r}(t_i) \times \mathbf{V}_i = \mathbf{r}(t_i) \times (\Delta\mathbf{H})_i \qquad (4.1.37)$$

and is expressed with the help of the jump of the momentum (or of the jump of the velocity at the moment of discontinuity).

Taking into account relation (4.1.36), we may write

$$\frac{\mathrm{d}}{\mathrm{d}t}\mathbf{K}_0(t) = \mathbf{M}_0\mathbf{F}(t) = \mathbf{M}_0\tilde{\mathbf{F}}(t) + \mathbf{M}_0\mathbf{F}_c(t), \qquad (4.1.38)$$

where the moment of the complementary force is expressed by

$$\mathbf{M}_0\mathbf{F}_c(t) = \sum_{i=1}^{n} \mathbf{r}(t_i) \times (\Delta\mathbf{H})_i\delta(t - t_i) = \sum_{i=1}^{n} (\Delta\mathbf{K})_i\delta(t - t_i). \tag{4.1.39}$$

We now state the following

Theorem 4.1.4. *The derivative with respect to time, in the sense of the theory of distributions, of the moment of momentum of a particle is equal to the torque of the generalized force acting on that particle about the origin of the same reference frame. This is the theorem of moment of momentum.*

We shall introduce also the magnitudes

$$\int_{t'}^{t''} \mathbf{M}_0\mathbf{F}(t)\, dt, \int_{t'}^{t''} \mathbf{M}_0\tilde{\mathbf{F}}(t)\, dt, \int_{t'}^{t''} \mathbf{M}_0\mathbf{F}_c(t)\, dt, \tag{4.1.40}$$

which represent the *angular impulse of the generalized force*, corresponding to the time interval $[t', t'']$, *the angular impulse of the ordinary force* and *the angular impulse of the complementary force*, respectively. The first and third of these integrals are meaningless from the point of view of the theory of distributions; however, for the sake of uniformity, we have adopted these classical notations.

Starting from relation (4.1.17) we may write

$$\int_{t'}^{t''} \mathbf{M}_0\mathbf{F}(t)\, dt = \int_{t'}^{t''} \mathbf{M}_0\tilde{\mathbf{F}}(t)\, dt + \int_{t'}^{t''} \mathbf{M}_0\mathbf{F}_c(t)\, dt, \tag{4.1.41}$$

a relation which, taking into account relation (4.1.36), leads to

$$\int_{t'}^{t''} \mathbf{M}_0\mathbf{F}(t)\, dt = \int_{t'}^{t''} \mathbf{M}_0\tilde{\mathbf{F}}(t)\, dt + \sum_{i=1}^{n} (\Delta\mathbf{K})_i$$

$$= \int_{t'}^{t''} m\, \frac{d}{dt}\left[\mathbf{r}(t) \times \mathbf{v}(t)\right] dt + \sum_{i=1}^{n} (\Delta\mathbf{K})_i \tag{4.1.41'}$$

and to the following

Theorem 4.1.4'. *The angular impulse of the generalized force acting on a particle, in a given time interval, is equal to the angular impulse of the ordinary force acting on that particle in the same time interval, to which is added the sum of the jumps of the moment of momentum of the particle, corresponding to the moments of discontinuity.*

In the case of a particle subject to constraints, we introduce the constraint forces (4.1.27) which permit us to write the theorem of moment of momentum (4.1.38) in the form

$$\frac{d}{dt}\mathbf{K}_0(t) = \mathbf{M}_0\mathbf{F}(t) + \mathbf{M}_0\mathbf{R}(t). \tag{4.1.42}$$

In the case of the problem of subsection 4.1.1.1 (Figure 4.1) the result is

$$\int_0^{t_0} \mathbf{r}(t) \times \mathbf{F}(t)\,\mathrm{d}t = \int_0^{t_0} \mathbf{r}(t) \times \tilde{\mathbf{F}}(t)\,\mathrm{d}t + \mathbf{r}(t_0) \times [m(\mathbf{v}_2 - \mathbf{v}_1)] = \overrightarrow{AB} \times (m\mathbf{v}_2), \quad (4.1.43)$$

where the point A has been taken as the origin of the co-ordinate axes.

4.1.1.5. Mechanical work. Kinetic energy

The mechanical work (or the *work*) effected by the force $\mathbf{F}(t)$ in the time interval $[t', t'']$ is

$$W = \int_{t'}^{t''} \mathbf{F} \cdot \mathrm{d}\mathbf{r} = \int_{t'}^{t''} m\,\frac{\mathrm{d}\mathbf{v}(t)}{\mathrm{d}t} \cdot \mathbf{v}(t)\,\mathrm{d}t = \int_{t'}^{t''} \frac{\mathrm{d}}{\mathrm{d}t} T(t)\,\mathrm{d}t, \quad (4.1.44)$$

where the sign "." signifies the dot product and

$$T = \frac{1}{2}\,mv^2(t) \quad (4.1.45)$$

is the *kinetic energy* of the particle.

In the classical case, relation (4.1.44) leads to the theorem of kinetic energy in the ordinary sense), in the form

$$W = T(t'') - T(t'), \quad (4.1.46)$$

where the work too, is considered in the usual sense.

If t_i is a moment of discontinuity, we may write for the kinetic energy the relations

$$\lim_{t \to t_i - 0} T(t) = \frac{1}{2}\,m[\lim_{t \to t_i - 0} v(t)]^2 = \frac{1}{2}\,mv^2(t_i - 0),$$

$$(4.1.47)$$

$$\lim_{t \to t_i + 0} T(t) = \frac{1}{2}\,m[\lim_{t \to t_i + 0} v(t)]^2 = \frac{1}{2}\,mv^2(t_i + 0);$$

this proves that the moments of discontinuity of the velocity are also moments of discontinuity of the kinetic energy.

On the other hand, taking into account formula (1.2.40) we may write

$$\frac{\mathrm{d}}{\mathrm{d}t} T(t) = \frac{\tilde{\mathrm{d}}}{\mathrm{d}t} T(t) + \sum_{i=1}^{n} (\Delta T)_i \delta(t - t_i), \quad (4.1.48)$$

where the jump of the kinetic energy corresponding to the moment of discontinuity t_i is given by

$$(\Delta T)_i = \frac{1}{2} m[v^2(t_i + 0) - v^2(t_i - 0)]. \qquad (4.1.49)$$

We introduce the notations

$$W_F = \int_{t'}^{t''} \mathbf{F} \cdot d\mathbf{r}, \; W_{\widetilde{F}} = \int_{t'}^{t''} \widetilde{\mathbf{F}} \cdot d\mathbf{r}, \; W_{F_c} = \int_{t'}^{t''} \mathbf{F}_c \cdot d\mathbf{r}, \qquad (4.1.50)$$

where W_F is the *work of the generalized force* $\mathbf{F}(t)$, $W_{\widetilde{F}}$ is the *work of the ordinary force* $\widetilde{\mathbf{F}}(t)$ and W_{F_c} is the *work of the complementary force* $\mathbf{F}_c(t)$. The first and third of these integrals are meaningless in the theory of distributions; however, for the sake of uniformity, we adopt these classical notations.

We may write

$$W_F = W_{\widetilde{F}} + W_{F_c} \qquad (4.1.51)$$

and therefore we may state the following

Theorem 4.1.5. *The work of the generalized force acting on a particle in a given time interval is equal to the sum of the work of the force in the usual sense acting on the particle in the same time interval, and the work of the complementary force acting on the particle in the time interval considered.*

Relation (4.1.51) may be written in the form

$$W_F = \int_{t'}^{t''} \frac{\widetilde{d}}{dt} T(t) \, dt + \sum_{i=1}^{n} \int_{t'}^{t''} (\Delta T)_i \delta(t - t_i) \, dt$$

$$= T(t'') - T(t') + \sum_{i=1}^{n} (\Delta T)_i, \qquad (4.1.52)$$

and hence we may state the following

Theorem 4.1.5'. *The work of the generalized force acting on a particle in a given time interval is equal to the difference between the kinetic energy at the final moment and the kinetic energy at the initial moment to which is added the sum of the jumps of the kinetic energy of the particle corresponding to the moments of discontinuity.* This is the *theorem of kinetic energy.*

In the case of the problem considered in section 4.1.1.1 we have

$$W_F = \int_0^{t_0} mg \sin \alpha \, dr + (\Delta T)_0 = mg\overline{AB} \sin \alpha + \frac{1}{2} m(v_2^2 - v_1^2); \qquad (4.1.53)$$

if we take $k = 1$ (perfectly elastic collision), the jump of the kinetic energy vanishes.

4.1.2. Fundamental equation of Newtonian mechanics

In the following discussion we shall treat of a modified form of Newton's equation containing also the initial conditions of the problem; we shall consider first the case of the free particle and then the case of the particle with constraints.

4.1.2.1. Case of the free particle

We assume that at the initial moment $t = t_0$ the radius vector and the velocity are $\mathbf{r}_0$ and $\mathbf{v}_0$; if $t \in R$ the radius vector is given by

$$\boldsymbol{\rho}(t) = \begin{cases} 0 & \text{for } t < t_0 \\ \mathbf{r}_0 & \text{for } t = t_0 \\ \mathbf{r}(t) & \text{for } t > t_0, \end{cases} \qquad (4.1.54)$$

and the velocity is expressed by

$$\mathbf{V}(t) = \begin{cases} 0 & \text{for } t < t_0 \\ \mathbf{v}_0 & \text{for } t = t_0 \\ \mathbf{v}(t) & \text{for } t > t_0. \end{cases} \qquad (4.1.55)$$

This shows that

$$(\Delta\boldsymbol{\rho})_{t_0} = \mathbf{r}_0, \quad (\Delta\mathbf{V})_{t_0} = \mathbf{v}_0 \qquad (4.1.56)$$

represents the jump of the radius vector and of the velocity, respectively, at the initial moment $t = t_0$.

Differentiating successively the radius vector by formula (1.2.40) we obtain

$$\frac{d}{dt}\boldsymbol{\rho}(t) = \frac{\tilde{d}}{dt}\boldsymbol{\rho}(t) + (\Delta\boldsymbol{\rho})_{t_0}\delta(t - t_0), \qquad (4.1.57)$$

$$\frac{d^2}{dt^2}\boldsymbol{\rho}(t) = \frac{\tilde{d}^2}{dt^2}\boldsymbol{\rho}(t) + (\Delta\mathbf{V})_{t_0}\delta(t - t_0) + (\Delta\boldsymbol{\rho})_{t_0}\dot{\delta}(t - t_0); \qquad (4.1.57')$$

therefore, the velocity and acceleration will be expressed in distributions in the form

$$\mathbf{V}(t) = \frac{\tilde{d}}{dt}\boldsymbol{\rho}(t) + \mathbf{r}_0\delta(t - t_0), \qquad (4.1.58)$$

$$\frac{d}{dt}\mathbf{V}(t) = \frac{\tilde{d}^2}{dt^2}\boldsymbol{\rho}(t) + \mathbf{v}_0\delta(t - t_0) + \mathbf{r}_0\dot{\delta}(t - t_0), \qquad (4.1.58')$$

where velocity and acceleration in the usual sense are indicated.

Remarking that in the second law of mechanics (4.1.1') acceleration occurs in the usual sense, relation (4.1.58') leads to a modified form of the fundamental equation of mechanics which includes also the initial conditions. This form of the fundamental equation may be useful for solving certain boundary-value problems. Therefore the *fundamental equation of mechanics* will take the form

$$m \frac{\mathrm{d}}{\mathrm{d}t} \mathbf{V}(t) = \tilde{\mathbf{F}}(t) + m\mathbf{v}_0 \delta(t - t_0) + m\mathbf{r}_0 \dot{\delta}(t - t_0). \tag{4.1.59}$$

If, besides the moment of discontinuity $t = t_0$, other moments of discontinuity $t = t_i$ $(i = 1, 2, \ldots, n)$ occur, then the fundamental equation takes the form

$$m \frac{\mathrm{d}}{\mathrm{d}t} \mathbf{V}(t) = \tilde{\mathbf{F}}(t) + m \sum_{i=0}^{n} \mathbf{v}_i \delta(t - t_i) + m\mathbf{r}_0 \dot{\delta}(t - t_0). \tag{4.1.60}$$

For example, if the force $\tilde{\mathbf{F}}(t) = \mathbf{F}$ is constant we may integrate equation (4.1.59) member by member and we obtain

$$m\mathbf{V} = (t - t_0)\mathbf{F} + m\mathbf{v}_0 \theta(t - t_0) + m\mathbf{r}_0 \delta(t - t_0), \tag{4.1.61}$$

$$m\boldsymbol{\rho} = \frac{1}{2}(t - t_0)^2 \mathbf{F} + m\mathbf{v}_0(t - t_0)_+ + m\mathbf{r}_0 \theta(t - t_0), \tag{4.1.61'}$$

where

$$(t - t_0)_+ = \begin{cases} 0 & \text{for } t < t_0 \\ t - t_0 & \text{for } t \geq t_0 \end{cases} \tag{4.1.62}$$

is the positive part of the function $(t - t_0)$ and where we have introduced the Heaviside function

$$\theta(t - t_0) = \begin{cases} 0 & \text{for } t < t_0 \\ 1 & \text{for } t \geq t_0. \end{cases} \tag{4.1.62'}$$

The results correspond, for example, to the motion of a particle in vacuum.

4.1.2.2. Case of a particle with constraints

Let us consider a particle of mass m and of radius vector $\mathbf{r}(x, y, z; t)$, on which acts a given force $\mathbf{F} = \mathbf{F}(t)$; we assume that this is a generalized force, which may include both the usual and the complementary force. We shall consider that the particle is subject to a bilateral holonomic constraint expressed in the form

$$f(x, y, z; t) = 0, \tag{4.1.63}$$

where the function $f(x, y, z; t)$ is defined piecewise by

$$f(x, y, z; t) = f_i(x, y, z; t) \text{ for } t_{i-1} < t < t_i \qquad (i = 1, 2, ..., n). \tag{4.1.63'}$$

Applying the axiom of the replacement of constraints, the problem reduces to a problem of free particle; by introducing the generalized constraint force $\mathbf{R}(t)$, the equation of motion of a free particle becomes

$$m \frac{d^2\mathbf{r}(t)}{dt^2} = \mathbf{F}(t) + \mathbf{R}(t). \tag{4.1.64}$$

The generalized constraint $\mathbf{R}(t)$ may be written in the form

$$\mathbf{R}(t) = \sum_{i=1}^{n} \mathbf{R}_i(t), \tag{4.1.65}$$

where $\mathbf{R}_i(t)$ are the generalized constraints given by

$$\mathbf{R}_i(t) = \tilde{\mathbf{R}}_i(t) + \mathbf{R}_{ic}(t). \tag{4.1.65'}$$

Here $\tilde{\mathbf{R}}_i(t)$ are the constraint forces in the usual sense, corresponding to the moments $t \in (t_{i-1}, t_i)$ $(i = 1, 2, \ldots, n)$ and expressed by

$$\tilde{\mathbf{R}}_i(t) = \lambda_i(t) \overset{\cdot}{\mathbf{V}} f_i = \lambda_i(t) \operatorname{grad} f_i, \tag{4.1.65''}$$

where $\lambda_i(t)$ is a scalar, and $\mathbf{V}$ is the *nabla* operator of Hamilton; it is assumed that the functions $f_i(x, y, z; t)$ are of class C^1 and $\mathbf{R}_{ic}(t)$ are the complementary constraint forces corresponding to the moments of discontinuity $t_i(i = 0, 1, 2, \ldots, n)$.

These constraint forces in the usual sense are directed, in the intervals of definition, along the normals to the surfaces $f_i(x, y, z; t) = 0$ considered rigid at the moment $t \in (t_{i-1}, t_i)$. At the joining points of the surfaces, i.e. at the points corresponding to the moments t_i, the constraint forces cannot be determined by using expressions (4.1.65''). In order to obtain a unitary expression of the constraint forces, $\mathbf{R}_i(t)$, we shall make use of the equation of motion (4.1.64) which we shall write in the form

$$\mathscr{F}(t) \equiv \mathbf{F}(t) + \mathbf{R}(t) - m \frac{d^2\mathbf{r}(t)}{dt^2} = 0; \tag{4.1.64'}$$

this equation is valid everywhere except at the points corresponding to the moments t_i.

In order to introduce also the moments if discontinuity in the calculation it is sufficient to replace equation (4.1.64') by the equation

$$\prod_{i=1}^{n} (t - t_i)^{m_i+1} \mathcal{F}(t) = 0, \qquad (4.1.66)$$

which obviously, holds for $t \in R$. Within the frame of ordinary functions the solution of equation (4.1.66) is (4.1.64'); for a general solution we admit that $\mathcal{F}(t)$ may be a distribution and we consider equation (4.1.66) in the sense of the theory of distributions. Thus, the solution in distributions of equation (4.1.66) will include as a particular case also solution (4.1.64').

Taking into account theorem 1.2.2 of section 1.2.1.5, we obtain

$$\mathbf{R}_c = \sum_{i=1}^{n} \sum_{j=0}^{m_i} \boldsymbol{\alpha}_{ij} \delta^{(j)}(t - t_i) \qquad (4.1.67)$$

for the complementary constraint force and

$$\mathbf{R}(t) = \sum_{i=1}^{\tilde{n}} \lambda_i(t) \nabla f_i + \sum_{i=1}^{n} \sum_{j=0}^{m_i} \boldsymbol{\alpha}_{ij} \delta^{(j)}(t - t_i) \qquad (4.1.67')$$

for the generalized constraint force, where the solution in the case of ordinary functions is also indicated. Hence we may state the following

Theorem 4.1.6. *The generalized constraint force which acts on a particle is equal to the sum of the ordinary constraint force at the moments for which it is defined, and the complementary constraint force due to the moments of discontinuity and expressed by the aid of Dirac distributions and its derivatives.*

This result shows that the ∇ operator applied in the sense of the theory of distributions leads to a formula of the form

$$\nabla f = \tilde{\nabla} f + \sum_{i=1}^{n} \sum_{j=0}^{m_i} \boldsymbol{\beta}_{ij} \delta^{(j)}(t - t_i), \qquad (4.1.68)$$

where $f(x, y, z; t)$ is a function of the class C^1 except at the points $t = t_i$ $(i = 1, 2, \ldots, n)$ where the function has discontinuities of the first species and the symbol $\tilde{\nabla}$ corresponds to the ∇ operator in the usual sense.

We remark that in the formulae (4.1.67) and (4.1.67') the number m_i is indeterminate; the same remark applies to the vectors $\boldsymbol{\alpha}_{ij}$. The signification of these magnitudes is clearly seen when we consider the motion of the particle. Thus, assuming that the radius vector $\mathbf{r}(t)$ has a discontinuity of the first species for $t = t_0$, we may write

$$\frac{d}{dt}\mathbf{r}(t) = \frac{\tilde{d}}{dt}\mathbf{r}(t) + (\Delta\mathbf{r})_{t_0}\delta(t - t_0), \tag{4.1.69}$$

$$\frac{d^2}{dt^2}\mathbf{r}(t) = \frac{\tilde{d}^2}{dt^2}\mathbf{r}(t) + \left(\Delta\left(\frac{\tilde{d}\mathbf{r}}{dt}\right)\right)_{t_0}\delta(t - t_0) + (\Delta\mathbf{r})_{t_0}\dot{\delta}(t - t_0), \tag{4.1.69'}$$

where $(\Delta\mathbf{r})_{t_0}$ and $\left(\Delta\left(\dfrac{d\mathbf{r}}{dt}\right)\right)_{t_0}$ represent the jump of the radius vector and of the velo-

city, respectively, at the moment t_0. Comparing this with equation (4.1.64) and taking into account relation (4.1.67), we deduce that $m_i = 1$; also the constant vectors $\boldsymbol{\alpha}_{ij}$ represent the jump of the static moment $m\mathbf{r}$ ($j = 1$), and the jump of the momentum $m\mathbf{v}$ ($j = 0$), respectively, corresponding to the moment of discontinuity t_i.

Taking into account theorem 4.1.5' of kinetic energy, stated in section 4.1.1.5, we deduce that the mechanical work effected by the complementary force $m\left(\Delta\left(\dfrac{\tilde{d}\mathbf{r}}{dt}\right)\right)_{t_0}\delta(t - t_0)$ is precisely the kinetic energy jump, expressed by

$$(\Delta T)_{t_0} = \frac{1}{2}mv^2(t_0 + 0) - \frac{1}{2}mv^2(t_0 - 0). \tag{4.1.70}$$

Returning now to the example considered in section 4.1.1.1 (Figure 4.1), the fundamental equation of mechanics (4.1.60) allows us to write

$$m\frac{d^2}{dt^2}\mathbf{r}(t) = m\mathbf{g} + (\Delta\mathbf{H})_{t_0}\delta(t - t_0), \tag{4.1.71}$$

where the jump of the momentum at the point B is given by

$$(\Delta\mathbf{H})_{t_0} = m[\mathbf{v}(t_0 + 0) - \mathbf{v}(t_0 - 0)]; \tag{4.1.72}$$

the expression $(\Delta\mathbf{H})_{t_0}\delta(t - t_0)$ represents the complementary force, namely the constraint force corresponding to the moment of discontinuity t_0. The work corresponding to this complementary force is expressed by

$$W_{F_c} = \frac{1}{2}m\left[v^2(t_0 + 0) - v^2(t_0 - 0)\right]. \tag{4.1.73}$$

The total work will be

$$W = \frac{1}{2}m(v_C^2 - v_A^2) + W_{F_c}, \tag{4.1.74}$$

where v_c and v_A represent the magnitudes of the velocities at the points C and A, respectively. In the case where the magnitude of the velocity $v(t_0 - 0)$ when reaching the point B is equal to the magnitude of the velocity $v(t_0 + 0)$ when leaving that point, we have $W_{F_c} = 0$, and hence

$$W = \frac{1}{2} m(v_C^2 - v_A^2).$$ (4.1.74')

We remark that usually this simplifying assumption is resorted to although it is not always necessary and often not justified either.

We also remark that in order to ensure the uniqueness of the solution of equation (4.1.71) it is necessary to know, besides the initial conditions at the point A, also the velocity on the right at the point B; otherwise, the motion on the segment BC remains indeterminate. By Cauchy's theorem concerning the uniqueness of the solutions of ordinary differential equations, the force which acts on the particle must be of the class C^1; it is precisely this condition which is not satisfied at the point B. In order that the motion on the segment BC be determinate, the initial conditions at the point B must be given; hence we must know $v(t_0 + 0)$ (the direction of the velocity is known). Usually it is assumed that there exists a relation between the velocities on the right and on the left of point B, relation which is of an experimental character; only under these conditions the equation (4.1.71) has a unique solution also for the segment BC.

4.1.2.3. Linear oscillator

We shall now consider the forced oscillations of a particle M acted on by an elastic force of atraction $F(x) = -kx(k > 0)$ and a perturbing force $Q(t) = mq(t)$ $(t \geqslant 0)$ (Figure 4.4), where m is the mass of the particle. The differential equation of motion is

$$m \frac{d^2 x(t)}{dt^2} + kx(t) = mq(t);$$ (4.1.75)

using the notation

$$\omega^2 = \frac{k}{m},$$ (4.1.76)

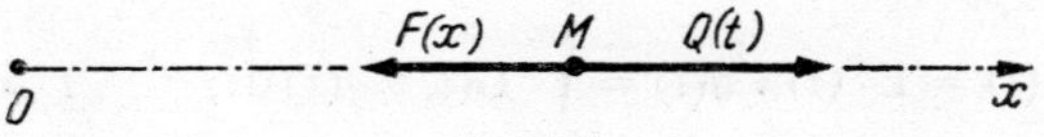

$$\overset{\bullet}{\underset{0}{\vphantom{|}}} \text{-------------} \quad \overset{F(x) \quad M \quad Q(t)}{\longleftarrow \bullet \longrightarrow} \text{------} \quad \overset{\bullet}{\underset{x}{\vphantom{|}}}$$

Fig. 4.4

where ω is the angular frequency of the particle, we may write

$$\frac{d^2 x(t)}{dt^2} + \omega^2 x(t) = q(t).$$ (4.1.75')

In accordance with the considerations of section 3.2.1.3 we shall deal with the general solution of the homogeneous equation

$$x(t) = C_1 \cos \omega t + C_2 \sin \omega t;$$ (4.1.77)

setting the initial conditions

$$x(0) = 0, \quad \frac{dx(0)}{dt} = 1,$$ (4.1.78)

we obtain

$$C_1 = 0, \quad C_2 = \frac{1}{\omega},$$ (4.1.79)

whence

$$x(t) = \frac{1}{\omega} \sin \omega t.$$ (4.1.77′)

Therefore the particular fundamental solution is

$$E^+(t) = \theta(t)x(t) = \frac{1}{\omega} \theta(t) \sin \omega t$$ (4.1.80)

and the equation

$$\frac{d^2}{dt^2} E^+(t) + \omega^2 E^+(t) = \delta(t)$$ (4.1.75″)

is satisfied.

Green's function corresponding to equation (4.1.75′) is given by

$$G(t, \tau) = E^+(t - \tau) = \frac{1}{\omega} \theta(t - \tau) \sin \omega(t - \tau).$$ (4.1.81)

Assuming that $q(t)$ is a function integrable on $[0, \infty)$, we may write a particular solution of equation (4.1.75′) in the form

$$x(t) = E^+(t) * q(t) = \int_0^\infty G(t, \tau)q(\tau)\, d\tau, \qquad t \geqslant 0.$$ (4.1.82)

Assuming the initial conditions

$$x(0) = x_0, \quad \frac{dx(0)}{dt} = \dot{x}_0,$$ (4.1.83)

formula (3.2.70′) leads to the general solution of equation (4.1.75′) namely

$$x(t) = E^+(t) * \theta(t)q(t) + \dot{x}_0 E^+(t) + x_0 \frac{\tilde{\mathrm{d}}}{\mathrm{d}t} E^+(t), \qquad t \geqslant 0, \qquad (4.1.84)$$

or in another form

$$x(t) = \frac{1}{\omega} \int_0^t \sin \omega(t - \tau)q(\tau)\,\mathrm{d}\tau + \frac{\dot{x}_0}{\omega} \sin \omega t + x_0 \cos \omega t, \ t \geqslant 0. \quad (4.1.84')$$

4.2. Stieltjes integral. Application to the geometry of masses

In the expression of the different laws of mechanics the elements of the geometry of masses play an important part; beside certain geometrical elements such as point, curve, surface, the structure of certain magnitudes includes also masses which may have a continuous distribution or may be concentrated.

The mathematical expression of static moments or moments of inertia is not possible in every case by means of the *Riemann integral;* that is why the *Stieltjes integral* plays an important part since it permits to express these magnitudes mathematically in a unique form, regardless of whether the masses are distributed or concentrated.

In the following discussion, expressions in distributions of static moments and of moments of inertia as well as some extensions of the Stieltjes integral are given; one-, two- and three-dimensional bodies will be dealt with in turn.

4.2.1. One-dimensional bodies

We shall present first an introductory treatment of the Stieltjes integral and then apply the results to the computation of certain elements of the geometry of masses.

4.2.1.1. The simple Stieltjes integral

Let $f(x)$ and $g(x)$ be two functions bounded on the interval $[a, b]$ and

$$a = x_0 < x_1 < x_2 < \cdots < x_i < x_{i+1} < \cdots < x_n = b \qquad (4.2.1)$$

a division of this interval; if ξ_i is an arbitrary point of the subinterval $[x_i, x_{i+1}]$ $(i = 0, 1, 2, \ldots, n - 1)$ then the expression

$$\sigma = \sum_{i=0}^{n-1} f(\xi_i)[g(x_{i+1}) - g(x_i)] \qquad (4.2.2)$$

represents the Stieltjes integral sum.

The finite limit of the Stieltjes sum σ, when $\lambda = \max (x_{i+1} - x_i)$ tends to zero, represents the *Stieltjes integral* of the function $f(x)$ with respect to $g(x)$ and is denoted by

$$(S)\int_a^b f(x)\,\mathrm{d}g(x) = \lim_{\lambda \to 0} \sum_{i=0}^{n-1} f(\xi_i)[g(x_{i+1}) - g(x_i)]. \tag{4.2.3}$$

We may state the following

Theorem 4.2.1. *If $f(x)$ is a function continuous on $[a, b]$ and $g(x)$ is a function with an integrable derivative on $[a, b]$, the Stieltjes integral (4.2.3) exists and its computation reduces to that of a Riemann integral*

$$(S)\int_a^b f(x)\,\mathrm{d}g(x) = (R)\int_a^b f(x)g'(x)\,\mathrm{d}x. \tag{4.2.4}$$

It is important to note that in the expression (4.2.3) of the Stieltjes integral $\mathrm{d}g(x)$ does not represent the differential of the function $g(x)$; the fact that the function has an integrable derivative under the conditions of the above theorem permits us to interpret $\mathrm{d}g(x)$ as the differential of $g(x)$ in the formula (4.2.4), thus realizing an easy passage from the Stieltjes to the Riemann integral.

If $f(x)$ and $g(x)$ are distributions and the Stieltjes integral exists, then $\mathrm{d}g(x)$ can be again interpreted as a differential in the sense of the theory of distributions. Thus we may state the following

Theorem 4.2.2. *If $f(x)$ is a continuous function on $[a, b]$ and $g(x)$ has a derivative $g'(x)$ absolutely integrable on the interval $[a, b]$, except at a finite number of points*

$$a = c_0 < c_1 < c_2 < \cdots < c_n = b, \tag{4.2.5}$$

where it has discontinuities of the first species, then the Stieltjes integral exists and we have the relation

$$(S)\int_a^b f(x)\,\mathrm{d}g(x) = (R)\int_a^b f(x)g'(x)\,\mathrm{d}x + f(a)[g(a + 0) - g(a)]$$

$$+ \sum_{k=1}^{m-1} f(c_k)[g(c_k + 0) - g(c_k - 0)] + f(b)[g(b) - g(b - 0)]. \tag{4.2.6}$$

Instead of the functions $f(x)$ and $g(x)$ defined on $[a, b]$ we shall consider the distributions defined on $x \in R$ by the functions

$$\bar{f}(x) = \begin{cases} f(a) & \text{for } x < a \\ f(x) & \text{for } x \in [a, b] \\ f(b) & \text{for } x > b, \end{cases} \tag{4.2.7}$$

$$\bar{g}(x) = \begin{cases} g(a) & \text{for } x < a \\ g(x) & \text{for } x \in [a, b] \\ g(b) & \text{for } x > b. \end{cases} \tag{4.2.7'}$$

In this way we obtain at the points of discontinuity a and b the jumps of the functions $\bar{g}(x)$ in the form

$$\Delta \bar{g}(a) = g(a + 0) - g(a - 0) = g(a + 0) - g(a), \qquad (4.2.8)$$

$$\Delta \bar{g}(b) = g(b + 0) - g(b - 0) = g(b) - g(b - 0). \qquad (4.2.8')$$

Differentiating the distribution $\bar{g}(x)$ in the sense of the theory of distributions we obtain

$$\frac{d\bar{g}(x)}{dx} = \frac{\widetilde{d\bar{g}(x)}}{dx} + \Delta \bar{g}(a)\delta(x - a)$$

$$+ \sum_{k=1}^{m-1} \Delta \bar{g}(c_k)\delta(x - c_k) + \Delta \bar{g}(b)\delta(x - b), \qquad (4.2.9)$$

where the derivative of $g(x)$ in the usual sense is given by

$$\frac{\widetilde{d\bar{g}(x)}}{dx} = \begin{cases} g'(x) \text{ for } x \in [a, b] \text{ and } x \neq a, b, c_k \ (k = 1, 2, ..., m - 1) \\ 0 \text{ in the rest.} \end{cases} \qquad (4.2.10)$$

Based on the continuity of $\bar{f}(x)$ we have

$$\bar{f}(x)\delta(x - x_0) = \bar{f}(x_0)\delta(x - x_0), \qquad (4.2.11)$$

so that multiplying relation (4.2.9) by $\bar{f}(x)$ and taking into account (4.2.10) and (4.2.11), we obtain

$$\bar{f}(x) \frac{d\bar{g}(x)}{dx} = f(x)g'(x) + f(a)\Delta \bar{g}(a)\delta(x - a)$$

$$+ \sum_{k=1}^{m-1} f(c_k)\Delta \bar{g}(c_k)\delta(x - c_k) + f(b)\Delta \bar{g}(b)\delta(x - b). \qquad (4.2.12)$$

If $\varphi(x) \in K$ is a fundamental function (possessing derivatives of all orders and having a compact support), we may write

$$\left(\bar{f}(x) \frac{d\bar{g}(x)}{dx}, \varphi(x) \right) = (f(x)g'(x), \varphi(x)) + f(a)\Delta \bar{g}(a)(\delta(x - a), \varphi(x))$$

$$+ \sum_{k=1}^{m-1} f(c_k)\Delta \bar{g}(c_k)(\delta(x - c_k), \varphi(x)) + f(b)\Delta \bar{g}(b)(\delta(x - b), \varphi(x)). \qquad (4.2.12')$$

Remarking that

$$(\delta(x - \xi), \varphi(x)) = \varphi(\xi)$$

and taking the fundamental function $\varphi(x)$ such that it is equal to unity on the interval $[a, b]$, which is always possible (since the function $\varphi(x)$ is arbitrary), relation (4.2.12′) takes the form

$$\left(\overline{f}(x) \frac{d\overline{g}(x)}{dx}, \varphi(x) \right) = (f(x)g'(x), \varphi(x)) + f(a)\Delta\overline{g}(a)$$

$$+ \sum_{k=1}^{m-1} f(c_k)\Delta\overline{g}(c_k) + f(b)\Delta\overline{g}(b). \tag{4.2.12″}$$

Since by (4.2.10), $g'(x)$ is defined only on $[a, b]$, we have

$$(f(x)g'(x), \varphi(x)) = (R)\int_a^b f(x)g'(x)\,dx, \tag{4.2.13}$$

so that formula (4.2.12″) becomes

$$\left(\overline{f}(x) \frac{d\overline{g}(x)}{dx}, \varphi(x) \right) = (R)\int_a^b f(x)g'(x)\,dx + f(a)\Delta\overline{g}(a)$$

$$+ \sum_{k=1}^{m-1} f(c_k)\Delta\overline{g}(c_k) + f(b)\Delta\overline{g}(b). \tag{4.2.12‴}$$

We remark that the second member of the above relation coincides with that of formula (4.2.6), thus proving theorem (4.2.2); hence in the Stieltjes integral the symbol $dg(x)$ may be interpreted as a differential in the sense of the theory of distributions.

Moreover, in many calculation involving the Stieltjes integral use of the functional $\overline{f}(x) \dfrac{d\overline{g}(x)}{dx}$ is made; the values of the integral are but mappings of this functional onto the space K of the fundamental functions. From relation (4.2.12‴) we deduce that the existence of the functional $\overline{f}(x) \dfrac{d\overline{g}(x)}{dx}$ is sufficient for the existence of the Stieltjes integral.

Let us consider the case where $g(x) = \delta(x)$. Since

$$f(x)\delta(x) = f(0)\delta(x), \tag{4.2.14}$$

we obtain by differentiation

$$f(x)\delta'(x) = f(0)\delta'(x) - f'(0)\delta(x), \tag{4.2.14'}$$

whence

$$(f(x)\delta'(x), \varphi(x)) = -f(0)\varphi'(0) - f'(0)\varphi(0). \tag{4.2.14''}$$

Choosing $\varphi(x)$ such as to be at the origin equal to unity and such that $\varphi'(0) = 0$, with the support on $[a, b]$, we obtain

$$(S)\int_a^b f(x)\,\mathrm{d}\delta(x) = -f'(0). \tag{4.2.15}$$

4.2.1.2. Mass, static moment and moment of inertia of a one-dimensional body

If $g(x) = m(x)$ represents the sum of the masses distributed and concentrated on the segment $[a, x]$, and if $g(a) = m(a) = 0$, then the *total mass* of the one-dimensional body represented by the segment $[a, b]$ is given by

$$M = (S)\int_a^b \mathrm{d}m(x). \tag{4.2.16}$$

The static moment about the origin O is

$$S_O = (S)\int_a^b x\,\mathrm{d}m(x) \tag{4.2.17}$$

and the *moment of inertia* about the same point is

$$I_O = (S)\int_a^b x^2\,\mathrm{d}m(x). \tag{4.2.18}$$

The derivative of $\overline{m}(x)$ in the sense of the theory of distributions is the *density*

$$\rho(x) = \frac{\mathrm{d}\overline{m}(x)}{\mathrm{d}x} = \tilde{\rho}(x) + \sum_{k=1}^n m_k \delta(x - x_k); \tag{4.2.19}$$

applying theorem 4.2.2, we may write also

$$M = (R)\int_a^b \tilde{\rho}(x)\,dx + \sum_{k=1}^n m_k, \qquad (4.2.16')$$

$$S_O = (R)\int_a^b x\tilde{\rho}(x)\,dx + \sum_{k=1}^n x_k m_k, \qquad (4.2.17')$$

$$I_O = (R)\int_a^b x^2\tilde{\rho}(x)\,dx + \sum_{k=1}^n x_k^2 m_k, \qquad (4.2.18')$$

where m_k $(k = 1, 2, \ldots, n)$ represent the jumps of the function $m(x)$ at the points x_k, being equal to the masses concentrated at these points, and $\tilde{\rho}(x)$ is the density in the usual sense.

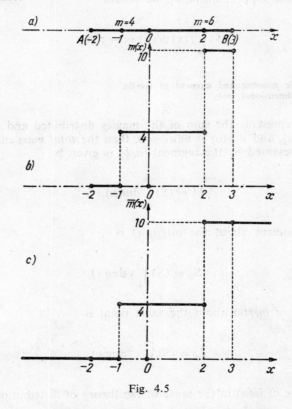

Fig. 4.5

Let us now consider the bar AB with $A(-2)$ and $B(3)$ (Figure 4.5a); we assume that at the point $C(-1)$ there is a concentrated mass $m = 4$ and at the point $D(2)$ a concentrated mass $m = 6$. The repartition function of the mass (Figure 4.5b) (neglecting the mass of the bar) will be

$$m(x) = \begin{cases} 0 & \text{for } -2 \leqslant x \leqslant -1 \\ 4 & \text{for } -1 < x \leqslant 2 \\ 10 & \text{for } 2 < x \leqslant 3, \end{cases} \tag{4.2.20}$$

and the function $\overline{m}(x)$ obtained by extending $m(x)$ is expressed by (Figure 4.5c)

$$\overline{m}(x) = \begin{cases} 0 & \text{for } -\infty < x \leqslant 1 \\ 4 & \text{for } -1 < x \leqslant 2 \\ 10 & \text{for } 2 < x < \infty; \end{cases} \tag{4.2.20'}$$

hence the total mass is

$$M = 10. \tag{4.2.20''}$$

We remark that the points of discontinuity of the function $\overline{m}(x)$ are (-1) and (2); the corresponding jumps will be

$$\Delta\overline{m}(-1) = m(-1 + 0) - m(-1 - 0) = 4,$$
$$\Delta\overline{m}(2) = m(2 + 0) - m(2 - 0) = 6 \tag{4.2.20'''}$$

and the specific mass is expressed by

$$\rho(x) = 4\delta(x + 1) + 6\delta(x - 2). \tag{4.2.21}$$

Therefore the static moment is

$$S_O = (-1) \cdot 4 + 2 \cdot 6 = 8 \tag{4.2.22}$$

and the moment of inertia

$$I_O = (-1)^2 \cdot 4 + 2^2 \cdot 6 = 28. \tag{4.2.23}$$

4.2.2. Two-dimensional bodies

Proceeding as in the preceding section, we shall treat first the double Stieltjes integral by introducing the two-dimensional differential and derivative; then we shall apply the results to the computation of the static moments and moments of inertia of two-dimensional bodies.

4.2.2.1. The double Stieltjes integral

Let $f(x, y)$ and $g(x, y)$ be two functions defined on the rectangular domain $D \equiv \{a_1, b_1; a_2, b_2\}^*$.

If $f(x, y)$ is a continuous function on the domain D and $g(x, y)$ a non-decreasing function with regard to each variable and such that

$$\Delta_2 g(x, y) \equiv g(x + h_1, y + h_2) - g(x + h_1, y)$$

$$- g(x, y + h_2) + g(x, y) \geqslant 0 \qquad (4.2.24)$$

for all $h_1, h_2 > 0$, at the point (x, y), then the double Stieltjes integral exists and represents the limit of the sum

$$\sigma = \sum_{j=1}^{m} \sum_{k=1}^{n} f(\xi_j, \eta_k)[g(x_j, y_k) - g(x_{j-1}, y_k) - g(x_j, y_{k-1}) + g(x_{j-1}, y_{k-1})], \quad (4.2.25)$$

when $\lambda \to 0$,

$$(S) \iint_D f(x, y) D_2 g(x, y) = \lim_{\lambda \to 0} \sigma, \qquad (4.2.25')$$

where λ is the greatest diameter of the subdomains into which the given domain has been resolved and where we have assumed

$$a_1 = x_0 < x_1 < \ldots < x_m = b_1,$$

$$a_2 = y_0 < y_1 < \ldots < y_n = b_2, \qquad (4.2.25'')$$

$$x_{j-1} \leqslant \xi_j \leqslant x_j, \, y_{k-1} \leqslant \eta_k \leqslant y_k \qquad (j = 1, 2, \ldots, m; k = 1, 2, \ldots, n). \qquad (4.2.25''')$$

We can state the following

Theorem 4.2.3. *If $f(x, y)$ is a continuous function in D and if the function $g(x)$ has the property (4.2.24) and admits continuous partial derivatives of the second order, then the computation of the double Stieltjes integral reduces to the computation of a double Riemann integral*

$$(S) \iint_D f(x, y) D_2 g(x, y) = (R) \iint_D f(x, y) g''_{xy}(x, y) \, dx \, dy. \qquad (4.2.26)$$

---

* In general we may take instead of the domain D an arbitrary domain which can be resolved into rectangles.

We remark that the Stieltjes integral sum (4.2.25) may be written in the form

$$\sigma = \sum_{j=1}^{m} \sum_{k=1}^{n} f(\xi_j, \eta_k)\Delta_2 g(x_{j-1}, y_{k-1}), \tag{4.2.27}$$

where we have introduced the two-dimensional variation of the function $g(x, y)$ at the point (x_j, y_k), namely

$$\Delta_2 g(x_{j-1}, y_{k-1}) = g(x_j, y_k) - g(x_{j-1}, y_k) - g(x_j, y_{k-1}) + g(x_{j-1}, y_{k-1}); \tag{4.2.28}$$

for a point (x, y) we shall use expression (4.2.24).

With the help of the auxilliary function

$$V(x) = g(x, y + h_2) - g(x, y) \tag{4.2.29}$$

the two-dimensional variation may be written in the form

$$\Delta_2 g(x, y) = V(x + h_1) - V(x); \tag{4.2.30}$$

since the function $g(x, y)$ is of class C^2, we may apply twice a mean value theorem which gives

$$\Delta_2 g(x, y) = h_1 V'_x(x + \theta_1 h_1) = h_1[g'_x(x + \theta_1 h_1, y + h_2) - g'_x(x + \theta_1 h_1, y)]$$

$$= h_1 h_2 g''_{xy}(x + \theta_1 h_1, y + \theta_2 h_2), \tag{4.2.30'}$$

where θ_1 and θ_2 are real numbers $(0 < \theta_1, \theta_2 < 1)$.

Thus, the two-dimensional variation at the point (x_j, y_k) will be written

$$\Delta_2 g(x_j, y_k) = \Delta x_j \Delta y_k g''_{xy}(\bar{\xi}_j, \bar{\eta}_k), \tag{4.2.30''}$$

where $x_{j-1} < \bar{\xi}_j < x_j$ and $y_{k-1} < \bar{\eta}_k < y_k$, and the Stieltjes integral sum becomes

$$\sigma = \sum_{j=1}^{m} \sum_{k=1}^{n} f(\xi_j, \eta_k) g''_{xy}(\bar{\xi}_j, \bar{\eta}_k)\Delta x_j \Delta y_k. \tag{4.2.27'}$$

Since ξ_j and η_k are arbitrary we may choose them such as to have $\xi_j = \bar{\xi}_j$ and $\eta_k = \bar{\eta}_k$; we remark that under these conditions the Stieltjes integral sum is a Riemann integral sum and therefore

$$\lim_{\lambda \to 0} \sigma = (R) \iint_D f(x, y)g''_{xy}(x, y) \, dx \, dy,$$

which proves theorem 4.2.3.

We remark that theorem 4.2.3 is still valid if instead of the continuity of the mixed derivative $g''_{xy}(x, y)$ only its integrability is required.

4.2.2.2. Two-dimensional differential and derivative

Theorem 4.2.3 points out the fact that the passage from the Stieltjes double integral to the Riemann double integral may be effected directly if instead of the symbol $D_2 g(x, y)$ we consider the expression $g''_{xy}(x, y) \, dx \, dy$. In fact, the symbol $D_2 g(x,y)$ represents a differential operator specific to the Stieltjes double integral which is really the *two-dimensional differential* of the function $g(x, y)$; also we remark that $g''_{xy}(x, y)$ represents the *two-dimensional derivative* of that function. We may write

$$D_2 g(x, y) = g''_{xy}(x, y) \, dx \, dy. \tag{4.2.31}$$

Thus we may write

$$\lim_{\substack{\Delta x_j \to 0 \\ \Delta y_k \to 0}} \frac{\Delta_2 g(x_{j-1}, y_{k-1})}{\Delta x_j \Delta y_k} = g''_{xy}(x_{j-1}, y_{k-1}). \tag{4.2.32}$$

It is interesting to note that relation (4.2.31) may be considered under wider conditions than those required by theorem 4.2.3.

In practice, an important case is that where the function $g(x, y)$ has everywhere continuous mixed derivatives of the second order except at a finite number of points where it has discontinuities of the first species. In this connection we shall consider the Heaviside function

$$\theta(x - x_0, y - y_0) = \begin{cases} 1 \text{ for } x \geqslant x_0, y \geqslant y_0 \\ \\ 0 \text{ in the rest,} \end{cases} \tag{4.2.33}$$

which obviously has discontinuities of the first species (Figure 4.6); we can state the following

Theorem 4.2.4. *If the function $f(x, y)$ defined on D is continuous, then*

$$(S) \iint_D f(x, y) D_2 \theta(x - x_0, y - y_0) = f(x_0, y_0). \tag{4.2.34}$$

Indeed, for the calculation of the Stieltjes integral sum (4.2.27) we shall consider the domain D divided into sub-domains such that the straight lines $x = x_0$ and $y = y_0$, representing the lines of discontinuity of the function $\theta(x - x_0, y - y_0)$, be comprised between the straight lines $x = x_{j-1}$ and $x = x_j$, $y = y_{k-1}$ and $y = y_k$, respectively; it may be easily seen that, by the definition (4.2.33), the integral sum (4.2.27) reduces to a single term, namely that corresponding to the sub-domain $[x_{j-1}, x_j] \times [y_{k-1}, y_k]$, i.e.

$$\sigma = f(\xi_j, \eta_k), \tag{4.2.35}$$

where $x_{j-1} < \xi_j < x_j$ and $y_{k-1} < \eta_k < y_k$.

When the sub-domains tend to zero, the sub-domain $[x_{j-1}, x_j] \times [y_{k-1}, y_k]$ approaches the point (x_0, y_0) so that $\xi_j \to x_0$, $\eta_k \to y_0$ and we have

$$(S) \iint_D f(x, y) D_2 \theta(x - x_0, y - y_0) = \lim_{\substack{\xi_j \to x_0 \\ \eta_k \to y_0}} f(\xi_j, \eta_k),$$

which proves the theorem.

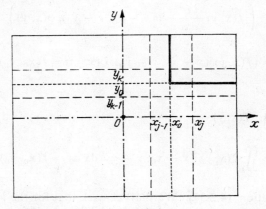

Fig. 4.6

We remark that formula (4.2.34) may be obtained from formula (4.2.26) although the function $\theta(x - x_0, y - y_0)$ does not satisfy the requirements of theorem 4.2.3 (it has not continuous mixed derivatives of the second order). For this reason, relation (4.2.32) shall be considered in the sense of the theory of distributions. Then we may write

$$\frac{\partial^2}{\partial x \partial y} \theta(x - x_0, y - y_0) = \delta(x - x_0) \times \delta(y - y_0) = \delta(x - x_0, y - y_0), \qquad (4.2.36)$$

where "$\times$" represents the direct product; there follows

$$D_2 \theta(x - x_0, y - y_0) = \delta(x - x_0, y - y_0) \, dx \, dy. \qquad (4.2.37)$$

Since the function $f(x, y)$ is continuous at (x_0, y_0) we have

$$f(x, y)\delta(x - x_0, y - y_0) = f(x_0, y_0)\delta(x - x_0, y - y_0)$$

and relation (4.2.37) becomes

$$f(x, y)D_2 \theta(x - x_0, y - y_0) = f(x_0, y_0)\delta(x - x_0, y - y_0) \, dx \, dy. \qquad (4.2.37')$$

Introducing the fundamental function $\varphi(x, y) \in K$, we may write

$$(\delta(x - x_0, y - y_0), \varphi(x, y)) = \varphi(x_0, y_0); \tag{4.2.38}$$

this function is arbitrary and it may be chosen such as to have D as support and to take the value 1 at the point (x_0, y_0). Thus, we obtain

$$\left(f(x, y) \frac{\partial^2}{\partial x \partial y} \theta(x - x_0, y - y_0), \varphi(x, y) \right)$$

$$= (f(x, y) \delta(x - x_0, y - y_0), \varphi(x, y)) = f(x_0, y_0) \tag{4.2.39}$$

or

$$\iint_{R^2} f(x, y) \frac{\partial^2}{\partial x \partial y} \theta(x - x_0, y - y_0) \varphi(x, y) \, dx \, dy$$

$$= \iint_D f(x, y) \delta(x - x_0, y - y_0) \, dx \, dy = f(x_0, y_0), \tag{4.2.39'}$$

which justifies relation (4.2.37). We shall remark that for the Heaviside function $\theta(x - x_0, y - y_0)$ we may again use formula (4.2.26), provided that the differentiations are performed in the sense of the theory of distributions.

Let us now consider functions $g(x, y)$ of the form

$$g(x, y) = \tilde{g}(x, y) + \sum_{k=1}^{n} g_k \theta(x - x_k, y - y_k), \tag{4.2.40}$$

where g_k are constants and $\tilde{g}(x, y)$ is the continuous part of the function $g(x, y)$ on D; such functions with discontinuities of the first species are particularly important in problems related to the geometry of masses. It can be shown that if the function $f(x, y)$ is continuous on D and the function $g(x, y)$ has the form (4.2.40) then the Stieltjes integral of the function $f(x, y)$ with regard to the function $g(x, y)$ exists.

According to relation (4.2.35) and to the property of additivity of the Stieltjes integral we have

$$(S) \iint_D f(x, y) D_2 g(x, y) = (R) \iint_D f(x, y) \, d\tilde{g}(x, y) + \sum_{k=1}^{n} g_k f(x_k y_k), \tag{4.2.41}$$

and, if the function $\tilde{g}(x, y)$ has continuous derivatives of the second order, we may write

$$(S) \iint_D f(x, y) D_2 g(x, y) = (R) \iint_D f(x, y) \frac{\partial^2 \tilde{g}(x, y)}{\partial x \partial y} \, dx \, dy + \sum_{k=1}^{n} g_k f(x_k, y_k). \tag{2.4.41'}$$

Formula (4.2.41') is the analogue of formula (4.2.6) for the simple Stieltjes integral; therefore we may write

$$D_2 g(x, y) = D_2 \tilde{g}(x, y) + \sum_{k=1}^{n} g_k D_2 \theta(x - x_k, y - y_k). \qquad (4.2.42)$$

From the structure of the function $g(x, y)$ defined by (4.2.40) one may easily deduce that the jumps of that function on the straight lines $x = x_k$, $y = y_k$ $(k = 1, 2, \ldots, n)$ are the same at the point (x_k, y_k), which is an intersection of the respective lines. Hence, if g_k represents the jump of the function $g(x, y)$ on the straight line $y = y_k$ at the point (x_k, y_k), we may write

$$g_k = g(x_k + 0, y_k + 0) - g(x_k - 0, y_k + 0) \qquad (4.2.43)$$

and

$$g(x_k + 0, y_k - 0) = g(x_k - 0, y_k - 0) = 0; \qquad (4.2.43')$$

in that case

$$g_k = g(x_k + 0, y_k + 0) - g(x_k - 0, y_k + 0)$$

$$- g(x_k + 0, y_k - 0) + g(x_k - 0, y_k - 0), \qquad (4.2.44)$$

i.e.

$$g_k = (\Delta g(x, y))_k = (\Delta g)_k, \qquad (4.2.44')$$

which represents the two-dimensional jump of the function $g(x, y)$ at the point of discontinuity (x_k, y_k).

Formula (4.2.42) may be written also in the form

$$D_2 g(x, y) = \tilde{D}_2 g(x, y) + \sum_{k=1}^{n} (\Delta g)_k \delta(x - x_k, y - y_k) \, dx \, dy, \qquad (4.2.42')$$

where $D_2 g(x, y)$ and $\tilde{D}_2 g(x, y)$ represent the two-dimensional differential in the sense of the theory of distributions and in the usual sense, respectively.

4.2.2.3. Mass, static moment and moment of inertia of a two-dimensional body

Function (4.2.40) permits us to express the repartition law of the sum of the masses of a plate, where $\tilde{m}(x, y)$ is the continuous part of that sum and m_k are the masses concentrated at the points (x_k, y_k) $(k = 1, 2, \ldots, n)$ of the plate; we have

$$m(x, y) = \tilde{m}(x, y) + \sum_{k=1}^{n} m_k \theta(x - x_k, y - y_k). \qquad (4.2.45)$$

The *density* will be expressed in that case by

$$D_2 m(x, y) = \rho(x, y)\, dx\, dy, \tag{4.2.46}$$

where the two-dimensional differential is considered in the sense of the theory of distributions; hence, taking into account formula (4.2.42′), we may write

$$\rho(x, y) = \tilde{\rho}(x, y) + \sum_{k=1}^{n} m_k \delta(x - x_k, y - y_k), \tag{4.2.46′}$$

where the *density in the usual sense* is expressed by

$$\tilde{D}_2 m(x, y) = \tilde{\rho}(x, y)\, dx\, dy. \tag{4.2.46″}$$

In that case the *total mass* of the two-dimensional body is given by

$$M = (S) \iint_D D_2 m(x, y) = (R) \iint_D \tilde{\rho}(x, y)\, dx\, dy + \sum_{k=1}^{n} m_k. \tag{4.2.47}$$

The *static moment* about the Oy axis is

$$S_{Oy} = (S) \iint_D x D_2 m(x, y) = (S) \iint_D x \rho(x, y)\, dx\, dy$$

$$= (R) \iint_D x \tilde{\rho}(x, y)\, dx\, dy + \sum_{k=1}^{n} x_k m_k, \tag{4.2.48}$$

and the moment of inertia about the origin O is

$$I_O = (S) \iint_D (x^2 + y^2) D_2 m(x, y) = (S) \iint_D (x^2 + y^2)\rho(x, y)\, dx\, dy$$

$$= (R) \iint_D (x^2 + y^2)\tilde{\rho}(x, y)\, dx\, dy + \sum_{k=1}^{n} (x_k^2 + y_k^2) m_k. \tag{4.2.49}$$

Other magnitudes may be defined in a similar way, for example the moment of inertia about the Oy axis is

$$I_{Oy} = (S) \iint_D x^2 D_2 m(x, y) = (S) \iint_D x^2 \rho(x, y)\, dx\, dy$$

$$= (R) \iint_D x^2 \tilde{\rho}(x, y)\, dx\, dy + \sum_{k=1}^{n} x_k^2 m_k. \tag{4.2.50}$$

It should be noted that the classical relations between the various magnitudes introduced are still valid; for example, the relation of invariance

$$I_O = I_{Ox} + I_{Oy} \qquad (4.2.51)$$

is easily verified.

4.2.3. Three-dimensional bodies

In the following discussion we shall introduce the triple Stieltjes integral, pointing out the three-dimensional derivative and differential; this will permit the computation of certain magnitudes pertaining to the geometry of masses of three-dimensional bodies.

4.2.3.1. The triple Stieltjes integral

Let us consider a function $g(x, y, z)$ defined on a rectangular parallelepiped $\Delta \equiv \{a_1, b_1; a_2, b_2; a_3, b_3\}^*$; it is assumed that the function does not decrease with regard to each variable and that it verifies the inequality

$$\Delta_3 g(x, y, z) \equiv g(x + h_1, y + h_2, z + h_3) - g(x, y + h_2, z + h_3)$$

$$- g(x + h_1, y, z + h_3) - g(x + h_1, y + h_2, z) + g(x + h_1, y, z)$$

$$+ g(x, y + h_2, z) + g(x, y, z + h_3) - g(x, y, z) \geqslant 0 \qquad (4.2.52)$$

for all x, y, z and $h_1, h_2, h_3 > 0$. With the help of the function $g(x, y, z)$ we may construct an additive function of the three-dimensional interval D, which is used for defining the three-dimensional Stieltjes integral. It may be shown that if $f(x, y, z)$ is a continuous function on D and $g(x, y, z)$ has the property (4.2.52), then the Stieltjes integral exists and may be defined as the limit of the integral sum when λ, which is the greatest diameter of the sub-domains considered, tends to zero; that is

$$\lim_{\lambda \to 0} \sum_{i=1}^{m} \sum_{j=1}^{n} \sum_{k=1}^{p} f(\xi_i, \eta_j, \zeta_k) \Delta_3 g(x_{i-1}, y_{j-1}, z_{k-1}) = (S) \iiint_D f(x, y, z) D_3 g(x, y, z),$$

$$(4.2.53)$$

---

* We may consider, as in 4.2.2.1., any domain that can be resolved into such parallelepipeds.

where

$$a_1 = x_0 < x_1 < \ldots < x_m = b_1,$$

$$a_2 = y_0 < y_1 < \ldots < y_n = b_2, \tag{4.2.54}$$

$$a_3 = z_0 < z_1 < \ldots < z_p = b_3,$$

$$x_{j-1} \leqslant \xi_i \leqslant x_i,\, y_{j-1} \leqslant \eta_j \leqslant y_j,\, z_{k-1} \leqslant \zeta_k \leqslant z_k, \tag{4.2.54'}$$

and the three-dimensional variation of the function $g(x, y, z)$ at the point $g(x_i, y_j, z_k)$ is given by

$$\Delta_3 g(x_{i-1}, y_{j-1}, z_{k-1}) = g(x_i, y_j, z_k) - g(x_{i-1}, y_j, z_k) - g(x_i, y_{j-1}, z_k)$$

$$- g(x_i, y_j, z_{k-1}) + g(x_i, y_{j-1}, z_{k-1}) + g(x_{i-1}, y_j, z_{k-1})$$

$$+ g(x_{i-1}, y_{j-1}, z_k) - g(x_{i-1}, y_{j-1}, z_{k-1}). \tag{4.2.55}$$

Taking into account (4.2.52) we may write

$$\Delta_3 g(x, y, z) = V(z + h_3) - V(z), \tag{4.2.56}$$

where we have introduced the auxiliary function

$$V(z) = g(x + h_1, y + h_2, z) - g(x, y + h_2, z)$$

$$- g(x + h_1, y, z) + g(x, y, z). \tag{4.2.56'}$$

On the other hand, we may write

$$V(z) = U(x + h_1, z) - U(x, z), \tag{4.2.57}$$

with the help of a new auxiliary function

$$U(x, z) = g(x, y + h_2, z) - g(x, y, z). \tag{4.2.57'}$$

Taking into account relations (4.2.56) — (4.2.57') and assuming that $g(x, y, z)$ possesses continuous partial derivatives of the third order, we may apply a mean-value formula; we obtain

$$\Delta_3 g(x, y, z) = h_3 V_z'(z + \theta_3 h_3) = h_3 \Delta_2 g_z'(x, y, z + \theta_3 h_3)$$

$$= h_1 h_2 h_3 g_{xyz}'''(x + \theta_1 h_1, y + \theta_2 h_2, z + \theta_3 h_3), \tag{4.2.58}$$

where $0 < \theta_i < 1$ $(i = 1, 2, 3)$ and Δ_2 represents the two-dimensional variation with respect to the variables x and y. Thus we have

$$\lim_{\substack{h_1 \to 0 \\ h_2 \to 0 \\ h_3 \to 0}} \frac{\Delta_3 g(x, y, z)}{h_1 h_2 h_3} = g'''_{xyz}(x, y, z). \tag{4.2.59}$$

We remark that expression (4.2.58) may be written also in the form

$$\Delta_3 g(x, y, z) = h_1 h_2 h_3 [g'''_{xyz}(x, y, z) + \varepsilon], \tag{4.2.58'}$$

where

$$\lim_{\substack{h_1 \to 0 \\ h_2 \to 0 \\ h_3 \to 0}} \varepsilon = 0; \tag{4.2.58''}$$

we state now the following

Theorem 4.2.5. *If the function $f(x, y, z)$ is a continuous function on D and $g(x,y,z)$ is a function possessing continuous derivatives of the third order and having the property* (4.2.52), *then the Stieltjes integral* (4.2.53) *reduces to a Riemann integral*

$$(S) \iiint_D f(x, y, z) D_3 g(x, y, z) = (R) \iiint_D f(x, y, z) g'''_{xyz}(x, y, z) \, dx \, dy \, dz. \tag{4.2.60}$$

We remark that

$$\Delta_3 g(x_{i-1}, y_{j-1}, z_{k-1}) = \Delta x_i \Delta y_j \Delta z_k [g'''_{xyz}(x_{i-1}, y_{j-1}, z_{k-1}) + \varepsilon], \tag{4.2.61}$$

where

$$\Delta x_i = x_i - x_{i-1}, \Delta y_j = y_j - y_{j-1}, \Delta z_k = z_k - z_{k-1}, \tag{4.2.61'}$$

$$\lim_{\substack{\Delta x_i \to 0 \\ \Delta y_j \to 0 \\ \Delta z_k \to 0}} \varepsilon = 0; \tag{4.2.61''}$$

the Stieltjes integral sum may be expressed in the form

$$\sum_{i=1}^{m} \sum_{j=1}^{n} \sum_{k=1}^{p} f(\xi_i, \eta_j, \zeta_k) \Delta_3 g(x_{i-1}, y_{j-1}, z_{k-1})$$

$$= \sum_{i=1}^{m} \sum_{j=1}^{n} \sum_{k=1}^{p} f(\xi_i, \eta_j, \zeta_k) g'''_{xyz}(x_{i-1}, y_{j-1}, z_{k-1}) \Delta x_i \Delta y_j \Delta z_k$$

$$+ \sum_{i=1}^{m} \sum_{j=1}^{n} \sum_{k=1}^{p} \varepsilon f(\xi_i, \eta_j, \zeta_k) \Delta x_i \Delta y_j \Delta z_k.$$

The first sum of the right-hand-side represents a Riemann integral sum whereas the second sum tends to zero owing to the boundedness of the continuous function $f(x, y, z)$ and to property (4.2.61'') of ε; passing to the limit we obtain formula (4.2.60).

4.2.3.2 Three-dimensional diferential and derivative

Formula (4.2.60) shows that, under the conditions stated, the passage from the Stieltjes to the Riemann integral may be performed directly if we admit the relation

$$D_3 g(x, y, z) = g'''_{xyz}(x, y, z)\, dx\, dy\, dz; \tag{4.2.62}$$

here $D_3 g(x, y, z)$ represents the *three-dimensional differential* of the function $g(x,y,z)$ and $g'''_{xyz}(x, y, z)$ the *three-dimensional derivative* of the function. As in the case of the one- and two-dimensional Stieltjes integral, this interpretation is always possible assuming that $f(x, y, z)$ and $g(x, y, z)$ are distributions for which the product $f(x, y, z)\, D_3 g(x, y, z)$ has a significance.

Introducing the Heaviside function,

$$\theta(x - x_0, y - y_0, z - z_0) = \begin{cases} 1 \text{ for } x \geqslant x_0, y \geqslant y_0, z \geqslant z_0 \\[2mm] 0 \text{ in the rest,} \end{cases} \tag{4.2.63}$$

it may be shown that

$$(S)\iiint_D f(x, y, z) D_3\theta(x - x_0, y - y_0, z - z_0) = f(x_0, y_0, z_0), \tag{4.2.64}$$

where $f(x, y, z)$ is a continuous function in D.

Since we have, in the sense of theory of distributions

$$\frac{\partial^3}{\partial x \partial y \partial z}\theta(x - x_0, y - y_0, z - z_0) = \delta(x - x_0, y - y_0, z - z_0), \tag{4.2.65}$$

it follows that formula (4.2.60) is valid in that case too, although the function $\theta(x - x_0,\ y - y_0,\ z - z_0)$ has discontinuities of the first species.

Let us now consider the function

$$g(x, y, z) = \tilde{g}(x, y, z) + \sum_{k=1}^{n} g_k\theta(x - x_k, y - y_k, z - z_k), \tag{4.2.66}$$

where g_k are constants and $\tilde{g}(x, y, z)$ is the continuous part of the function; as in section 4.2.2.2, it may be shown that

$$(S)\iiint_D f(x, y, z)D_3g(x, y, z) = (R)\iiint_D f(x, y, z)g'''_{xyz}(x, y, z)\,dx\,dy\,dz$$

$$+ \sum_{k=1}^n g_k f(x_k, y_k, z_k), \tag{4.2.67}$$

which extends the range of formula (4.2.60).

It follows also that

$$D_3g(x, y, z) = D_3\tilde{g}(x, y, z) + \sum_{k=1}^n g_k D_3\theta(x - x_k, y - y_k, z - z_k), \tag{4.2.68}$$

which may be written

$$D_3g(x, y, z) = \widetilde{D_3g}(x, y, z) + \sum_{k=1}^n (\Delta g)_k\delta(x - x_k, y - y_k, z - z_k)\,dx\,dy\,dz, \tag{4.2.68'}$$

where the three-dimensional differential in the sense of the theory of distributions and in the usual sense, as well as the three-dimensional jump of the function $g(x,y,z)$ at the point (x_k, y_k, z_k) are marked.

4.2.3.3 Mass, static moment and moment of inertia of a three-dimensional body

To express the distribution law $m(x, y, z)$ of the sum of the masses of an arbitrary three-dimensional body, where $\tilde{m}(x, y, z)$ is the continuous part of the sum and m_k are the masses concentrated at the points (x_k, y_k, z_k) $(k = 1, 2, \ldots, n)$, we use a function of the form (4.2.66); we may write

$$m(x, y, z) = \tilde{m}(x, y, z) + \sum_{k=1}^n m_k\theta(x - x_k, y - y_k, z - z_k). \tag{4.2.69}$$

The *density* is expressed by the relation

$$D_3m(x, y, z) = \rho(x, y, z)\,dx\,dy\,dz, \tag{4.2.70}$$

where the three-dimensional differential must be taken in the sense of the theory of distributions; taking into account relation (4.2.68'), we obtain

$$\rho(x, y, z) = \tilde{\rho}(x, y, z) + \sum_{k=1}^n m_k\delta(x - x_k, y - y_k, z - z_k), \tag{4.2.70'}$$

where the density in the ordinary sense is given by

$$\tilde{D}_3 m(x, y, z) = \tilde{\rho}(x, y, z)\,dx\,dy\,dz. \tag{4.2.70''}$$

Hence the *total mass* of the three-dimensional body will be

$$M = (S)\iiint_D D_3 m(x, y, z) = (R)\iiint_D \tilde{\rho}(x, y, z)\,dx\,dy\,dz + \sum_{k=1}^{n} m_k. \tag{4.2.71}$$

The *static moment* about the plane Oxy is

$$S_{Oxy} = (S)\iiint_D z D_3 m(x, y, z) = (S)\iiint_D z\rho(x, y, z)\,dx\,dy\,dz$$

$$= (R)\iiint_D z\tilde{\rho}(x, y, z)\,dx\,dy\,dz + \sum_{k=1}^{n} z_k m_k, \tag{4.2.72}$$

and the *moment of inertia* about the origin O is

$$I_O = (S)\iiint_D (x^2 + y^2 + z^2)D_3 m(x, y, z)$$

$$= (S)\iiint_D (x^2 + y^2 + z^2)\rho(x, y, z)\,dx\,dy\,dz$$

$$= (R)\iiint_D (x^2 + y^2 + z^2)\tilde{\rho}(x, y, z)\,dx\,dy\,dz + \sum_{k=1}^{n} (x_k^2 + y_k^2 + z_k^2)m_k. \tag{4.2.73}$$

Other magnitudes related to the geometry of masses may be introduced in a similar manner.

5

Theory of concentrated loads

A complete mathematical model of Newtonian mechanics should include the representation of loads acting on an arbitrary continuous body or on a system of particles. In the mechanics of rigid solids it is sufficient to consider that the loads are just forces which may be represented by sliding vectors, while in the case of systems of particles the loads are still forces which, however, are represented by bound vectors; the situation is different in the case of a deformable continuum. In that case, the loads are represented by bound vectors but they can be no longer replaced by systems of static equivalent loads. Owing to the deformability of the body, the loads have at the same time a local and an overall effect; the manner of representing the vector fields which correspond to various loads is particularly important. In this problem the *theory of distributions* plays a very important part.

In the following discussion we shall treat of the representation of arbitrary loads in general and especially of the representation of concentrated loads. Also, we shall give a general classification of concentrated loads.

5.1. Mathematical representation of loads

We shall first treat of the representation of concentrated forces; these are concentrated loads which will be considered as *fundamental* since starting from them we may construct any other concentrated loads. Also, with the aid of concentrated forces we may construct an arbitrary distributed load, the representation of which will be considered in the following.

5.1.1. Mathematical representation of concentrated forces

A category of loads the mathematical representation of which is simple and includes all its characteristics is that of *concentrated forces*. We shall first consider a single concentrated force and then a system of concentrated forces.

5.1.1.1. Mathematical representation of a concentrated force

Let A_0 be a point inside the arbitrary non-rigid domain Ω_ε dependent on the parameter $\varepsilon > 0$; we assume that when $\varepsilon \to 0$ the domain tends to the point A_0.

Also, we assume that in the domain Ω_ε there is a field of loads $\mathbf{Q}_\varepsilon(x, y, z)$ depending on the same parameter and having the projections

$$F_x g_\varepsilon'(x, y, z), \ F_y g_\varepsilon''(x, y, z), \ F_z g_\varepsilon'''(x, y, z) \tag{5.1.1}$$

on the co-ordinate axes, where

$$\lim_{\varepsilon \to 0} g_\varepsilon'(x, y, z) = \lim_{\varepsilon \to 0} g_\varepsilon''(x, y, z) = \lim_{\varepsilon \to 0} g_\varepsilon'''(x, y, z) =$$

$$= \delta(x - x_0, y - y_0, z - z_0); \tag{5.1.2}$$

the limits are considered in the sense of convergence in the space K.

If the field of loads reduces to the point A_0, we obtain the resultant force

$$\mathbf{R}_\varepsilon = \iiint_{\Omega_\varepsilon} \mathbf{Q}_\varepsilon \, dx \, dy \, dz \tag{5.1.3}$$

and the resultant moment

$$\mathbf{M}_\varepsilon = \iiint_{\Omega_\varepsilon} (\mathbf{r} - \mathbf{r}_0) \times \mathbf{Q}_\varepsilon \, dx \, dy \, dz, \tag{5.1.3'}$$

where $\mathbf{r}_0$ is the radius vector of the point A_0 and $\mathbf{r}$ is the radius vector of the point of application of the load $\mathbf{Q}_\varepsilon$. In general the field of loads $\mathbf{Q}_\varepsilon(x, y, z)$ is not equivalent to the torsor $(\mathbf{R}_\varepsilon, \mathbf{M}_\varepsilon)$ since the medium within the domain Ω_ε is non-rigid. Hence the field of forces $\mathbf{Q}_\varepsilon$ can be replaced only approximately by the torsor $(\mathbf{R}_\varepsilon, \mathbf{M}_\varepsilon)$; the approximation becomes closer as the dimensions of the domain Ω_ε grow smaller. In the case of such an approximation the medium enclosed in the domain Ω_ε has the *characteristics of a rigid solid*.

Since the domain Ω_ε tends to the point A_0 when $\varepsilon \to 0$ we may assume that, when passing to the limit, the field of loads $\mathbf{Q}_\varepsilon$ is equivalent to the torsor $(\mathbf{R}_\varepsilon, \mathbf{M}_\varepsilon)$. Applying theorem 1.2.4 of section 1.2.2.1, we may write

$$\lim_{\varepsilon \to 0} \mathbf{R}_\varepsilon = \mathbf{F}, \ \lim_{\varepsilon \to 0} \mathbf{M}_\varepsilon = 0, \tag{5.1.4}$$

which shows that the limit of the field of loads $\mathbf{Q}_\varepsilon$ is a concentrated force at the point A_0. Effecting the passage to the limit in the sense of the theory of distributions for the field $\mathbf{Q}_\varepsilon$, we obtain

$$\mathbf{Q}(x, y, z) = \lim_{\varepsilon \to 0} \mathbf{Q}_\varepsilon(x, y, z) = \mathbf{F}\delta(x - x_0, y - y_0, z - z_0) = \mathbf{F}\delta(\mathbf{r} - \mathbf{r}_0), \quad (5.1.5)$$

which is the mathematical representation of the concentrated force $\mathbf{F}$ of components F_x, F_y, F_z, applied at the point $A_0(x_0, y_0, z_0)$; we note that in defining the concentrated force the domain Ω_ε, with A_0 as an internal point, as well as the functions $q'_\varepsilon(x, y, z)$, $q''_\varepsilon(x, y, z)$, $q'''_\varepsilon(x, y, z)$, have played an auxiliary part only.

The mathematical representation (5.1.5) includes in a synthetic form all the characteristics of the concentrated force $\mathbf{F}$ considered as a *bound vector*, namely: *direction, modulus and location (point of application)*.

For example, the representative δ sequence

$$q_\varepsilon(x) = \begin{cases} \dfrac{1}{\varepsilon}\left(1 + \dfrac{x}{\varepsilon}\right) & \text{for } -\varepsilon \leqslant x \leqslant 0 \\[2mm] \dfrac{1}{\varepsilon}\left(1 - \dfrac{x}{\varepsilon}\right) & \text{for } 0 \leqslant x \leqslant \varepsilon \\[2mm] 0 & \text{for } |x| > \varepsilon \end{cases} \quad (5.1.6)$$

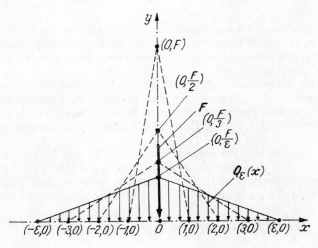

Fig. 5.1

permits the representation of a concentrated force as the limit of a sequence of loads distributed in a triangular form (Figure 5.1); we have

$$\mathbf{Q}_\varepsilon(x) = -F\mathbf{j}q_\varepsilon(x), \quad (5.1.7)$$

so that

$$\mathbf{R}_\varepsilon = -F\mathbf{j}, \quad \mathbf{M}_\varepsilon = 0, \tag{5.1.7'}$$

where $\mathbf{j}$ is the unit vector of the Oy axis (we have assumed that the concentrated force $\mathbf{F}$ acts at the point O in the direction of the Oy axis and has been obtained as the limit of loads acting in the plane Oxy).

5.1.1.2. Composition of concentrated forces

By the composition of concentrated forces of the type defined above we obtain forces of the same type; thus, the forces $\mathbf{F}_1, \mathbf{F}_2, \ldots, \mathbf{F}_n$ applied at the point A_0 are equivalent to the force $\mathbf{F} = \mathbf{F}_1 + \mathbf{F}_2 + \ldots + \mathbf{F}_n$ applied at the same point.

Indeed, the n forces are equivalent to the loads

$$\mathbf{Q}_i(x, y, z) = \mathbf{F}_i \delta(x - x_0, y - y_0, z - z_0) \qquad (i = 1, 2, \ldots, n), \tag{5.1.8}$$

by the composition of which we obtain

$$\mathbf{Q}(x, y, z) = \sum_{i=1}^{n} \mathbf{Q}_i(x, y, z) = \sum_{i=1}^{n} \mathbf{F}_i \delta(x - x_0, y - y_0, z - z_0). \tag{5.1.8'}$$

Also, by multiplying a bound force by a real number the point of application is not altered, a fact which is maintained in the mathematical representation (5.1.5); this shows that the representation of a concentrated force with the aid of the Dirac distribution is correct.

Although the composition of bound forces has a meaning, only when these have the same point of application, we may perform, in a certain sense, the composition of concentrated forces with different points of application; such operations are based on theorem 1.1.4 of section 1.1.2.7 concerning changes of variables in distributions.

Let us now assume that the concentrated forces $\mathbf{F}_1$ and $\mathbf{F}_2$ are *parallel and have the same direction* and that they are acting in the plane Oxy at the points $A_1(a_1, b)$ and $A_2(a_2, b)$, respectively (Figure 5.2). Denoting by $\mathbf{u}$ the unit vector of the two forces, their action is equivalent to the load $\mathbf{Q}$ the expression of which is

$$\mathbf{Q} = \mathbf{u}[F_1 \delta(x - a_1, y - b) + F_2 \delta(x - a_2, y - b)]. \tag{5.1.9}$$

We shall now consider the relation

$$F_1 \delta[\alpha_1(x - a_1), y - b] + F_2 \delta[\alpha_2(x - a_2), y - b]$$

$$= \frac{F_1}{\alpha_1} \delta(x - a_1, y - b) + \frac{F_2}{\alpha_2} \delta(x - a_2, y - b) \tag{5.1.10}$$

and we shall determine the numbers α_1, α_2 such as to satisfy the equality

$$\frac{F_1}{\alpha_1} = \frac{F_2}{\alpha_2}. \tag{5.1.11}$$

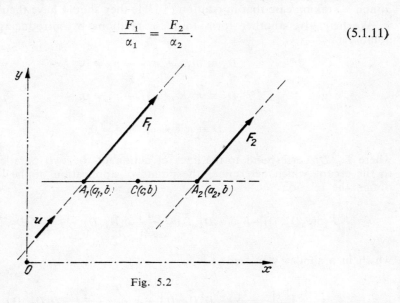

Fig. 5.2

If $C(c, b)$ is the centre of the two parallel forces, relation (5.1.11) expresses the equality of the moduli of the moments of the two forces about the point C, where α_1 and α_2 are the corresponding arms; the abscissa of the point C is given by

$$c = \frac{a_1 F_1 + a_2 F_2}{F_1 + F_2}, \tag{5.1.12}$$

so that

$$\alpha_1 = a_2 - c = \frac{(a_2 - a_1)F_1}{F_1 + F_2}, \quad \alpha_2 = c - a_1 = \frac{(a_2 - a_1)F_2}{F_1 + F_2}. \tag{5.1.12'}$$

Using the results of section 1.1.2.7, and relation (5.1.10), we obtain the equality

$$\mathbf{F}_1 \delta[\alpha_1(x - a_1), y - b] + \mathbf{F}_2 \delta[\alpha_2(x - a_2), y - b]$$

$$= (\mathbf{F}_1 + \mathbf{F}_2)\delta[(x - a_1)(x - a_2), y - b], \tag{5.1.13}$$

where the numbers α_1, α_2 satisfy relation (5.1.11). This formula constitutes a composition law for parallel forces considered as bound vectors; the composition law coincides with relation (5.1.9) only if $F_1 = F_2$, $\alpha_1 = \alpha_2 = 1$.

Since the Dirac distribution is even, we may consider also negative values for α_1 and α_2, taking care that in relation (5.1.11) they should have their absolute value.

We may give another form to these relations by introducing the following notations

$$D_1 \equiv m_1 x + n_1 y + p_1 = 0,$$

$$D_2 \equiv m_2 x + n_2 y + p_2 = 0, \qquad (5.1.14)$$

$$D \equiv mx + ny + p = 0,$$

where D_1, D_2 correspond to the lines of action of the two parallel forces and D to the secant which determines the points of application of the forces. We thus obtain the relation

$$F_1 \delta(\alpha_1 D_1, D) + F_2 \delta(\alpha_2 D_2, D) = \frac{F_1}{|\alpha_1|} \delta(D_1, D) + \frac{F_2}{|\alpha_2|} \delta(D_2, D), \qquad (5.1.15)$$

which in a similar way leads to

$$\mathbf{F}_1 \delta\left(\frac{\alpha_1}{m_2 - m_1} D_1 \bar{D}_2, D\right) + \mathbf{F}_2 \delta\left(\frac{\alpha_2}{m_2 - m_1} \bar{D}_1 D_2, D\right)$$

$$= (\mathbf{F}_1 + \mathbf{F}_2)\, \delta(D_1 D_2, D), \qquad (5.1.15')$$

where we have introduced the notations

$$\bar{D}_1 = m_1 a_2 + n_1 b + p,$$

$$\bar{D}_2 = m_2 a_1 + n_2 b + p. \qquad (5.1.14')$$

Let us now consider the case of two bound forces, *parallel and equal, of opposite directions* and of magnitude F (Figure 5.3). Their action is equivalent to

$$\mathbf{Q} = \mathbf{u} F[\delta(x - a_1, y - b) - \delta(x - a_2, y - b)]. \qquad (5.1.16)$$

Starting from relation (1.1.76') we obtain

$$(2x - a_1 - a_2)\, \delta[(x - a_1)(x - a_2),\, y - b]$$

$$= \operatorname{sgn}(a_1 - a_2)[\delta(x - a_1, y - b) - \delta(x - a_2, y - b)]; \qquad (5.1.17)$$

relation (5.1.16) becomes

$$\mathbf{Q} = \text{sgn}(a_1 - a_2)(2x - a_1 - a_2)F\mathbf{u}\delta[(x - a_1)(x - a_2), y - b], \qquad (5.1.18)$$

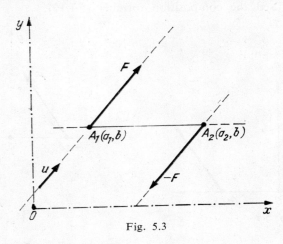

Fig. 5.3

which gives the composition formula

$$\mathbf{F}\delta(x - a_1, y - b) - \mathbf{F}\delta(x - a_2, y - b)$$

$$= 2\left(\frac{a_1 + a_2}{2} - x\right)\mathbf{F}\delta[(x - a_1)(x - a_2), y - b], \qquad (5.1.19)$$

where we have assumed $a_2 > a_1$.

Considering the more general case where $F_1 \neq F_2$, we obtain a relation of the form

$$\mathbf{F}_1\delta\left[\frac{\alpha_1}{(a_2 - a_1)^2}D_1\bar{D}_2, D\right] - \mathbf{F}_2\delta\left[\frac{\alpha_2}{(a_2 - a_1)^2}\bar{D}_1D_2, D\right]$$

$$= 2(\mathbf{F}_1 + \mathbf{F}_2)\left(\frac{a_1 + a_2}{2} - x\right)\delta(\Delta_1\Delta_2, \Delta). \qquad (5.1.20)$$

We remark that we can establish similar relations for the case where the forces are not in the plane Oxy; however, that is an unessential generalization.

In the particular case of Figure 5.4, formula (5.1.13) with $\alpha_1 = \alpha_2 = a$, $a > 0$, gives

$$\mathbf{F}\delta(x + a, y) + \mathbf{F}\delta(x - a, y) = 2\mathbf{F}a\delta(x^2 - a^2, y). \qquad (5.1.21)$$

Similarly, in the particular case of Figure 5.5 we may write

$$\mathbf{F}\delta(x + a, y) - \mathbf{F}\delta(x - a, y) = - 2\mathbf{F}x\delta(x^2 - a^2, y), \qquad (5.1.22)$$

where we have used the composition formula (5.1.19).

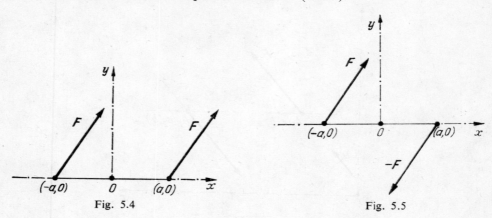

Fig. 5.4   Fig. 5.5

5.1.2. Mathematical representation of continuously distributed loads

In the following discussion we shall treat loads continuously distributed on certain varieties, especially loads represented by vectorial fields distributed on curves and surfaces. We shall show first how to represent, in general, such loads and then we will give a few important examples.

5.1.2.1. Arbitrary loads

Let

$$x = f(t), \quad y = g(t), \quad z = h(t) \qquad (5.1.23)$$

be the parametric equations of a curve, where the functions $f(t)$, $g(t)$ and $h(t)$ are assumed to be of class C^∞ and for which we may write

$$\lim_{t \to \pm \infty} [f^2(t) + g^2(t) + h^2(t)] = \infty. \qquad (5.1.24)$$

We consider also the vectorial field defined by

$$\tau(t) = \begin{cases} \mathbf{V}(t) \text{ for } t \in [a, b] \\ \\ 0 \quad \text{ for } t \notin [a, b]. \end{cases} \qquad (5.1.25)$$

In order to find the equivalent load of the vectorial field considered we divide the interval $[a, b]$ into n partial intervals of the form

$$a = t_0 < t_1 < t_2 < \ldots < t_{i-1} < t_i < t_{i+1} < \ldots < t_n = b \qquad (5.1.26)$$

and denote by $A \equiv M_0$, $B \equiv M_n$ and M_i the points on the curve (C) corresponding to the values $a = t_0$, $b = t_n$ and t_i $(i = 1, 2, \ldots, n - 1)$ of the parameter t, respectively.

Let us now associate to every arc $\overarc{M_{i-1}M_i}$ $(i = 1, 2, \ldots, n)$ of the length Δs_i, the concentrated load $\mathbf{F}_i$, expressed by

$$\mathbf{F}_i(x, y, z) = \Delta s_i \mathbf{V}(\overline{t_i}) \delta[x - f(\overline{t_i}), y - g(\overline{t_i}), z - h(\overline{t_i})], \qquad (5.1.27)$$

where $\overline{t_i}$ is an internal point of the interval $[t_{i-1}, t_i]$.

From the correspondence thus defined, it follows that to the arc $\overarc{AB}$ there corresponds the load

$$\mathbf{Q}_n(x, y, z) = \sum_{i=1}^{n} \Delta s_i \mathbf{V}(t_i) \delta[x - f(\overline{t_i}), y - g(\overline{t_i}), z - h(\overline{t_i})], \qquad (5.1.28)$$

which constitutes an approximation of the vector field $\boldsymbol{\tau}(t)$, defined by (5.1.25).

In order to obtain a representation of this vector field, we shall consider the limit in the sense of the theory of distributions of the load $\mathbf{Q}_n$; then we can write

$$\mathbf{Q}(x, y, z) = \lim_{\substack{n \to \infty \\ \Delta s_i \to 0}} \sum_{i=1}^{n} \Delta s_i \mathbf{V}(\overline{t_i}) \, \delta[x - f(\overline{t_i}), y - g(\overline{t_i}), z - h(\overline{t_i})]. \qquad (5.1.29)$$

We remark from (5.1.23) that if $\varphi(x, y, z)$ is a fundamental function, then

$$\psi(t) = \varphi[f(t), g(t), h(t)]$$

is a fundamental function too. We may write

$$(\mathbf{Q}_n(x, y, z), \varphi(x, y, z)) = \sum_{i=1}^{n} (\Delta s_i \mathbf{V}(\overline{t_i}) \, \delta[x - f(\overline{t_i}), y - g(\overline{t_i}), z - h(\overline{t_i})], \varphi(x, y, z))$$

$$= \sum_{i=1}^{n} \Delta s_i \mathbf{V}(\overline{t_i}) \, \varphi[f(\overline{t_i}), g(\overline{t_i}), h(\overline{t_i})],$$

from which it follows that the limit (5.1.29) exists and we have

$$\lim_{n \to \infty} (\mathbf{Q}_n(x, y, z), \varphi(x, y, z)) = (\mathbf{Q}(x, y, z), \varphi(x, y, z))$$

$$= \int_a^b \mathbf{V}(t) \sqrt{f'^2(t) + g'^2(t) + h'^2(t)} \, \varphi[f(t), g(t), h(t)] \, dt. \tag{5.1.30}$$

Introducing the function

$$\theta_1(t) = \begin{cases} 1 \text{ for } t \in [a, b] \\ \\ 0 \text{ for } t \notin [a, b], \end{cases} \tag{5.1.31}$$

we may write this relation in the form

$$(\mathbf{Q}(x, y, z), \quad \varphi(x, y, z))$$

$$= (\theta_1(t) \mathbf{V}(t) \sqrt{f'^2(t) + g'^2(t) + h'^2(t)}, \quad \varphi[f(t), g(t), h(t)]). \tag{5.1.30'}$$

We introduce now the auxilliary function

$$\boldsymbol{\alpha}_t(x, y, z) = \theta_1(t) \mathbf{V}(t) \sqrt{f'^2(t) + g'^2(t) + h'^2(t)} \, \theta[x - f(t)] \times \theta[y - g(t)] \times \theta[z - h(t)], \tag{5.1.32}$$

where $\theta(u)$ is the Heaviside function; since for regular functionals the direct product coincides with the usual product, we have

$$\boldsymbol{\alpha}_t(x, y, z) = \theta_t(x, y, z) \mathbf{V}(t) \sqrt{f'^2(t) + g'^2(t) + h'^2(t)}, \tag{5.1.32'}$$

where

$$\theta_t(x, y, z) = \theta_1(t) \theta[x - f(t)] \theta[y - g(t)] \theta[z - h(t)]$$

$$= \begin{cases} 1 \text{ for } x \geqslant f(t), \ y \geqslant g(t), z \geqslant h(t), \ t \in [a, b] \\ \\ 0 \text{ for } x < f(t) \text{ or } y < g(t) \text{ or } z < h(t) \text{ or } t \notin [a, b]. \end{cases} \tag{5.1.33}$$

Differentiating functions (5.1.32) or (5.1.32') in the sense of the theory of distributions, we obtain

$$\frac{\partial^3 \boldsymbol{\alpha}_t(x, y, z)}{\partial x \, \partial y \, \partial z} = \theta_1(t) \mathbf{V}(t) \sqrt{f'^2(t) + g'^2(t) + h'^2(t)} \, \delta[x - f(t), y - g(t), z - h(t)]. \tag{5.1.34}$$

Since the function $\alpha_t(x, y, z)$ is summable, when the variables x, y, z run over a bounded set and the parameter t runs over a bounded or unbounded set, we may write, based on the theorems of the integration of distributions depending on a parameter,

$$(\mathbf{Q}(x, y, z), \varphi(x, y, z)) = \int_{-\infty}^{\infty} \theta_1(t)\mathbf{V}(t)\sqrt{f'^2(t) + g'^2(t) + h'^2(t)}\,\varphi[f(t), g(t), h(t)]\,dt$$

$$= \int_{-\infty}^{\infty} (\theta_1(t)\mathbf{V}(t)\sqrt{f'^2(t) + g'^2(t) + h'^2(t)}\,\delta[x - f(t), y - g(t), z - h(t)], \varphi(x, y, z))\,dt$$

$$= \int_{-\infty}^{\infty} \left(\frac{\partial^3 \alpha_t(x, y, z)}{\partial x\,\partial y\,\partial z}, \varphi(x, y, z)\right)dt = \left(\int_{-\infty}^{\infty} \frac{\partial^3 \alpha_t(x, y, z)}{\partial x\,\partial y\,\partial z}\,dt, \varphi(x, y, z)\right)$$

$$= \left(\frac{\partial^3}{\partial x\,\partial y\,\partial z}\int_{-\infty}^{\infty} \alpha_t(x, y, z)\,dt, \varphi(x, y, z)\right);$$

from the definition of the equality of two distributions we obtain

$$\mathbf{Q}(x, y, z) = \frac{\partial^3}{\partial x\,\partial y\,\partial z}\int_{-\infty}^{\infty} \theta_t(x, y, z)\mathbf{V}(t)\sqrt{f'^2(t) + g'^2(t) + h'^2(t)}\,dt. \quad (5.1.35)$$

Formula (5.1.35) gives the *equivalent load of the vector field* $\tau(t)$ considered; this formula is the mathematical representation of the vector field $\tau(t)$ by the equivalent load $\mathbf{Q}(x, y, z)$.

If we denote by $V_x(t), V_y(t), V_z(t)$ the projections of the vector field $\mathbf{V}(t)$ defined by relation (5.1.25), then the *projections* $Q_x(x, y, z), Q_y(x, y, z), Q_z(x, y, z)$ of the *load* $\mathbf{Q}(x, y, z)$ *equivalent* to this vector field will be given by

$$Q_x(x, y, z) = \frac{\partial^3}{\partial x\,\partial y\,\partial z}\int_{-\infty}^{\infty} \theta_t(x, y, z)\,V_x(t)\sqrt{f'^2(t) + g'^2(t) + h'^2(t)}\,dt,$$

$$Q_y(x, y, z) = \frac{\partial^3}{\partial x\,\partial y\,\partial z}\int_{-\infty}^{\infty} \theta_t(x, y, z)\,V_y(t)\sqrt{f'^2(t) + g'^2(t) + h'^2(t)}\,dt, \quad (5.1.35')$$

$$Q_z(x, y, z) = \frac{\partial^3}{\partial x\,\partial y\,\partial z}\int_{-\infty}^{\infty} \theta_t(x, y, z)\,V_z(t)\sqrt{f'^2(t) + g'^2(t) + h'^2(t)}\,dt.$$

In particular, for a *plane curve* defined by the equations

$$x = f(t), \quad y = g(t) \quad\quad\quad (5.1.36)$$

and satisfying the condition

$$\lim_{t \to \pm \infty} [f^2(t) + g^2(t)] = \infty, \tag{5.1.37}$$

the expressions of the load which is equivalent to the vector field considered take the form

$$Q_x(x, y) = \frac{\partial^2}{\partial x\, \partial y} \int_{-\infty}^{\infty} \overline{\theta_t}(x, y)\, V_x(t)\, \sqrt{f'^2(t) + g'^2(t)}\, dt,$$

$$Q_y(x, y) = \frac{\partial^2}{\partial x\, \partial y} \int_{-\infty}^{\infty} \overline{\theta_t}(x, y)\, V_y(t)\, \sqrt{f'^2(t) + g'^2(t)}\, dt, \tag{5.1.38}$$

$$Q_z(x, y) = \frac{\partial^2}{\partial x\, \partial y} \int_{-\infty}^{\infty} \overline{\theta_t}(x, y)\, V_z(t)\, \sqrt{f'^2(t) + g'^2(t)}\, dt,$$

where

$$\overline{\theta_t}(x, y) = \theta_1(t)\, \theta\, [x - f(t)]\, \theta\, [y - g(t)]$$

$$= \begin{cases} 1 \;\; \text{for} \;\; x \geqslant f(t),\, y \geqslant g(t), \qquad t \in [a, b], \\[2mm] 0 \;\; \text{for} \;\; x < f(t) \;\; \text{or} \;\; y < g(t) \;\; \text{or} \;\; t \notin [a, b]. \end{cases} \tag{5.1.39}$$

In the case where the plane curve is defined by the equations

$$x = t, \; y = g(t) \tag{5.1.40}$$

and the condition

$$\lim_{t \to \pm \infty} [t^2 + g^2(t)] = \infty \tag{5.1.41}$$

is satisfied, the projections of the load equivalent to the vector field will be

$$Q_x(x, y) = \frac{\partial^2}{\partial x\, \partial y} \int_{-\infty}^{\infty} \overline{\overline{\theta_t}}(x, y)\, V_x(t)\, \sqrt{1 + g'^2((t)}\, dt,$$

$$Q_y(x, y) = \frac{\partial^2}{\partial x\, \partial y} \int_{-\infty}^{\infty} \overline{\overline{\theta_t}}(x, y)\, V_y(t)\, \sqrt{1 + g'^2(t)}\, dt, \tag{5.1.42}$$

$$Q_z(x, y) = \frac{\partial^2}{\partial x\, \partial y} \int_{-\infty}^{\infty} \overline{\overline{\theta_t}}(x, y)\, V_z(t)\, \sqrt{1 + g'^2(t)}\, dt,$$

where

$$\overline{\overline{\theta}}_t(x, y) = \theta_1(t)\, \theta(x - t)\, \theta[y - g(t)]$$

$$= \begin{cases} 1 \text{ for } x \geqslant t,\, y \geqslant g(t),\, t \in [a, b], \\\\ 0 \text{ for } x < t \text{ or } y < g(t) \text{ or } t \notin [a, b]. \end{cases} \qquad (5.1.43)$$

5.1.2.2. Vector fields defined on surfaces

Let

$$x = f(u, v), \quad y = g(u, v), \quad z = h(u, v) \qquad (5.1.44)$$

be the parametric equations of a surface, where the functions $f(u, v)$, $g(u, v)$, $h(u, v)$ are of class C^∞ and where the condition

$$\lim_{u^2 + v^2 \to \infty} [f^2(u, v) + g^2(u, v) + h^2(u, v)] = \infty \qquad (5.1.45)$$

is satisfied; also, let

$$\tau(u, v) = \begin{cases} \mathbf{V}(u, v) \text{ for } (u, v) \in D \\\\ 0 \quad \text{ for } (u, v) \notin D \end{cases} \qquad (5.1.46)$$

be a vector field, where D is the domain of definition of the parameters u and v.

Proceeding as before, we find that this *vector field* is *equivalent to the load*

$$\mathbf{Q}(x, y, z) = \frac{\partial^3}{\partial x\, \partial y\, \partial z} \int_{-\infty}^{\infty} \int_{-\infty}^{\infty} \theta_{uv}(x, y, z)\, \mathbf{V}(u, v)\, \sqrt{E(u, v)\, G(u, v) - F^2(u, v)}\, du\, dv,$$

$$(5.1.47)$$

where

$$\theta_{uv}(x, y, z) = \theta_1(u, v)\, \theta[x - f(u, v)]\, \theta[y - g(u, v)]\, \theta[z - h(u, v)], \qquad (5.1.48)$$

with

$$\theta_1(u, v) = \begin{cases} 1 \text{ for } (u, v) \in D \\\\ 0 \text{ for } (u, v) \notin D; \end{cases} \qquad (5.1.49)$$

$\theta(w)$ is the Heaviside function and $E(u, v)$, $F(u, v)$, $G(u, v)$ are the differential parameters

$$E(u, v) = f_u'^2(u, v) + g_u'^2(u, v) + h_u'^2(u, v),$$

$$F(u, v) = f_u'(u, v) f_v'(u, v) + g_u'(u, v) g_v'(u, v) + h_u'(u, v) h_v'(u, v), \qquad (5.1.50)$$

$$G(u, v) = f_v'^2(u, v) + g_v'^2(u, v) + h_v'^2(u, v).$$

5.1.2.3. Examples

Let there be a plane curve defined by the equations

$$x = t, \, y = y_0 = \text{const} \qquad (5.1.51)$$

and the vector field defined by (Figure 5.6)

$$V_x(t) = \frac{F_x}{2c}, \quad V_y(t) = \frac{F_y}{2c}, \quad V_z(t) = \frac{F_z}{2c}, t \in [-c, c], \qquad (5.1.52)$$

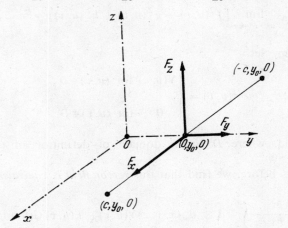

Fig. 5.6

where F_x, F_y, F_z are constant magnitudes. The expressions (5.1.42) take the form

$$Q_x(x, y; c) = \frac{\partial^2}{\partial x \partial y} \int_{-c}^{c} \frac{F_x}{2c} \theta(x - t)\theta(y - y_0) \, dt,$$

$$Q_y(x, y; c) = \frac{\partial^2}{\partial x \partial y} \int_{-c}^{c} \frac{F_y}{2c} \theta(x - t)\theta(y - y_0) \, dt, \qquad (5.1.53)$$

$$Q_z(x, y; c) = \frac{\partial^2}{\partial x \partial y} \int_{-c}^{c} \frac{F_z}{2c} \theta(x - t)\theta(y - y_0) \, dt.$$

Passing to the limit in the sense of the theory of distributions for $c \to 0$ and taking into account relation (1.3.15) we may write

$$Q_x(x, y) = \lim_{c \to 0} Q_x(x, y; c) = \frac{\partial^2}{\partial x \partial y}[F_x \theta(x) \times \theta(y - y_0)] = F_x \delta(x, y - y_0),$$

$$Q_y(x, y) = \lim_{c \to 0} Q_y(x, y; c) = \frac{\partial^2}{\partial x \partial y}[F_y \theta(x) \times \theta(y - y_0)] = F_y \delta(x, y - y_0), \quad (5.1.54)$$

$$Q_z(x, y) = \lim_{c \to 0} Q_z(x, y; c) = \frac{\partial^2}{\partial x \partial y}[F_z \theta(x) \times \theta(y - y_0)] = F_z \delta(x, y - y_0),$$

which is the expression of the *concentrated load* **F** applied at the point $A(0, y_0)$.

Let us now consider in the plane Oxy, a square $ABCD$, with the side $2a$, the centre at the origin and the sides parallel to the co-ordinate axes (Figure 5.7a).

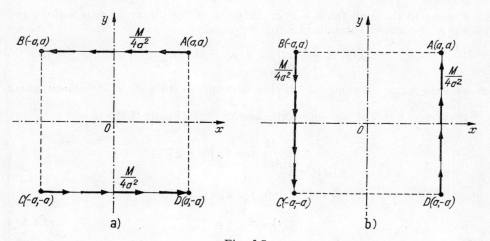

Fig. 5.7

We define on the side AB a vector field

$$V_x(t) = \begin{cases} -\dfrac{M}{4a^2} & \text{for } t \in [-a, a] \\[2ex] 0 & \text{for } t \notin [-a, a] \end{cases} \qquad (5.1.55)$$

$$V_y(t) = V_z(t) = 0; \qquad (5.1.55')$$

for the CD side we assume that

$$\overline{V}_x(t) = - V_x(t), \tag{5.1.56}$$

$$\overline{V}_y(t) = \overline{V}_z(t) = 0. \tag{5.1.56'}$$

The vector field defined on the sides AB and CD is equivalent to the load (component different from zero along Ox axis only)

$$Q_x(x, y; a) = \frac{\partial^2}{\partial x \partial y} \left\{ \frac{M}{4a^2} \left[\theta(y + a) - \theta(y - a) \right] \int_{-a}^{a} \theta(x - t) \, dt \right\}, \tag{5.1.57}$$

which may be written also

$$Q_x(x, y; a) = \frac{M}{2a} \left[\delta(y + a) - \delta(y - a) \right] \times \int_{-a}^{a} \frac{\delta(x - t)}{2a} \, dt. \tag{5.1.57'}$$

Passing to the limit for $a \to 0$ in the sense of the theory of distributions and taking into account formula (1.3.16), we obtain

$$Q_x(x, y) = \lim_{a \to 0} Q_x(x, y; a) = M\delta_y'(x, y); \tag{5.1.58}$$

as we shall see in section 5.2.1.1, this corresponds to a *directed moment* along the Ox axis.

If on the side DA we define the vector field (Figure 5.7b)

$$V_y(t) = \begin{cases} \dfrac{M}{4a^2} \text{ for } t \in [-a, a] \\[2mm] 0 \text{ for } t \notin [-a, a], \end{cases} \tag{5.1.59}$$

$$V_x(t) = V_z(t) = 0 \tag{5.1.59'}$$

and on the side BC the vector field

$$\overline{V}_y(t) = - V_y(t), \tag{5.1.60}$$

$$\overline{V}_x(t) = \overline{V}_z(t) = 0, \tag{5.1.60'}$$

we obtain by passing to the limit, for $a \to 0$, the load

$$Q_y(x, y) = \lim_{a \to 0} (x, y; a) = - M\delta_x'(x, y), \tag{5.1.61}$$

which corresponds to a directed moment along the Ox axis

By the superposition of effects, we find that if a vector field is defined on the sides of a square $ABCD$ as above (Figure 5.8), we obtain, when passing to the limit, for $a \to 0$, the loads

$$Q_x(x, y) = M\delta'_y(x, y), \quad Q_y(x, y) = -M\delta'_x(x, y); \tag{5.1.62}$$

this corresponds to a *concentrated moment (centre of rotation)* which acts in the positive direction of rotation at the origin of the co-ordinate axes.

We can also show that the vector field defined in Figure 5.7a may be expressed in the form

$$Q_x(x, y; a) = \frac{\partial}{\partial y}\left[\frac{M}{4a^2} h(x, y)\right], \tag{5.1.63}$$

where

$$h(x, y) = \begin{cases} 1 \text{ for } x \in [-a, a], y \in [-a, a] \\ \\ 0 \text{ for } x \notin [-a, a] \text{ or } y \notin [-a, a]; \end{cases} \tag{5.1.64}$$

formula (5.1.63) is a consequence of the general formula (5.1.35′). If we take into account the representative δ sequence (1.2.109′) we find again formula (5.1.58). We may proceed likewise for the formula (5.1.61).

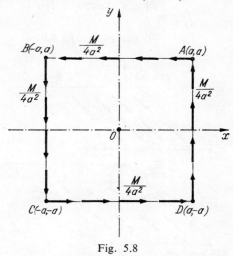

Fig. 5.8

Let us consider a simply connected domain Ω in the plane $z = z_0$ and the locally integrable vector field defined by

$$\tau(x, y, z_0) = \begin{cases} V(x, y, z_0) \text{ for } (x, y, z_0) \in \Omega \\ \\ 0 \qquad \text{for } (x, y, z_0) \notin \Omega; \end{cases} \tag{5.1.65}$$

applying the same reasoning as in subsection 5.1.2.2, we obtain the equivalent load

$$\mathbf{Q}(x, y, z) = \frac{\partial}{\partial z}[\theta_1(x, y, z)\mathbf{V}(x, y, z_0)], \tag{5.1.66}$$

where

$$\theta_1(x, y, z) = \bar{\theta}(x, y)\theta(z - z_0), \tag{5.1.67}$$

with

$$\bar{\theta}(x, y) = \begin{cases} 1 \text{ for } (x, y) \in \bar{\Omega} \\[2mm] 0 \text{ for } (x, y) \notin \bar{\Omega}, \end{cases} \tag{5.1.68}$$

where $\theta(w)$ is the Heaviside function and $\bar{\Omega}$ the projection of the domain Ω on the plane Oxy.

In particular, we consider a cube $ABCDA'B'C'D'$ with the side $2a$, the faces parallel to the co-ordinate axes and the centre at the origin O (Figure 5.9).

If the vector field

$$\tau(x, y, a) = \begin{cases} P\mathbf{i} \text{ for } (x, y, a) \in ABCD \\[2mm] 0 \text{ for } (x, y, a) \notin ABCD, \end{cases} \tag{5.1.69}$$

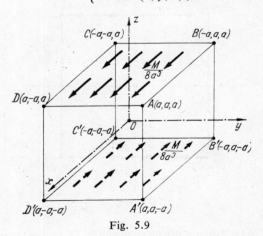

Fig. 5.9

where $\mathbf{i}$ is the unit vector on the Ox axis, acts on the face $ABCD$, we obtain for the equivalent load

$$\mathbf{Q}(x, y, z) = \frac{\partial}{\partial z}[\theta_1(x, y, z)P\mathbf{i}], \tag{5.1.70}$$

i.e.

$$Q_x(x, y, z) = \frac{\partial}{\partial z}[P\theta_1(x, y, z)], Q_y(x, y, z) = Q_z(x, y, z) = 0. \quad (5.1.70')$$

If a vector field, equal but of opposite direction, acts on the face $A'B'C'D'$ (Figure 5.9) and if we superpose the effects $\left(\text{we replace the magnitude } P \text{ by the magnitude } \dfrac{M}{8a^3}\right)$, we obtain finally

$$Q_x(x, y, z; a) = -\frac{\partial}{\partial z}\left[\frac{M}{8a^3} h(x, y, z)\right], \quad (5.1.71)$$

where

$$h(x, y, z) = \begin{cases} 1 \text{ for } x \in [-a, a], y \in [-a, a], z \in [-a, a] \\ 0 \text{ for } x \notin [-a, a] \text{ or } y \notin [-a, a] \text{ or } z \notin [-a, a]. \end{cases} \quad (5.1.72)$$

Passing to the limit, for $a \to 0$, in the sense of the theory of distributions, and taking into account formula (1.2.110'), we obtain

$$Q_x(x, y, z) = \lim_{a \to 0} Q_x(x, y, z; a) = -M\delta_z'(x, y, z), \quad (5.1.73$$

which leads again to a *directed moment* along the Ox axis.

We may obtain in a similar way a representation of a moment directed along the Oz axis and of a *rotational concentrated moment* (centre of rotation) in the plane Ozx; thus we see that the two representations are different as we consider a two- or a three-dimensional space.

5.2. Mathematical representation of concentrated loads

In the following paragraph we shall treat the case of concentrated loads; such loads play an important part by themselves and they also help in solving problems occurring in any other type of loading. Therefore, we shall first present the case of directed concentrated moments, with the help of which one can construct rotational concentrated moments (centres of rotation); then we shall consider the case of concentrated moments of dipole type.

5.2.1. Directed concentrated moments

Starting from the notion of concentrated force, we can construct directed concentrated moments of the first order as well as directed concentrated moments of a higher order; this will allow the introduction of directed moments, continuously distributed over a curve. Finally, using the notion of a directed concentrated moment, we shall study the problem of the equivalence of the action of forces; also, we shall be able to point out the highest order of a directed concentrated moment that may occur, depending on the nature of the body on which it is acting.

5.2.1.1. Directed concentrated moments of the first order

Usually, in the theory of sliding vectors, by a couple we mean a set of two vectors of the same modulus but with opposite directions; the lines of action of the two vectors must not coincide (if they do, the couple vanishes). The couple is characterized by its moment, which is a vector, normal to the plane determined by the lines of action of the two component vectors; its magnitude is equal to the area of the parallelogram formed by one of the vectors and the arm of the couple which connects the points of application of the two vectors (irrespective of their location on the lines of action); its direction is such that it corresponds to a positive rotation in the plane of the two vectors (Figure 5.10). Such a couple is characteristic of a rigid body.

In the case of a continuously deformable body, however, we have to consider sets of bound vectors; in that case a couple which results from a process involving a passage to the limit (the arm of the couple tends to zero) vanishes no longer. We are thus led to introduce the notion of *directed concentrated moment*.

Let $[-\mathbf{F}(-F_x, -F_y, -F_z)]$ be a concentrated force acting at the fixed point $A(x_0, y_0, z_0)$, and $\mathbf{F}_1(F_x, F_y, F_z)$ a concentrated force acting at the variable point

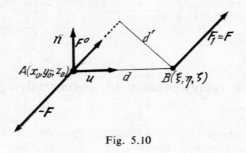

Fig. 5.10

$B(\xi, \eta, \zeta)$; evidently we can write $\mathbf{F}_1 = \mathbf{F}$ as free vectors. We denote by $\mathbf{d} = \overrightarrow{AB}$ the arm of the corresponding couple, by $\mathbf{F}^0$ the unit vector of $\mathbf{F}$, and by $\mathbf{u}$ the unit vector of the arm of the couple (Figure 5.10).

Let us introduce the following

Definition 5.2.1. *The direct concentrated moment at the point A is the limit, in the sense of the theory of distributions, of the set of concentrated forces $(-\mathbf{F}, \mathbf{F}_1)$, when the arm of the couple d tends to zero. The point A is considered fixed and the point B variable; we admit that the unit vectors $\mathbf{F}^0$ and $\mathbf{u}$ are constant and that the magnitude M of the moment is constant.*

The distance d' between the two lines of action is

$$d' = d|\mathbf{u} \times \mathbf{F}^0| \tag{5.2.1}$$

and the magnitude M of the moment is

$$M = Fd'. \tag{5.2.2}$$

The loads $\mathbf{Q}'$ and $\mathbf{Q}''$, equivalent to the concentrated forces applied at the points A and B, are

$$\mathbf{Q}'(x, y, z) = -\mathbf{F}\delta(x - x_0, y - y_0, z - z_0),$$

$$\mathbf{Q}''(x, y, z) = \mathbf{F}\delta(x - \xi, y - \eta, z - \zeta); \tag{5.2.3}$$

hence the set $(-\mathbf{F}, \mathbf{F}_1)$ will be equivalent to the load

$$\mathbf{Q}(x, y, z; d) = \mathbf{Q}'(x, y, z) + \mathbf{Q}''(x, y, z)$$

$$= \mathbf{F}[\delta(x - \xi, y - \eta, z - \zeta) - \delta(x - x_0, y - y_0, z - z_0)]$$

$$= \frac{M\mathbf{F}^0}{|\mathbf{u} \times \mathbf{F}^0|} \frac{\delta(x - \xi, y - \eta, z - \zeta) - \delta(x - x_0, y - y_0, z - z_0)}{d}, \tag{5.2.3'}$$

where account has been taken of relations (5.2.1) and (5.2.2).

The load $\mathbf{Q}$ corresponding to the directed moment will therefore be

$$\mathbf{Q}(x, y, z) = \lim_{d \to 0} \mathbf{Q}(x, y, z; d). \tag{5.2.4}$$

Denoting by $\cos \alpha_1$, $\cos \alpha_2$, $\cos \alpha_3$ the projections of the unit vector $\mathbf{u}$, we may introduce the *directional derivative in the direction of* $\mathbf{u}$ *of the Dirac distribution* by the relation

$$\lim_{d \to 0} \frac{\delta(x - x_0, y - y_0, z - z_0) - \delta(x - \xi, y - \eta, z - \zeta)}{d}$$

$$= \frac{\partial}{\partial u} \delta(x - x_0, y - y_0, z - z_0) = \cos \alpha_1 \delta'_x(x - x_0, y - y_0, z - z_0)$$

$$+ \cos \alpha_2 \delta'_y(x - x_0, y - y_0, z - z_0) + \cos \alpha_3 \delta'_z(x - x_0, y - y_0, z - z_0). \tag{5.2.5}$$

In that case, *the load corresponding to the directed concentrated moment* is expressed by

$$\mathbf{Q}(x, y, z) = -\frac{M\mathbf{F}^0}{|\mathbf{u} \times \mathbf{F}^0|} \frac{\partial}{\partial u} \delta(x - x_0, y - y_0, z - z_0);\qquad (5.2.4')$$

the load components are

$$Q_x(x, y, z) = -\frac{M \cos \beta_1}{|\mathbf{u} \times \mathbf{F}^0|} \frac{\partial}{\partial u} \delta(x - x_0, y - y_0, z - z_0),$$

$$Q_y(x, y, z) = -\frac{M \cos \beta_2}{|\mathbf{u} \times \mathbf{F}^0|} \frac{\partial}{\partial u} \delta(x - x_0, y - y_0, z - z_0),\qquad (5.2.4'')$$

$$Q_z(x, y, z) = -\frac{M \cos \beta_3}{|\mathbf{u} \times \mathbf{F}^0|} \frac{\partial}{\partial u} \delta(x - x_0, y - y_0, z - z_0),$$

where $\cos \beta_1$, $\cos \beta_2$, $\cos \beta_3$ are the projections of the unit vector $\mathbf{F}^0$. It is assumed that $\mathbf{u}$ and $\mathbf{F}^0$ are not collinear, hence

$$\mathbf{u} \times \mathbf{F}^0 \neq 0.\qquad (5.2.4''')$$

From relations (5.2.4') and (5.2.4'') we see that the directed concentrated moment is characterized by:
(*i*) the location $A(x_0, y_0, z_0)$;
(*ii*) the unit vector $\mathbf{F}^0$ of the forces which generate the moment;
(*iii*) the unit vector $\mathbf{u}$ of the direction in which the passage to the limit is effected (one observes that different results are obtained, depending on the choice of the location B on the line of action of the force $\mathbf{F}$);
(*iv*) the magnitude M of the moment.
The direction of the moment results from the above data and is specified by the unit vector $\mathbf{n}$ of the normal to the plane in which the directed concentrated moment is acting; the unit vector is chosen such that the mixed product $(\mathbf{u}, \mathbf{F}^0, \mathbf{n}) > 0$. The directed concentrated moment thus defined is positive since the rotation, as seen from the unit vector $\mathbf{n}$, occurs in the positive direction.
An important particular case is that where the unit vectors $\mathbf{u}$ and $\mathbf{F}^0$ are perpendicular to each other ($\mathbf{u} \cdot \mathbf{F}^0 = 0$); the corresponding equivalent load is

$$\mathbf{Q}(x, y, z) = -M\mathbf{F}^0 \frac{\partial}{\partial u} \delta(x - x_0, y - y_0, z - z_0).\qquad (5.2.6)$$

For example, if we take the origin of the co-ordinate axes as point of application and $\mathbf{F}^0 = -\mathbf{i}$, $\mathbf{u} = \mathbf{j}$, where $\mathbf{i}$, $\mathbf{j}$, $\mathbf{k}$ are the unit vectors of the co-ordinate axes, we obtain a directed concentrated moment, which induces a positive rotation in the plane Oxy (Figure 5.11) and is expressed by the equivalent load

$$\mathbf{Q}(x, y, z) = M\mathbf{i}\delta_y'(x, y, z). \tag{5.2.6'}$$

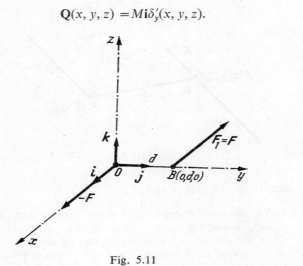

Fig. 5.11

The directed concentrated moment introduced above is a *moment of the first order;* therefore, its magnitude and its equivalent load will be designated, when necessary, by the superscript index 1 (written $M^{(1)}$ and $Q^{(1)}$).

5.2.1.2. Directed concentrated moments of higher order

In the definition of the directed concentrated moment of the first order, an important part is played by the set of forces $(-\mathbf{F}, \mathbf{F}_1)$; if these forces are replaced by directed concentrated moments of the first order, we can obtain by a similar procedure a *directed concentrated moment of the second order*.

Let $[-\mathbf{Q}^{(1)}]$ be the equivalent load acting at the fixed point $A(x_0, y_0, z_0)$ and $\mathbf{Q}_1^{(1)}$ the equivalent load acting at the variable point $B_2(\xi, \eta, \zeta)$; obviously, we can write $\mathbf{Q}_1^{(1)} = \mathbf{Q}^{(1)}$ as free vectors (Figure 5.12). Let $\mathbf{F}^0$ be the unit vector of the force $\mathbf{F}$, with the help of which we have defined the equivalent load $\mathbf{Q}^{(1)}$, and let $\mathbf{u}_1$ be the unit vector of the respective arm of the couple. We shall denote by $\mathbf{d}_2 = \overrightarrow{AB_2}$ the arm of the couple $(-\mathbf{Q}^{(1)}, \mathbf{Q}_1^{(1)})$ and by $\mathbf{u}_2$ its unit vector; the distance d_2' between the two lines of action is given by*

$$d_2' = d_2|\mathbf{u}_2 \times \mathbf{F}^0| \tag{5.2.7}$$

---

* Directed concentrated moments of the first order are designated by the superscribed index 1.

and the magnitude of the moment of the second order is

$$M^{(2)} = M^{(1)}d_2'. \qquad (5.2.8)$$

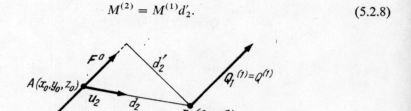

Fig. 5.12

We now introduce the following

Definition 5.2.2. *The directed concentrated moment of the second order at a point A is the limit, in the sense of the theory of distributions, of the set of equivalent loads* $(-\mathbf{Q}^{(1)}, \mathbf{Q}_1^{(1)})$, *when the arm of the couple d_2 tends to zero; it is assumed that the point A is fixed, the point B_2 is variable, the unit vectors $\mathbf{F}^0$ and $\mathbf{u}_2$ are constant, and the magnitude $M^{(2)}$ of the moment is constant.*

The set of loads $(-\mathbf{Q}^{(1)}, \mathbf{Q}_1^{(1)})$ leads to the equivalent load

$$\mathbf{Q}^{(2)}(x, y, z; d_2)$$

$$= -\frac{M^{(1)}\mathbf{F}^0}{|\mathbf{u}_1 \times \mathbf{F}^0|}\frac{\partial}{\partial u_1}\left[\delta(x - \xi, y - \eta, z - \zeta) - \delta(x - x_0, y - y_0, z - z_0)\right], \qquad (5.2.9)$$

which, by passing to the limit, gives

$$\mathbf{Q}^{(2)}(x, y, z) = \lim_{d_2 \to 0} \mathbf{Q}^{(2)}(x, y, z; d_2). \qquad (5.2.10)$$

We remark that from the limit

$$\lim_{d_2 \to 0}\frac{1}{d_2}\frac{\partial}{\partial u_1}\left[\delta(x - x_0, y - y_0, z - z_0) - \delta(x - \xi, y - \eta, z - \zeta)\right]$$

$$= \frac{\partial^2}{\partial u_1 \partial u_2}\delta(x - x_0, y - y_0, z - z_0) \qquad (5.2.11)$$

there results a *mixed directional derivative of the second order*, the directions being defined by the unit vectors $\mathbf{u}_1$ and $\mathbf{u}_2$; hence

$$\mathbf{Q}^{(2)}(x, y, z) = \frac{M^{(2)}\mathbf{F}^0}{|\mathbf{u}_1 \times \mathbf{F}^0|\,|\mathbf{u}_2 \times \mathbf{F}^0|}\; \frac{\partial^2}{\partial u_1 \partial u_2}\, \delta(x - x_0, y - y_0, z - z_0). \quad (5.2.10')$$

Thus we obtain the *equivalent load of a directed concentrated moment of the second order.*

Similarly, let

$$M^{(n)} = F d_1' d_2' \dots d_n' \qquad (5.2.12)$$

be the magnitude, of the *moment of order n*, expressed by the arms of couple corresponding to the moments of an order lower than n; we assume that the moment is applied at the point $A(x_0, y_0, z_0)$, which is considered fixed, and we denote by $\mathbf{u}_i (i = 1, \dots, n)$ the unit vectors defining the directions in which the passages to the limit take place.

Let us apply the equivalent load $[-\mathbf{Q}^{(n-1)}]$ at the fixed point $A(x_0, y_0, z_0)$ and the equivalent load $\mathbf{Q}_1^{(n-1)}$ at the variable point $B_n(\xi, \eta, \zeta)$, so that we may write $\mathbf{Q}_1^{(n-1)} = \mathbf{Q}^{(n-1)}$ from the viewpoint of free vectors, the arm of the couple being $\mathbf{d}_n = \overrightarrow{AB_n}$ (Figure 5.13); then we may introduce

Definition 5.2.3. *The directed concentrated moment of order n at the point A is the limit, in the sense of the theory of distributions, of the set of equivalent loads $(-\mathbf{Q}^{(n-1)}, \mathbf{Q}_1^{(n-1)})$, when d_n tends to zero, the point A being fixed and the point B_n variable; it is assumed that the unit vectors $\mathbf{F}^0$ and $\mathbf{u}_i (i = 1, 2, \dots, n)$ are constant and that the magnitude $M^{(n)}$ of the moment is constant.*

Then we can write

$$\mathbf{Q}^{(n)}(x, y, z) = \lim_{d_n \to 0} \mathbf{Q}^{(n)}(x, y, z; d_n) = \lim_{d_n \to 0} (-\mathbf{Q}^{(n-1)}(x, y, z), \mathbf{Q}_1^{(n-1)}(x, y, z)); \quad (5.2.13)$$

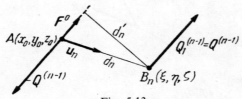

Fig. 5.13

applying the method of complete induction and using the foregoing results, we obtain the *equivalent load of a directed concentrated moment of order n*

$$\mathbf{Q}^{(n)}(x, y, z) = (-1)^n \frac{M^{(n)}\mathbf{F}^0}{\displaystyle\prod_{i=1}^{n} |\mathbf{u}_i \times \mathbf{F}^0|}\; \frac{\partial^n}{\partial u_1 \partial u_2 \dots \partial u_n}\, \delta(x - x_0, y - y_0, z - z_0), \quad (5.2.13')$$

where $\mathbf{u}_i$ are the unit vectors, corresponding to the directional derivatives mentioned above.

In particular, if we assume $\mathbf{u}_i = \mathbf{u}$, the expression of the equivalent load of the directed concentrated moment of order n becomes

$$\mathbf{Q}^{(n)}(x, y, z) = (-1)^n \frac{M^{(n)}\mathbf{F}^0}{|\mathbf{u} \times \mathbf{F}^0|^n} \frac{\partial^n}{\partial u^n} \delta(x - x_0, y - y_0, z - z_0); \qquad (5.2.14)$$

if we have also $\mathbf{u} = \mathbf{j}$ and $\mathbf{F}^0 = -\mathbf{i}$, we obtain

$$\mathbf{Q}^{(n)}(x, y, z) = (-1)^{n-1} M^{(n)}\mathbf{i} \frac{\partial^n}{\partial y^n} \delta(x, y, z), \qquad (5.2.15)$$

where the point of application is the origin of the co-ordinate axes.

We have given a general definition of a mixed directed concentrated moment of order n; the particular cases considered correspond to directed concentrated moments resulting from a single direction. They may prove very useful in many problems occurring in computation practice.

5.2.1.3. Directed moments continuously distributed over a curve

Let (C) be a curve (Figure 5.14) defined by the parameter equations

$$x = f(s), y = g(s), z = h(s), \qquad (5.2.16)$$

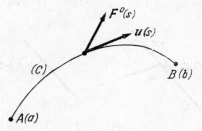

Fig. 5.14

where the functions $f(s)$, $g(s)$, $h(s)$ are assumed to be of class C^∞ with respect to the parameter s, which represents an arc of the curve, and such that

$$\lim_{s \to +\infty} [f^2(s) + g^2(s) + h^2(s)] = \infty. \qquad (5.2.17)$$

We assume that on the unit length of the arc $\overset{\frown}{AB}$, corresponding to the point of abscissa $s \in [a, b]$, there are acting directed concentrated moments of the first order, whose equivalent load is

$$\mathbf{Q}^{(1)}(s) = -\frac{M\mathbf{F}^0}{|\mathbf{u} \times \mathbf{F}^0|} \frac{\partial}{\partial s} \delta[x - f(s), y - g(s), z - h(s)], \qquad (5.2.18)$$

where M and $\mathbf{F}^0$ are known functions of s and $\mathbf{u} = \mathbf{u}(s)$ is the unit vector of the tangent to the arc of the curve. In that case we may associate the arc $\overset{\frown}{AB}$ with the equivalent load

$$\mathbf{Q}(x, y, z) = \int_a^b \mathbf{Q}^{(1)}(s) \, ds. \qquad (5.2.19)$$

If $\varphi(x, y, z)$ is a fundamental function we may write

$$(\mathbf{Q}(x, y, z), \varphi(x, y, z))$$

$$= \int_a^b \left(-\frac{M\mathbf{F}^0}{|\mathbf{u} \times \mathbf{F}^0|} \frac{\partial}{\partial s} \delta[x - f(s), y - g(s), z - h(s)], \varphi(x, y, z) \right) ds; \quad (5.2.20)$$

taking into account

$$\left(\frac{\partial}{\partial s} \delta[x - f(s), y - g(s), z - h(s)], \varphi(x, y, z) \right)$$

$$= -\frac{\partial}{\partial s} \varphi[f(s), g(s), h(s)],$$

it follows that

$$(\mathbf{Q}(x, y, z), \varphi(x, y, z)) = \int_a^b \frac{M\mathbf{F}^0}{|\mathbf{u} \times \mathbf{F}^0|} \frac{\partial}{\partial s} \varphi[f(s), g(s), h(s)] \, ds. \qquad (5.2.20')$$

Integrating by parts, we have

$$(\mathbf{Q}(x, y, z), \varphi(x, y, z)) = \left[\frac{M\varphi[f(s), g(s), h(s)]}{|\mathbf{u} \times \mathbf{F}^0|} \mathbf{F}^0 \right]_a^b$$

$$- \int_a^b \varphi[f(s), g(s), h(s)] \frac{\partial}{\partial s} \frac{M\mathbf{F}^0}{|\mathbf{u} \times \mathbf{F}^0|} \, ds, \qquad (5.2.21)$$

and introducing the function

$$\theta_1(s) = \begin{cases} 1 \text{ for } s \in [a, b] \\ \\ 0 \text{ for } s \notin [a, b], \end{cases} \tag{5.2.22}$$

we obtain

$$(\mathbf{Q}(x, y, z), \varphi(x, y, z)) = \left[\frac{M\varphi[f(s), g(s), h(s)]}{|\mathbf{u} \times \mathbf{F}^0|} \mathbf{F}^0 \right]_a^b$$

$$- \left(\theta_1(s) \frac{\partial}{\partial s} \frac{M\mathbf{F}}{|\mathbf{u} \times \mathbf{F}^0|}, \varphi[f(s), g(s), h(s)] \right). \tag{5.2.21'}$$

Finally, we obtain

$$\mathbf{Q}(x, y, z) = \frac{M(b)\mathbf{F}^0(b)}{|\mathbf{u} \times \mathbf{F}^0|_{s=b}} \delta(x - x_B, y - y_B, z - z_B)$$

$$- \frac{M(a)\mathbf{F}^0(a)}{|\mathbf{u} \times \mathbf{F}^0|_{s=a}} \delta(x - x_A, y - y_A, z - z_A)$$

$$- \frac{\partial^3}{\partial x \partial y \partial z} \int_{-\infty}^{\infty} \bar{\theta}_1(s) \frac{\partial}{\partial s} \frac{M\mathbf{F}^0}{|\mathbf{u} \times \mathbf{F}^0|} ds, \tag{5.2.21''}$$

where

$$\bar{\theta}_1(s) = \theta_1(s)\theta[x - f(s), y - g(s), z - h(s)]; \tag{5.2.22'}$$

$\theta(u)$ is Heaviside's function.

In particular, if the arc s is taken as a curvilinear co-ordinate, the expression of the equivalent load of the directed moment of the first order, concentrated at the point s_0, will be

$$\mathbf{Q}^{(1)}(s) = - \frac{M(s)\mathbf{F}^0(s)}{|\mathbf{u}(s) \times \mathbf{F}^0(s)|} \frac{\partial}{\partial s} \delta(s - s_0). \tag{5.2.18'}$$

Therefore the expression (5.2.21'') of *the equivalent load of the directed moment continuously distributed over the curve* (C) becomes

$$\mathbf{Q}(s) = \frac{M(b)\mathbf{F}^0(b)}{|\mathbf{u}(b) \times \mathbf{F}^0(b)|} \delta(s - b) - \frac{M(a)\mathbf{F}^0(a)}{|\mathbf{u}(a) \times \mathbf{F}^0(a)|} \delta(s - a)$$

$$- \theta_1(s) \frac{\partial}{\partial s} \frac{M(s)\mathbf{F}^0(s)}{|\mathbf{u}(s) \times \mathbf{F}^0(s)|}. \tag{5.2.21'''}$$

For example, if uniformly-distributed directed moments of magnitude M are acting on a segment AB of the axis Ox, and the unit vector $\mathbf{F}^0$ is constant and inclined on the axis Ox (Figure 5.15a) by the angle α, formula (5.2.21''') gives the equivalent load

$$\mathbf{Q}(x) = \frac{M\mathbf{F}^0}{\sin \alpha}[\delta(x - b) - \delta(x - a)]. \tag{5.2.23}$$

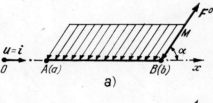

a)

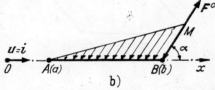

b)

Fig. 5.15

Likewise, if a directed moment with a linear variation, such that $M(a) = 0$ and $M(b) = M$, is acting on the segment AB of the axis Ox and the unit vector $\mathbf{F}^0$ is constant and inclined on the axis Ox by the angle α (Figure 5.15b), we obtain

$$\mathbf{Q}(x) = \frac{M\mathbf{F}^0}{\sin \alpha}\left[\delta(x - b) - \frac{\theta_1(x)}{b - a}\right]. \tag{5.2.24}$$

In a similar way, we may introduce directed moments of higher order continuously distributed over a curve.

5.2.1.4. Equivalence of the action of forces

Let $\mathbf{F}$ be a concentrated force acting at the point $A(x_0, y_0, z_0)$; we want to displace the force at the origin O of the co-ordinate axes, so that its effect should remain unaltered (Figure 5.16).

The loads corresponding to the concentrated force $\mathbf{F}$, considered applied at the point A or O, respectively are

$$\mathbf{Q}_A(x, y, z) = \mathbf{F}\delta(x - x_0, y - y_0, z - z_0), \tag{5.2.25}$$

$$\mathbf{Q}_0(x, y, z) = \mathbf{F}\delta(x, y, z). \tag{5.2.25'}$$

Applying Taylor's expansion to relation (5.2.25), we obtain

$$\mathbf{Q}_A(x, y, z) = \mathbf{F}\delta(x, y, z) - \left(x_0 \frac{\partial}{\partial x} + y_0 \frac{\partial}{\partial y} + z_0 \frac{\partial}{\partial z} \right) \mathbf{F}\delta(x, y, z)$$

$$+ \frac{1}{2!} \left(x_0 \frac{\partial}{\partial x} + y_0 \frac{\partial}{\partial y} + z_0 \frac{\partial}{\partial z} \right)^{(2)} \mathbf{F}\delta(x, y, z) + \dots$$

$$+ \frac{(-1)^{n-1}}{(n-1)!} \left(x_0 \frac{\partial}{\partial x} + y_0 \frac{\partial}{\partial y} + z_0 \frac{\partial}{\partial z} \right)^{(n-1)} \mathbf{F}\delta(x, y, z) + \mathbf{F}R_n(x, y, z), \quad (5.2.26)$$

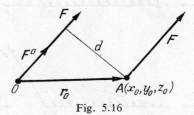

Fig. 5.16

where the remainder is expressed by

$$R_n(x, y, z) = \frac{(-1)^n}{n!} \left(x_0 \frac{\partial}{\partial x} + y_0 \frac{\partial}{\partial y} + z_0 \frac{\partial}{\partial z} \right)^{(n)} \delta(x - \theta x_0, y - \theta y_0, z - \theta z_0)$$

$$(5.2.27)$$

and $0 < \theta < 1$. Let $\mathbf{u}$ be the unit vector of the position vector $\overrightarrow{OA} = \mathbf{r}_0$; we may introduce directional derivatives and write expression (5.2.26) in the form

$$\mathbf{Q}_A(x, y, z) = \mathbf{F}\delta(x, y, z) - Fr_0\mathbf{F}^0 \frac{\partial}{\partial u} \delta(x, y, z)$$

$$+ \frac{1}{2!} Fr_0^2 \mathbf{F}^0 \frac{\partial^2}{\partial u^2} \delta(x, y, z) + \dots + \frac{(-1)^{n-1}}{(n-1)!} Fr_0^{n-1} \mathbf{F}^0 \frac{\partial^{n-1}}{\partial u^{n-1}} \delta(x, y, z)$$

$$+ \frac{(-1)^n}{n!} Fr_0^n \mathbf{F}^0 \frac{\partial^n}{\partial u^n} \delta(x - \theta x_0, y - \theta y_0, z - \theta z_0). \quad (5.2.26')$$

Denoting by d the distance between the point A and the line of action of the force $\mathbf{F}$ supposed to be applied at the point O, and by $\mathbf{F}^0$ the unit vector of that force, we obtain

$$\mathbf{Q}_A(x, y, z) = \mathbf{Q}_O(x, y, z) + \sum_{i=1}^{n-1} \frac{(-1)^i}{i!} \frac{Fd^i}{|\mathbf{u} \times \mathbf{F}^0|^i} \mathbf{F}^0 \frac{\partial^i}{\partial u^i} \delta(x, y, z)$$

$$+ \frac{(-1)^n}{n!} \frac{Fd^n}{|\mathbf{u} \times \mathbf{F}^0|^n} \mathbf{F}^0 \frac{\partial^n}{\partial u^n} \delta(x - \theta x_0, y - \theta y_0, z - \theta z_0). \quad (5.2.26'')$$

The magnitude of the moment of order n about the point A of the force $\mathbf{F}$, applied at the point O is, by definition,

$$M^{(i)} = Fd^i (i = 1, 2, ..., n);\tag{5.2.28}$$

in that case, taking into account expression (5.2.14) of a directed concentrated moment of order n and assuming that the direction in which the passage to the limit is effected remains the same, we obtain

$$\mathbf{Q}_A(x, y, z) = \mathbf{Q}_0(x, y, z) + \frac{1}{1!}\mathbf{Q}^{(1)}(x, y, z) + \frac{1}{2!}\mathbf{Q}^{(2)}(x, y, z) + \ldots$$

$$+ \frac{1}{(n-1)!}\mathbf{Q}^{(n-1)}(x, y, z) + \frac{1}{n!}\mathbf{Q}_1^{(n)}(x, y, z),\tag{5.2.29}$$

where

$$\mathbf{Q}_1^{(n)}(x, y, z) = \frac{(-1)^n}{|\mathbf{u} \times \mathbf{F}^0|^n}\, M^{(n)}\, \mathbf{F}^0\, \frac{\partial^n}{\partial u^n}\, \delta(x - \theta x_0, y - \theta y_0, z - \theta z_0).\tag{5.2.30}$$

Expression (5.2.29) gives the *law of the equivalence of the action of the force* $\mathbf{F}$ applied at the points A, and O, respectively. Thus, we can state the following

Theorem 5.2.1. *The action of a concentrated force $\mathbf{F}$ applied at the point A is equivalent to the action of the same force applied at the point O, to which are added the actions of $(n-1)$ directed concentrated moments of the first, second, ..., $(n-1)$th order applied at the point O, and whose equivalent loads $\mathbf{Q}^{(i)}$ are multiplied by* $\dfrac{1}{i!}$ ($i = 1, 2, ..., n-1$), *as well as the action of a directed concentrated moment of the nth order applied at a point located on the segment OA, and whose equivalent load $\mathbf{Q}_1^{(n)}$ is multiplied by* $\dfrac{1}{n!}$.

5.2.1.5. The highest order of directed concentrated moments

Depending on the physical properties of the medium on which the force $\mathbf{F}$ is acting and on the mathematical model selected for it, we can determine *the highest order of the directed concentrated moments* which may occur in the problem of the equivalence of forces, i.e. we can determine the number n.

Thus, in the case of a *rigid solid*, characterized by the fact that the distance between any two of its points is invariant in time, we remark that concentrated forces may be represented by sliding vectors; this results from the law of equivalence of the actions of two equal concentrated forces applied at two different points of the rigid solid on the same line of action. In that case, the highest order of the moments is 1 (hence the study of the equivalence of the actions of forces requires only moments of the first order). In other works, the law of equivalence takes the following form for rigid solids

$$\mathbf{Q}_A(x, y, z) = \mathbf{Q}_0(x, y, z) + \mathbf{Q}^{(1)}(x, y, z). \tag{5.2.31}$$

In the case of a *continuously deformable body* there may occur moments of higher order. Thus, in section 8.1.1. we shall see that in the bending of a bar, taking into consideration a simplified mathematical model, the highest order of the directed concentrated moment is $n = 3$; hence, in that case we have

$$\mathbf{Q}_A(x, y, z) = \mathbf{Q}_O(x, y, z) + \mathbf{Q}^{(1)}(x, y, z)$$

$$+ \frac{1}{2}\mathbf{Q}^{(2)}(x, y, z) + \frac{1}{6}\mathbf{Q}^{(3)}(x, y, z). \tag{5.2.32}$$

5.2.2. Rotational concentrated moments (centres of rotation)

Proceeding from the notion of directed concentrated moment, we can define another type of concentrated moment: the rotational moment; in the following we give some general results, and then we show how we can construct such a moment with the help of two directed concentrated moments.

5.2.2.1. General results

We have seen that a directed concentrated moment may be represented by means of a load, depending on the point of application $A(x_0, y_0, z_0)$, on the unit vector $\mathbf{F}^0$ of the generating concentrated forces, on the unit vector $\mathbf{u}$ of the direction in which the passage to the limit is effected, and on the magnitude M of the moment.

Let (P) be a plane passing through the point A and $\mathbf{n}(\cos\gamma_1, \cos\gamma_2, \cos\gamma_3)$ (Figure 5.17a) be the unit vector of its normal. We denote by $\{\mathbf{Q}\}$ the set of directed concentrated moments $\mathbf{Q}_i(x, y, z)$, applied at the point A, having the unit vectors $\mathbf{u}_i$ and $\mathbf{F}_i^0$ ($i = 1, 2, \ldots, n$) and belonging to the plane (P). We now introduce

Definition 5.2.4. *The sum of two or more elements of the set $\{\mathbf{Q}\}$ is called the rotational concentrated moment (or centre of rotation) corresponding to the point A and the plane (P), when the sum does not depend on the unit vectors $\mathbf{u}_i$ and $\mathbf{F}_i^0$.*

We shall denote the set thus defined by $\{\mathbf{R}\}$, and an element of it will be designated by (the equivalent load of a moment of magnitude M) (Figure 5.17b)

$$\mathbf{R}(x, y, z) = \mathbf{Q}_1(x, y, z) + \mathbf{Q}_2(x, y, z) + \ldots + \mathbf{Q}_n(x, y, z), n \geqslant 2, \quad (5.2.33)$$

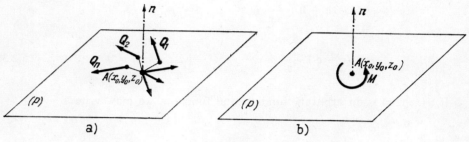

Fig. 5.17

where

$$\mathbf{Q}_i(x, y, z) = -\frac{M_i \mathbf{F}_i^0}{|\mathbf{u}_i \times \mathbf{F}_i^0|} \frac{\partial}{\partial u_i} \delta(x - x_0, y - y_0, z - z_0) \ (i = 1, 2, \ldots, n). \quad (5.2.34)$$

Replacing (5.2.34) in (5.2.33) and complying with the condition that the element $\mathbf{R}$ should not depend on the unit vectors $\mathbf{u}_i$ and $\mathbf{F}_i^0$ $(i = 1, 2, \ldots, n)$, we obtain

$$\mathbf{R}(x, y, z) = \mathbf{a} \frac{\partial}{\partial x} \delta(x - x_0, y - y_0, z - z_0)$$

$$+ \mathbf{b} \frac{\partial}{\partial y} \delta(x - x_0, y - y_0, z - z_0) + \mathbf{c} \frac{\partial}{\partial z} \delta(x - x_0, y - y_0, z - z_0), \quad (5.2.35)$$

where $\mathbf{a}, \mathbf{b}, \mathbf{c}$ are constant vectors depending only on the magnitudes M_i; we shall now show that the representation of the rotational concentrated moment in the form (5.2.35) is unique.

Let us assume that the moment can be written in the form

$$\mathbf{R}(x, y, z) = \mathbf{a}' \frac{\partial}{\partial x} \delta(x - x_0, y - y_0, z - z_0)$$

$$+ \mathbf{b}' \frac{\partial}{\partial y} \delta(x - x_0, y - y_0, z - z_0) + \mathbf{c}' \frac{\partial}{\partial z} \delta(x - x_0, y - y_0, z - z_0); \quad (5.2.35')$$

then, by comparing relations (5.2.35) and (5.2.35′), it follows that

$$(\mathbf{a} - \mathbf{a}') \frac{\partial}{\partial x} \delta(x - x_0, y - y_0, z - z_0)$$

$$+ (\mathbf{b} - \mathbf{b}') \frac{\partial}{\partial y} \delta(x - x_0, y - y_0, z - z_0)$$

$$+ (\mathbf{c} - \mathbf{c}') \frac{\partial}{\partial z} \delta(x - x_0, y - y_0, z - z_0) = 0. \tag{5.2.36}$$

If $\varphi(x, y, z)$ is an arbitrary fundamental function, we may write

$$\left((\mathbf{a} - \mathbf{a}') \frac{\partial}{\partial x} \delta(x - x_0, y - y_0, z - z_0) \right.$$

$$+ (\mathbf{b} - \mathbf{b}') \frac{\partial}{\partial y} \delta(x - x_0, y - y_0, z - z_0)$$

$$+ \left. (\mathbf{c} - \mathbf{c}') \frac{\partial}{\partial z} \delta(x - x_0, y - y_0, z - z_0), \varphi(x, y, z) \right) = 0; \tag{5.2.36′}$$

whence

$$(\mathbf{a} - \mathbf{a}') \frac{\partial}{\partial x} \varphi(x_0, y_0, z_0) + (\mathbf{b} - \mathbf{b}') \frac{\partial}{\partial y} \varphi(x_0, y_0, z_0)$$

$$+ (\mathbf{c} - \mathbf{c}') \frac{\partial}{\partial z} \varphi(x_0, y_0, z_0) = 0, \tag{5.2.36″}$$

a relation which, owing to its linearity with respect to the derivatives of the arbitrary fundamental function $\varphi(x, y, z)$, can occur only if

$$\mathbf{a} = \mathbf{a}', \ \mathbf{b} = \mathbf{b}', \ \mathbf{c} = \mathbf{c}'. \tag{5.2.37}$$

Hence, the *representation (5.2.35) of the rotational concentrated moment is unique,* which justifies the definition given above.

We shall now show that the *set* $\{\mathbf{R}\}$ *is non-void;* this and the fact that the representation (5.2.35) is unique allow the computation of the coefficients $\mathbf{a}, \mathbf{b}, \mathbf{c}$.

Let the directed concentrated moments, represented by the loads $\mathbf{Q}_1$ and $\mathbf{Q}_2$ be such that the following conditions are fulfilled (Figure 5.18)

$$M_1 = M_2 = \frac{M}{2},\qquad (5.2.38)$$

$$\mathbf{u}_1 \cdot \mathbf{u}_2 = \mathbf{u}_1 \cdot \mathbf{F}_1^0 = \mathbf{u}_2 \cdot \mathbf{F}_2^0 = 0,\qquad (5.2.38')$$

$$\mathbf{u}_1 \times \mathbf{F}_1^0 = \mathbf{u}_2 \times \mathbf{F}_2^0 = \mathbf{n}.\qquad (5.2.38'')$$

We remark that between the unit vectors occurring in the computation we have the relations

$$\mathbf{F}_1^0 = \mathbf{u}_2\,(\cos \beta_1, \cos \beta_2, \cos \beta_3),\ \mathbf{F}_2^0 = -\mathbf{u}_1\,(\cos \alpha_1, \cos \alpha_2, \cos \alpha_3),\quad (5.2.39)$$

involving also their projections which are connected, in accordance with (5.2.38') by the relation

$$\cos \alpha_1 \cos \beta_1 + \cos \alpha_2 \cos \beta_2 + \cos \alpha_3 \cos \beta_3 = 0.\qquad (5.2.40)$$

We remark further that we may write

$$\mathbf{u}_1 \times \mathbf{u}_2 = \mathbf{n},\qquad (5.2.41)$$

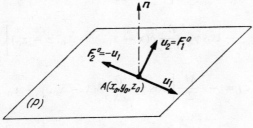

Fig. 5.18

that is

$$\cos \gamma_1 = \cos \alpha_2 \cos \beta_3 - \cos \alpha_3 \cos \beta_2,$$

$$\cos \gamma_2 = \cos \alpha_3 \cos \beta_1 - \cos \alpha_1 \cos \beta_3,\qquad (5.2.41')$$

$$\cos \gamma_3 = \cos \alpha_1 \cos \beta_2 - \cos \alpha_2 \cos \beta_1.$$

Taking into account relation (5.2.34), it follows that

$$\mathbf{Q}_1(x, y, z) = -\frac{M}{2}\mathbf{F}_1^0\left[\cos\alpha_1\frac{\partial}{\partial x}\delta(x - x_0, y - y_0, z - z_0)\right.$$

$$+\cos\alpha_2\frac{\partial}{\partial y}\delta(x - x_0, y - y_0, z - z_0)$$

$$\left.+\cos\alpha_3\frac{\partial}{\partial z}\delta(x - x_0, y - y_0, z - z_0)\right],$$

$$(5.2.42)$$

$$\mathbf{Q}_2(x, y, z) = -\frac{M}{2}\mathbf{F}_2^0\left[\cos\beta_1\frac{\partial}{\partial x}\delta(x - x_0, y - y_0, z - z_0)\right.$$

$$+\cos\beta_2\frac{\partial}{\partial y}\delta(x - x_0, y - y_0, z - z_0)$$

$$\left.+\cos\beta_3\frac{\partial}{\partial z}\delta(x - x_0, y - y_0, z - z_0)\right],$$

and hence the components of the equivalent load of the rotational concentrated moment are

$$Q_x(x, y, z) = -\frac{M}{2}\left[\cos\gamma_2\frac{\partial}{\partial z}\delta(x - x_0, y - y_0, z - z_0)\right.$$

$$\left.-\cos\gamma_3\frac{\partial}{\partial y}\delta(x - x_0, y - y_0, z - z_0)\right],$$

$$Q_y(x, y, z) = -\frac{M}{2}\left[\cos\gamma_3\frac{\partial}{\partial x}\delta(x - x_0, y - y_0, z - z_0)\right.$$

$$(5.2.43)$$

$$\left.-\cos\gamma_1\frac{\partial}{\partial z}\delta(x - x_0, y - y_0, z - z_0)\right],$$

$$Q_z(x, y, z) = -\frac{M}{2}\left[\cos\gamma_1\frac{\partial}{\partial y}\delta(x - x_0, y - y_0, z - z_0)\right.$$

$$\left.-\cos\gamma_2\frac{\partial}{\partial x}\delta(x - x_0, y - y_0, z - z_0)\right].$$

Comparing the representations (5.2.35) and (5.2.43) and taking into account the uniqueness of the representation (5.2.35), we obtain the constants

$$\mathbf{a}\left(0, -\frac{M}{2}\cos\gamma_3, \frac{M}{2}\cos\gamma_2\right), \mathbf{b}\left(\frac{M}{2}\cos\gamma_3, 0, -\frac{M}{2}\cos\gamma_1\right),$$

(5.2.44)

$$\mathbf{c}\left(-\frac{M}{2}\cos\gamma_2, \frac{M}{2}\cos\gamma_1, 0\right).$$

We remark that the relations

$$\mathbf{a}\cos\gamma_1 + \mathbf{b}\cos\gamma_2 + \mathbf{c}\cos\gamma_3 = 0, \tag{5.2.45}$$

$$\mathbf{a}\cdot\mathbf{n} = \mathbf{b}\cdot\mathbf{n} = \mathbf{c}\cdot\mathbf{n} = 0 \tag{5.2.45'}$$

are satisfied and hence the *vectors* **a**, **b**, **c** *are coplanar* and located on the plane (P). Ultimately, the *equivalent load of the rotational concentrated moment* has the expression

$$\mathbf{Q}(x, y, z) = -\frac{1}{2}M\mathbf{n} \times \operatorname{grad} \delta(x - x_0, y - y_0, z - z_0). \tag{5.2.46}$$

Thus we see that the rotational concentrated moment is characterized by:
(*i*) the location $A(x_0, y_0, z_0)$;
(*ii*) the plane (P) determined by the unit vector **n** of the normal to the plane at the point A;
(*iii*) the magnitude M of the moment.
The rotational concentrated moment defined above is positive since the rotation in the plane (P), seen from **n** (Figure 5.17), is counter-clockwise. In this way, the direction of the moment results from the data specified above and is indicated by the unit vector **n**.

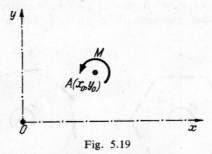

Fig. 5.19

In particular, in a two-dimensional case, the components of the equivalent load of a moment of magnitude M applied at the point $A(x_0, y_0)$ in the plane Oxy are (Figure 5.19)

$$\mathbf{Q}_x(x, y) = \frac{1}{2}M\frac{\partial}{\partial y}\delta(x - x_0, y - y_0),$$

(5.2.47)

$$\mathbf{Q}_y(x, y) = -\frac{1}{2}M\frac{\partial}{\partial x}\delta(x - x_0, y - y_0).$$

5.2.2.2. Construction of a rotational concentrated moment by means of two directed concentrated moments

Taking into account the examples given in subsection 5.1.2.3 and the considerations presented in the previous subsection, we may use as a *canonical represen-*

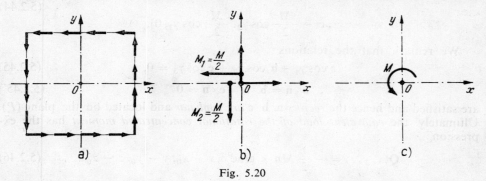

Fig. 5.20

tation of a rotational concentrated moment the representation obtained by the superposition of two directed concentrated moments of the same direction, same magnitude, and the component forces of which are normal to each other.

Thus, the system of concentrated forces acting on the sides of a square (Figure 5.20a) leads to the superposition of two directed concentrated moments (Figure 5.20b), hence to the rotational concentrated moment (Figure 5.20c). We remark that the result is the same for any position of a square of which centre is A. The fact that we have considered the plane Oxy particularizes in no way the problem from a physical point of view. It should also be mentioned that this points out a method of *constructing experimentally*, of *modelling* a rotational concentrated moment.

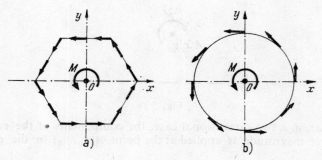

Fig. 5.21

It can be shown that starting from loads which are tangential to and uniformly distributed over the sides of a regular polygon (Figure 5.21a), or from loads tangential to and uniformly distributed over a circle (Figure 5.21b), we obtain again a rotational concentrated moment.

5.2.3. Concentrated moments of dipole type

In section 5.2.1.1 we have defined the directed concentrated moment under the assumption that the unit vector $\mathbf{F}^0$ of the forces which make up the moment, and the unit vector $\mathbf{u}$, which gives the direction in which the passage to the limit takes place, are not collinear ($\mathbf{u} \times \mathbf{F}^0 \neq 0$). For the case where they are, we introduce a new type of concentrated moment: *the concentrated moment of dipole type*.

In the following section we shall deal with concentrated moments of linear, plane, and spatial dipole type as well as concentrated moments of sectorial (plane and spatial) dipole type.

5.2.3.1. Concentrated moments of linear dipole type

Let $[-\mathbf{F}(-F_x, -F_y, -F_z)]$ (Figure 5.22) be a concentrated force applied at the fixed point $A(x_0, y_0, z_0)$ and let $\mathbf{F}_1 = \mathbf{F}(F_x, F_y, F_z)$ (the equality is considered in the sense of free vectors) be another concentrated force applied at the variable point $B(\xi, \eta, \zeta)$. Let $\mathbf{u}$ be the unit vector of the force $\mathbf{F}$. We assume that the two concentrated forces have the same line of action; hence the unit vector of the vector $\mathbf{AB}$ will be again $\mathbf{u}$. We set $\overline{AB} = d$ and introduce the *magnitude of the dipole moment*

$$D = Fd. \tag{5.2.48}$$

We can introduce the following

Definition 5.2.5. *The concentrated moment of linear dipole type (dipole of concentrated forces) at the point A is the limit, in the sense of the theory of distributions, of the set of concentrated forces $(-\mathbf{F}, \mathbf{F}_1)$ when the arm of the couple d tends to zero. The point A is considered fixed and the point B variable; we assume that the two concentrated forces have the same line of action, and that the unit vector $\mathbf{u}$ and the magnitude D of the dipole moment are constant.*

The common line of action of the two concentrated forces is the *line of action of the dipole*, which is considered to be positive if it tends to further separate the points of application of the forces (Figure 5.22) and negative in the opposite case.

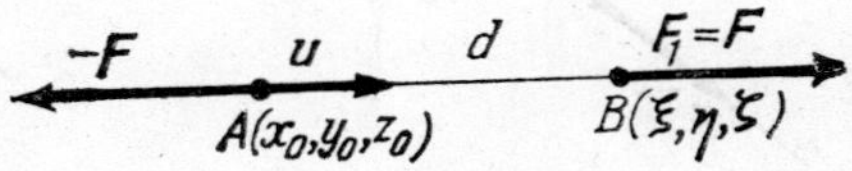

Fig. 5.22

The equivalent load of the set of two concentrated forces may be written

$$\mathbf{Q}(x, y, z; d) = F\mathbf{u}[\delta(x - \xi, y - \eta, z - \zeta) - \delta(x - x_0, y - y_0, z - z_0)]; \tag{5.2.49}$$

taking into account formula (5.2.5), which introduces the directional derivative, and the moment (5.2.48), we may write

$$\mathbf{Q}(x, y, z) = \lim_{d \to 0} \mathbf{Q}(x, y, z; d) = -D\mathbf{u}\frac{\partial}{\partial u}\delta(x - x_0, y - y_0, z - z_0), \quad (5.2.50)$$

where the directional derivative has been introduced in the direction defined by the unit vector **u**. This is the *equivalent load of a concentrated moment of linear dipole type* which must be introduced in the exceptional case where we cannot define the directed concentrated moment.

From relation (5.2.50) we deduce that the concentrated moment of linear dipole type is characterized by:

    (*i*) the location $A(x_0, y_0, z_0)$;
    (*ii*) the unit vector **u** of the forces which generate the moment;
    (*iii*) the magnitude D of the dipole moment.

The direction of the dipole is specified by the unit vector $\mathbf{F}^0$ of the force **F**; if $\mathbf{F}^0 = \mathbf{u}$, which is the case considered above, then the dipole is positive, while if $\mathbf{F}^0 = -\mathbf{u}$, it is negative and the sign of relation (5.2.50) must be changed.

An important particular case is that where the point of application is the origin of the co-ordinate axes and $\mathbf{u} = -\mathbf{i}$, where **i** is the unit vector of the Ox axis (Figure 5.23); we thus obtain a dipole of concentrated forces, which is expressed by the equivalent load $\left(\text{we note that }\dfrac{\partial}{\partial u} = -\dfrac{\partial}{\partial x}\right)$

$$\mathbf{Q}(x, y, z) = -D\mathbf{i}\,\delta'_x(x, y, z). \quad (5.2.51)$$

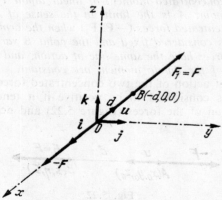

Fig. 5.23

The concentrated moment of linear dipole type introduced above is a *moment of the first order;* for this reason its magnitude and equivalent load will include when necessary the superscript 1 (written $D^{(1)}$ and $Q^{(1)}$).

We can introduce, as we did in section 5.2.1.2, concentrated moments of linear dipole type of a higher order. Thus we may write the *equivalent load of a concentrated moment of linear dipole type of order n* in the form

$$\mathbf{Q}^{(n)}(x, y, z) = (-1)^n D^{(n)} \mathbf{u} \frac{\partial^n}{\partial u^n} \delta(x - x_0, y - y_0, z - z_0), \qquad (5.2.52)$$

where the magnitude of the moment is given by

$$D^{(n)} = F d_1 d_2 \ldots d_n ; \qquad (5.2.53)$$

$d_1, d_2, \ldots, d_n$ are the distances between the points of application of the forces acting at the variable points on the line of action of the unit vector $\mathbf{u}$ and the location of the concentrated force acting at the fixed point $A(x_0, y_0, z_0)$. In particular, if $d_1 = d_2 = \ldots = d_n = d$, we have

$$D^{(n)} = F d^n. \qquad (5.2.53')$$

Now let $ABCD$ be a square in the plane Oxy, having the centre at the origin and the sides parallel to the co-ordinate axes and equal to $2a$ (Figure 5.24); we assume that normal loads of magnitude $\dfrac{D}{4a^2}$ are acting on the unit length of the

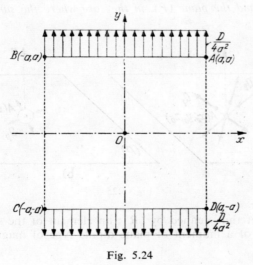

Fig. 5.24

sides AB and CD. Using the results of section 5.1.2.3, it can be shown that the corresponding equivalent loads may be expressed in the form

$$Q_x(x, y; a) = 0, \quad Q_y(x, y; a) = -\frac{\partial}{\partial y} \left[\frac{D}{4a^2} h(x, y) \right], \qquad (5.2.54)$$

where $h(x, y)$ is the characteristic function corresponding to the square $ABCD$ and given by relation (5.1.64).

Passing to the limit when $a \to 0$, in the sense of the theory of distributions, we obtain

$$Q_x(x, y) = 0, \quad Q_y(x, y) = - D\delta_y'(x, y) \tag{5.2.55}$$

and hence the loading case shown in Figure 5.24 leads to a dipole of concentrated forces applied at the origin of the co-ordinate axes and having the line of action in the direction of the Oy axis.

5.2.3.2. Concentrated moments of plane dipole type

Let (P) be a plane passing through the point $A(x_0, y_0, z_0)$ and $\mathbf{n}(\cos \gamma_1, \cos \gamma_2, \cos \gamma_3)$ (Figure 5.25a) be the unit vector of the normal to the plane. We denote by $\{\mathbf{Q}\}$ the set of moments of linear dipole type $\mathbf{Q}_i(x, y, z)$ applied at the point A, with the unit vectors $\mathbf{u}_i$ $(i = 1, 2, \dots, n)$ belonging to the plane (P). We now introduce the following

Definition 5.2.6. *The sum of two or more elements of the set* $\{\mathbf{Q}\}$ *is called a concentrated moment of the plane dipole type* (or *centre of plane dilatation*), *corresponding to the point A and the plane (P), in the case where the sum does not depend on the unit vectors* $\mathbf{u}_i$.

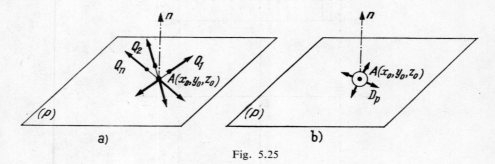

a)   b)

Fig. 5.25

We denote the set thus defined by $\{\mathbf{R}\}$; an element of the set is expressed by (the equivalent load of a moment of plane dipole type of magnitude D_p) (Figure 5.25b)

$$\mathbf{R}(x, y, z) = \sum_{i=1}^{n} \mathbf{Q}_i(x, y, z), \quad n \geqslant 2, \tag{5.2.56}$$

where

$$\mathbf{Q}_i(x, y, z) = - D_i \mathbf{u}_i \frac{\partial}{\partial u} \delta(x - x_0, y - y_0, z - z_0). \tag{5.2.57}$$

Replacing (5.2.57) in (5.2.56) and imposing the condition that **R** should not depend on the unit vectors $\mathbf{u}_i$ $(i = 1, 2, \ldots, n)$, we find that an element of the set may be expressed in the form (5.2.35), where **a**, **b**, **c** are constant vectors de-

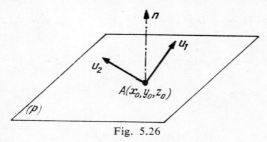

Fig. 5.26

pending only on the magnitudes D_i; we can show, as we did in section 5.2.2.1, that the *representation* of a concentrated moment of plane dipole type in the form (5.2.35) is *unique*, which justifies the definition given above.

In that case it is sufficient to consider in the plane (P) two concentrated moments of linear dipole type, given by

$$\mathbf{Q}_1(x, y, z) = - D_1 \mathbf{u}_1 [\cos \alpha_1 \, \delta'_x(x - x_0, y - y_0, z - z_0)$$

$$+ \cos \alpha_2 \, \delta'_y(x - x_0, y - y_0, z - z_0) + \cos \alpha_3 \, \delta'_z(x - x_0, y - y_0, z - z_0)],$$

$$\mathbf{Q}_2(x, y, z) = - D_2 \mathbf{u}_2 [\cos \beta_1 \, \delta'_x(x - x_0, y - y_0, z - z_0)$$

$$+ \cos \beta_2 \, \delta'_y(x - x_0, y - y_0, z - z_0) + \cos \beta_3 \, \delta'_z(x - x_0, y - y_0, z - z_0)],$$

$$(5.2.58)$$

where the lines of action are given by the unit vectors $\mathbf{u}_1(\cos \alpha_1, \cos \alpha_2, \cos \alpha_3)$ and $\mathbf{u}_2(\cos \beta_1, \cos \beta_2, \cos \beta_3)$; we assume that these lines of action are perpendicular to each other (Figure 5.26) and that the relations

$$\mathbf{u}_1 \cdot \mathbf{u}_2 = 0, \quad \mathbf{u}_1 \times \mathbf{u}_2 = \mathbf{n} \tag{5.2.59}$$

are satisfied; therefore the projections of these unit vectors will satisfy the conditions of orthogonality

$$\cos^2 \alpha_1 + \cos^2 \beta_1 + \cos^2 \gamma_1 = \cos^2 \alpha_2 + \cos^2 \beta_2 + \cos^2 \gamma_2$$

$$= \cos^2 \alpha_3 + \cos^2 \beta_3 + \cos^3 \gamma_3 = 1, \tag{5.2.60}$$

$$\cos \alpha_2 \cos \alpha_3 + \cos \beta_2 \cos \beta_3 + \cos \gamma_2 \cos \gamma_3$$

$$= \cos \alpha_3 \cos \alpha_1 + \cos \beta_3 \cos \beta_1 + \cos \gamma_3 \cos \gamma_1$$

$$= \cos \alpha_1 \cos \alpha_2 + \cos \beta_1 \cos \beta_2 + \cos \gamma_1 \cos \gamma_2 = 0. \tag{5.2.60'}$$

Taking into account relations (5.2.60), (5.2.60′) we obtain the components of the *equivalent load of the concentrated moment of plane dipole type* in the form

$$Q_x(x, y, z) = -\frac{1}{2} D_p[(1 - \cos^2 \gamma_1)\delta_x'(x - x_0, y - y_0, z - z_0)$$

$$- \cos \gamma_1 \cos \gamma_2\, \delta_y'(x - x_0, y - y_0, z - z_0)$$

$$- \cos \gamma_3 \cos \gamma_1\, \delta_z'(x - x_0, y - y_0, z - z_0)],$$

$$Q_y(x, y, z) = -\frac{1}{2} D_p[-\cos \gamma_1 \cos \gamma_2\, \delta_x'(x - x_0, y - y_0, z - z_0)$$

$$+ (1 - \cos^2 \gamma_2)\delta_y'(x - x_0, y - y_0, z - z_0) \quad (5.2.61)$$

$$- \cos \gamma_2 \cos \gamma_3\, \delta_z'(x - x_0, y - y_0, z - z_0)],$$

$$Q_z(x, y, z) = -\frac{1}{2} D_p[-\cos \gamma_3 \cos \gamma_1\, \delta_x'(x - x_0, y - y_0, z - z_0)$$

$$- \cos \gamma_2 \cos \gamma_3\, \delta_y'(x - x_0, y - y_0, z - z_0)$$

$$+ (1 - \cos^2 \gamma_3)\delta_z'(x - x_0, y - y_0, z - z_0)],$$

where the *magnitude* D_p of the dipole is given by

$$D_1 = D_2 = \frac{1}{2} D_p. \quad (5.2.62)$$

We remark that the equivalent load may be written, using vectorial notations, in the form

$$\mathbf{Q}(x, y, z) = -\frac{1}{2} D_p \operatorname{grad} \delta(x - x_0, y - y_0, z - z_0)$$

$$+ \frac{1}{2} D_p \mathbf{n}[\mathbf{n} \cdot \operatorname{grad} \delta(x - x_0, y - y_0, z - z_0)], \quad (5.2.61')$$

where

$$\mathbf{n} \cdot \operatorname{grad} \delta(x - x_0, y - y_0, z - z_0) = \frac{\partial}{\partial n} \delta(x - x_0, y - y_0, z - z_0); \quad (5.2.63)$$

by introducing the double vectorial product we can also write

$$\mathbf{Q}(x, y, z) = \frac{1}{2} D_p \mathbf{n} \times [\mathbf{n} \times \operatorname{grad} \delta(x - x_0, y - y_0, z - z_0)]. \quad (5.2.61'')$$

The concentrated moment of plane dipole type is considered to be positive when it is obtained from positive linear dipoles; in that case the concentrated moment of plane dipole type will be a *centre of plane dilatation*. Conversely (if we use negative linear dipoles), we obtain a *negative* concentrated moment of plane dipole type, which represents a *centre of plane* concentration, while relations (5.2.61), (5.2.61'), (5.2.61'') must be written with a changed sign.

The concentrated moment of plane dipole type is characterized by:
(*i*) the location $A(x_0, y_0, z_0)$;
(*ii*) the unit vector $\mathbf{n}$ of the normal to the plane whereon the moment is acting;
(*iii*) the magnitude D_p of the dipole moment.

The direction of the plane dipole is determined as specified above.

Taking into account formula (5.2.46) which expresses the rotational concentrated moment and assuming $M = D_p$, we may write

$$\mathbf{Q}_{D_p}(x, y, z) + \mathbf{n} \times \mathbf{Q}_M(x, y, z) = 0. \quad (5.2.64)$$

If the plane (P) coincides with the plane Oxy, expression (5.2.61') leads to

$$\mathbf{Q}(x, y, z) = -\frac{1}{2} D_p [\mathbf{i} \delta'_x(x - x_0, y - y_0, z - z_0) + \mathbf{j} \delta'_y(x - x_0, y - y_0, z - z_0)],$$

$$(5.2.65)$$

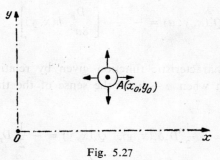

Fig. 5.27

where $\mathbf{i}$ and $\mathbf{j}$ are the unit vectors of the co-ordinate axes; in particular, in the plane case (the variable z is missing) we write (Figure 5.27)

$$\mathbf{Q}(x, y) = -\frac{1}{2} D_p \operatorname{grad} \delta(x - x_0, y - y_0). \quad (5.2.65')$$

Now let $ABCD$ be a square in the plane Oxy with the center at the origin and the sides parallel to the co-ordinate axes and equal to $2a$ (Figure 5.28a); we assume that normal loads equal to $\dfrac{D_p}{8a^2}$ per unit length are acting on all four

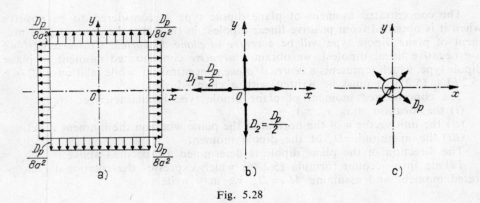

Fig. 5.28

sides of the square. Using the results of section 5.1.2.3 it can be shown that the corresponding equivalent loads are

$$Q_x(x, y; a) = -\frac{\partial}{\partial x}\left[\frac{D_p}{8a^2}\, h(x, y)\right],$$

$$Q_y(x, y; a) = -\frac{\partial}{\partial y}\left[\frac{D_p}{8a^2}\, h(x, y)\right],$$

(5.2.66)

where $h(x, y)$ is the characteristic function given by relation (5.1.64).

Passing to the limit when $a \to 0$, in the sense of the theory of distributions, we obtain

$$Q_x(x, y) = -\frac{1}{2} D_p \delta_x'(x, y), \quad Q_y(x, y) = -\frac{1}{2} D_p \delta_y'(x, y),$$

(5.2.67)

a representation which corresponds to two concentrated moments of linear dipole type; these moments are orthogonal, of the same magnitude and sign (Figure 5.28b); we shall consider that this representation by means of two linear dipoles is the *canonical representation of a concentrated moment of plane dipole type* (Figure 5.28c).

We further remark that we obtain the same result if we start from a circle (Figure 5.29), over the circumference of which uniformly distributed normal loads are acting; in fact, this leads to the usual graphical representation of a concentrated moment of plane dipole type.

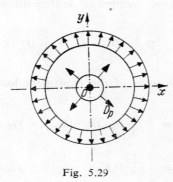

Fig. 5.29

5.2.3.3. Concentrated moments of spatial dipole type

Let $\{\mathbf{Q}\}$ be the set of the moments of linear dipole type $\mathbf{Q}_i(x, y, z)$ applied at the point $A(x_0, y_0, z_0)$ and let $\mathbf{u}_i$ $(i = 1, 2, \ldots, n)$ be their respective unit vectors. We introduce the following

Definition 5.2.7. *The sum of three or more elements of the set* $\{\mathbf{Q}\}$ *is called a concentrated moment of spatial dipole type* (or *centre of spatial dilatation*) *corresponding to the point* A, *in the case where the sum does not depend on the unit vectors* $\mathbf{u}_i$.

An element of the set $\{\mathbf{R}\}$ thus defined may be written in the form (the equivalent load of a moment of spatial dipole type of magnitude D_s)

$$\mathbf{R}(x, y, z) = \sum_{i=1}^{n} \mathbf{Q}_i(x, y, z), \qquad n \geqslant 3, \tag{5.2.68}$$

where the loads $\mathbf{Q}_i(x, y, z)$ are given by relation (5.2.57). Replacing (5.2.57) in (5.2.68) and imposing the condition that $\mathbf{R}$ should not depend on the unit vectors $\mathbf{u}_i$ $(i = 1, 2, \ldots, n)$, we may write the element in the form (5.2.35), where $\mathbf{a}, \mathbf{b}, \mathbf{c}$ are constant vectors depending on the magnitudes D_i; following the same procedure as in section 5.2.2.1, it can be shown that this *representation* is *unique*, which justifies the definition given above.

Let us now consider three concentrated moments of linear dipole type, given by

$$\mathbf{Q}_i(x, y, z) = -D_i \mathbf{u}_i \frac{\partial}{\partial u_i} \delta(x - x_0, y - y_0, z - z_0) \qquad (i = 1, 2, 3) \tag{5.2.69}$$

and let $\mathbf{u}_1(\cos\alpha_1, \cos\alpha_2, \cos\alpha_3)$, $\mathbf{u}_2(\cos\beta_1, \cos\beta_2, \cos\beta_3)$, $\mathbf{u}_3(\cos\gamma_1, \cos\gamma_2, \cos\gamma_3)$ be the unit vectors of their respective lines of action; we assume that the lines of action are normal to each other (Figure 5.30) and that the projections of the corresponding unit vectors satisfy the conditions (5.2.60), (5.2.60′). If the *magnitude D_s* of the concentrated moment of spatial dipole type is given by

$$D_1 = D_2 = D_3 = \frac{1}{3} D_s, \tag{5.2.70}$$

we can express the components of the equivalent load in the form

$$Q_x(x, y, z) = -\frac{1}{3} D_s \delta'_x(x - x_0, y - y_0, z - z_0),$$

$$Q_y(x, y, z) = -\frac{1}{3} D_s \delta'_y(x - x_0, y - y_0, z - z_0), \tag{5.2.71}$$

$$Q_z(x, y, z) = -\frac{1}{3} D_s \delta'_z(x - x_0, y - y_0, z - z_0);$$

in vectorial notation one has

$$\mathbf{Q}(x, y, z) = -\frac{1}{3} D_s \operatorname{grad} \delta(x - x_0, y - y_0, z - z_0). \tag{5.2.71′}$$

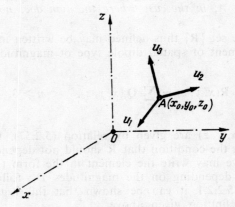

Fig. 5.30

A concentrated moment of spatial dipole type is considered to be *positive* if it has been obtained from positive linear dipoles; in that case it will be a *centre of spatial dilatation*. If the component linear dipoles are negative, we obtain a *nega-*

tive concentrated moment of spatial dipole type; this will represent a *centre of spatial contraction*, and relations (5.2.71), (5.2.71') will change sign.

We remark that a concentrated moment of spatial dipole type is characterized by:

(*i*) the location $A(x_0, y_0, z_0)$;

(*ii*) the magnitude D_s of the dipole moment.

The direction of a spatial dipole is determined as specified above.

Let us consider a cube of side $2a$, with the center at the origin and the faces parallel to the coordinate planes; we assume that normal loads of magnitude $\dfrac{D_s}{24a^3}$ per unit area are acting on all the faces. Using the results of section 5.1.2.3 we may write the corresponding equivalent loads in the form

$$Q_x(x, y, z; a) = - \frac{\partial}{\partial x}\left[\frac{D_s}{24\,a^3}\, h(x, y, z) \right],$$

$$Q_y(x, y, z; a) = - \frac{\partial}{\partial y}\left[\frac{D_s}{24\,a^3}\, h(x, y, z) \right], \qquad (5.2.72)$$

$$Q_z(x, y, z; a) = - \frac{\partial}{\partial z}\left[\frac{D_s}{24\,a^3}\, h(x, y, z) \right],$$

where the characteristic function $h(x, y, z)$ of the cube is given by relation (5.1.72).

Passing to the limit when $a \to 0$, in the sense of the theory of distributions, we obtain

$$Q_x(x, y, z) = - \frac{1}{3}\, D_s \delta_x'(x, y, z),$$

$$Q_y(x, y, z) = - \frac{1}{3}\, D_s \delta_y'(x, y, z), \qquad (5.2.73)$$

$$Q_z(x, y, z) = - \frac{1}{3}\, D_s \delta_z'(x, y, z);$$

this representation corresponds to three concentrated moments of linear dipole type, which are orthogonal and of the same magnitude and sign (Figure 5.31). We shall consider that this representation by means of three linear dipoles is the *canonical representation of a concentrated moment of spatial dipole type.*

We observe that we obtain the same result if we consider uniformly distributed normal loads acting over the surface of a sphere.

It is interesting to remark that the concentrated moment of the spatial dipole type introduces a singularity of a different type than of the concentrated force for example. Indeed, the concentrated force may be obtained with the aid of δ representative sequences, whereas the centre of spatial expansion is approximated by δ' representative sequences.

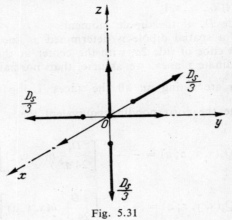

Fig. 5.31

5.2.3.4. Concentrated moments of sectorial dipole type

Let us consider in the plane Oxy a set of concentrated moments of linear dipole type applied, for simplicity, at the origin of the co-ordinate axes (Figure 5.32). We introduce the following

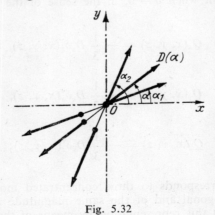

Fig. 5.32

Definition 5.2.8. *The concentrated load represented by the equivalent load*

$$\mathbf{Q}(x, y) = \int_{\alpha_1}^{\alpha_2} \mathbf{Q}(\alpha)\, d\alpha, \qquad (5.2.74)$$

where the equivalent loads

$$\mathbf{Q}(\alpha) = - D(\alpha)\mathbf{u}(\alpha) \frac{\partial}{\partial u} \delta(x, y) \qquad (5.2.75)$$

correspond to concentrated moments of linear dipole type, their lines of action making the angle α with the Ox axis, is said to be a concentrated moment of plane sectorial dipole type, applied at the origin of the co-ordinate axes.

Thus, the expression of a concentrated moment of plane sectorial type has the form

$$\mathbf{Q}(x, y) = \mathbf{a}\delta_x'(x, y) + \mathbf{b}\delta_y'(x, y), \qquad (5.2.74')$$

where **a** and **b** are constant vectors given by

$$\mathbf{a} = - \int_{\alpha_1}^{\alpha_2} D(\alpha)\mathbf{u}(\alpha) \cos \alpha \, d\alpha,$$

$$\qquad (5.2.76)$$

$$\mathbf{b} = - \int_{\alpha_1}^{\alpha_2} D(\alpha)\mathbf{u}(\alpha) \sin \alpha \, d\alpha.$$

In particular, if $D = \text{const}$ and $\alpha_1 = 0$, $\alpha_2 = \pi$, we obtain the concentrated moment of plane dipole type considered in section 5.2.3.2.

Similarly, we consider in the space a set of concentrated moments of linear dipole type applied at the origin of the co-ordinate axes; we assume that the extremity of the unit vector $\mathbf{u}(\cos \alpha, \cos \beta, \cos \gamma)$ applied at the origin describes a domain D on the surface of the sphere of unit radius, whose equations are

$$x = \cos \alpha, \, y = \cos \beta, \, z = \cos \gamma = \sqrt{1 - \cos^2 \alpha - \cos^2 \beta}. \qquad (5.2.77)$$

Under these conditions we give the following

Definition 5.2.9. *The concentrated load represented by the equivalent load*

$$\mathbf{Q}(x, y, z) = \iint_D \mathbf{Q}(\alpha, \beta) \, dS, \qquad (5.2.78)$$

where the equivalent loads

$$\mathbf{Q}(\alpha, \beta) = - D(\alpha, \beta)\mathbf{u}(\alpha, \beta) \frac{\partial}{\partial u} \delta(x, y, z) \qquad (5.2.79)$$

correspond to concentrated moments of linear dipole type, their lines of action being determined by the angles α and β, is said to be a concentrated moment of spatial sectorial dipole type.

The surface element dS is given by

$$dS = \frac{\sin^2\alpha \sin^2\beta}{1 - \cos^2\alpha - \cos^2\beta} \sqrt{1 - \cot^2\alpha \cot^2\beta}\, d\alpha\, d\beta. \qquad (5.2.80)$$

Therefore, the equivalent load of a concentrated moment of spatial sectorial dipole type may be expressed in the form

$$\mathbf{Q}(x, y, z) = \mathbf{a}\delta_x'(x, y, z) + \mathbf{b}\delta_y'(x, y, z) + \mathbf{c}\delta_z'(x, y, z), \qquad (5.2.78')$$

where the constant vectors $\mathbf{a}$, $\mathbf{b}$, $\mathbf{c}$ are given by

$$\mathbf{a} = -\iint_D D(\alpha, \beta)\mathbf{u}(\alpha, \beta)\cos\alpha\, dS,$$

$$\mathbf{b} = -\iint_D D(\alpha, \beta)\mathbf{u}(\alpha, \beta)\cos\beta\, dS, \qquad (5.2.81)$$

$$\mathbf{c} = -\iint_D D(\alpha, \beta)\mathbf{u}(\alpha, \beta)\cos\gamma\, dS.$$

In particular, if $D(\alpha,\beta) = $ const and the angles α, β, γ vary in the interval $[0,\ \pi]$, we obtain the concentrated moment of spatial dipole type considered in subsection 5.2.3,3.

5.3. Tensor properties of concentrated loads

We have seen in the previous paragraph how we may obtain directed concentrated moments and concentrated moments of linear dipole type, starting from the notion of concentrated force. By the superposition of two directed concentrated moments, which are orthogonal and of the same magnitude and sign (and for which the arm of the couple is normal to the direction of the component concentrated forces), we obtain a rotational concentrated moment (center of rotation), which has lost its directional effect and leads to a *mechanical phenomenon with axial anti-symmetry;* in fact, this is the essential property of the concentrated moment.

In the same way, we can introduce a quadripole of concentrated forces, obtained by superposing the effects of two orthogonal concentrated moments of linear dipole type, of the same magnitude and sign; thus we obtain a concentrated moment of plane dipole type (centre of plane dilatation) which, likewise, has no directional effect any longer and leads to a *mechanical phenomenon with axial symmetry.*

By the superposition of effects of three orthogonal concentrated moments of linear dipole type of the same magnitude and sign, we obtain a concentrated moment of spatial dipole type (centre of spatial dilatation) which leads to *a mechanical phenomenon with central symmetry.*

Results of this kind have been pointed out as early as the beginning of the 20th century, but without any further and deeper investigation of the matter; in the sequel we shall justify the use of these concentrated loads and emphasize their *tensorial aspect.*

5.3.1. General tensor properties

In the following discussion we shall see that, from a certain point of view, concentrated loads exibit interesting tensorial properties which justify the properties mentioned above. Also, this will aid in classifying the concentrated loads.

In order to emphasize the different tensorial properties of concentrated loads, we shall consider a deformable solid* on which these loads are acting, thus inducing in the body a state of stress and strain. We shall refer to a body on which we may apply the principle of the *local superposition of effects*, for example, to a *linearly elastic body.*

In the case of an elastic space (or an elastic plane), the state of stress and the state of strain do not show any particular features depending on the direction of the line of action of an internal concentrated load (concentrated force, directed concentrated moment, concentrated moment of linear dipole type, etc.). But in the case of a domain whose boundary is, in some part at least, at a finite distance we have to deal with favourite directions which depend on the boundary.

5.3.1.1. Concentrated loads of the first order

Let f_x, f_y, f_z be concentrated forces equal to unity and acting at the point (x_0, y_0, z_0) in the direction of the co-ordinate axes Ox, Oy, Oz (Figure 5.33); it is easy to see that they are components of a vector (tensor of the first order). Denoting by $f(f_x)$, $f(f_y)$, $f(f_z)$ a component of the displacement vector, stress tensor, or strain tensor at the point (x, y, z), due to the action of the force f_x, f_y, f_z respectively, we remark that $f(f_x)$, $f(f_y)$, $f(f_z)$ are also the components of a tensor of the first order. We give the following

Definition 5.3.1. *Concentrated loads, which behave like a tensor of the first order from the point of view of the state of stress and strain they induce in a body, are called concentrated loads of the first order.*

Therefore, concentrated forces are concentrated loads of the first order.

We can also construct other concentrated loads of the first order, for example *dipoles of centres of spatial dilatation.* Such concentrated loads are obtained by a process of passing to the limit, in the sense of the theory of distributions, when

---

* Details are given in chapter 7.

a variable point approaches a fixed point; it is assumed that a centre of spatial concentration acts at the fixed point and a centre of spatial dilatation at the variable point.

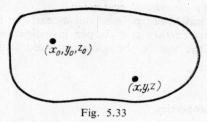

Fig. 5.33

5.3.1.2. Concentrated loads of the second order

Let M_{yz}, M_{zy}, M_{zx}, M_{xz}, M_{xy}, M_{yx} be the direct concentrated moments corresponding to a rectangular trihedron $Oxyz$ (we assume that the arm of the couple of each directed concentrated moment is at a right angle to the common direction of the component concentrated forces); let D_{xx}, D_{yy}, D_{zz} be the concentrated moments of linear dipole type corresponding to the co-ordinate axes. The directed concentrated moments are positive if they correspond to a positive rotation on the plane wherein they act. In the above notation the first index refers to the direction of the component forces and the second index to the direction of the arm of the couple.

Taking into account the conventions regarding the signs and the fact that the magnitude of these concentrated moments is obtained as a product of the components of a force and the components of a position vector (two tensors of the first order), we obtain a tensor of the second order whose components are

$$\left\{ \begin{array}{ccc} D_{xx} & -M_{xy} & M_{xz} \\ M_{yx} & D_{yy} & -M_{yz} \\ -M_{zx} & M_{zy} & D_{zz} \end{array} \right\}. \tag{5.3.1}$$

Let m_{yz}, m_{zy}, m_{zx}, m_{xz}, m_{xy}, m_{yx} be the corresponding directed concentrated moments of magnitude equal to unity, and d_{xx}, d_{yy}, d_{zz} the corresponding concentrated moments of linear dipole type, equal to unity, applied at the point (x_0, y_0, z_0). Denoting by $f(m_{yz})$, $f(m_{zy})$, ..., $f(d_{zz})$ a geometrical or mechanical magnitude, induced at the point (x, y, z) by a single concentrated load, and admitting the principle of the local superposition of effects, we obtain a tensor of the second order of the form

$$T_f \equiv \left\{ \begin{array}{ccc} f(d_{xx}) & -f(m_{xy}) & f(m_{xz}) \\ f(m_{yx}) & f(d_{yy}) & -f(m_{yz}) \\ -f(m_{zx}) & f(m_{zy}) & f(d_{zz}) \end{array} \right\}. \tag{5.3.2}$$

Thus we may formulate the following

Definition 5.3.2. *Concentrated loads which are behaving like a tensor of the second order from the point of view of the state of stress and strain they induce are called concentrated loads of the second order.*

Directed concentrated moments and concentrated moments of linear dipole type are concentrated loads of the second order.

The antisymmetric part of the tensor T_f leads to a vector whose components are

$$f(m_x) = \frac{1}{2}\left[f(m_{yz}) + f(m_{zy})\right], f(m_y) = \frac{1}{2}\left[f(m_{zx}) + f(m_{xz})\right],$$

$$\hspace{8cm}(5.3.3)$$

$$f(m_z) = \frac{1}{2}\left[f(m_{xy}) + f(m_{yx})\right];$$

thus we obtain a *rotational concentrated moment* which is a concentrated load of the first order.

5.3.1.3. Concentrated loads of the zero order

By a similar procedure we introduce the following

Definition 5.3.3. *Concentrated loads which behave like a scalar (tensor of zero order) from the point of view of the state of stress and strain they induce are called concentrated loads of the zero order.*

The first invariant of the tensor T_f, which is a scalar, leads to a centre of spatial dilatation, expressed by

$$f(d_s) = \frac{1}{3}\left[f(d_{xx}) + f(d_{yy}) + f(d_{zz})\right];\hspace{2cm}(5.3.4)$$

hence the centres of spatial dilatation are concentrated loads of the zero order.

5.3.1.4. Concentrated loads of the nth order

We have seen how by proceeding from a concentrated force (concentrated load of the first order) and passing to the limit, in the sense of the theory of distributions, to a neighbouring point, we obtain a concentrated load of the second order. By a similar procedure, starting from a concentrated load of the $(n-1)$th order, one obtains a *concentrated load of the nth order*.

We remark further that by a *tensor contraction* we may obtain also other concentrated loads.

It is interesting to note that in the case of bodies of *classical type* all the concentrated loads which may occur are constructed by proceeding from a concentrated force (which may be a *concentrated body force*); in the case of bodies of *non-classical type* (e.g. bodies of Cosserat type, where a particle has six degrees of freedom), the concentrated body force is no longer sufficient for defining concentrated loads. In that case, besides the concentrated body force there also occurs as a fundamental concentrated load the *concentrated body moment* (different from the center of rotation); obviously in that case the concentrated loads obtained are more complicated.

We remark that, in accordance with the above classification, directed concentrated moments of order n as well as concentrated moments of linear dipole type of order n are concentrated loads of the order $(n + 1)$.

5.3.2. Plane case

In the plane Oxy tensor T_f has the form

$$
T_f \equiv \left\{ \begin{array}{cc} f(d_{xx}) & -f(m_{xy}) \\[2mm] f(m_{yx}) & f(d_{yy}) \end{array} \right\}, \tag{5.3.2'}
$$

which leads to the *centre of plane dilatation*

$$
f(d_p) = \frac{1}{2}\left[f(d_{xx}) + f(d_{yy}) \right] \tag{5.3.5}
$$

and the *rotational concentrated moment*

$$
f(m) = \frac{1}{2}\left[f(m_{xy}) + f(m_{yx}) \right]. \tag{5.3.6}
$$

The last relation is also an invariant in the plane case; therefore, the rotational concentrated moment behaves like a load of the zero order.

In the following we shall emphasize some properties related to the principal directions of the tensor T_f; we shall also examine the case where a concentrated load is acting on the boundary of the domain considered.

5.3.2.1. Principal directions

In various classical textbooks the influence of a rotational concentrated moment is often introduced with the aid of a single directed concentrated moment; this is possible only in very particular cases. The problem may be associated with that of the *principal directions of the tensor T_f'*.

For example, in the plane case the principal directions are expressed by

$$\tan 2\alpha_1 = -\frac{f(m_{xy}) - f(m_{yx})}{f(d_{xx}) - f(d_{yy})}, \tag{5.3.7}$$

where α_1 is the angle made with the Ox axis; this leads to

$$f_{\text{extr}}(d_{\alpha_1}) = f(d_p) \pm \frac{1}{2}\sqrt{[f(d_{xx}) - f(d_{yy})]^2 + [f(m_{xy}) - f(m_{yx})]^2} \tag{5.3.8}$$

and to

$$f(m_{\alpha_1}) = f(m). \tag{5.3.8'}$$

Similarly the principal directions expressed by

$$\tan 2\alpha_2 = \frac{f(d_{xx}) - f(d_{yy})}{f(m_{xy}) - f(m_{yx})} \tag{5.3.9}$$

bisect the preceding principal directions and lead to

$$f_{\text{extr}}(m_{\alpha_2}) = f(m) \pm \frac{1}{2}\sqrt{[f(d_{xx}) - f(d_{yy})]^2 + [f(m_{xy}) - f(m_{yx})]^2} \tag{5.3.10}$$

and to

$$f(d_{\alpha_2}) = f(d_p). \tag{5.3.10'}$$

Formulae (5.3.8') and (5.3.10') show that the effect of a rotational concentrated moment or of a centre of plane dilatation cannot be correctly introduced by means of a directed concentrated moment or a concentrated moment of linear dipole type, respectively, unless the direction of the latter loads is a principal direction for the concentrated moments of linear dipole type or for the directed concentrated moments respectively.

We observe that we can also write the relation

$$f_{\text{extr}}(m_{\alpha_1}) - f_{\text{extr}}(d_{\alpha_2}) = f(m) - f(d_p). \tag{5.3.11}$$

As the functions $f(m_{xy})$, $f(m_{yx})$, $f(d_{xx})$, $f(d_{yy})$ are themselves the components of a tensor of the first or second order, the problem arises of finding a group of two directions (one at the point (x_0, y_0) and the other at the point (x, y)) for which the above functions admit a *maximum maximorum*.

5.3.2.2. Concentrated loads acting on the boundary of plane domains

The above results may be also used when the concentrated loads are acting on the boundary of a plane domain. An interesting problem is the passage from a

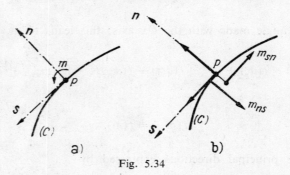

Fig. 5.34

point inside the domain to a point on the boundary; in fact, this is the only way of obtaining an arbitrary concentrated load on the boundary.

Let us consider first the case of a boundary consisting of a smooth curve (C). Let **s** be the direction of the tangent to the curve at point P and **n** that of the outer normal. Then a rotational concentrated moment (Figure 5.34a) equal to unity is expressed by a relation of the form (Figure 5.34b).

$$f(m) = \frac{1}{2}\left[f(m_{ns}) + f(m_{sn})\right] \tag{5.3.12}$$

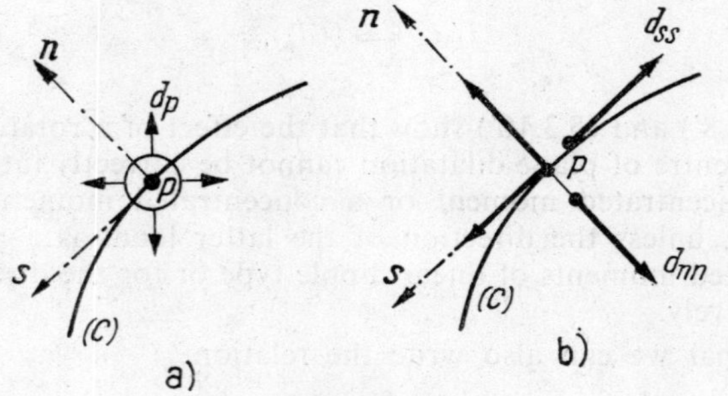

Fig. 5.35

and a centre of plane dilatation (Figure 5.35a) by a relation of the form (Figure 5.35b).

$$f(d_p) = \frac{1}{2}\left[f(d_{nn}) + f(d_{ss})\right]. \tag{5.3.13}$$

Likewise, the effect of a directed concentrated moment or of a concentrated moment of linear dipole type is given by

$$f(m_\beta) = f(m) - \frac{1}{2}\left[f(m_{ns}) - f(m_{sn})\right]\cos 2\beta - \frac{1}{2}\left[f(d_{nn}) - f(d_{ss})\right]\sin 2\beta, \quad (5.3.14)$$

$$f(d_\beta) = f(d_p) - \frac{1}{2}\left[f(d_{nn}) - f(d_{ss})\right]\cos 2\beta + \frac{1}{2}\left[f(m_{ns}) - f(m_{sn})\right]\sin 2\beta, \quad (5.3.14')$$

where β is the angle made with the tangent to the boundary curve.

In the case of a *cuspidal point* of the boundary of a plane domain we may introduce a rotational concentrated moment, using the formula

$$f(m) = \frac{1}{2}\left[f(m_{n_1 s_1}) + f(m_{n_2 s_2})\right] + \frac{1}{2}\left[f(d_{s_1 s_1}) - f(d_{s_2 s_2})\right]\cot \omega, \quad (5.3.15)$$

where the indices n_1, n_2 and s_1, s_2 correspond to the directions of the outer normals and of the tangents at the point, respectively (Figure 5.36); the indices 1 and 2 correspond to the two tangents making the angle ω.

The effect of a center of plane dilatation is introduced in a similar way with the aid of the relation

$$f(d_p) = \frac{1}{2}\left[f(d_{s_1 s_1}) + f(d_{s_2 s_2})\right] - \frac{1}{2}\left[f(m_{n_1 s_1}) - f(m_{n_2 s_2})\right]\cot \omega. \quad (5.3.16)$$

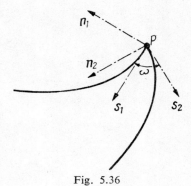

Fig. 5.36

The action of a directed concentrated moment is given by

$$f(m_\beta) = \frac{1}{2}\left[f(m_{n_1 s_1}) + f(m_{n_2 s_2})\right] - \frac{1}{2}\frac{\sin(\omega - 2\beta)}{\sin \omega}\left[f(m_{n_1 s_1}) - f(m_{n_2 s_2})\right]$$

$$+ \frac{\cos \alpha \cos(\omega - \beta)}{\sin \omega}\left[f(d_{s_1 s_1}) - f(d_{s_2 s_2})\right], \quad (5.3.17)$$

and that of a concentrated moment of linear dipole type is expressed in the form

$$f(d_\beta) = \frac{1}{2}\left[f(d_{s_1s_1}) + f(d_{s_2s_2})\right] + \frac{1}{2}\frac{\sin(\omega - 2\beta)}{\sin\omega}\left[f(d_{s_1s_1}) - f(d_{s_2s_2})\right]$$

$$+ \frac{\sin\alpha\sin(\omega - \beta)}{\sin\omega}\left[f(m_{n_1s_1}) - f(m_{n_2s_2})\right], \tag{5.3.17'}$$

where β is the angle made with the tangential direction s_1.

6

Applications

We shall apply the results obtained so far to the study of mechanical phenomena which exhibit discontinuities; in fact, the results prove their usefulness precisely in the case of such phenomena. Thus, we shall deal with the problem of collisions and that of the systems of particles of variable mass.

6.1. Collision

In the following we shall restrict ourselves to the study of the collision of two spheres considered as particles of mass m_1 and m_2. (Figure 6.1). Let $\mathbf{v}_1$, $\mathbf{v}_2$ and $\mathbf{u}_1$, $\mathbf{u}_2$ be their velocities before and after collision, respectively. In the absence of external forces acting on the particles (the constraint forces at the moment of collision are internal forces) the theorem of momentum, introduced in section 4.1.1.3

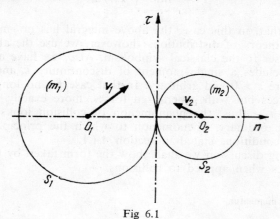

Fig 6.1

for the particular case of a single particle, permits us to write a theorem of conservation of momentum in the form

$$m_1\mathbf{v}_1 + m_2\mathbf{v}_2 = m_1\mathbf{u}_1 + m_2\mathbf{u}_2. \tag{6.1.1}$$

If during the collision no friction forces become manifest we may admit that the velocity components along the common tangent τ will satisfy the relations

$$u_{1\tau} = v_{1\tau}, \qquad u_{2\tau} = v_{2\tau}. \tag{6.1.2}$$

As regards the velocity components along the common normal $\mathbf{n}$ we introduce, in general, an experimental relation of the form

$$k = \frac{u_{2_n} - u_{1_n}}{v_{1_n} - v_{2_n}}, \tag{6.1.3}$$

where k is a coefficient of restitution ($0 \leqslant k \leqslant 1$). In the case $k=1$ we are dealing with *perfectly elastic collisions* while in the case $k = 0$ collisions are *perfectly plastic;* the intermediate case is that of the *elastic collisions* (arbitrary).

We shall give first some considerations concerning the elastic collisions and then we shall deal with plastic collisions.

6.1.1. Elastic collision

In connection with the problem of elastic collisions we shall recall the considerations presented in section 4.1.1 about the dynamics of a particle. Starting from the concept of generalized force and impulse of a generalized force (in the sense of the theory of distributions), we shall introduce the notion of *percussion* of a particle with the aid of the formula

$$\mathbf{P} = \lim_{t'' - t' \to 0} \int_{t'}^{t''} \mathbf{F}(t) \, dt. \tag{6.1.4}$$

It is obvious that, in this case, the above integral has no meaning from the viewpoint of the theory of distributions; however, we use the above notation in order to keep closer to the classical symbolism. Also, we have assumed that the interval $[t', t'']$ includes a single moment of discontinuity t_0 and it is such that $|t'' - t'| < \varepsilon$ where $\varepsilon > 0$ and arbitrary. In that case we no longer have to deal with ordinary forces but with generalized forces, more exactly there occurs the distribution $\delta(t - t_0)$; thus, the phenomenon of collision looks no longer as a special, but as an ordinary, phenomenon to which the principles of mechanics apply under the conditions stated in section 4.1.1.

In the following discussion we shall show the form taken by the general theorems of mechanics when applied to collisions.

6.1.1.1 Theorem of momentum

We remark first that by using a mean value theorem we may write

$$\lim_{t'' - t' \to 0} \int_{t'}^{t''} \widetilde{\mathbf{F}}(t) \, dt = 0, \tag{6.1.5}$$

i.e. the *impulse of the ordinary force may be neglected with respect to the impulse of the complementary force* owing to the discontinuities; hence

$$\lim_{t''-t'\to 0} \int_{t'}^{t''} \mathbf{F}(t)\,dt = m\mathbf{v}_0,$$

(6.1.6)

where $\mathbf{v}_0$ is the velocity jump corresponding to the moment of discontinuity t_0.
Theorem 4.1.3' takes the form

$$(\Delta\mathbf{H})_0 = \mathbf{P}$$

(6.1.7)

and we may state the following

Theorem 6.1.1. *The jump of the momentum of a particle at a moment of discontinuity is equal to the percussion acting on the particle at that moment*. This is the *theorem of momentum* in the case of collisions.

6.1.1.2 Theorem of moment of momentum

Just as in the previous case we may write

$$\lim_{t''-t'\to 0} \int_{t'}^{t''} \mathbf{M}_0\,\tilde{\mathbf{F}}(t) = 0,$$

(6.1.8)

i.e. *the angular impulse of the ordinary force may be neglected with respect to the angular impulse of the complementary force* owing to the discontinuities; we obtain

$$\lim_{t''-t'\to 0} \int_{t'}^{t''} \mathbf{M}_0\,\mathbf{F}(t)\,dt = \mathbf{r}_0 \times (m\mathbf{v}_0),$$

(6.1.9)

where $\mathbf{r}_0$ is the radius vector corresponding to the moment of discontinuity t_0.
Using theorem 4.1.4' and notation (6.1.4) we obtain

$$(\Delta\mathbf{K})_0 = \mathbf{r}_0 \times (\Delta\mathbf{H})_0 = \mathbf{r}_0 \times \mathbf{P}$$

(6.1.10)

and we may state the following

Theorem 6.1.2. *The jump of the moment of momentum of a particle* (which is equal to the *moment of the linear momenium jump* of that particle) *at a moment of discontinuity is equal to the moment of the percussion acting on the particle at that moment;* obviously, the two moments are taken about the same reference point. This represents the *theorem of moment of momentum* in the case of collision.

6.1.1.3 Theorems of kinetic energy

Starting from the theorem of momentum written in the form

$$m\mathbf{v}_0 = m(\mathbf{v}'' - \mathbf{v}') = \mathbf{P}, \tag{6.1.11}$$

where $\mathbf{v}'$ is the velocity of the particle before collision, and $\mathbf{v}''$ the velocity after collision, the dot product by $\mathbf{v}''$ gives

$$m\mathbf{v}''^2 - m\mathbf{v}' \cdot \mathbf{v}'' = \mathbf{P} \cdot \mathbf{v}'' \tag{6.1.12}$$

or

$$T'' - T' + T_0 = \mathbf{P} \cdot \mathbf{v}'', \tag{6.1.12'}$$

where

$$T' = \frac{1}{2}m\mathbf{v}'^2 \tag{6.1.13}$$

is *the kinetic energy before collision*,

$$T'' = \frac{1}{2}m\mathbf{v}''^2 \tag{6.1.13'}$$

is *the kinetic energy after collision*, and

$$T_0 = \frac{1}{2}m(\mathbf{v}'' - \mathbf{v}')^2 = \frac{1}{2}m\mathbf{v}_0^2 \tag{6.1.13''}$$

is *the kinetic energy of the lost velocity*.
  The variation of the kinetic energy is given by

$$(\Delta T)_0 = T'' - T'; \tag{6.1.14}$$

hence formula (6.1.12') may be written in the form

$$(\Delta T)_0 + T_0 = \mathbf{P} \cdot \mathbf{v}'' \tag{6.1.12''}$$

and we may state the following

Theorem 6.1.3. *The sum of the variation of the kinetic energy of a particle at a moment of discontinuity and the kinetic energy of the lost velocity is equal to the dot product of the percussion acting on the particle and its velocity after the moment of discontinuity.* This represents the *theorem of kinetic energy* in the case of collision.
  If one has

$$\mathbf{P} \cdot \mathbf{v}'' = 0, \tag{6.1.15}$$

which occurs, for instance, if the velocity of the particle vanishes after collision, one obtains the relation

$$(\Delta T)_0 + T_0 = 0 \qquad (6.1.16)$$

and we may state the following

Theorem 6.1.4. *If condition* (6.1.15) *is satisfied, then the sum of the variation of kinetic energy of a particle at a moment of discontinuity and the kinetic energy of the lost velocity at the same moment is equal to zero.* This represents *Carnot's theorem* in the case of collision.

Applying the dot product by $\mathbf{v}'$ to relation (6.1.11) we obtain

$$m\mathbf{v}' \cdot \mathbf{v}'' - m\mathbf{v}'^2 = \mathbf{P} \cdot \mathbf{v}' \qquad (6.1.17)$$

or

$$T'' - T' - T_0 = \mathbf{P} \cdot \mathbf{v}', \qquad (6.1.17')$$

whence

$$(\Delta T)_0 - T_0 = \mathbf{P} \cdot \mathbf{v}' ; \qquad (6.1.17'')$$

therefore we may state the following

Theorem 6.1.3'. *The difference between the variation of the kinetic energy of a particle at a moment of discontinuity and the kinetic energy of the lost velocity at the same moment is equal to the dot product of the percussion acting on the particle and the velocity of the latter before the moment of discontinuity.* This represents an *analogue of the theorem of kinetic energy* in the case of collision.

In particular, if we have

$$\mathbf{P} \cdot \mathbf{v}' = 0 \qquad (6.1.18)$$

there follows

$$(\Delta T)_0 = T_0 \qquad (6.1.19)$$

and we may state the following

Theorem 6.1.4'. *If condition* (6.1.18) *is satisfied, then the variation of the kinetic energy of a particle at a moment of discontinuity is equal to the kinetic energy of the lost velocity at the same moment.* This represents an *analogue of Carnot's theorem* in the case of collision.

Adding relations (6.1.12'') and (6.1.17'') member by member, we obtain

$$(\Delta T)_0 = \frac{1}{2} \mathbf{P} \cdot (\mathbf{v}' + \mathbf{v}'') \qquad (6.1.20)$$

and we may state the following

Theorem 6.1.5. *The variation of the kinetic energy of a particle at a moment of discontinuity is equal to the dot product of the percussion acting on that particle and the half-sum of the velocities of the particle before and after the moment of discontinuity.* This represents *Kelvin's theorem* in the case of collision.

Subtracting relations (6.1.12″) and (6.1.17″) member by member, we obtain

$$T_0 = \frac{1}{2}\mathbf{P} \cdot (\mathbf{v}'' - \mathbf{v}') = \frac{1}{2}\mathbf{P} \cdot \mathbf{v}_0 \qquad (6.1.21)$$

and we may state the following

Theorem 6.1.5′. *The kinetic energy of the lost velocity by a particle at a moment of discontinuity is equal to half the dot product of the percussion acting on that particle and the velocity jump at the moment of discontinuity.* This represents an *analogue of Kelvin's theorem* in the case of collision.

6.1.2. Plastic collision

There are some phenomena which can be better interpreted, and whose properties can be presented in a unitary form, if they are studied in a certain definite space which is adequate for the respective phenomenon. For example, in the case of plastic collision, it is particularly useful to introduce a certain space which will be termed the space of plastic collisions; in the following discussion we shall introduce it and emphasize its usefulness.

6.1.2.1 General considerations

We have seen that in the case of a plastic collision of two particles of masses m_1, m_2, the coefficient k of relation (6.1.3) is equal to zero; in other words, after collision the two particles (considered to be two spheres) adhere to each other; the mass of the body thus obtained (a system of two particles) will be (m_1+m_2) and its velocity is

$$\mathbf{u} = \frac{m_1\mathbf{v}_1 + m_2\mathbf{v}_2}{m_1 + m_2}. \qquad (6.1.22)$$

We remark that, from a mathematical point of view, the plastic collision of two particles leads to a *composition law of mass and velocity*.

An important part is played by the theorems of kinetic energy. Thus, we remark that Carnot's theorem extended to two particles is still valid; we denote by

$$(\Delta T)_0 = T'' - T' = \frac{1}{2}(m_1 + m_2)\,\mathbf{u}^2 - \left(\frac{1}{2}m_1\mathbf{v}_1^2 + \frac{1}{2}m_2\mathbf{v}_2^2\right), \qquad (6.1.23)$$

$$T_0 = \frac{1}{2}m_1(\mathbf{u} - \mathbf{v}_1)^2 + \frac{1}{2}m_2(\mathbf{u} - \mathbf{v}_2)^2 = \frac{m_1m_2}{2(m_1 + m_2)}(\mathbf{v}_1 - \mathbf{v}_2)^2 \qquad (6.1.23')$$

the variation of the kinetic energy and the kinetic energy of the lost velocity. Whence

$$T'' \leqslant T' \qquad (6.1.23'')$$

and Carnot's theorem becomes

Theorem 6.1.6. *The kinetic energy of two particles after a plastic collision is smaller than, or at most equal to the kinetic energy before collision.*

It follows that the variation of the kinetic energy is *negative*.

6.1.2.2 Space of plastic collisions

Let M be a particle of mass m and velocity $\mathbf{v}$; we shall denote the mass-velocity pair which characterizes the moving particle by

$$S = (m; \mathbf{v}). \qquad (6.1.24)$$

We assume that mass and velocity are functions of time of class C^1 and moreover that the mass $m(t)$ is a positive function $(m(t) > 0)$. We shall denote by $\mathscr{M}$ the set consisting of the elements S defined by relation (6.1.24), i.e.

$$\mathscr{M} = \{S\}. \qquad (6.1.24')$$

In the set thus obtained we introduce a relation of equivalence, denoted by K and specified by the following

Definition 6.1.1. *Two elements S_1, $S_2 \in \mathscr{M}$ are said to be equivalent and we shall denote this fact by $S_1 = S_2$ if the two elements have the same momentum*

$$m_1 \mathbf{v}_1 = m_2 \mathbf{v}_2 ; \qquad (6.1.25)$$

this will be written

$$K(S_1) = K(S_2). \qquad (6.1.25')$$

It is easily seen that relation K, defined on $\mathscr{M}$, is an *equivalence relation* since it is *reflexive, symmetrical and transitive*.

The equivalence relation K introduced in the set M defines a *partition* of $\mathscr{M}$ into *equivalence classes*. We shall denote by $\mathscr{M}/K$ the set of these equivalence classes which represents the *quotient set* of $\mathscr{M}$ by K.

In this manner, by relation K, to each element $S \in \mathscr{M}$ there corresponds an equivalence class $\tilde{S}$, i.e.

$$S \xrightarrow{K} \tilde{S} \in \mathscr{M}/K. \qquad (6.1.26)$$

We shall define in the set of equivalence classes $\mathcal{M}/K$ a *law of internal composition*, using an *additive* notation

$$(\tilde{S}_1, \tilde{S}_2) \rightarrow \tilde{S}_1 + \tilde{S}_2, \tag{6.1.27}$$

and a *law of external composition*, using a *multiplicative* notation

$$(\alpha \tilde{S}) \rightarrow \alpha \tilde{S}, \tag{6.1.27'}$$

where α is an arbitrary real number.

Let there be $\tilde{S}_1 = (m_1; \mathbf{v}_1)$, $\tilde{S}_2 = (m_2; \mathbf{v}_2)$ with $\tilde{S}_1, \tilde{S}_2 \in \mathcal{M}/K$; we define the law of internal composition as follows

$$\tilde{S}_1 + \tilde{S}_2 = \left(m_1 + m_2; \frac{m_1 \mathbf{v}_1 + m_2 \mathbf{v}_2}{m_1 + m_2} \right). \tag{6.1.28}$$

The *neutral element* $\tilde{0} \in \mathcal{M}/K$ is defined by the relation

$$\tilde{0} = (m; 0), \tag{6.1.29}$$

in other words by $K(\tilde{0}) = 0$.

The property of *commutativity*

$$\tilde{S}_1 + \tilde{S}_2 = \tilde{S}_2 + \tilde{S}_1 \tag{6.1.30}$$

and the property of *associativity*

$$(\tilde{S}_1 + \tilde{S}_2) + \tilde{S}_3 = \tilde{S}_1 + (\tilde{S}_2 + \tilde{S}_3) \tag{6.1.31}$$

are easily verified.

From the definition (6.1.29) of the neutral element it follows that

$$\tilde{S}_1 + \tilde{0} = (m_1; \mathbf{v}_1) + (m; 0) = \left(m_1 + m; \frac{m_1 \mathbf{v}_1}{m_1 + m} \right) = \tilde{S}_1 ; \tag{6.1.32}$$

the relation

$$K(\tilde{S}_1 + \tilde{0}) = m_1 \mathbf{v}_1 = K(\tilde{S}_1) \tag{6.1.32'}$$

points out the property of *null effect* of the neutral element $\tilde{0}$.

Thus, the law of internal composition defines an *Abelian group* on $\mathcal{M}/K$.

We shall define the law of external composition in the form

$$\alpha \tilde{S} = (m; \alpha \mathbf{v}), \tag{6.1.33}$$

where $\tilde{S} = (m; \mathbf{v})$ and $\alpha \in R$. The *element $\tilde{S}'$ opposed* to the element $\tilde{S}$, is specified by

$$\tilde{S}' = (m; -\mathbf{v}) = -(m; \mathbf{v}) = -\tilde{S} ; \qquad (6.1.34)$$

obviously we have

$$\tilde{S} + \tilde{S}' = (m; \mathbf{v}) + (m; -\mathbf{v}) = (2m; 0) = \tilde{0}, \qquad (6.1.35)$$

so that $K(\tilde{S} + \tilde{S}') = 0$ and we may write

$$\tilde{S}' = -\tilde{S}. \qquad (6.1.35')$$

For the real number $\alpha = 1$ we may write the property

$$1\tilde{S} = (m; \mathbf{v}) = \tilde{S}; \qquad (6.1.36)$$

also, the property of *associativity with respect to the numbers* α, $\beta \in R$ follows immediately

$$\alpha(\beta\tilde{S}) = \beta(\alpha \tilde{S}) = (\alpha\beta)\tilde{S} = (m; \alpha\beta\mathbf{v}). \qquad (6.1.37)$$

In a similar way we may easily prove a property of *distributivity for the addition of the real numbers* α, $\beta \in R$

$$(\alpha + \beta)\tilde{S} = \alpha\tilde{S} + \beta\tilde{S} \qquad (6.1.38)$$

and a property of *distributivity for the law of internal composition*

$$\alpha(\tilde{S}_1 + \tilde{S}_2) = \alpha\tilde{S}_1 + \alpha\tilde{S}_2. \qquad (6.1.39)$$

Thus we have proved that the quotient set $\mathscr{M}/K$ where internal and external composition laws have been defined, represents a *vector space on the field of real numbers R*.

It is important to note that the internal composition law introduced above corresponds to the plastic collision, since the mass of the sum element is equal to the sum of the masses of the component elements, while the resultant velocity corresponds to that given by formula (6.1.22). Also, the quotient set $\mathscr{M}/K$ obtained with the aid of the equivalence law $K(S) = m\mathbf{v}$ corresponds to the physical meaning of collision; indeed, if the elements S_1 and S_2 are equivalent they have the same momentum. This justifies the name of vector space given to the *space of plastic collisions* thus obtained.

Applying the second principle of mechanics to two equivalent elements S_1 and S_2, we may write

$$\frac{d}{dt}(m_1\mathbf{v}_1) = \frac{d}{dt}(m_2\mathbf{v}_2) = \mathbf{F}_1 = \mathbf{F}_2 ; \qquad (6.1.40)$$

hence, the differential equation of the motion of two elements of the same class of equivalence is the same and this illustrates the mechanical sense of the equivalence K.

We may write

$$\alpha_1 \tilde{S}_1 + \alpha_2 \tilde{S}_2 + \alpha_3 \tilde{S}_3 = \left(m_1 + m_2 + m_3; \frac{\alpha_1 m_1 \mathbf{v}_1 + \alpha_2 m_2 \mathbf{v}_2 + \alpha_3 m_3 \mathbf{v}_3}{m_1 + m_2 + m_3} \right);$$

(6.1.41)

if we take into account the definition (6.1.29) of the neutral element, it follows that relation

$$K(\alpha_1 \tilde{S}_1 + \alpha_2 \tilde{S}_2 + \alpha_3 \tilde{S}_3) = K(\tilde{0}) = 0$$

(6.1.41')

is verified if

$$(\alpha_1 m_1)\mathbf{v}_1 + (\alpha_2 m_2)\mathbf{v}_2 + (\alpha_3 m_3)\mathbf{v}_3 = 0.$$

(6.1.41'')

In this way, relation

$$\alpha_1 \tilde{S}_1 + \alpha_2 \tilde{S}_2 + \alpha_3 \tilde{S}_3 = 0, \quad \alpha_1, \alpha_2, \alpha_3 \in R,$$

(6.1.42)

cannot occur for arbitrary $\tilde{S}_1$, $\tilde{S}_2$, $\tilde{S}_3$, unless $\alpha_1 = \alpha_2 = \alpha_3 = 0$.

On the other hand, we may have a relation of the form

$$\sum_{i=1}^{n} \alpha_i \tilde{S}_i = 0, \quad \alpha_i \in R \quad (i = 1, 2, ..., n)$$

(6.1.43)

without all α_i being zero for all $n > 3$; it follows that the *space of plastic collisions is three-dimensional*.

6.1.2.3 Metrization of the space of plastic collisions. Carnot's theorem

We shall now consider the mapping $\tilde{S} \to |\tilde{S}|$ of $\mathcal{M}/K$ onto the real positive half-axis R defined by relation

$$|\tilde{S}| = \frac{1}{2} m \mathbf{v}^2 ;$$

(6.1.44)

we remark that in this way we have introduced the *kinetic energy* of the element.

It may be shown that the mapping defined by relation (6.1.44) satisfies the following properties:

$$|\tilde{S}| \geqslant 0 \quad \text{(the equality implies } \tilde{S} = \tilde{0} \in \mathcal{M}/K\text{)},$$

(6.1.45)

$$|\tilde{S}_1 + \tilde{S}_2| \leqslant |\tilde{S}_1| + |\tilde{S}_2|, \tilde{S}_1, \tilde{S}_2 \in \mathcal{M}/K,$$

(6.1.45')

$$|\alpha \tilde{S}| = \alpha^2 |\tilde{S}|, \alpha \in R.$$

(6.1.45'')

Indeed, properties (6.1.45) and (6.1.45″) follow immediately from relation (6.1.44). We remark further that

$$|\tilde{S}_1| + |\tilde{S}_2| = \frac{1}{2} m_1 \mathbf{v}_1^2 + \frac{1}{2} m_2 \mathbf{v}_2^2, \tag{6.1.46}$$

$$|\tilde{S}_1 + \tilde{S}_2| = \frac{1}{2} \frac{(m_1 \mathbf{v}_1 + m_2 \mathbf{v}_2)^2}{m_1 + m_2} ; \tag{6.1.46'}$$

here $|\tilde{S}_1| + |\tilde{S}_2|$ is the kinetic energy T' of the system of two particles before plastic collision, while $|\tilde{S}_1 + \tilde{S}_2|$ is the kinetic energy T'' after plastic collision. Thus, Carnot's theorem 6.1.6 leads to the property (6.1.45′).

We remark that the space of plastic collisions might have been normed by introducing a norm defined by the relation

$$\|\tilde{S}\| = |m\,\mathbf{v}| ; \tag{6.1.47}$$

in that case the distance in the respective space is given by

$$\mathrm{d}(\tilde{S}_1, \tilde{S}_2) = |m_1 \mathbf{v}_1 - m_2 \mathbf{v}_2|. \tag{6.1.47'}$$

Although the mapping (6.1.44) does not constitute a norm in the space of plastic collisions, it permits the introduction of the notion of *distance* in the set $\mathcal{M}/K$ and thus the space of plastic collisions becomes *a metric space*. Therefore we shall use

Definition 6.1.2. *The number*

$$\mathrm{d}(\tilde{S}_1, \tilde{S}_2) = |\tilde{S}_1 - \tilde{S}_2| \tag{6.1.48}$$

is said to be the distance in the set $\mathcal{M}/K$; *whence we obtain*

$$\mathrm{d}(\tilde{S}_1, \tilde{S}_2) = \frac{1}{2} \frac{(m_1 \mathbf{v}_1 - m_2 \mathbf{v}_2)^2}{m_1 + m_2}, \tag{6.1.48'}$$

an expression which may be written in different ways with the aid of the kinetic energy.

Taking into account the mapping (6.1.44), it may be easily seen that the distance as defined above, satisfies the following conditions:

$$\mathrm{d}(\tilde{S}_1, \tilde{S}_2) = 0 \text{ if } \tilde{S}_1 = \tilde{S}_2, \tag{6.1.49}$$

$$\mathrm{d}(\tilde{S}_1, \tilde{S}_2) = \mathrm{d}(\tilde{S}_2, \tilde{S}_1) \text{ for all } \tilde{S}_1, \tilde{S}_2 \in \mathcal{M}/K, \tag{6.1.49'}$$

$$\mathrm{d}(\tilde{S}_1, \tilde{S}_3) \leqslant \mathrm{d}(\tilde{S}_1, \tilde{S}_2) + \mathrm{d}(\tilde{S}_2, \tilde{S}_3). \tag{6.1.49''}$$

6.1.2.4. Differentiation in the space of plastic collisions.
The analogue of Carnot's theorem for the energy of acceleration

Let $\tilde{S} = (m(t), \mathbf{v}(t))$ be an element of $\mathcal{M}/K$; we assume that $m(t)$ and $\mathbf{v}(t)$ are functions of class C^1 with respect to the time t. If t' is a value belonging to the domain of definition, there follows

$$\tilde{S}(t') - \tilde{S}(t) = \left(m(t') + m(t); \frac{m(t')\mathbf{v}(t') - m(t)\mathbf{v}(t)}{m(t') + m(t)} \right). \qquad (6.1.50)$$

Multiplying by $\dfrac{1}{t' - t}$ we obtain

$$\frac{\tilde{S}(t') - \tilde{S}(t)}{t' - t} = \left(m(t') + m(t); \frac{1}{m(t') + m(t)} \frac{\Delta(m\mathbf{v})}{\Delta t} \right), \qquad (6.1.50')$$

where $\Delta(m\mathbf{v}) = m(t')\mathbf{v}(t') - m(t)\mathbf{v}(t)$ and $\Delta t = t' - t$; passing to the limit we have

$$\frac{d\tilde{S}}{dt} = \left(2m; \frac{1}{2m} \frac{d}{dt}(m\mathbf{v}) \right). \qquad (6.1.51)$$

Taking into account the second law of mechanics we can write also

$$\frac{d\tilde{S}}{dt} = \left(2m; \frac{\mathbf{F}}{2m} \right). \qquad (6.1.51')$$

The equivalence relation corresponding to the derivative $\dfrac{d\tilde{S}}{dt}$ is

$$K\left(\frac{d\tilde{S}}{dt} \right) = 2m \frac{\mathbf{F}}{2m} = \mathbf{F} \qquad (6.1.52)$$

and leads to the force acting on the particle, as was to be expected considering Newton's second law.

Introducing the mapping (6.1.44) for the element (6.1.51'), we obtain

$$2 \left| \frac{d\tilde{S}}{dt} \right| = \frac{1}{2} m \left(\frac{\mathbf{F}}{m} \right)^2, \qquad (6.1.53)$$

where the *acceleration energy* has been marked out (corresponding to the case $m = \text{const}$).

We remark that the derivative is a linear operator in the space of plastic collisions and verifies the relations

$$\frac{d}{dt}(\alpha \tilde{S}) = \alpha \frac{d\tilde{S}}{dt} + \frac{d\alpha}{dt} \tilde{S}, \tag{6.1.54}$$

$$\frac{d}{dt}(\tilde{S}_1 + \tilde{S}_2) = \frac{d\tilde{S}_1}{dt} + \frac{d\tilde{S}_2}{dt}. \tag{6.1.54'}$$

It two elements $\tilde{S}_1$ and $\tilde{S}_2$ are acted on by the forces $\mathbf{F}_1$ and $\mathbf{F}_2$ the corresponding energies of acceleration are

$$\mathcal{T}_1 = \frac{1}{2} m_1 \left(\frac{\mathbf{F}_1}{m_1} \right)^2, \ \mathcal{T}_2 = \frac{1}{2} m_2 \left(\frac{\mathbf{F}_2}{m_2} \right)^2, \tag{6.1.55}$$

and the acceleration energy before collision is

$$\mathcal{T}' = \mathcal{T}_1 + \mathcal{T}_2; \tag{6.1.55'}$$

after a plastic collision it results the acceleration energy

$$\mathcal{T}'' = \frac{1}{2}(m_1 + m_2) \left(\frac{\mathbf{F}_1 + \mathbf{F}_2}{m_1 + m_2} \right)^2. \tag{6.1.55''}$$

The property (6.1.45') in the space of plastic collisions leads to the relation

$$\mathcal{T}'' \leqslant \mathcal{T}', \tag{6.1.56}$$

thus permitting to state

Theorem 6.1.6'. *The acceleration energy after the plastic collision of two particles is smaller or at most equal to the acceleration energy before collision. This is an analogue of Carnot's theorem 6.1.6* and may be written also in the form

$$\frac{1}{2} \frac{(\mathbf{F}_1 + \mathbf{F}_2)^2}{m_1 + m_2} \leqslant \frac{1}{2} \frac{\mathbf{F}_1^2}{m_1} + \frac{1}{2} \frac{\mathbf{F}_2^2}{m_2}. \tag{6.1.56'}$$

6.1.2.5. Primitives in the space of plastic collisions

We cannot introduce the notion of Riemann integral in the set $\mathcal{M}/K$ but we can introduce the concept of *primitive* by the following

Definition 6.1.3. $\tilde{S}^* \in \mathcal{M}/K$ *is said to be a primitive for* $\tilde{S} \in \mathcal{M}/K$ *if*

$$\frac{d\tilde{S}^*}{dt} = \tilde{S}; \tag{6.1.57}$$

this may be written also

$$\tilde{S}^* = \int \tilde{S}\, dt. \qquad (6.1.57')$$

We also give the following

Definition 6.1.4. *The element* $\tilde{S} = (m, \mathbf{v}) \in \mathcal{M}/K$ *is said to be constant if its momentum is constant, i.e.*

$$K(\tilde{S}) = m\mathbf{v} = \mathbf{c}. \qquad (6.1.58)$$

Thus, the element

$$\tilde{S}_0 = \left(m; \frac{\mathbf{c}}{m}\right) \qquad (6.1.58')$$

is constant in $\mathcal{M}/K$.

Since

$$\frac{d\tilde{S}_0}{dt} = \left(2m; \frac{1}{2m}\frac{d\mathbf{c}}{dt}\right),$$

we deduce that

$$K\left(\frac{d\tilde{S}_0}{dt}\right) = \frac{d\mathbf{c}}{dt} = 0,$$

which shows that we may write for a constant element $\tilde{S}_0$

$$\frac{d\tilde{S}_0}{dt} = \tilde{0}. \qquad (6.1.58'')$$

Also, we can write

$$\frac{d}{dt}(\tilde{S}^* + \tilde{S}_0) = \frac{d\tilde{S}^*}{dt} + \frac{d\tilde{S}_0}{dt} = \tilde{S} + \tilde{0} = \tilde{S}, \qquad (6.1.59)$$

and therefore we may state

Theorem 6.1.7. *If* $\tilde{S}^*$ *is a primitive for* $\tilde{S}$, *then* $(\tilde{S}^* + \tilde{S}_0)$ *is a primitive for* $\tilde{S}$ *too.*

Let us now consider

$$\tilde{S} = (m(t); \mathbf{v}(t)), \qquad (6.1.60)$$

$$\tilde{S}^* = (m_1(t); \mathbf{v}_1(t)), \qquad (6.1.60')$$

where $m(t)$ and $\mathbf{v}(t)$ are functions of class C^1; from relation

$$\frac{\mathrm{d}\tilde{S}^*}{\mathrm{d}t} = \tilde{S} \tag{6.1.61}$$

we obtain

$$\left(2m_1; \frac{1}{2m_1}\frac{\mathrm{d}}{\mathrm{d}t}(m_1\mathbf{v}_1)\right) = (m; \mathbf{v}), \tag{6.1.61'}$$

so that, by the equality of momenta, we may write

$$\frac{\mathrm{d}}{\mathrm{d}t}(m_1\mathbf{v}_1) = m\mathbf{v}. \tag{6.1.62}$$

Integrating we obtain

$$m_1\mathbf{v}_1 = \int m\mathbf{v}\,\mathrm{d}t, \tag{6.1.62'}$$

therefore *the general form of the primitive $\tilde{S}^*$* is expressed by

$$\tilde{S}^* = \int \tilde{S}\,\mathrm{d}t = \left(m_1(t), \frac{1}{m_1(t)}\int m\mathbf{v}\,\mathrm{d}t\right), \tag{6.1.63}$$

where $m_1(t)$ is an arbitrary positive function.

6.2. Bodies of variable mass

The problem of bodies of variable mass is closely related to the problem of collision; we may consider *bodies with a continuously variable mass,* but we may have also to deal with bodies with a *discontinuously variable mass*; from a mathematical point of view, such a phenomenon may be described in the same manner as a collision. Therefore, we shall use in the following discussion the results of the preceding section.

We shall first study the case of a particle and then the case of a system of particles; the case of a continuous body may be studied in a similar way.

6.2.1. Particle of variable mass

We shall establish Meščerski's equation for a particle of variable mass using the space of plastic collisions introduced in the preceding section. We shall first consider the case of a particle with a continuously variable mass and then the case of a particle with a discontinuously variable mass.

6.2.1.1. Meščerski's equation for a particle with continuously variable mass

Let $\tilde{S} = (m(t), \mathbf{v}(t))$ be an arbitrary element of $\mathcal{M}/K$, the momentum of which is $K(\tilde{S}) = m\mathbf{v}$. Assuming that the functions $m(t)$ and $\mathbf{v}(t)$ are of class C^1, we may express the derivative of S by relation (6.1.51); the momentum of this derivative is the force $\mathbf{F}$ acting on the element $\tilde{S}$, if we take into account the second law of mechanics in the form

$$\frac{\mathrm{d}}{\mathrm{d}t}\left[m(t)\mathbf{v}(t)\right] = \mathbf{F}(t), \tag{6.2.1}$$

corresponding to a particle of variable mass.

If $\mathbf{u}$ is the velocity and $\dfrac{\mathrm{d}m}{\mathrm{d}t}$ the mass of a particle attached to $\tilde{S}$, we may write

$$\tilde{S}' = \left(\frac{\mathrm{d}m}{\mathrm{d}t}; -\mathbf{u}\right); \tag{6.2.2}$$

the velocity $\mathbf{u}$ has been taken with the sign $(-)$ because the particle is directed toward the particle of mass m and becomes attached to the latter (as it has been assumed).

Composing $\dfrac{\mathrm{d}\tilde{S}}{\mathrm{d}t}$ and $\tilde{S}'$ we obtain

$$\frac{\mathrm{d}\tilde{S}}{\mathrm{d}t} + \tilde{S}' = \left(2m + \frac{\mathrm{d}m}{\mathrm{d}t}; \frac{1}{2m + \dfrac{\mathrm{d}m}{\mathrm{d}t}}\left[\frac{\mathrm{d}(m\mathbf{v})}{\mathrm{d}t} - \frac{\mathrm{d}m}{\mathrm{d}t}\mathbf{u}\right]\right), \tag{6.2.3}$$

the momentum of the new element being equal to the force $\mathbf{F}$ which is acting on it

$$K\left(\frac{\mathrm{d}\tilde{S}}{\mathrm{d}t} + \tilde{S}'\right) = \frac{\mathrm{d}}{\mathrm{d}t}(m\mathbf{v}) - \frac{\mathrm{d}m}{\mathrm{d}t}\mathbf{u} = \mathbf{F}; \tag{6.2.4}$$

hence, we may write

$$\frac{\mathrm{d}}{\mathrm{d}t}(m\mathbf{v}) = \mathbf{F} + \frac{\mathrm{d}m}{\mathrm{d}t}\mathbf{u}. \tag{6.2.5}$$

This is *Meščerski's equation for a particle of continuously variable mass* in the case where *new particles are attached* to the initial particle.

When a particle is detached from the particle of mass $m(t)$ we may introduce it in the form

$$\tilde{S}'' = \left(- \frac{dm}{dt}; \mathbf{u} \right), \tag{6.2.6}$$

where the mass $m(t)$ is decreasing; proceeding as before we find again equation (6.2.5) for a particle with continuously variable mass in the case where *particles are detached* from the initial particle.

Denoting the velocity of the detached particle with respect to the initial particle by

$$\mathbf{v}_r = \mathbf{u} - \mathbf{v}, \tag{6.2.7}$$

we may write the equation of motion in the form

$$m \frac{d\mathbf{v}}{dt} = \mathbf{F} + \mathbf{\Phi}, \tag{6.2.8}$$

where

$$\mathbf{\Phi} = \frac{dm}{dt} \mathbf{v}_r \tag{6.2.9}$$

represents the *reactive force*.

Let us now consider a particle the initial mass of which is

$$m(t) = m_1(t) + m_2(t), \tag{6.2.10}$$

where $m_1(t)$ and $m_2(t)$ represent the *attached* and *the detached mass,* respectively, at the initial moment; the element $\tilde{S} = (m; \mathbf{v})$ corresponds to the particle, and $\mathbf{F}$ is the force acting on it. We shall also consider the particles

$$\tilde{S}_1 = \left(\frac{dm_1}{dt}; -\mathbf{u}_1 \right), \quad \tilde{S}_2 = \left(- \frac{dm_2}{dt}; \mathbf{u}_2 \right), \tag{6.2.11}$$

which correspond to an increase or decrease of the mass, respectively; proceeding as before we obtain the equation of motion in the form

$$\frac{d}{dt}(m\mathbf{v}) = \mathbf{F} + \frac{dm_1}{dt} \mathbf{u}_1 + \frac{dm_2}{dt} \mathbf{u}_2, \tag{6.2.12}$$

which represents a generalization of Meščerski's equation (6.2.5).

6.2.1.2. Meščerski's equation for a particle of discontinuous variable mass

We shall assume that the force $\mathbf{F}$ considered in the preceding section is conservative, i.e. we have

$$\mathbf{F} = \operatorname{grad} U, \tag{6.2.13}$$

where the derivatives are considered in the sense of the theory of distributions; also, we assume that the masses m_1 and m_2 are dependent on the time variable t only. In that case, equation (6.2.12) may be obtained on condition that the functional

$$I(x, y, z) = \int_{t'}^{t''} \left[\frac{1}{2} m(x^2 + y^2 + z^2) + U(x, y, z; t) \right.$$

$$\left. + U_1(x, y, z; t) + U_2(x, y, z; t) \right] dt \qquad (6.2.14)$$

be stationary; we have introduced the functions

$$U_1 = \frac{dm_1}{dt} \mathbf{u}_1 \cdot \mathbf{r}, \; U_2 = \frac{dm_2}{dt} \mathbf{u}_2 \cdot \mathbf{r}, \qquad (6.2.15)$$

where $\mathbf{r} = \mathbf{r}(x, y, z)$ is the radius vector.

Assuming that $x(t)$, $y(t)$, $z(t)$, $U(t)$, $U_1(t)$ and $U_2(t)$ are functions of class C^0 while $m_1(t)$, $m_2(t)$ are functions of class C^1 everywhere except at a finite number of points of discontinuity of the first species, it may be stated that the operations indicated in the functional (6.2.14) have a meaning in the theory of distributions. In that case the generalized Meščerski's equation (6.2.12) may be considered in the sense of the theory of distributions under the conditions stated for the abovementioned functions.

Effecting the operations in the sense of the theory of distributions, we may write

$$\frac{d}{dt}(m\mathbf{v}) = \frac{\tilde{d}}{dt}(m\mathbf{v}) + \sum_{i=1}^{n} (\Delta \mathbf{H})_{t_i} \delta(t - t_i), \qquad (6.2.16)$$

since the momentum $\mathbf{H} = m\mathbf{v}$ has discontinuities of the first species only; we assume that the moments of discontinuity are $t_i (i = 1, 2, \ldots, n)$. The jump of the momentum is

$$(\Delta \mathbf{H})_{t_i} = m(t_i + 0)\mathbf{v}(t_i + 0) - m(t_i - 0)\mathbf{v}(t_i - 0). \qquad (6.2.17)$$

Assuming that of the n moments of discontinuity the first n_1 correspond to the discontinuities of the mass $m_1(t)$ and the last $n_2 = n - n_1$ to the discontinuities of the mass $m_2(t)$ we may write

$$\frac{dm_1}{dt} = \frac{\tilde{d}m_1}{dt} + \sum_{j=1}^{n_1} (\Delta m_1)_{t_j} \delta(t - t_j), \qquad (6.2.18)$$

$$\frac{dm_2}{dt} = \frac{\tilde{d}m_2}{dt} + \sum_{k=n_1+1}^{n} (\Delta m_2)_{t_k} \delta(t - t_k), \qquad (6.2.18')$$

where we have introduced the jumps

$$(\Delta m_1)_{t_j} = m_1(t_j + 0) - m_1(t_j - 0) \qquad (j = 1, 2, ..., n_1), \qquad (6.2.19)$$

$$(\Delta m_2)_{t_k} = m_2(t_k + 0) - m_2(t_k - 0) \qquad (k = n_1 + 1, n_1 + 2, ..., n). \quad (6.2.19')$$

Therefore equation (6.2.12) takes the form

$$\frac{\tilde{d}}{dt}(m\mathbf{v}) + \sum_{i=1}^{n} (\Delta \mathbf{H})_{t_i} \delta(t - t_i) = \mathbf{F} + \frac{\tilde{d}m_1}{dt} \mathbf{u}_1 + \frac{\tilde{d}m_2}{dt} \mathbf{u}_2$$

$$+ \sum_{j=1}^{n_1} (\Delta m_1)_{t_j} \mathbf{u}_1 \delta(t - t_j) + \sum_{k=n_1+1}^{n} (\Delta m_2)_{t_k} \mathbf{u}_2 \delta(t - t_k); \qquad (6.2.20)$$

this equation may be replaced by the equation

$$\frac{\tilde{d}}{dt}(m\mathbf{v}) = \mathbf{F} + \frac{\tilde{d}m_1}{dt} \mathbf{u}_1 + \frac{\tilde{d}m_2}{dt} \mathbf{u}_2, \qquad (6.2.21)$$

corresponding to the points of continuity and where the derivatives are taken in the ordinary sense, and by the relations

$$(\Delta \mathbf{H})_{t_j} = (\Delta m_1)_{t_j} \mathbf{u}_1 \qquad (j = 1, 2, ..., n_1),$$

$$(\Delta \mathbf{H})_{t_k} = (\Delta m_2)_{t_k} \mathbf{u}_2 \qquad (k = n_1 + 1, n_1 + 2, ..., n), \qquad (6.2.21')$$

corresponding to the discontinuities of the mass. In this way, we obtain *the generalized equation of Meščerski for a particle with a discontinuous variable mass*.

If we consider the form (4.1.60) of the equation of motion of a particle, we remark that the effect of initial conditions may be introduced in equation (6.2.20), in a similar manner.

6.2.2. Systems of particles of variable mass

We shall establish Lagrange's equations for systems of particles of variable mass; proceeding as before we shall consider first a continuous and then a discontinuous variation.

6.2.2.1. Lagrange's equations for a system of particles of continuous variable mass

Let there be a system of n particles of mass $m_i(t)$ $(i = 1, 2, ..., n)$; the generalized co-ordinates which determine the position of the system are q_k $(k = 1, 2, ..., h)$. Obviously, it is assumed that we have to deal with *holonomic constraints* only.

We mark by $\mathbf{u}_i'(t)$ and $\mathbf{u}_i''(t)$ the velocities of the particles *captured* or *emitted,* respectively, by the particle of mass $m_i(t)$, and we denote by $m_i'(t)$ and $m_i''(t)$ the corresponding masses at the initial moment, so as to have

$$m_i(t) = m_i'(t) + m_i''(t). \tag{6.2.22}$$

We assume further that the generalized forces Q_k are derivatives of a potential $U = U(q_1, q_2, \ldots, q_h; t)$, i.e. we have

$$Q_k = \frac{\partial U}{\partial q_k} \qquad (k = 1, 2, \ldots, h), \tag{6.2.23}$$

where the derivative is considered in the sense of the theory of distributions; also, it may be seen easily that the generalized forces Q_k', Q_k'', corresponding to capture and emmision, respectively, are derivatives of the potentials

$$U_1 = \sum_{i=1}^{n} \frac{\mathrm{d}m_i'}{\mathrm{d}t}\,\mathbf{u}_i'(t)\cdot\mathbf{r}_i(t), \quad U_2 = \sum_{i=1}^{n} \frac{\mathrm{d}m_i''}{\mathrm{d}t}\,\mathbf{u}_i''(t)\cdot\mathbf{r}_i(t), \tag{6.2.24}$$

where $\mathbf{r}_i(t)$ are the corresponding radius vectors of the particles, and may be written in the form

$$Q_k' = \frac{\partial U_1}{\partial q_k}, \quad Q_k'' = \frac{\partial U_2}{\partial q_k} \qquad (k = 1, 2, \ldots, h). \tag{6.2.24'}$$

We remark that the potentials U_1 and U_2 have significance in the theory of distributions since the velocities $\mathbf{u}_i'(t)$, $\mathbf{u}_i''(t)$ are of the class C^0, $\mathbf{r}_i(t)$ is of class C^1 with respect to the generalized co-ordinates, and $m_i'(t)$, $m_i''(t)$ are everywhere of class C^1 except at a finite number of moments of discontinuity of the first species. The kinetic energy of the system $T = T(q_1, q_2, \ldots, q_h; \dot{q}_1, \dot{q}_2, \ldots, \dot{q}_h; t)$ is also introduced.

Under the above conditions the equations of motion of a system of particles of variable mass are obtained by writing that the functional

$$I(q_1, q_2, \ldots, q_h) = \int_{t'}^{t''} (T + U + U_1 + U_2)\,\mathrm{d}t \tag{6.2.25}$$

is stationary in the sense of the theory of distributions and the admissible lines might be distributions.

Thus we obtain equations

$$\frac{\mathrm{d}}{\mathrm{d}t}\left(\frac{\partial T}{\partial \dot{q}_k}\right) - \frac{\partial T}{\partial q_k} = Q_k + Q_k' + Q_k'' \qquad (k = 1, 2, \ldots, h), \tag{6.2.26}$$

which represent *Lagrange's equations of motion of a system of particles of continuous variable mass*. If the forces $\mathbf{F}_i$ are not conservative, we may write

$$Q_k = \sum_{i=1}^{n} \mathbf{F}_i \cdot \frac{\partial \mathbf{r}_i}{\partial q_k} . \tag{6.2.27}$$

Denoting by

$$\mathscr{L} = T + U + U_1 + U_2 \tag{6.2.28}$$

a distribution analogous to the *kinetic potential of Lagrange,* the equations (6.2.26) become

$$\frac{d}{dt}\left(\frac{\partial \mathscr{L}}{\partial \dot{q}_k}\right) - \frac{\partial \mathscr{L}}{\partial q_k} = 0 \qquad (k = 1, 2, ..., h). \tag{6.2.26'}$$

Obviously, equations (6.2.26) are still valid in the case where the system of particles is subject to certain percussion forces (complementary forces) $\mathscr{F}_i \delta(t - t_i)$ $(i = 1, 2, ..., n)$; in that case the generalized force Q_k is expressed by

$$Q_k = \sum_{i=1}^{n} \mathscr{F}_i \cdot \frac{\partial \mathbf{r}_i}{\partial q_k} \delta(t - t_i). \tag{6.2.27'}$$

6.2.2.2. Lagrange's equations for a system of particles of discontinuous variable mass

If the masses $m_i'(t)$ have discontinuities of the first species corresponding to the moments $t_j (j = 1, 2, ..., n')$ and the masses $m_i''(t)$ have discontinuities of the first species corresponding to the moments $t_l (l = n' + 1, n' + 2, ..., n'')$ we may write the relations

$$\frac{dm_i'}{dt} = \frac{\widetilde{d}m_i'}{dt} + \sum_{j=1}^{n'} (\Delta m_i')_{t_j} \delta(t - t_j), \tag{6.2.29}$$

$$\frac{dm_i''}{dt} = \frac{\widetilde{d}m_i''}{dt} + \sum_{l=n'+1}^{n''} (\Delta m_i'')_{t_l} \delta(t - t_l) \qquad (i = 1, 2, ..., n), \tag{6.2.29'}$$

where the jumps of the masses are

$$(\Delta m_i')_{t_j} = m_i'(t_j + 0) - m_i'(t_j - 0) \qquad (j = 1, 2, ..., n'), \tag{6.2.30}$$

$$(\Delta m_i'')_{t_l} = m_i''(t_l + 0) - m_i''(t_l - 0) \qquad (l = n' + 1, n' + 2, ..., n''). \tag{6.2.30'}$$

Under the above assumptions, the kinetic energy T and the Lagrangian $\mathscr{L}$ will have discontinuities of the first species only, so that we may write

$$\frac{\mathrm{d}}{\mathrm{d}t}\left(\frac{\partial T}{\partial \dot{q}_k}\right) = \frac{\mathrm{d}}{\mathrm{d}t}\left(\frac{\partial \mathscr{L}}{\partial \dot{q}_k}\right) = \frac{\tilde{\mathrm{d}}}{\mathrm{d}t}\left(\frac{\partial T}{\partial \dot{q}_k}\right) + \sum_{i=1}^{\tilde{n}''}(\Delta p_k)_{t_i}\delta(t - t_i) \quad (k = 1, 2, ..., h),$$

(6.2.31)

where $p_k = \dfrac{\partial \mathscr{L}}{\partial \dot{q}_k}$ is the *generalized momentum*.

Theorem 1.1.3. of section 1.1.2.6 permits us to write the equation describing the motion of the system in the domain of continuity, namely

$$\frac{\tilde{\mathrm{d}}}{\mathrm{d}t}\left(\frac{\partial T}{\partial \dot{q}_k}\right) - \frac{\partial T}{\partial q_k} = Q_k + \tilde{Q}'_k + \tilde{Q}''_k \qquad (k = 1, 2, ..., h), \qquad (6.2.32)$$

where we introduce the generalized forces in the ordinary sense

$$\tilde{Q}'_k = \sum_{i=1}^{n}\frac{\tilde{\mathrm{d}}m'_i}{\mathrm{d}t}\mathbf{u}'_i \cdot \frac{\partial \mathbf{r}_i}{\partial q_k}, \quad \tilde{Q}''_k = \sum_{i=1}^{n}\frac{\tilde{\mathrm{d}}m''_i}{\mathrm{d}t}\mathbf{u}''_i \cdot \frac{\partial \mathbf{r}_i}{\partial q_k} \qquad (k = 1, 2, ..., h); \qquad (6.2.33)$$

the conditions for jumps will be of the form

$$(\Delta p_k)_{t_j} = \sum_{i=1}^{n}(\Delta m'_i)_{t_j}\mathbf{u}'_i \cdot \frac{\partial \mathbf{r}_i}{\partial q_k}, \quad (\Delta p_k)_{t_l} = \sum_{i=1}^{n}(\Delta m''_i)_{t_l}\mathbf{u}''_k \cdot \frac{\partial \mathbf{r}_i}{\partial q_k}$$

$$(j = 1, 2, ..., n', l = n' + 1, n' + 2, ..., n''). \qquad (6.2.34)$$

These equations correspond to the *motion of a system of particles of discontinuous variable mass*.

6.2.2.3. Lagrange's equations for collision

If in equation (6.2.26) we put

$$Q'_k = Q''_k = 0 \qquad (6.2.35)$$

we obtain *Lagrange's equations corresponding to a system of particles of constant mass* (classical equations).

Assuming that $t \in [t', t'']$, we may write

$$\int_{t'}^{t''}\frac{\mathrm{d}}{\mathrm{d}t}\left(\frac{\partial T}{\partial \dot{q}_k}\right)\mathrm{d}t - \int_{t'}^{t''}\frac{\partial T}{\partial q_k}\mathrm{d}t = \int_{t'}^{t''}Q_k\,\mathrm{d}t. \qquad (6.2.36)$$

Proceeding as in section 6.1.1, we assume that in that interval we have a moment of discontinuity $t = t_0$. Also, we shall use the above notations although they have no significance in the theory of distributions, but in this way we maintain the classical symbolism.

Applying a mean value theorem we may write

$$\lim_{t''-t'\to 0} \int_{t'}^{t''} \frac{\partial T}{\partial q_k}\, dt = 0; \qquad (6.2.37)$$

also

$$\lim_{t''-t'\to 0} \int_{t'}^{t''} \frac{d}{dt}\left(\frac{\partial T}{\partial \dot{q}_k}\right) dt = (\Delta p_k)_0, \qquad (6.2.38)$$

where we have introduced the jump of the generalized momentum.

We introduce the generalized percussion in the form

$$\mathscr{P}_k = \lim_{t''-t'\to 0} \int_{t'}^{t''} Q_k\, dt. \qquad (6.2.39)$$

In that case equations (6.2.36) become

$$(\Delta p_k)_0 = \mathscr{P}_k \qquad (k = 1, 2, ..., h), \qquad (6.2.40)$$

and we may state the following

Theorem 6.2.1. *The jump of the generalized momentum of a system of particles corresponding to a moment of discontinuity, is equal to the generalized percussion corresponding to the same moment.* This represents *the analogue of the momentum theorem* in the case of collision.

Part 3

Applications of the theory of distributions in mechanics of solids

7

Mathematical model of the mechanics of solids

In order to present some applications of the theory of distributions to the mechanics of solids it is necessary to show first how to construct a *mathematical model* of such bodies.

The considerations which follow refer in general to arbitrary deformable solids. However, considering the applications we shall deal with, we shall restrict our study to the case of homogeneuos, isotropic, linearly elastic bodies, subject to small deformations and rotations.

7.1. General equations of deformable solids

First, we shall point out briefly the *geometrical* and *mechanical* aspect of the problem; then we shall examine the *constitutive law* of deformable solids which emphasizes the *physical* properties of the body and links the geometrical to the mechanical aspects of the problem.

Evidently, all these considerations will be made within the frame of the mathematical model of Newtonian mechanics. Also, certain fundamental assumptions will be introduced, namely:

(*i*) The solid body investigated is considered to be at rest with respect to a fixed reference frame and is acted upon by *external loads in equilibrium*; if the body is in motion, we introduce the inertia forces, hence the external loads will be in *dynamical equilibrium*. Also, every part of the body will be acted upon by loads in equilibrium (a load system equivalent to zero).

(*ii*) The solid body is considered to be a *continuous medium* (without voids, microscopic internal cracks, etc). This makes us assume that the deformation of the body will also proceed in a continuous manner, hence that stresses and strains will be represented mathematically by functions which are at least continuous. Certain points of the body (singular points) where stresses and strains tend to infinity will be treated separately; besides, the introduction of distributions eliminates such points. Also, bodies with internal voids (continuous bodies corresponding to *multiply connected domains*) and bodies with internal cuts (particular cases of multiple connection), must be treated separately.

(*iii*) In the solid body investigated there are no *initial stresses* which may be due to some initial deformations of the material; these initial deformations might be caused by the processing of the material (rolling, etc.), during erection, or by phenomena occurring before the material is acted upon by external loads, by concrete contraction, etc. Although such phenomena exist and cannot be eliminated, it will be assumed for the sake of simplicity that there are no initial stresses. Hence, we shall assume that in a body which is not acted upon by external loads the state of stress is zero. If the internal stresses cannot be neglected, additional calculation are performed which take into account their effect.

Other hypotheses will be introduced later as required.

7.1.1. Geometrical and mechanical aspects of the problem

In the following discussion we shall first deal with the geometrical aspect of the problem which is in fact the first of which we become aware; then we shall emphasize the mechanical aspect of the problem connected to the internal forces which appear.

7.1.1.1. Geometry and kinematics of deformation

Under the action of external loads the particles which make up a solid body change (eventually in time) the position (with respect to a fixed reference point) which they had before they were acted upon by the loads. If, by a translation and a rotation, we can bring the particles of the body to the same position which they had before being acted upon, the *motion* is said to be that *of a rigid body*. Otherwise, the body undergoes a *deformation*. The totality of the deformations to which a particle of the body has been subjected constitutes the state of deformation at a point (centre of mass of the particle). The totality of the states of deformation corresponding to all the points (particles) of the solid body constitutes its *state of deformation*.

Together with the notion of deformation we introduce the notion of *displacement*. The totality of displacements corresponding to a particle of the solid body make up the state of displacement at a point. The totality of the states of displacement corresponding to all the points of the solid body constitutes the *state of displacement* of the body. Frequently, the state of deformation means also the state of displacement, because the final aim is to determine the components of the displacement vector. In general, we shall use the notion of state of deformation.

Bodies which can be subjected to rigid body displacements only are said to be *rigid solids*; all the other solid bodies are termed *deformable solids*.

In the following discussion we shall make a simplifying hypothesis: we shall assume that we have to deal with small deformations and rotations only, which are negligible when referred to unity.

We shall use an orthogonal system of co-ordinate axes $Oxyz^*$.
The state of deformation is characterized by the *strain tensor*

$$
T_\varepsilon \equiv \left\{
\begin{matrix}
\varepsilon_x & \dfrac{1}{2}\gamma_{xy} & \dfrac{1}{2}\gamma_{zx} \\[2mm]
\dfrac{1}{2}\gamma_{xy} & \varepsilon_y & \dfrac{1}{2}\gamma_{yz} \\[2mm]
\dfrac{1}{2}\gamma_{zx} & \dfrac{1}{2}\gamma_{yz} & \varepsilon_z
\end{matrix}
\right\},
\tag{7.1.1}
$$

which is a symmetric tensor of the second order; here $\varepsilon_x = \varepsilon_x(x, y, z)$, $\varepsilon_y = \varepsilon_y(x, y, z)$, $\varepsilon_z = \varepsilon_z(x, y, z)$ are *linear strains* (the index specifies the direction along which the deformation is measured), and $\gamma_{yz} = \gamma_{yz}(x, y, z)$, $\gamma_{zx} = \gamma_{zx}(x, y, z)$, $\gamma_{xy} = \gamma_{xy}(x, y, z)$ are *angular strains* (the indices specify the angular element whose deformation is considered).

The state of displacement is given by the *displacement vector* **u**, the components of which are $u = u(x, y, z)$, $v = v(x, y, z)$, $w = w(x, y, z)$.

In the linear case considered we have between the components of the tensor T_ε and the components of the vector **u** Cauchy's relations

$$
\varepsilon_x = \frac{\partial u}{\partial x}, \quad \varepsilon_y = \frac{\partial v}{\partial y}, \quad \varepsilon_z = \frac{\partial w}{\partial z},
\tag{7.1.2}
$$

$$
\gamma_{yz} = \frac{\partial v}{\partial z} + \frac{\partial w}{\partial y}, \quad \gamma_{zx} = \frac{\partial w}{\partial x} + \frac{\partial u}{\partial z}, \quad \gamma_{xy} = \frac{\partial u}{\partial y} + \frac{\partial v}{\partial x}.
\tag{7.1.2'}
$$

Equations (7.1.2), (7.1.2'), considered as a system of equations of the unknown functions u, v, w, must satisfy a set of compatibility conditions. The conditions, which are at the same time *conditions of continuity of deformations,* have been stated by B. de Saint-Venant in the form

$$
\frac{\partial^2 \varepsilon_y}{\partial z^2} + \frac{\partial^2 \varepsilon_z}{\partial y^2} = \frac{\partial^2 \gamma_{yz}}{\partial y \partial z}, \quad \frac{\partial^2 \varepsilon_z}{\partial x^2} + \frac{\partial^2 \varepsilon_x}{\partial z^2} = \frac{\partial^2 \gamma_{zx}}{\partial z \partial x}, \quad \frac{\partial^2 \varepsilon_x}{\partial y^2} + \frac{\partial^2 \varepsilon_y}{\partial x^2} = \frac{\partial^2 \gamma_{xy}}{\partial x \partial y},
\tag{7.1.3}
$$

$$
\frac{\partial}{\partial x}\left(-\frac{\partial \gamma_{yz}}{\partial x} + \frac{\partial \gamma_{zx}}{\partial y} + \frac{\partial \gamma_{xy}}{\partial z}\right) = 2\frac{\partial^2 \varepsilon_x}{\partial y \partial z},
$$

$$
\frac{\partial}{\partial y}\left(-\frac{\partial \gamma_{zx}}{\partial y} + \frac{\partial \gamma_{xy}}{\partial z} + \frac{\partial \gamma_{yz}}{\partial x}\right) = 2\frac{\partial^2 \varepsilon_y}{\partial z \partial x},
\tag{7.1.3'}
$$

$$
\frac{\partial}{\partial z}\left(-\frac{\partial \gamma_{xy}}{\partial z} + \frac{\partial \gamma_{yz}}{\partial x} + \frac{\partial \gamma_{zx}}{\partial y}\right) = 2\frac{\partial^2 \varepsilon_z}{\partial x \partial y};
$$

---

* Owing to the occurrence of privileged directions in the applications which will be considered, we shall not use the tensorial symbolism in the following discussion.

these conditions are necessary and sufficient in the case of solid bodies occupying a simply connected domain, but they are necessary only in the case of bodies occupying a multiply connected domain.

If the strains satisfy these continuity conditions, we can integrate the system of equations (7.1.2), (7.1.2′), and the state of displacement will be given by Cesàro's formulae

$$u_1(x_1, y_1, z_1) = u_0 - (y_1 - y_0)\,\omega_z^0 + (z_1 - z_0)\,\omega_y^0$$

$$+ \int_{\widehat{P_0 P_1}} \left\{ \left[\varepsilon_x + (y_1 - y)\left(\frac{\partial \varepsilon_x}{\partial y} - \frac{1}{2}\frac{\partial \gamma_{xy}}{\partial x} \right) + (z_1 - z)\left(\frac{\partial \varepsilon_x}{\partial z} - \frac{1}{2}\frac{\partial \gamma_{zx}}{\partial x} \right) \right] dx \right.$$

$$+ \left[\frac{1}{2}\gamma_{xy} - (y_1 - y)\left(\frac{\partial \varepsilon_y}{\partial x} - \frac{1}{2}\frac{\partial \gamma_{xy}}{\partial y} \right) + \frac{1}{2}(z_1 - z)\left(\frac{\partial \gamma_{xy}}{\partial z} - \frac{\partial \gamma_{yz}}{\partial x} \right) \right] dy$$

$$+ \left. \left[\frac{1}{2}\gamma_{zx} + \frac{1}{2}(y_1 - y)\left(\frac{\partial \gamma_{zx}}{\partial y} - \frac{\partial \gamma_{yz}}{\partial x} \right) - (z_1 - z)\left(\frac{\partial \varepsilon_z}{\partial x} - \frac{1}{2}\frac{\partial \gamma_{zx}}{\partial z} \right) \right] dz \right\},$$

$$v_1(x_1, y_1, z_1) = v_0 - (z_1 - z_0)\,\omega_x^0 + (x_1 - x_0)\,\omega_z^0$$

$$+ \int_{\widehat{P_0 P_1}} \left\{ \left[\frac{1}{2}\gamma_{xy} + \frac{1}{2}(z_1 - z)\left(\frac{\partial \gamma_{xy}}{\partial z} - \frac{\partial \gamma_{zx}}{\partial y} \right) - (x_1 - x)\left(\frac{\partial \varepsilon_x}{\partial y} - \frac{1}{2}\frac{\partial \gamma_{xy}}{\partial x} \right) \right] dx \right.$$

$$+ \left[\varepsilon_y + (z_1 - z)\left(\frac{\partial \varepsilon_y}{\partial z} - \frac{1}{2}\frac{\partial \gamma_{yz}}{\partial y} \right) + (x_1 - x)\left(\frac{\partial \varepsilon_y}{\partial x} - \frac{1}{2}\frac{\partial \gamma_{xy}}{\partial y} \right) \right] dy \tag{7.1.4}$$

$$+ \left. \left[\frac{1}{2}\gamma_{yz} - (z_1 - z)\left(\frac{\partial \varepsilon_z}{\partial y} - \frac{1}{2}\frac{\partial \gamma_{yz}}{\partial z} \right) + \frac{1}{2}(x_1 - x)\left(\frac{\partial \gamma_{yz}}{\partial x} - \frac{\partial \gamma_{zx}}{\partial y} \right) \right] dz \right\},$$

$$w_1(x_1, y_1, z_1) = w_0 - (x_1 - x_0)\,\omega_y^0 + (y_1 - y_0)\,\omega_x^0$$

$$+ \int_{\widehat{P_0 P_1}} \left\{ \left[\frac{1}{2}\gamma_{zx} - (x_1 - x)\left(\frac{\partial \varepsilon_x}{\partial z} - \frac{1}{2}\frac{\partial \gamma_{zx}}{\partial x} \right) + \frac{1}{2}(y_1 - y)\left(\frac{\partial \gamma_{zx}}{\partial y} - \frac{\partial \gamma_{xy}}{\partial z} \right) \right] dx \right.$$

$$+ \left[\frac{1}{2}\gamma_{yz} + \frac{1}{2}(x_1 - x)\left(\frac{\partial \gamma_{yz}}{\partial x} - \frac{\partial \gamma_{xy}}{\partial z} \right) - (y_1 - y)\left(\frac{\partial \varepsilon_y}{\partial z} - \frac{1}{2}\frac{\partial \gamma_{yz}}{\partial y} \right) \right] dy$$

$$+ \left. \left[\varepsilon_z + (x_1 - x)\left(\frac{\partial \varepsilon_z}{\partial x} - \frac{1}{2}\frac{\partial \gamma_{zx}}{\partial z} \right) + (y_1 - y)\left(\frac{\partial \varepsilon_z}{\partial y} - \frac{1}{2}\frac{\partial \gamma_{yz}}{\partial z} \right) \right] dz \right\};$$

here $u_0, v_0, w_0, \omega_x, \omega_y, \omega_z$ represent the displacements and rotations, respectively, corresponding to the *motion as a rigid body of the deformable solid*, written for the point $P_0(x_0, y_0, z_0)$.

We remark that the displacements must be functions of class C^3 whereas it is sufficient for the strains to be of class C^2.

In the dynamic case, these functions are also dependent on the time variable t, the displacements being thus of the form $u = u(x, y, z; t)$, $v = v(x, y, z; t)$, $w = w(x, y, z; t)$. Since small variations and rotations have been assumed, the *substantial derivative is equal to the partial derivative with respect to time* and therefore we shall consider the *displacement velocities* $\dot{u}$, $\dot{v}$, $\dot{w}$ and the *displacement accelerations* $\ddot{u}$, $\ddot{v}$, $\ddot{w}$.

We can also point out a *local rotation vector as a rigid body* expressed by relation

$$\boldsymbol{\omega} = \frac{1}{2}\operatorname{curl} \mathbf{u} \qquad (7.1.5)$$

or by the components

$$\omega_x = \frac{1}{2}\left(\frac{\partial w}{\partial y} - \frac{\partial v}{\partial z}\right),\quad \omega_y = \frac{1}{2}\left(\frac{\partial u}{\partial z} - \frac{\partial w}{\partial x}\right),\quad \omega_z = \frac{1}{2}\left(\frac{\partial v}{\partial x} - \frac{\partial u}{\partial y}\right). \quad (7.1.5')$$

In the case of *bodies of Cosserat type* we introduce an additional *vector of free rotation* and therefore a particle of such a body will have *six degrees of freedom*; however, we shall not deal with bodies of that type, although the methods of the theory of distributions which will be applied to bodies of the classical type may be equally used in that more general case.

It should be noted that by a *state of plane deformation* we mean a state of deformation where the points of the body are displaced on planes whose normal has a fixed direction. Such is the case, for example, of a cylinder of infinite length (practically finite), acted on its lateral surface by a load uniformly distributed along the generatrices and which has no component along their direction.

7.1.1.2. Mechanics of stresses

Owing to deformations, the static or dynamic equilibrium of the constraint forces acting between the particles of a body is disturbed, and additional internal forces are generated; the totality of these internal forces, termed as *stresses,* corresponding to a particle, makes up the state of stress at a point (centre of mass of the particle). The totality of the states of stress corresponding to all the points of the solid body constitutes the *state of stress* of the body.

The state of stress is characterized by the *stress tensor*

$$T_\sigma \equiv \left\{ \begin{array}{ccc} \sigma_x & \tau_{xy} & \tau_{zx} \\ \tau_{xy} & \sigma_y & \tau_{yz} \\ \tau_{zx} & \tau_{yz} & \sigma_z \end{array} \right\}, \qquad (7.1.6)$$

which is a symmetric tensor of the second order; $\sigma_x = \sigma_x(x, y, z)$, $\sigma_y = \sigma_y(x, y, z)$, $\sigma_z = \sigma_z(x, y, z)$ (the index specifies the external normal to the area element upon which the respective stress is acting) are *normal stresses,* while $\tau_{yz} = \tau_{yz}(x, y, z)$, $\tau_{zx} = \tau_{zx}(x, y, z)$, $\tau_{xy} = \tau_{xy}(x, y, z)$ are *tangential stresses* (the first index specifies the external normal to the element of area on which the stress is acting and the second index represents the direction along which the respective component is considered).

In the static case, the tensor components satisfy the system of partial differential equations *(equations of equilibrium)*

$$\frac{\partial \sigma_x}{\partial x} + \frac{\partial \tau_{xy}}{\partial y} + \frac{\partial \tau_{zx}}{\partial z} + X = 0,$$

$$\frac{\partial \tau_{xy}}{\partial x} + \frac{\partial \sigma_y}{\partial y} + \frac{\partial \tau_{yz}}{\partial z} + Y = 0, \qquad (7.1.7)$$

$$\frac{\partial \tau_{zx}}{\partial x} + \frac{\partial \tau_{yz}}{\partial y} + \frac{\partial \sigma_z}{\partial z} + Z = 0,$$

where $X = X(x, y, z)$, $Y = Y(x, y, z)$, $Z = Z(x, y, z)$ are the components of the body force; we remark that these components must be continuous functions. The stresses must be functions of class C^1; taking into account the constitutive law of a linear elastic field, for example, it follows that these functions must be of class C^2.

In the dynamic case, the components $\sigma_x = \sigma_x(x, y, z; t), \ldots, \tau_{xy} = \tau_{xy}(x, y, z; t)$ of the tensor T_σ must satisfy the *equations of motion*

$$\frac{\partial \sigma_x}{\partial x} + \frac{\partial \tau_{xy}}{\partial y} + \frac{\partial \tau_{zx}}{\partial z} + X = \rho \ddot{u},$$

$$\frac{\partial \tau_{xy}}{\partial x} + \frac{\partial \sigma_y}{\partial y} + \frac{\partial \tau_{yz}}{\partial z} + Y = \rho \ddot{v}, \qquad (7.1.8)$$

$$\frac{\partial \tau_{zx}}{\partial x} + \frac{\partial \tau_{yz}}{\partial y} + \frac{\partial \sigma_z}{\partial z} + Z = \rho \ddot{w},$$

where ρ is the *density* (we have admitted the hypothesis of small motions).

In the case of a *damping* proportional to the displacement velocity the differential operator $\rho \dfrac{\partial^2}{\partial t^2}$ is replaced by the operator $\left(\rho \dfrac{\partial^2}{\partial t^2} + k \dfrac{\partial}{\partial t} \right)$, where k is a *damping coefficient*.

In the case of Cosserat type bodies the stress tensor is asymmetric; besides, a *couple-stress tensor* also occurs, which is again an asymmetric tensor of the second order (couple-stresses or micromoments are stresses of the second order). *Body moments* also are in evidence in the above case.

It should be noted that by *state of plane stress* we mean a state of stress where no stresses occur on elements of area the normals of which have a fixed direction. For example, the case of thin plates acted upon under certain conditions. If we assume a thin plane plate is free of loads on both parallel faces and it is acted upon only on the lateral face, along the thickness of the plate, by loads parallel to the middle plane whose resultant is located in that plane, we find that not all the equations of linearly elastic solids are satisfied; in order to satisfy these equations we must consider the mean stresses on the thickness of the plate, thus pointing out a generalized state of plane stress.

7.1.2. Constitutive laws

The constitutive law of a solid emphasizes the physical properties of the material; from a mathematical viewpoint such a law furnishes the relations existing between the components of the tensor T_ε and the components of the tensor T_σ. In the following discussion we shall make some simplifying hypotheses which affect the constitutive law and will be used in connection with linearly elastic bodies. We shall make the following assumptions:

(*i*) The solid body is *isotropic*, i.e. it has the same mechanical and physical properties in any direction in the neighbourhood of any of its points. This property is expressed by the relation between stresses and strains. In the case where this hypothesis is not complied with, the body is said to be *anisotropic*.

(*ii*) The solid body is *homogeneous* if it possesses the same mechanical and physical properties at every point. Hence, the mechanical coefficients of the material which occur in the constitutive law are constant (with respect to the spatial variables). When the coefficients are variable the body is *non-homogeneous*.

The properties of isotropy and homogeneity are not dependent on each other. A solid body can be *homogeneous* and *isotropic* (as will be assumed generally), or *homogeneous* and *anisotropic* (i.e. it has the same properties at any point in the same direction), or *isotropic* and *non-homogeneous* (i.e. it has the same properties in any direction, but different from one point to another), or it may be even *non-homogeneous* and *anisotropic*. We remark that by changing the reference system of co-ordinates in the case of an anisotropic body we may obtain a body with properties of non-homogeneity. In that case we admit that an anisotropic body is non-homogeneous if it remains non-homogeneous in any reference system of curvilinear co-ordinates (otherwise it might be homogeneous and have a *curvilinear anisotropy*).

(*iii*) The solid under investigation is *perfectly elastic*. When it is acted upon by external loads the body undergoes deformation; when the action of the loads ceases, the body comes back to its initial position (in the case of static loads) and its initial form (without the occurrence of any hysteresis effect); the deformation is *reversible*. Hence the distorted shape of a solid body will be affected only by the external loads acting upon it at that moment. A one-to-one correspondence results between stress and strain; thus the state of stress at a point of the solid body will depend only on the state of strain in the neighbourhood of that point. In this

way we are in the realm of the theory of elasticity. In the case of many bodies currently used, when the stresses exceed the limit of elasticity, *remanent deformations* result
after removal of the load. Beyond the limit of elasticity the body has *elastoplastic
properties*. In that case, whether we may or may not neglect elastic deformations,
the computation of the state of stress and strain must be effected with the aid of the
theory of plasticity.

In general, we shall express the one-to-one connection between stresses and
strains by a linear relation (Hooke's law). This hypothesis often corresponds satis-
factorily to the physical phenomenon and leads, on the other hand, to important
simplifications of the mathematical calculations. Sometimes more complicated
mathematical relations are encountered involving a *non-linear* theory of elasticity
from a physical point of view. Also, special relations may be considered as in the
hypo- or hyper-elastic cases.

In general, the deformation of bodies is accompanied by a change of volume
(solid bodies are *compressible*). In particular, we shall also consider the case of
incompressible solids, which may be looked upon as a first approximation in calculations. The constitutive law is altered in the case of such bodies.

(*iv*) No account is taken of the velocity of deformation of the body; therefore
no account is taken of bodies with *rheological properties* (where the effect of *viscosity*
intervenes) and for which the constitutive law — in general integro-differential—
depends on the time variable too. Hence, no account is taken of *creep* and *relaxation*
effects.

In the following we shall present the constitutive law corresponding to a linearly
elastic solid in the general anisotropic and isotropic case.

7.1.2.1. Elastic potential. General anisotropic case

Let us denote the components of the tensors T_ε and T_σ by

$$\varepsilon_x = \varepsilon_1, \quad \varepsilon_y = \varepsilon_2, \quad \varepsilon_z = \varepsilon_3, \quad \gamma_{yz} = \varepsilon_4, \quad \gamma_{zx} = \varepsilon_5, \quad \gamma_{xy} = \varepsilon_6, \tag{7.1.9}$$

$$\sigma_x = \sigma_1, \quad \sigma_y = \sigma_2, \quad \sigma_z = \sigma_3, \quad \tau_{yz} = \sigma_4, \quad \tau_{zx} = \sigma_5, \quad \tau_{xy} = \sigma_6, \tag{7.1.9'}$$

thus replacing these tensors of the second order in a three-dimensional space E_3,
by first-order tensors in a six-dimensional space E_6. We postulate, therefore, the
existence of an *elastic potential* *

$$W = \frac{1}{2} c_{ij}\, \varepsilon_i\, \varepsilon_j, \tag{7.1.10}$$

---

* The dummy-index notation (Einstein) is used.

which plays the part of a potential function for the components of the stress tensor; the stresses are expressed by Green's formulae

$$\sigma_i = \frac{\partial W}{\partial \varepsilon_i} \qquad (i = 1, 2, ..., 6), \tag{7.1.11}$$

where c_{ij} are the *elastic constants* of the material.

The tensor c_{ij} is symmetric $(c_{ij} = c_{ji})$ so that the number of distinct elastic constants is 21 in the general case of anisotropy. Also, we assume that the *density of strain energy* (the elastic potential, W) is a positive definite quadratic form.

Similarly, we can write Castigliano's relations

$$\varepsilon_i = \frac{\partial W}{\partial \sigma_i} \qquad (i = 1, 2, ..., 6). \tag{7.1.11'}$$

For different particular cases of anisotropy we obtain different conditions for the elastic constants c_{ij} and the number of distinct constants is smaller.

The constitutive law (Hooke's law) will take the form

$$\sigma_i = c_{ij}\,\varepsilon_j. \tag{7.1.12}$$

Evidently, this law is easily transposed in the space E_3.

7.1.2.2. Isotropic case

In the isotropic case there are only two distinct elastic constants left. Returning to the space E_3 we may express this constitutive law in the form

$$\sigma_x = \lambda\varepsilon_v + 2\mu\varepsilon_x, \quad \sigma_y = \lambda\varepsilon_v + 2\mu\varepsilon_y, \quad \sigma_z = \lambda\varepsilon_v + 2\mu\varepsilon_z, \tag{7.1.13}$$

$$\tau_{yz} = \mu\gamma_{yz}, \quad \tau_{zx} = \mu\gamma_{zx}, \quad \tau_{xy} = \mu\gamma_{xy}, \tag{7.1.13'}$$

where λ and μ are Lamé's elastic constants; we have introduced the cubical dilatation

$$\varepsilon_v = \varepsilon_x + \varepsilon_y + \varepsilon_z. \tag{7.1.14}$$

Solving relations (7.1.13), (7.1.13') with respect to strains we obtain Hooke's law expressed by relations

$$\varepsilon_x = \frac{1}{E}[\sigma_x - v(\sigma_y + \sigma_z)], \quad \varepsilon_y = \frac{1}{E}[\sigma_y - v(\sigma_z + \sigma_x)],$$

$$\varepsilon_z = \frac{1}{E}[\sigma_z - v(\sigma_x + \sigma_y)], \tag{7.1.15}$$

$$\gamma_{yz} = \frac{1}{G}\tau_{yz}, \quad \gamma_{zx} = \frac{1}{G}\tau_{zx}, \quad \gamma_{xy} = \frac{1}{G}\tau_{xy}, \tag{7.1.15'}$$

where E and G are the moduli of longitudinal and transversal elasticity, respectively, and v is Poisson's ratio $\left(0 \leqslant v \leqslant \dfrac{1}{2}\right)$.

Of these elastic constants of the material only two are distinct, and we may write the relations

$$\lambda = \frac{vE}{(1 + v)(1 - 2v)}, \quad \mu = G = \frac{E}{2(1 + v)}. \tag{7.1.16}$$

We remark that relations (7.1.15) may be also written

$$\varepsilon_x = -\frac{v}{E}\,\Theta + \frac{1}{2G}\,\sigma_x, \quad \varepsilon_y = -\frac{v}{E}\,\Theta + \frac{1}{2G}\,\sigma_y, \quad \varepsilon_z = -\frac{v}{E}\,\Theta + \frac{1}{2G}\,\sigma_z, \tag{7.1.15''}$$

where the sum of normal stresses is expressed in the form

$$\Theta = \sigma_x + \sigma_y + \sigma_z. \tag{7.1.17}$$

7.2. Formulations of the problems of the theory of elasticity. Limiting conditions

In the following we shall briefly show how the fundamental problems of the theory of elasticity are stated in terms of stresses and displacements depending on how the limiting conditions are imposed; representations with the help of stress or displacement potentials will be also given. Further, we shall present some modified forms of the equations of the theory of elasticity which include the limiting conditions too.

It should be noted that the fundamental system of equations of the theory of elasticity will be considered in distributions, and therefore we shall seek solutions in distributions for the problems encountered.

7.2.1. Formulation in stresses of the problems of the theory of elasticity

Stating the problems of the theory of elasticity in terms of stresses involves the elimination of displacements and strains from the corresponding fundamental system of equations. In the following we shall treat of both the three-dimensional and the plane case.

7.2.1.1. Three-dimensional case

Eliminating the displacements from the equations (7.1.2), (7.1.2′) we obtain the equations of continuity (7.1.3), (7.1.3′).

Using Hooke's law (7.1.15), (7.1.15′) and the equations of equilibrium (7.1.7) we may replace equations (7.1.3), (7.1.3′) by the Beltrami-Michell equations

$$\Delta\sigma_x + \frac{1}{1+v}\frac{\partial^2\Theta}{\partial x^2} = -\frac{v}{1-v}\left(\frac{\partial X}{\partial x} + \frac{\partial Y}{\partial y} + \frac{\partial Z}{\partial z}\right) - 2\frac{\partial X}{\partial x},$$

$$\Delta\sigma_y + \frac{1}{1+v}\frac{\partial^2\Theta}{\partial y^2} = -\frac{v}{1-v}\left(\frac{\partial X}{\partial x} + \frac{\partial Y}{\partial y} + \frac{\partial Z}{\partial z}\right) - 2\frac{\partial Y}{\partial y}, \quad (7.2.1)$$

$$\Delta\sigma_z + \frac{1}{1+v}\frac{\partial^2\Theta}{\partial z^2} = -\frac{v}{1-v}\left(\frac{\partial X}{\partial x} + \frac{\partial Y}{\partial y} + \frac{\partial Z}{\partial z}\right) - 2\frac{\partial Z}{\partial z},$$

$$\Delta\tau_{yz} + \frac{1}{1+v}\frac{\partial^2\Theta}{\partial y\partial z} = -\left(\frac{\partial Y}{\partial z} + \frac{\partial Z}{\partial y}\right),$$

$$\Delta\tau_{zx} + \frac{1}{1+v}\frac{\partial^2\Theta}{\partial z\partial x} = -\left(\frac{\partial Z}{\partial x} + \frac{\partial X}{\partial z}\right), \quad (7.2.1')$$

$$\Delta\tau_{xy} + \frac{1}{1+v}\frac{\partial^2\Theta}{\partial x\partial y} = -\left(\frac{\partial X}{\partial y} + \frac{\partial Y}{\partial x}\right),$$

thus eliminating strains. By proceeding in this way we may state, *in the static case,* the problems of the theory of elasticity in terms of stresses; thus, the six components of the stress tensor T_σ will be determined by the system of equations (7.1.7), (7.2.1), (7.2.1′). It should be noted that this system of nine equations with six unknown functions is consistent, since it has been obtained with the aid of the compatibility conditions (7.1.3), (7.3.1′).

We also note that the sum of normal stresses must satisfy the equation

$$\Delta\Theta = -\frac{1+v}{1-v}\left(\frac{\partial X}{\partial x} + \frac{\partial Y}{\partial y} + \frac{\partial Z}{\partial z}\right); \quad (7.2.2)$$

the components of the tensor T_σ must satisfy, each of them separately, the biharmonic equation

$$\Delta\Delta\Phi(x, y, z) = 0 \quad (7.2.3)$$

in the absence of body forces.

A statement of the *second fundamental problem of the theory of elasticity* in terms of stresses is convenient when the *boundary conditions* are given *in terms of stresses;* the boundary conditions are written in the form

$$p_{nx} = \sigma_x \cos(n, x) + \tau_{xy} \cos(n, y) + \tau_{zx} \cos(n, z),$$

$$p_{ny} = \tau_{xy} \cos(n, x) + \sigma_y \cos(n, y) + \tau_{yz} \cos(n, z), \qquad (7.2.4)$$

$$p_{nz} = \tau_{zx} \cos(n, x) + \tau_{yz} \cos(n, y) + \sigma_z \cos(n, z),$$

where p_{nx}, p_{ny}, p_{nz} are the components of the surface loads acting on the boundary element whose external normal is **n**.

In the *dynamic case,* the Laplace operator is replaced by the d'Alembert operators $\Box_i$ $(i = 1, 2)$

$$\Box_i = \Delta - \frac{1}{c_i^2} \frac{\partial^2}{\partial t^2} \qquad (i = 1, 2), \qquad (7.2.5)$$

where the wave propagation velocity is given by

$$c_1^2 = \frac{1}{\rho}(\lambda + 2\mu), \quad c_2^2 = \frac{1}{\rho}\mu; \qquad (7.2.6)$$

it is important to note that between the d'Alembert and the Laplace operators we have the relation

$$2(1 - v)\Box_1 = \Delta + (1 - 2v)\Box_2. \qquad (7.2.7)$$

In that case the equations of the Beltrami type take the form

$$\Box_2 \sigma_x + \frac{1}{1+v}\left[\frac{\partial^2}{\partial x^2} + \frac{v}{2(1-v)c_2^2}\frac{\partial^2}{\partial t^2}\right]\Theta$$

$$= -\frac{v}{1-v}\left(\frac{\partial X}{\partial x} + \frac{\partial Y}{\partial y} + \frac{\partial Z}{\partial z}\right) - 2\frac{\partial X}{\partial x},$$

$$\Box_2 \sigma_y + \frac{1}{1+v}\left[\frac{\partial^2}{\partial y^2} + \frac{v}{2(1-v)c_2^2}\frac{\partial^2}{\partial t^2}\right]\Theta$$

$$= -\frac{v}{1-v}\left(\frac{\partial X}{\partial x} + \frac{\partial Y}{\partial y} + \frac{\partial Z}{\partial z}\right) - 2\frac{\partial Y}{\partial y}, \qquad (7.2.8)$$

$$\Box_2 \sigma_z + \frac{1}{1+v}\left[\frac{\partial^2}{\partial z^2} + \frac{v}{2(1-v)c_2^2}\frac{\partial^2}{\partial t^2}\right]\Theta$$

$$= -\frac{v}{1-v}\left(\frac{\partial X}{\partial x} + \frac{\partial Y}{\partial y} + \frac{\partial Z}{\partial z}\right) - 2\frac{\partial Z}{\partial z},$$

$$\Box_2 \tau_{yz} + \frac{1}{1+v} \frac{\partial^2 \Theta}{\partial y \partial z} = -\left(\frac{\partial Y}{\partial z} + \frac{\partial Z}{\partial y} \right),$$

$$\Box_2 \tau_{zx} + \frac{1}{1+v} \frac{\partial^2 \Theta}{\partial z \partial x} = -\left(\frac{\partial Z}{\partial x} + \frac{\partial X}{\partial z} \right), \qquad (7.2.8')$$

$$\Box_2 \tau_{xy} + \frac{1}{1+v} \frac{\partial^2 \Theta}{\partial x \partial y} = -\left(\frac{\partial X}{\partial y} + \frac{\partial Y}{\partial x} \right),$$

and the sum of the normal stresses satisfy the equation

$$\Box_1 \Theta = -\frac{1+v}{1-v} \left(\frac{\partial X}{\partial x} + \frac{\partial Y}{\partial y} + \frac{\partial Z}{\partial z} \right); \qquad (7.2.9)$$

in the absence of body forces, every component of the tensor T_σ must satisfy the double equation of waves

$$\Box_1 \Box_2 \Psi(x, y, z; t) = 0. \qquad (7.2.10)$$

Stating the dynamical problem in terms of stresses requires that the system (7.2.8), (7.2.8′) should be associated with the system of equations of motion (7.1.8) and the equations (7.1.2), (7.1.2′), (7.1.15), (7.1.15′); in the case of the first fundamental problem too, we must add to the boundary conditions set for a moment $t \geqslant t_0$, the *initial conditions*

$$\sigma_x(x, y, z; t_0) = \sigma_x^0(x, y, z), \quad \sigma_y(x, y, z; t_0) = \sigma_y^0(x, y, z),$$

$$\sigma_z(x, y, z; t_0) = \sigma_z^0(x, y, z), \qquad (7.2.11)$$

$$\tau_{yz}(x, y, z; t_0) = \tau_{yz}^0(x, y, z), \quad \tau_{zx}(x, y, z; t_0) = \tau_{zx}^0(x, y, z),$$

$$\tau_{xy}(x, y, z; t_0) = \tau_{xy}^0(x, y, z), \qquad (7.2.11')$$

$$\dot{\sigma}_x(x, y, z; t_0) = \dot{\sigma}_x^0(x, y, z), \quad \dot{\sigma}_y(x, y, z; t_0) = \dot{\sigma}_y^0(x, y, z),$$

$$\dot{\sigma}_z(x, y, z; t_0) = \dot{\sigma}_z^0(x, y, z), \qquad (7.2.12)$$

$$\dot{\tau}_{yz}(x, y, z; t_0) = \dot{\tau}_{yz}^0(x, y, z), \quad \dot{\tau}_{zx}(x, y, z; t_0) = \dot{\tau}_{zx}^0(x, y, z),$$

$$\dot{\tau}_{xy}(x, y, z; t_0) = \dot{\tau}_{xy}^0(x, y, z), \qquad (7.2.12')$$

corresponding to the initial moment $t = t_0$.

7.2.1.2. Plane case

The equilibrium equations are

$$\frac{\partial \sigma_x}{\partial x} + \frac{\partial \tau_{xy}}{\partial y} + X = 0,$$

$$\frac{\partial \tau_{xy}}{\partial x} + \frac{\partial \sigma_y}{\partial y} + Y = 0,$$

(7.2.13)

and the equations of motion are

$$\frac{\partial \sigma_x}{\partial x} + \frac{\partial \tau_{xy}}{\partial y} + X = \rho \ddot{u},$$

$$\frac{\partial \tau_{xy}}{\partial x} + \frac{\partial \sigma_y}{\partial y} + Y = \rho \ddot{v}.$$

(7.2.14)

The relations between strains and displacements become

$$\varepsilon_x = \frac{\partial u}{\partial x}, \qquad \varepsilon_y = \frac{\partial v}{\partial y},$$

(7.2.15)

$$\gamma_{xy} = \frac{\partial u}{\partial y} + \frac{\partial v}{\partial x}.$$

(7.2.15')

In the case of a *state of plane stress* Hooke's law takes the form

$$\varepsilon_x = \frac{1}{E}(\sigma_x - v\sigma_y), \qquad \varepsilon_y = \frac{1}{E}(\sigma_y - v\sigma_x),$$

(7.2.16)

$$\gamma_{xy} = \frac{2(1 + v)}{E} \tau_{xy};$$

(7.2.16')

in the case of a *state of plane strain* one may use the same relations but the elastic constants E and v must be replaced by the generalized elastic constants

$$E_0 = \frac{E}{1 - v^2}, \qquad v_0 = \frac{v}{1 - v}.$$

(7.2.17)

By eliminating the strains and the displacements we obtain in the static case (for a state of plane stress)

$$\Delta(\sigma_x + \sigma_y) = -(1 + v)\left(\frac{\partial X}{\partial x} + \frac{\partial Y}{\partial y}\right);$$

(7.2.18)

thus the three components $\sigma_x = \sigma_x(x, y)$, $\sigma_y = \sigma_y(x, y)$, $\tau_{xy} = \tau_{xy}(x, y)$ of the stress tensor will be given by the system of equations (7.2.13), (7.2.18).

Boundary conditions are expressed by

$$p_{nx} = \sigma_x \cos(n, x) + \tau_{xy} \cos(n, y),$$

$$p_{ny} = \tau_{xy} \cos(n, x) + \sigma_y \cos(n, y),$$

(7.2.19)

where p_{nx}, p_{ny} are the components of the external load acting upon the boundary element whose external normal is **n**.

The *dynamic case* may be stated in terms of stresses in a similar manner.

7.2.2. Formulation in displacements of the problems of the theory of elasticity

Stating the problems of the theory of elasticity in terms of displacements involves the elimination of stresses and strains from the equations of the corresponding fundamental system of equations. As in the preceding section we shall give results for both the three-dimensional and the plane cases.

7.2.2.1. Three-dimensional case

By successive eliminations we may obtain in the *static case* Lamé's equations

$$\mu \Delta u + (\lambda + \mu) \frac{\partial \varepsilon_v}{\partial x} + X = 0,$$

$$\mu \Delta v + (\lambda + \mu) \frac{\partial \varepsilon_v}{\partial y} + Y = 0,$$

(7.2.20)

$$\mu \Delta w + (\lambda + \mu) \frac{\partial \varepsilon_v}{\partial z} + Z = 0,$$

which, in a formulation in terms of displacements of the problems of elasticity, furnish the three components of the displacement vector.

The cubical dilatation must satisfy the equation

$$\Delta \varepsilon_v = - \frac{1}{\lambda + 2\mu} \left(\frac{\partial X}{\partial x} + \frac{\partial Y}{\partial y} + \frac{\partial Z}{\partial z} \right);$$

(7.2.21)

in the absence of body forces, every displacement must satisfy the biharmonic equation (7.2.3).

Such a formulation is convenient in the case of the *first fundamental problem of the theory of elasticity,* where *boundary conditions are given in terms of displacements*; these conditions are written in the form

$$u = u_n, \quad v = v_n, \quad w = w_n, \tag{7.2.22}$$

where u_n, v_n, w_n are the displacements corresponding to a point of the boundary where the external normal is **n**.

In the *dynamic case* Lamé's equations take the form

$$\mu \square_2 u + (\lambda + \mu) \frac{\partial \varepsilon_v}{\partial x} + X = 0,$$

$$\mu \square_2 v + (\lambda + \mu) \frac{\partial \varepsilon_v}{\partial y} + Y = 0, \tag{7.2.23}$$

$$\mu \square_2 w + (\lambda + \mu) \frac{\partial \varepsilon_v}{\partial z} + Z = 0,$$

and the cubical dilatation satisfies the equation

$$\square_1 \varepsilon_v = - \frac{1}{\lambda + 2\mu} \left(\frac{\partial X}{\partial x} + \frac{\partial Y}{\partial y} + \frac{\partial Z}{\partial z} \right); \tag{7.2.24}$$

also, in the absence of body forces every component of the displacement vector will satisfy the double equation of waves (7.2.10).

It should be noted that to the conditions on the boundary at a moment $t \geqslant t_0$, one must add the *initial conditions.*

$$u(x, y, z; t_0) = u_0(x, y, z), \quad v(x, y, z; t_0) = v_0(x, y, z),$$

$$w(x, y, z; t_0) = w_0(x, y, z), \tag{7.2.25}$$

$$\dot{u}(x, y, z; t_0) = \dot{u}_0(x, y, z), \quad \dot{v}(x, y, z; t_0) = \dot{v}_0(x, y, z),$$

$$\dot{w}(x, y, z; t_0) = \dot{w}_0(x, y, z), \tag{7.2.25'}$$

corresponding to the initial moment $t = t_0$.

Further, it should be noted that a formulation in terms of displacements is useful also in the case of the *mixed fundamental problem* where on a part of the boundary the conditions are given in terms of stresses, while on the rest of the contour they are given in terms of displacements.

7.2.2.2. Plane case

In the case of a *state of plane deformation* due to *static loads* Lamé's equations may be written in the form

$$\mu \Delta u + (\lambda + \mu)\frac{\partial \varepsilon_v}{\partial x} + X = 0,$$

$$\mu \Delta u + (\lambda + \mu)\frac{\partial \varepsilon_v}{\partial y} + Y = 0; \tag{7.2.26}$$

the conditions on the boundary, corresponding to a point where the external normal is **n**, are

$$u = u_n, \quad v = v_n. \tag{7.2.27}$$

In the *dynamic case* these equations become (still for a state of plane deformation)

$$\mu \square_2 u + (\lambda + \mu)\frac{\partial \varepsilon_v}{\partial x} + X = 0,$$

$$\mu \square_2 v + (\lambda + \mu)\frac{\partial \varepsilon_v}{\partial y} + Y = 0, \tag{7.2.28}$$

where we must also associate the initial conditions of the form

$$u(x, y; t_0) = u_0(x, y), \quad v(x, y; t_0) = v_0(x, y), \tag{7.2.29}$$

$$\dot{u}(x, y; t_0) = \dot{u}_0(x, y), \quad \dot{v}(x, y; t_0) = \dot{v}_0(x, y) \tag{7.2.29'}$$

for the initial moment $t = t_0$.

7.2.3. Modified forms of the equations of the theory of elasticity

The formulations of the preceding sections were deduced by assuming that the displacements were functions of class C^3 and the stresses were functions of class C^2. The functions which intervene in the limiting conditions (boundary and initial conditions) must be of a corresponding class and the body forces must be of class C^1.

However, in the applications which will follow, it is assumed that these equations are written in terms of *distributions* and that the limiting conditions are of a less restrictive character. In this connection, we shall show how to obtain the equilibrium equations in terms of distributions in the case of a linearly elastic homogeneous and anisotropic body; at the same time, we shall point out a method for including, in these equations, the conditions on the boundary, thus obtaining a modified form of the equations of equilibrium.

In the case of the dynamic problem we shall show how to include the initial conditions in Lamé's equations corresponding to that problem; otherwise we could not consider that the equations are written in terms of distributions since displacements are not defined for $t \in R$.

Also, we assume that the classical theorems of existence and uniqueness are valid for the fundamental system of equations written in terms of distributions (both in the static and the dynamic cases).

7.2.3.1. Modified form of the equations of equilibrium

Let D be a linearly elastic body, bounded by the surface $P(x_1, x_2, x_3) = 0$, of the class C^1; the respective domain is considered to be simply connected. We shall denote by X_i the components of the body forces acting on the linearly elastic medium and by p_{ni} the components of the external loads acting on the surface $P=0$; we use the notations of section 7.1.2.1 and Einstein's summation convention.

The *potential energy* of a linearly elastic body is given by the functional

$$\prod(u_1, u_2, u_3) = \iiint_D W \, dV - \left(\iiint_D X_i u_i \, dV + \iint_{P=0} p_{ni} u_i \, dS \right), \qquad (7.2.30)$$

where W is the density of strain energy given by relation (7.1.10), and u_i are the components of the displacement vector.

Noting by $\delta(P)$ the Dirac distribution concentrated on the surface $P = 0$ and defined by relation (1.4.10) we may express the surface integral of (7.2.30) by a volume integral of the form

$$\iint_{P=0} p_{ni} u_i \, dS = \iiint_D p_{ni} u_i \sqrt{\frac{\partial P}{\partial x_j} \frac{\partial P}{\partial x_j}} \, \delta(P) \, dV; \qquad (7.2.31)$$

in that case the functional (7.2.30) takes the form

$$\prod(u_1, u_2, u_3) = \iiint_D W \, dV - \iiint_D \left[X_i u_i + p_{ni} u_i \sqrt{\frac{\partial P}{\partial x_j} \frac{\partial P}{\partial x_j}} \, \delta(P) \right] dV. \qquad (7.2.30')$$

We remark that the expression under the sign of integral is a function which is continuous with respect to all its arguments; in order that the expression may have a significance when multiplication and differentiation are effected in the sense of the theory at distributions, it is sufficient to assume that the displacements u_i are functions of class C^0. Thus, even if the X_i are expressed by the Dirac distribution $\delta(x_1, x_2, x_3)$, the products $X_i u_i$ have a sense, since

$$(u_i(x_1, x_2, x_3)\delta(x_1, x_2, x_3), \quad \varphi(x_1, x_2, x_3))$$

$$= (\delta(x_1, x_2, x_3), \quad u_i(x_1, x_2, x_3)\varphi(x_1, x_2, x_3)) = u_i(0, 0, 0)\varphi(0, 0, 0), \qquad (7.2.32)$$

where $\varphi(x_1, x_2, x_3) \in K$ is a fundamental function owing precisely to the continuity of the function u_i.

Taking into account the principle of minimum potential energy and writing the conditions required for an extremum of the functional (7.2.30'), we obtain

$$\frac{\partial}{\partial x_j}\left(\frac{\partial W}{\partial\left(\dfrac{\partial u_i}{\partial x_j}\right)}\right) + X_i + p_{ni}\sqrt{\frac{\partial P}{\partial x_j}\frac{\partial P}{\partial x_j}}\,\delta(P) = 0 \quad (i = 1, 2, 3); \quad (7.2.33)$$

with the aid of relations (7.1.11), (7.1.11'), of the relations between strains and displacements, and using the notations (7.1.9), (7.1.9'), we obtain a *modified form of the equations of equilibrium*, where differentiation is considered in the sense of the theory of distributions and where we have returned to the notations of the orthogonal Cartesian co-ordinate system $Oxyz$

$$\frac{\partial\sigma_x}{\partial x} + \frac{\partial\tau_{xy}}{\partial y} + \frac{\partial\tau_{zx}}{\partial z} + X + p_{nx}\sqrt{\left(\frac{\partial P}{\partial x}\right)^2 + \left(\frac{\partial P}{\partial y}\right)^2 + \left(\frac{\partial P}{\partial z}\right)^2}\,\delta(P) = 0,$$

$$\frac{\partial\tau_{xy}}{\partial x} + \frac{\partial\sigma_y}{\partial y} + \frac{\partial\tau_{yz}}{\partial z} + Y + p_{ny}\sqrt{\left(\frac{\partial P}{\partial x}\right)^2 + \left(\frac{\partial P}{\partial y}\right)^2 + \left(\frac{\partial P}{\partial z}\right)^2}\,\delta(P) = 0, \quad (7.2.34)$$

$$\frac{\partial\tau_{zx}}{\partial x} + \frac{\partial\tau_{yz}}{\partial y} + \frac{\partial\sigma_z}{\partial z} + Z + p_{nz}\sqrt{\left(\frac{\partial P}{\partial x}\right)^2 + \left(\frac{\partial P}{\partial y}\right)^2 + \left(\frac{\partial P}{\partial z}\right)^2}\,\delta(P) = 0.$$

Since stresses, represented in general by the function $f = f(x, y, z)$ are functions with discontinuities of the first species when transversing the surface $P = 0$, we may write

$$f(x, y, z) = \begin{cases} \overline{f}(x, y, z) & \text{for } (x, y, z) \in D \\ 0 & \text{for } (x, y, z) \notin D; \end{cases} \quad (7.2.35)$$

also, the partial derivatives will be written in the form

$$\frac{\partial f}{\partial x} = \frac{\widetilde{\partial} f}{\partial x} + (\Delta_x f)_P\,\frac{\partial P}{\partial x}\,\delta(P), \quad (7.2.36)$$

where $(\Delta_x f)_P$ represents the jump of the function $f(x, y, z)$ on transversing the surface $P = 0$ along the direction of the Ox axis.

Applying differentiation formulae of the form (7.2.36) to the system of equations (7.2.34), and taking into account theorem 1.4.2 of section 1.4.1.2., we find again the equations of equilibrium

$$\frac{\tilde{\partial}\bar{\sigma}_x}{\partial x} + \frac{\tilde{\partial}\bar{\tau}_{xy}}{\partial y} + \frac{\tilde{\partial}\bar{\tau}_{zx}}{\partial z} + X = 0,$$

$$\frac{\tilde{\partial}\bar{\tau}_{xy}}{\partial x} + \frac{\tilde{\partial}\bar{\sigma}_y}{\partial y} + \frac{\tilde{\partial}\bar{\tau}_{yz}}{\partial z} + Y = 0, \qquad (7.2.37)$$

$$\frac{\tilde{\partial}\bar{\tau}_{zx}}{\partial x} + \frac{\tilde{\partial}\bar{\tau}_{yz}}{\partial y} + \frac{\tilde{\partial}\bar{\sigma}_z}{\partial z} + Z = 0,$$

where derivatives are taken in the ordinary sense and the conditions on the boundary are

$$\bar{\sigma}_x \cos(n, x) + \bar{\tau}_{xy} \cos(n, y) + \bar{\tau}_{zx} \cos(n, z) = p_{nx},$$

$$\bar{\tau}_{xy} \cos(n, x) + \bar{\sigma}_y \cos(n, y) + \bar{\tau}_{yz} \cos(n, z) = p_{ny}, \qquad (7.2.38)$$

$$\bar{\tau}_{zx} \cos(n, x) + \bar{\tau}_{yz} \cos(n, y) + \bar{\sigma}_z \cos(n, z) = p_{nz};$$

we remark that the jump of the stresses on the boundary are given just by the values of the stresses at the respective points, with a changed sign, and that we have relations of the form

$$\frac{\partial P}{\partial x} = \sqrt{\left(\frac{\partial P}{\partial x}\right)^2 + \left(\frac{\partial P}{\partial y}\right)^2 + \left(\frac{\partial P}{\partial z}\right)^2} \cos(n, x), \qquad (7.2.39)$$

corresponding to the external normal **n** at a point of the boundary $P = 0$ of the domain D. Thus, we find again the results given in section 7.1.1.2.

If the elastic body D has, besides the external surface $P_0 = P = 0$, other surfaces of discontinuity of the first species $P_i = 0$ $(i = 1, 2, \ldots, n)$ as regards stresses, then formula (7.2.36) will be written

$$\frac{\partial f}{\partial x} = \frac{\tilde{\partial} f}{\partial x} + \sum_{i=0}^{n} (\Delta_x f)_{P_i} \frac{\partial P_i}{\partial x} \delta(P_i), \qquad (7.2.36')$$

where $(\Delta_x f)_{P_i}$ represents the jump of the function $f(x, y, z)$ when it transverses the surface P_i along the direction of the Ox axis.

We remark that relation (7.2.39) takes the form

$$\frac{\partial P_i}{\partial x} = \sqrt{\left(\frac{\partial P_i}{\partial x}\right)^2 + \left(\frac{\partial P_i}{\partial y}\right)^2 + \left(\frac{\partial P_i}{\partial z}\right)^2} \cos(n_i, x), \qquad (7.2.39')$$

where $\mathbf{n}_i$ is the external normal corresponding to a point of the surface $P_i = 0$ $(i = 1, 2, \ldots, n)$.

In this way, we obtain by a similar procedure the *additional conditions of jump*

$$(\Delta_x\sigma_x)_{P_i}\cos(n_i, x) + (\Delta_y\tau_{xy})_{P_i}\cos(n_i, y) + (\Delta_z\tau_{zx})_{P_i}\cos(n_i, z) = 0,$$

$$(\Delta_x\tau_{xy})_{P_i}\cos(n_i, x) + (\Delta_y\sigma_y)_{P_i}\cos(n_i, y) + (\Delta_z\tau_{yz})_{P_i}\cos(n_i, z) = 0, \quad (7.2.40)$$

$$(\Delta_x\tau_{zx})_{P_i}\cos(n_i, x) + (\Delta_y\tau_{yz})_{P_i}\cos(n_i, y) + (\Delta_z\sigma_z)_{P_i}\cos(n_i, z) = 0,$$

which must be satisfied by the stresses on the surfaces of discontinuity $P_i = 0$ ($i = 1, 2, \ldots, n$).

7.2.3.2. A modified form of Lamé's equations in the dynamic case

In order to transcribe in the distributions space Lamé's equations in the dynamic case, we shall replace the body forces X, Y, Z and the displacements u, v, w defined for $t \geq 0$ only (for simplicity we assume $t_0 = 0$), by the distributions $\overline{X}, \overline{Y}, \overline{Z}$ and $\overline{u}, \overline{v}, \overline{w}$, defined by the relations *

$$\overline{X}(x, y, z; t) = \begin{cases} 0 & \text{for} \quad t < 0 \\[2mm] X & \text{for} \quad t \geq 0, \ldots, \end{cases} \qquad (7.2.41)$$

$$\overline{u}(x, y, z; t) = \begin{cases} 0 & \text{for} \quad t < 0 \\[2mm] u & \text{for} \quad t \geq 0, \ldots, \end{cases} \qquad (7.2.42)$$

hence, for all $t \in R$.

We remark that distributions (7.2.42) have discontinuities of the first species with respect to the time variable, and therefore their partial derivatives with respect to that variable will be of the form

$$\frac{\partial}{\partial t}\,\overline{u}(x, y, z; t) = \frac{\widetilde{\partial}}{\partial t}\,\overline{u}(x, y, z; t) + u_0(x, y, z)\,\delta(t), \ldots, \qquad (7.2.43)$$

$$\frac{\partial^2}{\partial t^2}\,\overline{u}(x, y, z; t) = \frac{\widetilde{\partial}^2}{\partial t^2}\,\overline{u}(x, y, z; t) + \dot{u}_0(x, y, z)\,\delta(t) + u_0(x, y, z)\,\dot{\delta}(t), \ldots, \qquad (7.2.43')$$

where account has been taken of the initial conditions (7.2.25), (7.2.25'). Also, these distributions have no discontinuities with regard to the variables x, y, z since

---

* We shall denote by "..." the fact that the other relations are obtained by circular permutations.

it is assumed that displacements u, v, w are functions of class C^2; therefore we may write

$$\frac{\partial^2}{\partial x^2}\, \bar{u}(x, y, z; t) = \frac{\tilde{\partial}^2}{\partial x^2}\, \bar{u}(x, y, z; t), \ldots , \qquad (7.2.44)$$

$$\frac{\partial^2}{\partial y \partial z}\, \bar{u}(x, y, z; t) = \frac{\tilde{\partial}^2}{\partial y \partial z}\, \bar{u}(x, y, z; t), \ldots . \qquad (7.2.44')$$

Replacing relations $(7.2.43) - (7.2.44')$ in Lamé's equations $(7.2.23)$ and introducing the distribution

$$\bar{\varepsilon}_v = \frac{\partial \bar{u}}{\partial x} + \frac{\partial \bar{v}}{\partial y} + \frac{\partial \bar{w}}{\partial z} \qquad (7.2.45)$$

we obtain a modified form of these equations

$$\mu \square_2 \bar{u} + (\lambda + \mu)\, \frac{\partial \bar{\varepsilon}_v}{\partial x} + \bar{X} + \rho \dot{u}_0 \delta(t) + \rho u_0 \dot{\delta}(t) = 0,$$

$$\mu \square_2 \bar{v} + (\lambda + \mu)\, \frac{\partial \bar{\varepsilon}_v}{\partial y} + \bar{Y} + \rho \dot{v}_0 \delta(t) + \rho v_0 \dot{\delta}(t) = 0, \qquad (7.2.46)$$

$$\mu \square_2 \bar{w} + (\lambda + \mu)\, \frac{\partial \bar{\varepsilon}_v}{\partial z} + \bar{Z} + \rho \dot{w}_0 \delta(t) + \rho w_0 \dot{\delta}(t) = 0;$$

this *modified form of Lamé's equations in the dynamic case* has the advantage of being written in terms of distributions and of including the initial conditions.

In the *plane case* we introduce in a similar way the distributions defined by the functions

$$\bar{X}(x, y; t) = \begin{cases} 0 & \text{for} \quad t < 0 \\ \\ X & \text{for} \quad t \geqslant 0, \end{cases} \qquad (7.2.47)$$

$$\bar{Y}(x, y; t) = \begin{cases} 0 & \text{for} \quad t < 0 \\ \\ Y & \text{for} \quad t \geqslant 0, \end{cases} \qquad (7.2.47')$$

and the distributions defined by the functions

$$\bar{u}(x, y; t) = \begin{cases} 0 & \text{for} \quad t < 0 \\ \\ u & \text{for} \quad t \geqslant 0, \end{cases} \qquad (7.2.48)$$

$$\bar{v}(x, y; t) = \begin{cases} 0 & \text{for} \quad t < 0 \\ \\ v & \text{for} \quad t \geqslant 0. \end{cases} \qquad (7.2.48')$$

By introducing the distribution

$$\bar{\varepsilon}_p = \frac{\partial \bar{u}}{\partial x} + \frac{\partial \bar{v}}{\partial y} \tag{7.2.49}$$

we obtain by a similar procedure the modified form of Lamé's equations in the dynamic case for a state of plane deformation

$$\mu \square_2 \bar{u} + (\lambda + \mu) \frac{\partial \bar{\varepsilon}_p}{\partial x} + \bar{X} + \rho \dot{u}_0 \delta(t) + \rho u_0 \dot{\delta}(t) = 0,$$

$$\mu \square_2 \bar{v} + (\lambda + \mu) \frac{\partial \bar{\varepsilon}_p}{\partial y} + \bar{Y} + \rho \dot{v}_0 \delta(t) + \rho v_0 \dot{\delta}(t) = 0, \tag{7.2.50}$$

which, likewise, includes the initial conditions.

In the case where the domain D, three- or two-dimensional, is infinite but has at least a part of its boundary at a finite distance, or if it is finite, the formulae (7.2.44), (7.2.44′) can no longer be applied. In that case we must use a method similar to that indicated by the formulae (7.2.35) concerning the spatial variables; the formulae (7.2.44), (7.2.44′) shall be replaced by formulae of the form (7.2.36) where there occur concentrated distributions on surfaces (three-dimensional problem) or concentrated distributions on curves (plane problem).

Thus we obtain, for the general problem, differential equations of the form

$$\mu \square_2 \bar{u} + (\lambda + \mu) \frac{\partial \bar{\varepsilon}_v}{\partial x} + f = 0,$$

$$\mu \square_2 \bar{v} + (\lambda + \mu) \frac{\partial \bar{\varepsilon}_v}{\partial y} + g = 0, \tag{7.2.51}$$

$$\mu \square_2 \bar{w} + (\lambda + \mu) \frac{\partial \bar{\varepsilon}_v}{\partial z} + h = 0,$$

where $f = f(x, y, z; t)$, $g = g(x, y, z; t)$ and $h = h(x, y, z; t)$ are known distributions or distributions which can be determined with the help of additional conditions; similar equations may be written for the plane case. The static problem may be studied by the same method.

The matrix

$$(U) \equiv \begin{pmatrix} u_{11} & u_{12} & u_{13} \\ u_{21} & u_{22} & u_{23} \\ u_{31} & u_{32} & u_{33} \end{pmatrix}, \tag{7.2.52}$$

with $u_{ij} = u_{ij}(x, y, z; t)$ $(i, j = 1, 2, 3)$, is termed the *fundamental solution of the elastodynamic problem* if its elements satisfy the system of equations

$$\mu \square_2 u_{11} + (\lambda + \mu) \frac{\partial \varepsilon_v'}{\partial x} + \delta(x, y, z) \delta(t) = 0,$$

$$\mu \square_2 u_{21} + (\lambda + \mu) \frac{\partial \varepsilon_v'}{\partial y} = 0, \tag{7.2.53}$$

$$\mu \square_2 u_{31} + (\lambda + \mu) \frac{\partial \varepsilon_v'}{\partial z} = 0, \dots ,$$

with the notations

$$\varepsilon_v' = \frac{\partial u_{11}}{\partial x} + \frac{\partial u_{21}}{\partial y} + \frac{\partial u_{31}}{\partial z},$$

$$\varepsilon_v'' = \frac{\partial u_{12}}{\partial x} + \frac{\partial u_{22}}{\partial y} + \frac{\partial u_{32}}{\partial z}, \tag{7.2.54}$$

$$\varepsilon_v''' = \frac{\partial u_{13}}{\partial x} + \frac{\partial u_{23}}{\partial y} + \frac{\partial u_{33}}{\partial z}.$$

From a mechanical point of view the elements of the first column of matrix (U) give the state of displacement corresponding to the elastic space acted on by a force which is concentrated at the origin, in the positive direction of the Ox axis, is equal to unity and is applied in the form of a shock at the initial moment, with *homogeneous initial conditions* (initial conditions equal to zero).

If we introduce, in addition, the matrices

$$(\bar{u}) \equiv \begin{pmatrix} \bar{u} \\ \bar{v} \\ w \end{pmatrix}, \qquad (Q) \equiv \begin{pmatrix} f \\ g \\ h \end{pmatrix} \tag{7.2.55}$$

we may write the state of displacement in the most general case, in the form

$$(\bar{u}) = (U) * (Q), \tag{7.2.56}$$

where the convolution product refers both to the space and to the time variables.

The passage from generalized to ordinary displacements is effected, for example, in the case of the elastic space, by formulae of the form

$$\bar{u}(x, y, z; t) = u(x, y, z; t) \theta(t), \dots , \tag{7.2.57}$$

i.e. by taking the positive part of the corresponding functions. Concerning numerical computation one calculates the values of the corresponding functions for distributions of the function type, while singular distributions are approximate with the aid of representative sequences.

It follows that by a translation to an arbitrary point of the space the fundamental solutions become *Green's distributions.*

The most important advantage offered by the methods of the theory of distributions is that all boundary value problems (including both contour and initial conditions) may be treated in a unitary form for every kind of external loads.

A similar formulation may be given to a solution of the problem expressed in terms of stresses.

7.2.4. Representation of the solution by stress and displacement potentials

We shall show how one may represent in the dynamic case a state of stress or of displacement with the help of potential functions. The starting point is a solution of the problem in terms of stresses.

7.2.4.1. Representation of the Beltrami-Finzi type

In the absence of body forces and considering the equations of motion (7.1.8) the state of stress may be represented in the form

$$\sigma_x = \frac{\partial^2 F_{yy}}{\partial z^2} + \frac{\partial^2 F_{zz}}{\partial y^2} - \frac{\partial^2 F_{yz}}{\partial y \partial z} + \rho \ddot{f}_{xx},$$

$$\sigma_y = \frac{\partial^2 F_{zz}}{\partial x^2} + \frac{\partial^2 F_{xx}}{\partial z^2} - \frac{\partial^2 F_{zx}}{\partial z \partial x} + \rho \ddot{f}_{yy}, \qquad (7.2.58)$$

$$\sigma_z = \frac{\partial^2 F_{xx}}{\partial y^2} + \frac{\partial^2 F_{yy}}{\partial x^2} - \frac{\partial^2 F_{xy}}{\partial x \partial y} + \rho \ddot{f}_{zz},$$

$$\tau_{yz} = \frac{\partial}{\partial x}\left(-\frac{\partial F_{yz}}{\partial x} + \frac{\partial F_{zx}}{\partial y} + \frac{\partial F_{xy}}{\partial z} \right) - \frac{\partial^2 F_{xx}}{\partial y \partial z} + \rho \ddot{f}_{yz},$$

$$\tau_{zx} = \frac{\partial}{\partial y}\left(-\frac{\partial F_{zx}}{\partial y} + \frac{\partial F_{xy}}{\partial z} + \frac{\partial F_{yz}}{\partial x} \right) - \frac{\partial^2 F_{yy}}{\partial z \partial x} + \rho \ddot{f}_{zx}, \qquad (7.2.58')$$

$$\tau_{xy} = \frac{\partial}{\partial z}\left(-\frac{\partial F_{xy}}{\partial z} + \frac{\partial F_{yz}}{\partial x} + \frac{\partial F_{zx}}{\partial y} \right) - \frac{\partial^2 F_{zz}}{\partial x \partial y} + \rho \ddot{f}_{xy},$$

and the state of displacement is given by

$$u = \frac{\partial f_{xx}}{\partial x} + \frac{\partial f_{xy}}{\partial y} + \frac{\partial f_{zx}}{\partial z},$$

$$v = \frac{\partial f_{xy}}{\partial x} + \frac{\partial f_{yy}}{\partial y} + \frac{\partial f_{yz}}{\partial z},\qquad (7.2.59)$$

$$w = \frac{\partial f_{zx}}{\partial x} + \frac{\partial f_{yz}}{\partial y} + \frac{\partial f_{zz}}{\partial z},$$

where we have introduced the symmetrical tensors of the second order

$$T_F \equiv \begin{Bmatrix} F_{xx} & F_{xy} & F_{zx} \\ F_{xy} & F_{yy} & F_{yz} \\ F_{zx} & F_{yz} & F_{zz} \end{Bmatrix}, \qquad T_f \equiv \begin{Bmatrix} f_{xx} & f_{xy} & f_{zx} \\ f_{xy} & f_{yy} & f_{yz} \\ f_{zx} & f_{yz} & f_{zz} \end{Bmatrix}, \qquad (7.2.60)$$

the components of which must be functions of class C^3; the tensor T_F is B. Finzi's tensor and the above representation is of the Beltrami-Finzi type. Such a *representation* is *complete* (any state of stress and displacement which satisfies the equations of motion may be represented in this form).

Putting in the above representation

$$F_{yz} = F_{zx} = F_{xy} = 0, \quad f_{yz} = f_{zx} = f_{xy} = 0, \qquad (7.2.61)$$

we obtain a representation of the Maxwell type; putting

$$F_{xx} = F_{yy} = F_{zz} = 0, \quad f_{xx} = f_{yy} = f_{zz} = 0 \qquad (7.2.61')$$

we find a representation of the Morera type.

7.2.4.2. Representation of the Schaefer type

If we choose the components of the Finzi tensor such that

$$F_{xx} = \Theta_{yy} + \Theta_{zz} - \Omega, \quad F_{yy} = \Theta_{zz} + \Theta_{xx} - \Omega, \quad F_{zz} = \Theta_{xx} + \Theta_{yy} - \Omega, \qquad (7.2.62)$$

$$F_{yz} = -\Theta_{yz}, \quad F_{zx} = -\Theta_{zx}, \quad F_{xy} = -\Theta_{xy} \qquad (7.2.62')$$

and the components of the tensor T_f such that

$$f_{xx} = \overline{\Theta}_{yy} + \overline{\Theta}_{zz} - \overline{\Omega}, \quad f_{yy} = \overline{\Theta}_{zz} + \overline{\Theta}_{xx} - \overline{\Omega}, \quad f_{zz} = \overline{\Theta}_{xx} + \overline{\Theta}_{yy} - \overline{\Omega}, \quad (7.2.63)$$

$$f_{yz} = -\overline{\Theta}_{yz}, \quad f_{zx} = -\overline{\Theta}_{zx}, \quad f_{xy} = -\overline{\Theta}_{xy}, \quad (7.2.63')$$

assuming that

$$\overline{\Theta}_{xx} = \frac{1}{2G}(\Theta_{xx} - \Theta_{yy} - \Theta_{zz}), \quad \overline{\Theta}_{yy} = \frac{1}{2G}(\Theta_{yy} - \Theta_{zz} - \Theta_{xx}),$$

$$(7.2.64)$$

$$\overline{\Theta}_{zz} = \frac{1}{2G}(\Theta_{zz} - \Theta_{xx} - \Theta_{yy}),$$

$$\overline{\Theta}_{yz} = \frac{1}{G}\Theta_{yz}, \quad \overline{\Theta}_{zx} = \frac{1}{G}\Theta_{zx}, \quad \overline{\Theta}_{xy} = \frac{1}{G}\Theta_{xy} \quad (7.2.64')$$

and

$$\overline{\Omega} + \frac{1}{2G}\Omega = 0, \quad (7.2.65)$$

we may write the state of stress in the form

$$\sigma_x = \frac{\partial^2 \Theta_{yy}}{\partial y^2} + \frac{\partial^2 \Theta_{zz}}{\partial z^2} - \frac{\partial^2 \Theta_{xx}}{\partial x^2} + 2\frac{\partial^2 \Theta_{yz}}{\partial y \partial z}$$

$$-\left(\frac{\partial^2}{\partial y^2} + \frac{\partial^2}{\partial z^2} - \frac{\rho}{2G}\frac{\partial^2}{\partial t^2}\right)\Omega,$$

$$\sigma_y = \frac{\partial^2 \Theta_{zz}}{\partial z^2} + \frac{\partial^2 \Theta_{xx}}{\partial x^2} - \frac{\partial^2 \Theta_{yy}}{\partial y^2} + 2\frac{\partial^2 \Theta_{zx}}{\partial z \partial x}$$

$$(7.2.66)$$

$$-\left(\frac{\partial^2}{\partial z^2} + \frac{\partial^2}{\partial x^2} - \frac{\rho}{2G}\frac{\partial^2}{\partial t^2}\right)\Omega,$$

$$\sigma_z = \frac{\partial^2 \Theta_{xx}}{\partial x^2} + \frac{\partial^2 \Theta_{yy}}{\partial y^2} - \frac{\partial^2 \Theta_{zz}}{\partial z^2} + 2\frac{\partial^2 \Theta_{xy}}{\partial x \partial y}$$

$$-\left(\frac{\partial^2}{\partial x^2} + \frac{\partial^2}{\partial y^2} - \frac{\rho}{2G}\frac{\partial^2}{\partial t^2}\right)\Omega,$$

$$\tau_{yz} = \frac{\partial}{\partial x} \left(\frac{\partial \Theta_{yz}}{\partial x} - \frac{\partial \Theta_{zx}}{\partial y} - \frac{\partial \Theta_{xy}}{\partial z} \right)$$

$$- \frac{\partial^2}{\partial y \partial z} (\Theta_{yy} + \Theta_{zz} - \Omega) - \frac{\rho}{G} \frac{\partial^2 \Theta_{yz}}{\partial t^2},$$

$$\tau_{zx} = \frac{\partial}{\partial y} \left(\frac{\partial \Theta_{zx}}{\partial y} - \frac{\partial \Theta_{xy}}{\partial z} - \frac{\partial \Theta_{yz}}{\partial x} \right) \qquad (7.2.66')$$

$$- \frac{\partial^2}{\partial z \partial x} (\Theta_{zz} + \Theta_{xx} - \Omega) - \frac{\rho}{G} \frac{\partial^2 \Theta_{zx}}{\partial t^2},$$

$$\tau_{xy} = \frac{\partial}{\partial z} \left(\frac{\partial \Theta_{xy}}{\partial z} - \frac{\partial \Theta_{yz}}{\partial x} - \frac{\partial \Theta_{zx}}{\partial y} \right)$$

$$- \frac{\partial^2}{\partial x \partial y} (\Theta_{xx} + \Theta_{yy} - \Omega) - \frac{\rho}{G} \frac{\partial^2 \Theta_{xy}}{\partial t^2};$$

the state of displacement is expressed by

$$2Gu = \frac{\partial \Omega}{\partial x} - \left(\frac{\partial \Theta_{xx}}{\partial x} + \frac{\partial \Theta_{xy}}{\partial y} + \frac{\partial \Theta_{zx}}{\partial z} \right),$$

$$2Gv = \frac{\partial \Omega}{\partial y} - \left(\frac{\partial \Theta_{xy}}{\partial x} + \frac{\partial \Theta_{yy}}{\partial y} + \frac{\partial \Theta_{yz}}{\partial z} \right), \qquad (7.2.67)$$

$$2Gw = \frac{\partial \Omega}{\partial z} - \left(\frac{\partial \Theta_{zx}}{\partial x} + \frac{\partial \Theta_{yz}}{\partial y} + \frac{\partial \Theta_{zz}}{\partial z} \right).$$

The components of H. Schaefer's tensor

$$T_\theta \equiv \begin{Bmatrix} \Theta_{xx} & \Theta_{xy} & \Theta_{zx} \\ \Theta_{xy} & \Theta_{yy} & \Theta_{yz} \\ \Theta_{zx} & \Theta_{yz} & \Theta_{zz} \end{Bmatrix} \qquad (7.2.68)$$

must satisfy the equation of transverse waves

$$\Box_2 \varkappa(x, y, z; t) = 0, \tag{7.2.69}$$

being of class C^4 with regard to the space variables and of class C^2 with respect to the time variable; the function Ω must satisfy the equation

$$\Box_1 \Omega = \frac{1}{1-v} \left[\frac{\partial^2 \Theta_{xx}}{\partial x^2} + \frac{\partial^2 \Theta_{yy}}{\partial y^2} + \frac{\partial^2 \Theta_{zz}}{\partial z^2} \right.$$

$$\left. + 2 \left(\frac{\partial^2 \Theta_{yz}}{\partial y \partial z} + \frac{\partial^2 \Theta_{zx}}{\partial z \partial x} + \frac{\partial^2 \Theta_{xy}}{\partial x \partial y} \right) \right], \tag{7.2.70}$$

being of class C^4 with respect to all the variables. Thus, the complete system of equations of elastodynamics, in the absence of body forces, is satisfied; the above representation is complete for a simply connected domain.

7.2.4.3. A new form of the representation of the Schaefer type

By introducing a vector potential $\boldsymbol{\Phi} = \boldsymbol{\Phi}(x, y, z; t)$ whose components are defined by the relations

$$\Phi_x = \frac{\partial \Theta_{xx}}{\partial x} + \frac{\partial \Theta_{xy}}{\partial y} + \frac{\partial \Theta_{zx}}{\partial z},$$

$$\Phi_y = \frac{\partial \Theta_{xy}}{\partial x} + \frac{\partial \Theta_{yy}}{\partial y} + \frac{\partial \Theta_{yz}}{\partial z}, \tag{7.2.71}$$

$$\Phi_z = \frac{\partial \Theta_{zx}}{\partial x} + \frac{\partial \Theta_{yz}}{\partial y} + \frac{\partial \Theta_{zz}}{\partial z},$$

we may express the state of stress in the form

$$\sigma_x = -\frac{\partial \Phi_x}{\partial x} + \frac{\partial \Phi_y}{\partial y} + \frac{\partial \Phi_z}{\partial z} - \left(\frac{\partial^2}{\partial y^2} + \frac{\partial^2}{\partial z^2} - \frac{\rho}{2G} \frac{\partial^2}{\partial t^2} \right) \Omega,$$

$$\sigma_y = \frac{\partial \Phi_x}{\partial x} - \frac{\partial \Phi_y}{\partial y} + \frac{\partial \Phi_z}{\partial z} - \left(\frac{\partial^2}{\partial z^2} + \frac{\partial^2}{\partial x^2} - \frac{\rho}{2G} \frac{\partial^2}{\partial t^2} \right) \Omega, \tag{7.2.72}$$

$$\sigma_z = \frac{\partial \Phi_x}{\partial x} + \frac{\partial \Phi_y}{\partial y} - \frac{\partial \Phi_z}{\partial z} - \left(\frac{\partial^2}{\partial x^2} + \frac{\partial^2}{\partial y^2} - \frac{\rho}{2G} \frac{\partial^2}{\partial t^2} \right) \Omega,$$

$$\tau_{yz} = \frac{\partial^2 \Omega}{\partial y \partial z} - \left(\frac{\partial \Phi_y}{\partial z} + \frac{\partial \Phi_z}{\partial y} \right),$$

$$\tau_{zx} = \frac{\partial^2 \Omega}{\partial z \partial x} - \left(\frac{\partial \Phi_z}{\partial x} + \frac{\partial \Phi_x}{\partial z} \right), \tag{7.2.72'}$$

$$\tau_{xy} = \frac{\partial^2 \Omega}{\partial x \partial y} - \left(\frac{\partial \Phi_x}{\partial y} + \frac{\partial \Phi_y}{\partial x} \right);$$

the state of displacement expressed in vector form will be

$$2G\mathbf{u} = \operatorname{grad} \Omega - 2\mathbf{\Phi}. \tag{7.2.73}$$

The components of the vector potential $\mathbf{\Phi}$ satisfy the equation of transverse waves (7.2.68), being functions of class C^3 with regard to the space variables and of class C^2 with regard to the time variable. Also, the function Ω must satisfy the equation

$$\square_1 \Omega = \frac{1}{1-v} \operatorname{div} \mathbf{\Phi}. \tag{7.2.74}$$

Introducing the vector potential $\mathbf{\Gamma} = \mathbf{\Gamma}(x, y, z; t)$ by the relation

$$\mathbf{\Phi} = -(1-v) \square_1 \mathbf{\Gamma} \tag{7.2.75}$$

and remarking that in that case we may write

$$\Omega = -\operatorname{div} \mathbf{\Gamma}, \tag{7.2.76}$$

we obtain the representation of C. Somigliana and M. Iacovache

$$2G\mathbf{u} = 2(1-v) \square_1 \mathbf{\Gamma} - \operatorname{grad} \operatorname{div} \mathbf{\Gamma}, \tag{7.2.77}$$

the components of the vector $\mathbf{\Gamma}$ being of class C^4 and satisfying the double equation of waves (7.2.10).

Thus, we pass from stress potentials to displacement potentials.

7.2.4.4. Non-zero body forces

In the case of non-zero body forces we may use the same representations, for example the representation (7.2.72), (7.2.72′), (7.2.73), the vector $\boldsymbol{\Phi}$ being given by equation

$$\square_2 \boldsymbol{\Phi} = \mathbf{F}, \tag{7.2.78}$$

where $\mathbf{F}$ is the body force vector; the function Ω will be a particular solution of equation

$$\square_1 \square_2 \Omega = \frac{1}{1-v} \operatorname{div} \mathbf{F}. \tag{7.2.79}$$

In particular, the influence of a *conservative body force*

$$\mathbf{F} = \operatorname{grad} \chi \tag{7.2.80}$$

is introduced in the form

$$\sigma_x = \frac{v}{1-2v} \Delta\omega + \frac{\partial^2\omega}{\partial x^2},$$

$$\sigma_y = \frac{v}{1-2v} \Delta\omega + \frac{\partial^2\omega}{\partial y^2}, \tag{7.2.81}$$

$$\sigma_z = \frac{v}{1-2v} \Delta\omega + \frac{\partial^2\omega}{\partial z^2},$$

$$\tau_{yz} = \frac{\partial^2\omega}{\partial y\partial z}, \quad \tau_{zx} = \frac{\partial^2\omega}{\partial z\partial x}, \quad \tau_{xy} = \frac{\partial^2\omega}{\partial x\partial y}, \tag{7.2.81′}$$

$$2G\mathbf{u} = \operatorname{grad} \omega, \tag{7.2.82}$$

where the potential $\omega = \omega(x, y, z; t)$ is given by the equation

$$\square_1\omega + \frac{1-2v}{1-v} \chi = 0. \tag{7.2.83}$$

Applying the results of section 3.3.2.3 we may write the particular solutions of the above equations

$$\Phi_x = -\frac{1}{4\pi} L^{-1}\left[\frac{1}{R}\, e^{-p\frac{R}{c_2}} * L[\bar{X}]\right], \ldots, \qquad (7.2.84)$$

$$\omega = \frac{1-2v}{4(1-v)\pi} L^{-1}\left[\frac{1}{R}\, e^{-p\frac{R}{c_1}} * L[\bar{\chi}]\right], \qquad (7.2.85)$$

$$\Omega = \frac{1}{2\pi}\, c_2^2\, L^{-1}\left[\frac{1}{p^2 R}\,(e^{-p\frac{R}{c_1}} - e^{-p\frac{R}{c_2}}) * L\left(\frac{\partial \bar{X}}{\partial x} + \frac{\partial \bar{Y}}{\partial y} + \frac{\partial \bar{Z}}{\partial z}\right)\right]; \qquad (7.2.86)$$

here we have introduced the distributions corresponding to the given body forces, using the formulae (7.2.4) and a similar formula for the potential χ.

The representations given here will be used in chapter 10.

8

Bars. Strings

8.1. Straight bar

Bodies which have one dimension (the length) considerably greater than the other two (the cross-section) are termed bars. The notion will be stated more precisely by giving a constructive definition.

Let Γ be an arc of a curve of finite length l *(bar axis)*. In the plane normal to the curve at an arbitrary point of the latter let us consider a closed curve C which bounds a plane domain D *(cross-section of the bar)*; the centre of gravity of the area of the domain is supposed to be on the curve Γ, at the point P. When the point P moves along the curve Γ the curve C, which may be deformable, will generate a surface which bounds a three-dimensional domain called a *bar* (Figure 8.1); the two dimensions (mean values) a and b of the cross-section should be of the same order of magnitude and satisfy the condition $a, b \ll l$.

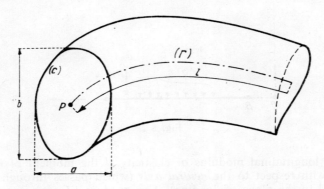

Fig. 8.1

According to the shape of the axis, the bars may be straight or curved. In the following we shall treat of *straight bars* and of problems concerning their bending and transverse vibrations.

8.1.1. Bending of a straight bar

We shall study the bending of a straight bar adopting the hypotheses used in the *strength of materials* (particularly Bernoulli's hypothesis concerning plane cross-sections) and applying the results obtained previously for ordinary differential equations. At the same time, we shall also point out the maximum number of directed concentrated moments of higher order which must be introduced in order to shift the action of a concentrated force from one point to another point of the straight bar[*].

**8.1.1.1. Generalized solution of the differential equation
              of the elastic line of a straight bar subjected to bending**

In the theory of strength of materials a straight bar reduces, as far as deflection is concerned, to the *medium fibre;* the deflected medium fibre is also called the *elastic line of the bar* (or the deflected curve). Assuming that a straight bar is subjected to beding due to a normal load which can be represented by a distribution $q(x)$ (Figure 8.2), and taking the axis of the bar BC as a co-ordinate axis Ox, we may write the differential equation in the form

$$\frac{d^4v(x)}{dx^4} = \frac{1}{EI}\, q(x) \; ; \tag{8.1.1}$$

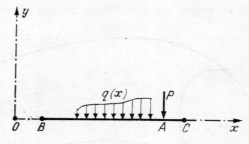

Fig. 8.2

here E is the longitudinal modulus of elasticity, I the moment of inertia of the cross-section with respect to the *neutral axis* (which passes through the bar axis and is normal to the plane of loading). The product EI represents the *flexural rigidity* of the bar. The load $q(x)$ is considered positive if it pulls the bar, and the displacement $v(x)$ is positive if it is directed in the positive direction of the Oy axis.

---

[*] See section 5.2.1.5.

Using the results given in section 3.2.1.3, we remark that the normal funda-mental system of solutions of the homogeneous equation corresponding to equation (8.1.1) is

$$v_1(x) = 1, \quad v_2(x) = x, \quad v_3(x) = \frac{1}{2}x^2, \quad v_4(x) = \frac{1}{6}x^3 ; \qquad (8.1.2)$$

in that case the particular fundamental solution $E^+(x)$ will be given by

$$E^+(x) = \frac{1}{EI} v_4^+(x) = \begin{cases} \dfrac{1}{6EI}x^3 & \text{for } x \geqslant 0 \\[2mm] 0 & \text{for } x < 0. \end{cases} \qquad (8.1.3)$$

Therefore, the general solution of equation (8.1.1) will be

$$EIv(x) = C_1 + C_2 x + \frac{1}{2}C_3 x^2 + \frac{1}{6}C_4 x^3 + \frac{1}{6}x_+^3 * q(x), \qquad (8.1.4)$$

where the constants C_1, C_2, C_3, C_4 will result from the *supporting conditions* of the bar.

We remark that the equation of the elastic line may be also written in the form

$$EIv(x) = \left(A_1 + A_2 x + \frac{1}{2}A_3 x^2 + \frac{1}{6}A_4 x^3 + \frac{1}{6}x_+^3 \right) * q(x), \qquad (8.1.4')$$

where A_1, A_2, A_3, A_4 are other constants to be determined in the same way.

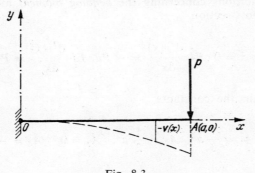

Fig. 8.3

For example, let there be a straight bar OA, of length a (Figure 8.3), having a built-in end at the point O and the other end free *(cantilever)*; we assume that a concentrated force P acts on the free end. The equivalent load will be

$$q(x) = -P \delta(x - a), \qquad (8.1.5)$$

and we may write the equation of the elastic line in the form

$$EIv(x) = C_1 + C_2x + \frac{1}{2}C_3x^2 + \frac{1}{6}C_4x^3 - \frac{P}{6}x_+^3 * \delta(x-a) ; \qquad (8.1.6)$$

since

$$x_+^3 * \delta(x-a) = (x-a)_+^3,$$

we may also write

$$EIv(x) = C_1 + C_2x + \frac{1}{2}C_3x^2 + \frac{1}{6}C_4x^3 - \frac{P}{6}(x-a)_+^3, \quad x \in [0, \infty). \qquad (8.1.6')$$

We remark that equation (8.1.6) may be also written

$$EIv(x) = -\left(A_1 + A_2x + \frac{1}{2}A_3x^2 + \frac{1}{6}A_4x^3 + \frac{1}{6}x_+^3 \right) * P\,\delta(x-a), \qquad (8.1.6'')$$

the constants of integration resulting from the *conditions at the built-in end*

$$x = 0: \quad v(x) = 0, \quad \frac{dv(x)}{dx} = 0 \qquad (8.1.7)$$

and from the conditions concerning the *bending moment* and the *shearing force* at the built-in cross-section

$$x = 0 : EI\frac{d^2v(x)}{dx^2} = -Pa, \quad EI\frac{d^3v(x)}{dx^3} = P. \qquad (8.1.7')$$

Thus, we obtain the constants

$$A_1 = \frac{1}{3}a^3, \quad A_2 = \frac{1}{2}a^2, \quad A_3 = 0, \quad A_4 = -1. \qquad (8.1.8)$$

8.1.1.2. Maximum order of directed concentrated moments

The equation of the elastic line of a straight bar subjected to bending may be written

$$v(x) = v_0(x) + E^+(x) * q(x), \qquad (8.1.9)$$

where $E^+(x)$ represents the particular fundamental solution given by (8.1.3) and the function $v_0(x)$ is expressed by

$$v_0(x) = \frac{1}{EI}\left(C_1 + C_2 x + \frac{1}{2} C_3 x^2 + \frac{1}{6} C_4 x^3 \right). \qquad (8.1.9')$$

We assume that a concentrated force P, expressed in the form $q(x) = Q_A(x)$ is applied at the point A of the bar (Figure 8.2); taking into account the representation (5.2.29), (5.2.30) we may write

$$v(x) = v_0(x) * \left[Q_0(x) + Q^{(1)}(x) + \frac{1}{2} Q^{(2)}(x) + \frac{1}{6} Q^{(3)}(x) \right], \, x \in [0, a], \quad (8.1.10)$$

since

$$v_0(x) * Q^{(i)}(x) = 0, \; i \geqslant 4. \qquad (8.1.11)$$

Hence the *maximum order of directed concentrated moments* will be in that case $n = 3$, thus proving relation (5.2.32).

Returning to the example given in the preceding subsection about a cantilever OA acted on by a concentrated force P at a point A (Figure 8.3), and taking into account the results presented in section 5.2.1.5, we may write

$$Q_0(x) = - P\delta(x), \; Q^{(1)}(x) = Pa\delta'(x),$$

$$Q^{(2)}(x) = - Pa^2\delta''(x), \; Q^{(3)}(x) = Pa^3\delta'''(x). \qquad (8.1.12)$$

Denoting

$$\mathscr{E}(x) = \frac{1}{3} a^3 + \frac{1}{2} a^2 x - \frac{1}{6} x^3, \qquad (8.1.13)$$

we may write for all $x \in [0, a]$

$$EIv(x) = - \mathscr{E}(x) * P\delta(x - a) = \mathscr{E}(x) * Q_0(x) + \mathscr{E}(x) * Q^{(1)}(x)$$

$$+ \frac{1}{2} \mathscr{E}(x) * Q^{(2)}(x) + \frac{1}{6} \mathscr{E}(x) * Q^{(3)}(x); \qquad (8.1.14)$$

whence

$$EIv(x) = - \left(\frac{1}{3} a^3 + \frac{1}{2} a^2 x - \frac{1}{6} x^3 \right) P$$

$$+ \left(\frac{1}{2} a^2 - \frac{1}{2} x^2 \right) Pa + \frac{1}{2} xPa^2 - \frac{1}{6} Pa^3. \qquad (8.1.15)$$

Finally, the equation of the elastic line is given, for all $x \in [0, a]$, by the expression

$$v(x) = -\frac{P}{6\,EI}x^2\,(3a - x). \tag{8.1.15$'$}$$

8.1.2. Free transverse vibrations of an infinite straight bar

We shall deal in the following, with the problem of *free transverse vibrations (eigenvibrations)* of a straight bar of infinite length; this will permit us to eliminate the conditions at the ends of the bar, and at the same time will exemplify a method for solving such problems.

In this connection, we shall first establish a modified form of the differential equation of motion, and then we shall integrate that equation with initial conditions using the method of integral transformations.

8.1.2.1. Modified form of the differential equation of motion

Taking the axis of a bar of infinite length as the Ox axis and considering the Oxy plane, we may write the differential equation of the free transverse vibrations of the bar in the form

$$\frac{\partial^4}{\partial x^4}\,v(x;t) + \frac{1}{c^2}\frac{\partial^2}{\partial t^2}\,v(x;t) = 0, \tag{8.1.16}$$

where the velocity of wave propagation is given by

$$c = \sqrt{\frac{EI}{\rho}}\,; \tag{8.1.17}$$

EI is the fundamental rigidity and ρ the density.

We set the initial conditions

$$v(x;0) = f_1(x), \quad \frac{\partial}{\partial t}v(x;0) = g(x) = c\,\frac{d^2 f_2(x)}{dx^2}, \tag{8.1.18}$$

where differentiation is effected in the sense of the theory of distributions and $f_1(x)$ and $f_2(x)$ are given distributions.

Since the function $v(x; t)$ is defined only for $t > 0$ we cannot replace the differential equation (8.1.16) by a differential equation in terms of distributions; to obtain the latter we shall introduce the generalized displacement

$$\bar{v}(x; t) = \begin{cases} 0 & \text{for } t < 0 \\[2mm] v(x; t) & \text{for } t \geqslant 0, \end{cases} \tag{8.1.19}$$

which involves a discontinuity of the first species for $t = 0$.

In that case we may write

$$\frac{\partial}{\partial t} \bar{v}(x; t) = \frac{\tilde{\partial}}{\partial t} \bar{v}(x; t) + \bar{v}(x; 0)\, \delta(t) = \frac{\tilde{\partial}}{\partial t} v(x; t) + f_1(x)\, \delta(t), \tag{8.1.20}$$

$$\frac{\partial^2}{\partial t^2} \bar{v}(x; t) = \frac{\tilde{\partial}^2}{\partial t} \bar{v}(x; t) + c\,\frac{\mathrm{d}^2 f_2(x)}{\mathrm{d}x^2}\, \delta(t) + f_1(x)\, \dot{\delta}(t), \tag{8.1.20'}$$

where the tilde sign corresponds to differentiation in the ordinary sense and the jumps at the moment of discontinuity $t = 0$, which correspond to the initial conditions, are pointed out.

On the other hand, we have

$$\frac{\partial^4}{\partial x^4} \bar{v}(x; t) = \frac{\tilde{\partial}^4}{\partial x^4} \bar{v}(x; t). \tag{8.1.21}$$

Thus, the differential equation (8.1.16) is expressed in the *modified form*

$$c^2 \frac{\partial^4}{\partial x^4} \bar{v}(x; t) + \frac{\partial^2}{\partial t^2} \bar{v}(x; t) = c\,\frac{\mathrm{d}^2 f_2(x)}{\mathrm{d}x^2}\, \delta(t) + f_1(x)\, \dot{\delta}(t), \tag{8.1.22}$$

which also includes the initial conditions, since it is written in terms of distributions.

It should be noted that the same procedure may be applied to other differential equations of motion of a similar form.

8.1.2.2. Integration of the differential equation of motion

To obtain the solution of the differential equation (8.1.22), we shall first apply the Laplace transform with respect to the time-variable t, using the results of sections 2.2.1.2 and 2.2.1.3. We obtain

$$c^2 \frac{\mathrm{d}^4}{\mathrm{d}x^4} L[\bar{v}(x; t)] + p^2 L[\bar{v}(x; t)] = c\,\frac{\mathrm{d}^2 f_2(x)}{\mathrm{d}x^2} + p f_1(x). \tag{8.1.23}$$

Then we apply the Fourier transform with respect to the spatial variable x and using the results given in paragraph 2.1 we have

$$c^2(-i\alpha)^4 F[L[\bar{v}(x;t)]] + p^2 F[L[\bar{v}(x;t)]]$$

$$= -\alpha^2 c F[f_2(x)] + p F[f_1(x)], \tag{8.1.23'}$$

whence

$$F[L[\bar{v}(x;t)]] = \frac{p}{p^2 + \alpha^4 c^2} F[f_1(x)] - \frac{\alpha^2 c}{p^2 + \alpha^4 c^2} F[f_2(x)]. \tag{8.1.24}$$

By applying the inverse Laplace transform we may write

$$F[\bar{v}(x;t)] = F[f_1(x)] L^{-1}\left[\frac{p}{p^2 + \alpha^4 c^2}\right] - F[f_2(x)] L^{-1}\left[\frac{\alpha^2 c}{p^2 + \alpha^4 c^2}\right]. \tag{8.1.25}$$

Since

$$L^{-1}\left[\frac{p}{p^2 + \alpha^4 c^2}\right] = \cos(\alpha^2 ct), \quad L^{-1}\left[\frac{\alpha^2 c}{p^2 + \alpha^4 c^2}\right] = \sin(\alpha^2 ct), \tag{8.1.26}$$

formula (8.1.25) takes the form

$$F[\bar{v}(x;t)] = F[f_1(x)] \cos(\alpha^2 ct) - F[f_2(x)] \sin(\alpha^2 ct). \tag{8.1.25'}$$

Applying now the inverse Fourier transform to relation (8.1.25') we obtain

$$\bar{v}(x;t) = F^{-1}[F[f_1(x)] \cos(\alpha^2 ct)] - F^{-1}[F[f_2(x)] \sin(\alpha^2 ct)]; \tag{8.1.27}$$

remarking that

$$F^{-1}[\cos(\alpha^2 ct)] = \frac{1}{2\sqrt{2\pi ct}}\left(\cos\frac{x^2}{4ct} + \sin\frac{x^2}{4ct}\right),$$

$$\tag{8.1.28}$$

$$F^{-1}[\sin(\alpha^2 ct)] = \frac{1}{2\sqrt{2\pi ct}}\left(\cos\frac{x^2}{4ct} - \sin\frac{x^2}{4ct}\right)$$

and taking into account formula (2.1.17) concerning the Fourier transform of a convolution product in the case where one of the factors is a temperate distribution and the other factor is a distribution with a bounded support, we may write the solution of the problem in the form

$$\bar{v}(x;t) = \frac{1}{2\sqrt{2\pi ct}} \left[f_1(x) * \left(\cos \frac{x^2}{4ct} + \sin \frac{x^2}{4ct} \right) \right.$$

$$\left. - f_2(x) * \left(\cos \frac{x^2}{4ct} - \sin \frac{x^2}{4ct} \right) \right]. \tag{8.1.27'}$$

8.2. Strings

We shall consider the constructive definition of a bar given at the beginning of the preceding section. If the dimensions of the cross-section of the bar are quite negligible with respect to the length of the bar, so that the latter becomes *perfectly flexible* (i.e. cannot be subjected to bending), we obtain a body called a *string*.

In the following discussion it will be assumed that we have to deal with inextensible strings only. We shall first consider the equilibrium of such a string acted on by loads which are expressed by a continuous function with discontinuities of the first species, then we shall consider the problem of *free transverse vibrations* of a string (the problem of a *vibrating string*).

8.2.1. Equilibrium of a string acted on by loads expressed by continuous functions with discontinuities of the first species

In the mechanics of continuously deformable material curves it is shown that the equilibrium equations of a *perfectly flexible and inextensible* string acted on by forces represented by a continuous function $\mathbf{F}(s)$, are

$$\frac{d\mathbf{T}(s)}{ds} + \mathbf{F}(s) = 0, \tag{8.2.1}$$

$$\mathbf{u}(s) \times \mathbf{T}(s) = 0, \tag{8.2.2}$$

where s is an arc of the curve and $\mathbf{u}(s)$ and $\mathbf{T}(s)$ represent the unit vector of the tangent to the equilibrium curve of the string and the *tension* in the string, respectively, at the point whose curvilinear abscissa is s.

In most of the problems which occur in practice the repartition of the forces along a string is expressed, from a mathematical point of view, by piecewise continuous functions or, more exactly, by a function which is continuous everywhere except at a finite number of points where it has discontinuities of the first species. Such a case occurs in the design of cableways and electrical networks.

A study of the equilibrium of strings acted on by such loads may be effected by applying the equations (8.2.1) and (8.2.2) to those parts of the string where external loads are expressed by continuous functions. If the string is acted on by concentrated loads only, the above equations can be applied no longer, but the equilibrium conditions must be written for every separate point where a concentrated load is acting. All these different cases of loading may be considered in a unitary form when we apply the theory of distributions.

We shall establish the equilibrium equation of a string expressed in terms of distributions followed by a few examples of various cases of loading.

8.2.1.1. Equations of equilibrium

Let AB be a perfectly flexible and inextensible string fixed at the points A and B (Figure 8.4). The arcs are measured with A as a reference point and the positive direction is from A toward B. We assume that the string is acted on by forces expressed by a vector function $\mathbf{F}(s)$ which is continuous everywhere with respect to the arc s, except at a finit number of points where it has discontinuities of the first species. Obviously, such loads also include concentrated forces since, if $\mathbf{F}_1$ represents a force acting on the string at the point P_0 defined by the arc $AP_0 = s_0$, it may be represented by the function

$$\mathbf{F}(s) = \begin{cases} \mathbf{F}_1 & \text{for } s = s_0 \\ \\ 0 & \text{for } s \neq s_0, \end{cases} \tag{8.2.3}$$

which has lateral limits at the point $s = s_0$.

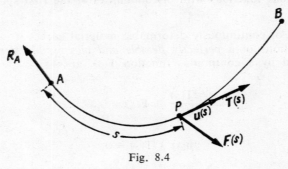

Fig. 8.4

In order to establish the equilibrium equations we cancel the constraint at the point A and replace it by the reaction $\mathbf{R}_A$, and we cut the string at a point P defined by the arc $\overset{\frown}{AP} = s$ and replace the action of the part $\overset{\frown}{PB}$ on the part $\overset{\frown}{AP}$ by the force $\mathbf{T}$ (tension) applied at that point and directed along the tangent to the equilibrium curve.

We shall denote by $\mathbf{R}(s)$ *the resultant of the given forces* acting on the part $\overgroup{AP}$. The necessary condition of equilibrium of the part $\overgroup{AP}$ is in that case

$$\mathbf{R}_A + \mathbf{R}(s) + \mathbf{T}(s) = 0. \tag{8.2.4}$$

From the hypothesis about the forces acting on the string we deduce that both the resultant $\mathbf{R}(s)$ and the tension $\mathbf{T}(s)$ will be functions which, together with the derivatives of the first order, are continuous everywhere, except at a finite number of points where they have discontinuities of the first species. It is important to note that the points of discontinuity of the tension and of the resultant are precisely the points of application of the concentrated forces; between the points of discontinuity the tension $\mathbf{T}(s)$ is directed, in accordance with equation (8.2.1), along the tangent to the equilibrium curve of the string.

If the forces acting on the string are expressed by a continuous function $\mathbf{F}(s)$, in which case the resultant $\mathbf{R}(s)$ is also a continuous function together with its derivative of the first order and having the expression

$$\mathbf{R}(s) = \int_0^s \mathbf{F}(s)\,\mathrm{d}s, \tag{8.2.5}$$

which is equivalent to

$$\frac{\mathrm{d}\mathbf{R}(s)}{\mathrm{d}s} = \mathbf{F}(s), \tag{8.2.5'}$$

then we obtain equation (8.2.1). The condition that the moment about an arbitrary point of all the forces acting on the arc $\overgroup{AP}$ be zero, leads to equation (8.2.2).

However, in general the resultant $\mathbf{R}(s)$ is not a function which is differentiable everywhere so that relation (8.2.5') does not hold everywhere. Since the resultant $\mathbf{R}(s)$ is an integrable function we may compute derivatives of any order within the frame of the theory of distributions. Thus, by differentiating relation (8.2.4) in the sense of the theory of distributions we obtain

$$\frac{\mathrm{d}\mathbf{T}(s)}{\mathrm{d}s} + \frac{\mathrm{d}\mathbf{R}(s)}{\mathrm{d}s} = 0. \tag{8.2.6}$$

We shall express the derivative in the sense of the theory of distributions of the resultant $\mathbf{R}(s)$ in the form (8.2.5'), and we shall call it a *generalized force* per unit length of string; in this way equation (8.2.1) retains the same form but is expressed in terms of distributions.

Hence, we may write the derivatives in the sense of the theory of distributions in the form

$$\frac{\mathrm{d}\mathbf{T}(s)}{\mathrm{d}s} = \frac{\widetilde{\mathrm{d}\mathbf{T}(s)}}{\mathrm{d}s} + \sum_{i=1}^{n} (\Delta\mathbf{T})_i\,\delta(s - s_i), \tag{8.2.7}$$

$$\frac{\mathrm{d}\mathbf{R}(s)}{\mathrm{d}s} = \frac{\widetilde{\mathrm{d}\mathbf{R}(s)}}{\mathrm{d}s} + \sum_{i=1}^{n} (\Delta\mathbf{R})_i\,\delta(s - s_i), \tag{8.2.7'}$$

where $(\Delta \mathbf{T})_i$ and $(\Delta \mathbf{R})_i$ represent the jump of the tension and of the resultant, respectively, corresponding to the point of discontinuity $s = s_i (i = 1, 2, \ldots, n)$. Equation (8.2.6) takes the form

$$\frac{\tilde{d}\mathbf{T}(s)}{ds} + \frac{\tilde{d}\mathbf{R}(s)}{ds} + \sum_{i=1}^{n} [(\Delta \mathbf{T})_i + (\Delta \mathbf{R})_i] \, \delta(s - s_i) = 0. \tag{8.2.6'}$$

By theorem 1.1.3 of section 1.1.2.6, relation (8.2.6') may take place if, and only if,

$$\frac{\tilde{d}\mathbf{T}(s)}{ds} + \frac{\tilde{d}\mathbf{R}(s)}{ds} = 0, \tag{8.2.8}$$

$$(\Delta \mathbf{T})_i + (\Delta \mathbf{R})_i = 0 \quad (i = 1, 2, \ldots, n). \tag{8.2.8'}$$

Introducing the force in the ordinary sense $\tilde{\mathbf{F}}(s)$, given by

$$\tilde{\mathbf{F}}(s) = \frac{\tilde{d}\mathbf{R}(s)}{ds}, \tag{8.2.9}$$

we may write equation (8.2.8) in the form

$$\frac{\tilde{d}\mathbf{T}(s)}{ds} + \tilde{\mathbf{F}}(s) = 0 ; \tag{8.2.8''}$$

the above equation is formally identical to equation (8.2.1) but it applies only to the parts between the points of discontinuity of the tension $\mathbf{T}(s)$.

Equation (8.2.8') shows that at every point of discontinuity the sum of the tension and of the resultant jumps is zero.

8.2.1.2. Applications

If, for example, a string is acted on at the point of curvilinear abscissa s_0 by a concentrated force $\mathbf{F}$, the resultant $\mathbf{R}(s)$ is expressed by

$$\mathbf{R}(s) = \begin{cases} 0 & \text{for } s < s_0 \\ \mathbf{F} & \text{for } s \geqslant s_0, \end{cases} \tag{8.2.10}$$

which may be also written

$$\mathbf{R}(s) = \mathbf{F} \, \theta(s - s_0), \tag{8.2.10'}$$

where we have introduced the Heaviside distribution; from relation (8.2.5′) we obtain the generalized force per unit length in the form

$$\mathbf{F}(s) = \mathbf{F}\,\delta(s - s_0). \tag{8.2.11}$$

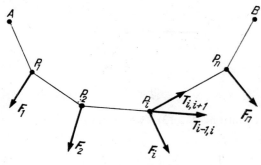

Fig. 8.5

In that case the equilibrium equation of a string acted on by a single concentrated force applied at the point $s = s_0$ becomes

$$\frac{d\mathbf{T}(s)}{ds} + \mathbf{F}\,\delta(s - s_0) = 0. \tag{8.2.12}$$

If the string is acted on by the concentrated forces $\mathbf{F}_i\ (i = 1, 2, \ldots, n)$ applied at the points $s_1 < s_2 < \ldots < s_i < \ldots < s_n$, it will take the form of a polygonal line (Figure 8.5); in that case the equation of equilibrium written in terms of distributions will be

$$\frac{d\mathbf{T}(s)}{ds} + \sum_{i=1}^{n} \mathbf{F}_i\,\delta(s - s_i) = 0. \tag{8.2.13}$$

This is equivalent to

$$\frac{\tilde{d}\mathbf{T}(s)}{ds} = 0, \tag{8.2.14}$$

$$(\Delta\mathbf{T})_i + \mathbf{F}_i = 0 \qquad (i = 1, 2, \ldots, n). \tag{8.2.14′}$$

The first equation shows that the stress is piecewise constant, i.e. that the equilibrium curve has indeed the shape of a polygonal line; on the other hand, noting that

$$(\Delta\mathbf{T})_i = \mathbf{T}_{i,\,i+1} - \mathbf{T}_{i-1,\,i}, \tag{8.2.15}$$

(where

$$\mathbf{T}_{i-1,\,i} + \mathbf{T}_{i,\,i-1} = 0, \tag{8.2.16}$$

it follows that equation (8.2.14′) corresponds to the equilibrium of the point P_i, acted on by the concentrated force $\mathbf{F}_i$ and by the tensions $\mathbf{T}_{i,\,i-1}$ and $\mathbf{T}_{i,\,i+1}$ (Figure 8.6).

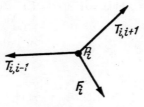

Fig. 8.6

By integrating equation (8.2.13), we obtain

$$\mathbf{T}(s) + \sum_{i=1}^{n} \mathbf{F}_i \theta(s - s_i) + \mathbf{C} = 0 \; ; \tag{8.2.17}$$

for $s = 0$ we may write

$$\mathbf{T}_A + \mathbf{C} = 0, \tag{8.2.18}$$

whence

$$\mathbf{C} = -\mathbf{T}_A = \mathbf{R}_A, \tag{8.2.18′}$$

where $\mathbf{T}_A$ and $\mathbf{R}_A$ are the tension and the reaction, respectively, corresponding to the supporting point A; therefore, we have

$$\mathbf{T}(s) = -\sum_{i=1}^{n} \mathbf{F}_i \, \theta(s - s_i) - \mathbf{R}_A. \tag{8.2.19}$$

Thus, we have obtained the tension at every point of the string except the points of discontinuity; besides, at these points too, the corresponding jumps are marked out.

If $\mathbf{r}(s)$ is the position vector of a point of the string then, from equation (8.2.2) we obtain

$$\frac{d\mathbf{r}(s)}{ds} = -\frac{1}{T}\left[\sum_{i=1}^{n} \mathbf{F}_i \theta(s - s_i) + \mathbf{R}_A\right]; \tag{8.2.20}$$

here, account has been taken of (8.2.19) and of

$$T = |\mathbf{T}| = \sqrt{\left[\mathbf{R}_A + \sum_{i=1}^{n} \mathbf{F}_i \theta(s - s_i)\right]^2}. \tag{8.2.21}$$

Since the unit vector of the tangent to the equilibrium curve $\dfrac{d\mathbf{r}}{ds}$ has discontinuities of the first species only, and considering the point A as the origin, we obtain by integration

$$\mathbf{r}(s) = \sum_{\substack{j=0 \\ (k\geqslant 1)}}^{k-1} \frac{\displaystyle\sum_{i=0}^{j} \mathbf{F}_i}{\sqrt{\left[\displaystyle\sum_{i=0}^{j} \mathbf{F}_i\right]^2}} (s_{j+1} - s_j) + \frac{\displaystyle\sum_{i=0}^{k} \mathbf{F}_i}{\sqrt{\left[\displaystyle\sum_{i=0}^{k} \mathbf{F}_i\right]^2}} (s - s_k), \qquad (8.2.22)$$

for $s_k \leqslant s \leqslant s_{k+1}$ $(k = 0, 1, 2, \ldots, n)$, with $s_0 = 0$ and $s_{n+1} = l$, where l is the total length of the string; also, we consider $\mathbf{F}_0 = \mathbf{R}_A$.

In this way we obtain the equation of equilibrium of a string acted on by n concentrated forces $\mathbf{F}_i (i = 1, 2, \ldots, n)$.

Let us now consider a string $\overset{\frown}{AB}$, of length l, acted on by its own weight (weight per unit volume γ) and by a concentrated force $\mathbf{F}$, applied at the point P_0 of curvilinear abscissa s_0 (Figure 8.7). The forces acting on the string being parallel (the direction of the force $\mathbf{F}$ is assumed to be parallel to the force of grav-

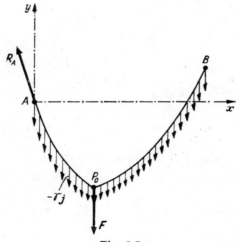

Fig. 8.7

ity) it follows that the equilibrium curve of the string is a plane curve. The differential equation of equilibrium will take the form

$$\frac{d\mathbf{T}(s)}{ds} - \gamma \mathbf{j} + \mathbf{F}\,\delta(s - s_0) = 0, \qquad (8.2.23)$$

where $\mathbf{j}$ is the unit vector of the Oy axis.

If $\mathbf{R}_A(X_A, Y_A)$ is the reaction at the point A, we obtain by integration

$$\mathbf{T}(s) = \gamma s \mathbf{j} - \mathbf{F}\,\theta(s - s_0) - \mathbf{R}_A ; \qquad (8.2.24)$$

projecting on the co-ordinate axes we have

$$\frac{dx}{ds} = -\frac{1}{T}X_A , \qquad (8.2.24')$$

$$\frac{dy}{ds} = \frac{1}{T}[\gamma s + F\,\theta(s - s_0) - Y_A], \qquad (8.2.24'')$$

where the magnitude of the tension in the string is given by

$$T = |\mathbf{T}| = \sqrt{X_A^2 + [\gamma s + F\,\theta(s - s_0) - Y_A]^2}. \qquad (8.2.25)$$

By integrating equation (8.2.24') we obtain

$$x = \begin{cases} -X_A \displaystyle\int_0^s \frac{ds}{\sqrt{X_A^2 + (\gamma s - Y_A)^2}} & \text{for } 0 \leqslant s \leqslant s_0 \\[4mm] -X_A \displaystyle\int_0^{s_0} \frac{ds}{\sqrt{X_A^2 + (\gamma s - Y_A)^2}} - X_A \displaystyle\int_{s_0}^s \frac{ds}{\sqrt{X_A^2 + (\gamma s + F - Y_A)^2}} & \text{for } s_0 \leqslant s \leqslant l ; \end{cases} \qquad (8.2.26)$$

using the substitutions

$$\sinh u = \frac{\gamma_s - Y_A}{|X_A|} \text{ for } 0 \leqslant s \leqslant s_0, \qquad (8.2.27)$$

$$\sinh v = \frac{\gamma s + F - Y_A}{|X_A|} \text{ for } s_0 \leqslant s \leqslant l, \qquad (8.2.27')$$

we can also write

$$x = \begin{cases} -\dfrac{X_A}{\gamma}(u - u_A) & \text{for } u_A \leqslant u \leqslant u_0 \\[4mm] -\dfrac{X_A}{\gamma}(u_0 - u_A + v - v_0) & \text{for } v_0 \leqslant v \leqslant v_B, \end{cases} \qquad (8.2.26')$$

where

$$u_A = u(s)|_{s=0}, \qquad v_B = v(s)|_{s=l},$$

$$u_0 = u(s)|_{s=s_0}, \qquad v_0 = v(s)|_{s=s_0}.$$

(8.2.28)

We can also write

$$y = \begin{cases} \dfrac{|X_A|}{\gamma}(\cosh u - \cosh u_A) \text{ for } u_A \leqslant u \leqslant u_0 \\[3mm] \dfrac{|X_A|}{\gamma}(\cosh u_0 - \cosh u_A + \cosh v - \cosh v_0) \text{ for } v_0 \leqslant v \leqslant v_B, \end{cases}$$

(8.2.29)

which states exactly the equilibrium equation of the string.

We remark that here, as in the preceding case, the reaction $\mathbf{R}_A$ must be specified by certain initial conditions.

Let us now consider a perfectly flexible and inextensible string of negligible weight, resting on a rough cylindrical surface wound along the plane section line $\overset{\frown}{ACB}$, with a discontinuity of the first species at the point C(Figure 8.8). We assume that the tensions at the two ends of the string are $\mathbf{T}_A$ and $\mathbf{T}_B$, respectively, and that the coefficients of friction between the string and the surface over the parts $\overset{\frown}{AC}$ and $\overset{\frown}{CB}$, are f_1 and f_2. Under these conditions we want to determine the tension at the point C so that the string be in equilibrium.

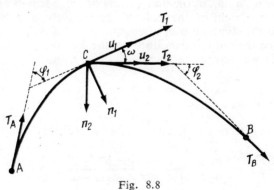

Fig. 8.8

The condition for a jump at the point C is

$$(\Delta \mathbf{T})_C + (\Delta \mathbf{R})_C = 0.$$

(8.2.30)

Assuming that the string tends to become displaced from A toward B and denoting by $\mathbf{T}_1$, $\mathbf{T}_2$ the tensions on the left and on the right, respectively, of the point C in the equilibrium position of the string, we may write L. Euler's relations

$$e^{-f_1\varphi_1} \leqslant \frac{T_1}{T_A} \leqslant e^{f_1\varphi_1}, \quad e^{-f_2\varphi_2} \leqslant \frac{T_B}{T_2} \leqslant e^{f_2\varphi_2}, \qquad (8.2.31)$$

where φ_1 is the angle between the tangent at the point A and the tangent to the left of the point C, while φ_2 is the angle between the tangent at the point B and the tangent to the right of the point C. Hence, we may also write

$$e^{-(f_1\varphi_1 + f_2\varphi_2)} \leqslant \frac{T_B}{T_A} \frac{T_1}{T_2} \leqslant e^{f_1\varphi_1 + f_2\varphi_2} . \qquad (8.2.31')$$

Remarking that

$$(\Delta\mathbf{T})_C = \mathbf{T}_2 - \mathbf{T}_1, \qquad (\Delta\mathbf{R})_C = \mathbf{R}_2 - \mathbf{R}_1, \qquad (8.2.32)$$

we may express relation (8.2.30) in the form

$$\mathbf{T}_2 - \mathbf{T}_1 + \mathbf{R}_2 - \mathbf{R}_1 = 0, \qquad (8.2.30')$$

where

$$\begin{aligned} \mathbf{R}_1 &= -\mu_1 (N_1\mathbf{n}_1 - f_1 N_1\mathbf{u}_1), \\ \mathbf{R}_2 &= -\mu_2 (N_2\mathbf{n}_2 + f_2 N_2\mathbf{u}_2) ; \end{aligned} \qquad (8.2.33)$$

we have denoted by $\mathbf{n}_1$, $\mathbf{n}_2$, $\mathbf{u}_1$, $\mathbf{u}_2$ the unit vectors of the normals and of the tangents on the left and on the right of the point C, respectively, and by μ_1, μ_2 two coefficients which have the dimension of a length.

Let ρ_1 and ρ_2 be the radii of curvature on the left and on the right of the point of discontinuity C; we may write

$$N_1 = \frac{T_1}{\rho_1}, \qquad N_2 = \frac{T_2}{\rho_2}. \qquad (8.2.34)$$

Projecting on the direction of the unit vector $\mathbf{v}_2$ we obtain

$$\frac{T_1}{T_2} = \frac{\rho_1}{\rho_2} \frac{\rho_2 - \mu_2 f_2}{(\rho_1 - \mu_1 f_1)\cos\omega - \mu_2 \sin\omega}, \qquad (8.2.35)$$

where ω is the angle between the unit vectors $\mathbf{u}_1$ and $\mathbf{u}_2$. Substituting this ratio in relation (8.2.31') we obtain the relation which must exist between $\mathbf{T}_A$ and $\mathbf{T}_B$ so that the string be in equilibrium.

The coefficients μ_1 and μ_2 specify the direction of the total reaction $\mathbf{R}_C$ at the point C

$$\mathbf{R}_C = \mathbf{R}_2 - \mathbf{R}_1. \qquad (8.2.33')$$

Denoting by β the angle between the total reaction $\mathbf{R}_C$ and the negative direction of the normal $\mathbf{n}_2$ we obtain

$$\frac{T_1}{T_2} = \frac{\cos \beta}{\cos \beta \cos \omega - \sin \beta} \; ; \tag{8.2.36}$$

as the ratio of the tensions is positive and since $0 \leqslant \beta \leqslant \dfrac{\pi}{2}$, it follows that

$$\tan \beta \leqslant \cos \omega, \tag{8.2.37}$$

which marks out the possible values of the angle β as a function of the angle ω, although the angle β remains indeterminate.

8.2.2. Vibrating string

As a dynamic problem we shall consider the case of the *free transverse vibrations of a string*, which is known in the literature as the *problem of the vibrating string*.
The differential equation of motion may be written in the form

$$\frac{\partial^2 v(x;t)}{\partial x^2} - \frac{1}{c^2} \frac{\partial^2 v(x;t)}{\partial t^2} = 0, \tag{8.2.38}$$

where the velocity of wave propagation is given by

$$c = \sqrt{\frac{T}{\rho A}} \; ; \tag{8.2.39}$$

T is the tension in the string, ρ is the density and A the area of the cross-section.
Just as in the case considered in section 8.1.2 we could complete the equation with initial conditions by extending the time variable t for values smaller than the initial moment and by applying successively the Laplace and the Fourier transforms.
In the following discussion we shall restrict our attention to the *fundamental solution* using the results of section 3.2.2.3 for a particular class of partial differential equations.
The characteristic curves of equation (8.2.38) are

$$\varphi(x; t) \equiv x - ct = C_1, \quad \psi(x; t) \equiv x + ct = C_2, \tag{8.2.40}$$

and the fundamental solution may be written in the form

$$E(x; t) = \frac{1}{2c} \left[\theta(x + ct) - \theta(x - ct) \right]. \tag{8.2.41}$$

Remarking that

$$\theta(x + ct) = \begin{cases} 1 \text{ for } x + ct \geqslant 0 \\ \\ 0 \text{ for } x + ct < 0, \end{cases} \tag{8.2.42}$$

$$\theta(x - ct) = \begin{cases} 1 \text{ for } x - ct \geqslant 0 \\ \\ 0 \text{ for } x - ct < 0, \end{cases} \tag{8.2.42'}$$

whence

$$\theta(x + ct) - \theta(x - ct) = \begin{cases} 1 \text{ for } |x| \leqslant ct \\ \\ 0 \text{ for } |x| > ct, \end{cases} \tag{8.2.42''}$$

the fundamental solution may be also written

$$E(x; t) = \begin{cases} \dfrac{1}{2c} \text{ for } |x| \leqslant ct \\ \\ 0 \quad \text{ for } |x| > ct. \end{cases} \tag{8.2.41'}$$

9

Plane problems of the theory of elasticity

In the following discussion we shall show how to apply the methods of the theory of distributions to some plane problems of the theory of elasticity. We shall consider an elastic plane acted on by static or dynamic loads and an elastic half-plane acted on by static loads. The methods presented may be extended to other plane domains.

9.1. Elastic plane

Let us consider an elastic plane acted on by the body forces $X = X(x, y)$, $Y = Y(x, y)$, expressed in terms of distributions (Figure 9.1). We assume that we are in the case of a state of plane strain (hence in the case of an elastic space acted on by body forces constant along a certain direction and having no components with respect to that direction).

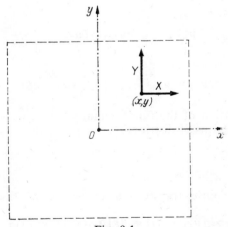

Fig. 9.1

We shall first show how to obtain the fundamental solution, and then use that solution for finding the solution corresponding to arbitrary body forces.

Also, we shall consider the elastic plane acted on by body forces $X=X(x, y; t)$, $Y = Y(x, y; t)$ variable in time.

9.1.1. Static problem. Fundamental solution

The differential equations of the problem expressed in terms of stresses are the equations (7.2.13) and the equation

$$\Delta(\sigma_x + \sigma_y) = -\frac{1}{1-v}\left(\frac{\partial X}{\partial x} + \frac{\partial Y}{\partial y}\right), \tag{9.1.1}$$

which is obtained from equation (7.2.18) using the generalized elastic constants (7.2.17).

The matrix

$$(U) \equiv \begin{pmatrix} u_{11}(x, y) & u_{12}(x, y) \\ u_{21}(x, y) & u_{22}(x, y) \\ u_{31}(x, y) & u_{32}(x, y) \end{pmatrix} \tag{9.1.2}$$

will be termed the *fundamental solution* corresponding to the elastic plane if its elements satisfy the equations

$$\frac{\partial u_{11}}{\partial x} + \frac{\partial u_{31}}{\partial y} = -\delta(x, y), \quad \frac{\partial u_{31}}{\partial x} + \frac{\partial u_{21}}{\partial y} = 0, \tag{9.1.3}$$

$$\Delta(u_{11} + u_{21}) = -\frac{1}{1-v}\delta'_x(x, y) \tag{9.1.3'}$$

and the equations

$$\frac{\partial u_{12}}{\partial x} + \frac{\partial u_{32}}{\partial y} = 0, \quad \frac{\partial u_{32}}{\partial x} + \frac{\partial u_{22}}{\partial y} = -\delta(x, y), \tag{9.1.4}$$

$$\Delta(u_{12} + u_{22}) = -\frac{1}{1-v}\delta'_y(x, y). \tag{9.1.4'}$$

For the determination of the fundamental solution we shall apply the double Fourier transform in terms of distributions.

9.1.1.1. Fourier transform

Applying the double Fourier transform to equations (9.1.3), (9.1.3'), we obtain

$$\begin{aligned} \alpha F[u_{11}(x, y)] + \beta F[u_{31}(x, y)] &= -i, \\ \alpha F[u_{31}(x, y)] + \beta F[u_{21}(x, y)] &= 0, \end{aligned} \tag{9.1.5}$$

$$F[u_{11}(x, y)] + F[u_{21}(x, y)] = -\frac{i}{1-v}\frac{\alpha}{\alpha^2 + \beta^2}, \tag{9.1.5'}$$

where we have denoted by F[   ] the double Fourier transform; α and β are complex variables.

Solving the system of algebraic equations (9.1.5), (9.1.5'), we obtain

$$F[u_{11}(x, y)] = -\,i\,\frac{\alpha}{\alpha^2 + \beta^2} - \frac{i}{1 - v}\,\frac{\alpha\beta^2}{(\alpha^2 + \beta^2)^2}\,,$$

$$F[u_{21}(x, y)] = i\,\frac{\alpha}{\alpha^2 + \beta^2} - \frac{i}{1 - v}\,\frac{\alpha^3}{(\alpha^2 + \beta^2)^2}\,, \qquad (9.1.6)$$

$$F[u_{31}(x, y)] = -\,i\,\frac{\beta}{\alpha^2 + \beta^2} + \frac{i}{1 - v}\,\frac{\alpha^2\beta}{(\alpha^2 + \beta^2)^2}\,.$$

If we apply the double Fourier transform to equations (9.1.4) and (9.1.4') we may write, in a similar way,

$$F[u_{12}(x, y)] = i\,\frac{\beta}{\alpha^2 + \beta^2} - \frac{i}{1 - v}\,\frac{\beta^3}{(\alpha^2 + \beta^2)^2}\,,$$

$$F[u_{22}(x, y)] = -\,i\,\frac{\beta}{\alpha^2 + \beta^2} - \frac{i}{1 - v}\,\frac{\alpha^2\beta}{(\alpha^2 + \beta^2)^2}\,, \qquad (9.1.6')$$

$$F[u_{32}(x, y)] = -\,i\,\frac{\alpha}{\alpha^2 + \beta^2} + \frac{i}{1 - v}\,\frac{\alpha\beta^2}{(\alpha^2 + \beta^2)^2}\,.$$

9.1.1.2. Inverse Fourier transform

Applying the double Fourier transform to relation (1.4.52) we obtain

$$F[\log{(x^2 + y^2)}] = -\,\frac{4\pi}{\alpha^2 + \beta^2}\,; \qquad (9.1.7)$$

it follows that

$$F\left[\frac{\partial}{\partial x}\log{(x^2 + y^2)}\right] = -\,i\alpha F\,[\log{(x^2 + y^2)}] = 4\pi i\,\frac{\alpha}{\alpha^2 + \beta^2}\,,$$

i.e.

$$\frac{1}{2\pi}\,F\left[\frac{x}{x^2 + y^2}\right] = i\,\frac{\alpha}{\alpha^2 + \beta^2}\,. \qquad (9.1.8)$$

Similarly, we may write

$$\frac{1}{2\pi} F\left[\frac{y}{x^2 + y^2}\right] = \mathrm{i}\,\frac{\beta}{\alpha^2 + \beta^2} \; . \tag{9.1.8'}$$

We denote by $F_x[\;\;]$, $F_y[\;\;]$ the simple Fourier transform with respect to the variable x and y, respectively; if $k > 0$ we may write

$$F\left[\frac{x}{x^2 + k^2 y^2}\right] = F_y\!\left[F_x\!\left[\frac{x}{x^2 + k^2 y^2}\right]\right] = F_y\!\left[\frac{\pi \mathrm{i}|\alpha|}{\alpha}\, e^{-k|\alpha||y|}\right]$$

$$= \frac{\pi \mathrm{i}|\alpha|}{\alpha}\, F_y[e^{-k|\alpha||y|}] = \frac{\pi \mathrm{i}|\alpha|}{\alpha} \cdot \frac{2k|\alpha|}{k^2\alpha^2 + \beta^2} \, ,$$

whence

$$\frac{1}{2\pi} F\left[\frac{x}{x^2 + k^2 y^2}\right] = \mathrm{i}\,\frac{k\alpha}{k^2\alpha^2 + \beta^2} \, , \qquad k > 0, \tag{9.1.9}$$

thus generalizing formula (9.1.8); likewise we have

$$\frac{1}{2\pi} F\left[\frac{y}{k^2 x^2 + y^2}\right] = \mathrm{i}\,\frac{k\beta}{\alpha^2 + k^2\beta^2} \, , \qquad k > 0. \tag{9.1.9'}$$

Differentiating relations (9.1.9) and (9.1.9') with respect to the parameter k, we obtain

$$\frac{1}{2\pi} F\left[\frac{\mathrm{d}}{\mathrm{d}k}\left(\frac{x}{x^2 + k^2 y^2}\right)\right] = -\,\mathrm{i}\,\frac{\alpha(k^2\alpha^2 - \beta^2)}{(k^2\alpha^2 + \beta^2)^2} \, , \tag{9.1.10}$$

$$\frac{1}{2\pi} F\left[\frac{\mathrm{d}}{\mathrm{d}k}\left(\frac{x}{k^2 x^2 + y^2}\right)\right] = \mathrm{i}\,\frac{\beta(\alpha^2 - k^2\beta^2)}{(\alpha^2 + k^2\beta^2)^2} \, . \tag{9.1.10'}$$

In particular, for $k = 1$ we may write

$$\frac{1}{\pi} F\left[\frac{xy^2}{(x^2 + y^2)^2}\right] = \mathrm{i}\,\frac{\alpha(\alpha^2 - \beta^2)}{(\alpha^2 + \beta^2)^2} \, , \tag{9.1.11}$$

$$\frac{1}{\pi} F\left[\frac{yx^2}{(x^2 + y^2)^2}\right] = -\,\mathrm{i}\,\frac{\beta(\alpha^2 - \beta^2)}{(\alpha^2 + \beta^2)^2} \, . \tag{9.1.11'}$$

From relations (9.1.8), (9.1.8′), (9.1.11′) we derive

$$\frac{1}{2\pi} F\left[\frac{x}{2(x^2 + y^2)} - \frac{xy^2}{(x^2 + y^2)^2} \right] = i \frac{\alpha\beta^2}{(\alpha^2 + \beta^2)^2}, \tag{9.1.12}$$

$$\frac{1}{2\pi} F\left[\frac{x}{2(x^2 + y^2)} + \frac{xy^2}{(x^2 + y^2)^2} \right] = i \frac{\alpha^3}{(\alpha^2 + \beta^2)^2}, \tag{9.1.12′}$$

$$\frac{1}{2\pi} F\left[\frac{y}{2(x^2 + y^2)} - \frac{x^2 y}{(x^2 + y^2)^2} \right] = i \frac{\alpha^2\beta}{(\alpha^2 + \beta^2)^2}, \tag{9.1.13}$$

$$\frac{1}{2\pi} F\left[\frac{y}{2(x^2 + y^2)} + \frac{x^2 y}{(x^2 + y^2)^2} \right] = i \frac{\beta^3}{(\alpha^2 + \beta^2)^2}, \tag{9.1.13′}$$

which give the components of the matrix (9.1.2) of the fundamental solution in the form

$$u_{11}(x, y) = F^{-1}[F[u_{11}(x, y)]]$$

$$= -\frac{1}{2\pi} \left\{ \frac{x}{x^2 + y^2} + \frac{1}{1 - v}\left[\frac{x}{2(x^2 + y^2)} - \frac{xy^2}{(x^2 + y^2)^2} \right] \right\}$$

$$-\frac{1}{4\pi(1 - v)} \frac{x}{x^2 + y^2}\left(1 - 2v + \frac{2x^2}{x^2 + y^2} \right),$$

$$u_{21}(x, y) = F^{-1}[F[u_{21}x, y)]]$$

$$= \frac{1}{2\pi} \left\{ \frac{x}{x^2 + y^2} - \frac{1}{1 - v}\left[\frac{x}{2(x^2 + y^2)} + \frac{xy^2}{(x^2 + y^2)^2} \right] \right\} \tag{9.1.14}$$

$$= \frac{1}{4\pi(1 - v)} \frac{x}{x^2 + y^2}\left(1 - 2v - \frac{2y^2}{x^2 + y^2} \right),$$

$$u_{31}(x, y) = F^{-1}[F[u_{31}(x, y)]]$$

$$= -\frac{1}{2\pi} \left\{ \frac{y}{x^2 + y^2} - \frac{1}{1 - v}\left[\frac{y}{2(x^2 + y^2)} - \frac{x^2 y}{(x^2 + y^2)^2} \right] \right\}$$

$$= -\frac{1}{4\pi(1 - v)} \frac{y}{x^2 + y^2}\left(1 - 2v + \frac{2x^2}{x^2 + y^2} \right),$$

$$u_{12}(x, y) = F^{-1}[F[u_{12}(x, y)]]$$

$$= \frac{1}{2\pi} \left\{ \frac{y}{x^2 + y^2} - \frac{1}{1 - v} \left[\frac{y}{2(x^2 + y^2)} + \frac{x^2 y}{(x^2 + y^2)^2} \right] \right\}$$

$$= \frac{1}{4\pi(1 - v)} \frac{y}{x^2 + y^2} \left(1 - 2v - \frac{2x^2}{x^2 + y^2} \right),$$

$$u_{22}(x, y) = F^{-1}[F[u_{22}(x, y)]]$$

$$= -\frac{1}{2\pi} \left\{ \frac{y}{x^2 + y^2} + \frac{1}{1 - v} \left[\frac{y}{2(x^2 + y^2)} - \frac{x^2 y}{(x^2 + y^2)^2} \right] \right\} \qquad (9.1.14')$$

$$= -\frac{1}{4\pi(1 - v)} \frac{y}{x^2 + y^2} \left(1 - 2v + \frac{2y^2}{x^2 + y^2} \right),$$

$$u_{32}(x, y) = F^{-1}[F[u_{32}(x, y)]]$$

$$= -\frac{1}{2\pi} \left\{ \frac{x}{x^2 + y^2} - \frac{1}{1 - v} \left[\frac{x}{2(x^2 + y^2)} - \frac{x y^2}{(x^2 + y^2)^2} \right] \right\}$$

$$= -\frac{1}{4\pi(1 - v)} \frac{x}{x^2 + y^2} \left(1 - 2v + \frac{2y^2}{x^2 + y^2} \right).$$

9.1.2. Static problem. Case of arbitrary loads

With the help of the matrix (U) of the fundamental solution and of the convolution product we may construct the solution corresponding to arbitrary body forces. Using that mode of expression we shall then give results for a few particular cases of loading.

9.1.2.1. Case of arbitrary body forces

Introducing the matrices

$$(\sigma) \equiv \begin{pmatrix} \sigma_x \\ \sigma_y \\ \tau_{xy} \end{pmatrix}, \qquad (Q) \equiv \begin{Bmatrix} X \\ Y \end{Bmatrix}, \qquad (9.1.15)$$

we may write

$$(\sigma) = (U) * (Q); \tag{9.1.16}$$

in this way the state of stress for arbitrary body forces may be expressed in the form

$$\sigma_x(x, y) = u_{11}(x, y) * X(x, y) + u_{12}(x, y) * Y(x, y),$$

$$\tag{9.1.17}$$

$$\sigma_y(x, y) = u_{21}(x, y) * X(x, y) + u_{22}(x, y) * Y(x, y),$$

$$\tau_{xy}(x, y) = u_{31}(x, y) * X(x, y) + u_{32}(x, y) * Y(x, y). \tag{9.1.17'}$$

Ignoring the proof given in the preceding section, it may be shown that the formulae (9.1.17), (9.1.17') give the generalized solution (in terms of distributions) of the system of equations (7.2.13) and (9.1.1), by a direct verification, which may be easily performed taking into account the results of subsection 1.4.2.2.

9.1.2.2. Examples

Let there be an elastic plane acted on by a concentrated force $\mathbf{F}(F_x, F_y)$ at the point $A(x_0, y_0)$ (Figure 9.2); the equivalent load of the concentrated force is ex-

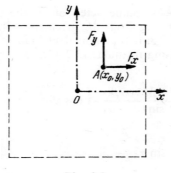

Fig. 9.2

pressed by

$$X(x, y) = F_x \delta(x - x_0, y - y_0),$$

$$\tag{9.1.18}$$

$$Y(x, y) = F_y \delta(x - x_0, y - y_0).$$

Using formulae (9.1.17) and (9.1.17′) we obtain the corresponding state of stress in the form

$$\sigma_x(x, y) = F_x u_{11}(x - x_0, y - y_0) + F_y u_{12}(x - x_0, y - y_0),$$

$$\tag{9.1.19}$$

$$\sigma_y(x, y) = F_x u_{21}(x - x_0, y - y_0) + F_y u_{22}(x - x_0, y - y_0),$$

$$\tau_{xy}(x, y) = F_x u_{31}(x - x_0, y - y_0) + F_y u_{32}(x - x_0, y - y_0). \tag{9.1.19′}$$

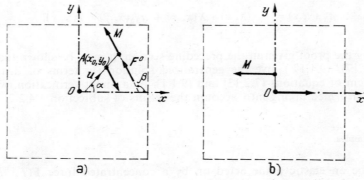

Fig. 9.3

Let us consider the elastic plane acted on at the point $A(x_0, y_0)$ by a directed concentrated moment of magnitude M, specified by the unit vectors $\mathbf{u}(\cos\alpha, \sin\alpha)$ and $\mathbf{F}^0(\cos\beta, \sin\beta)$ (Figure 9.3a); the equivalent load of this directed concentrated moment may be written, using formulae (5.2.4″), in the form

$$X(x, y) = -\frac{M\cos\beta}{|\sin(\beta - \alpha)|}\left[\cos\alpha \frac{\partial}{\partial x}\delta(x - x_0, y - y_0)\right.$$

$$\left. + \sin\alpha \frac{\partial}{\partial y}\delta(x - x_0, y - y_0)\right],$$

$$\tag{9.1.20}$$

$$Y(x, y) = -\frac{M\sin\beta}{|\sin(\beta - \alpha)|}\left[\cos\alpha \frac{\partial}{\partial x}\delta(x - x_0, y - y_0)\right.$$

$$\left. + \sin\alpha \frac{\partial}{\partial y}\delta(x - x_0, y - y_0)\right],$$

whence the state of stress given by

$$\sigma_x(x, y) = -\frac{M}{|\sin(\beta - \alpha)|}\left\{\cos\beta\left[\cos\alpha\frac{\partial}{\partial x}u_{11}(x - x_0, y - y_0)\right.\right.$$

$$+ \sin\alpha\frac{\partial}{\partial y}u_{11}(x - x_0, y - y_0)\bigg] + \sin\beta\left[\cos\alpha\frac{\partial}{\partial x}u_{12}(x - x_0, y - y_0)\right.$$

$$+ \sin\alpha\frac{\partial}{\partial y}u_{12}(x - x_0, y - y_0)\bigg]\bigg\},$$

$$(9.1.21)$$

$$\sigma_y(x, y) = -\frac{M}{|\sin(\beta - \alpha)|}\left\{\cos\beta\left[\cos\alpha\frac{\partial}{\partial x}u_{21}(x - x_0, y - y_0)\right.\right.$$

$$+ \sin\alpha\frac{\partial}{\partial y}u_{21}(x - x_0, y - y_0)\bigg] + \sin\beta\left[\cos\alpha\frac{\partial}{\partial x}u_{22}(x - x_0, y - y_0)\right.$$

$$+ \sin\alpha\frac{\partial}{\partial y}u_{22}(x - x_0, y - y_0)\bigg]\bigg\},$$

$$\tau_{xy}(x, y) = -\frac{M}{|\sin(\beta - \alpha)|}\left\{\cos\beta\left[\cos\alpha\frac{\partial}{\partial x}u_{31}(x - x_0, y - y_0)\right.\right.$$

$$+ \sin\alpha\frac{\partial}{\partial y}u_{31}(x - x_0, y - y_0)\bigg] + \sin\beta\left[\cos\alpha\frac{\partial}{\partial x}u_{32}(x - x_0, y - y_0)\right. \quad (9.1.21')$$

$$+ \sin\alpha\frac{\partial}{\partial y}u_{32}(x - x_0, y - y_0)\bigg]\bigg\}.$$

In the particular case where $\mathbf{u} = \mathbf{j}$ and $\mathbf{F_0} = -\mathbf{i}$, $\mathbf{i}$ and $\mathbf{j}$ are the unit vectors of the co-ordinate axes, we obtain a directed concentrated moment which causes a positive rotation in the plane Oxy (Figure 9.3b), is assumed to be applied at the origin and is expressed by the equivalent load

$$X(x, y) = M\delta'_y(x, y), \quad Y(x, y) = 0; \quad (9.1.22)$$

in that case the state of stress will be written

$$\sigma_x(x, y) = M \frac{\partial}{\partial y} u_{11}(x, y)$$

$$= \frac{M}{2\pi(1 - v)} \frac{xy}{(x^2 + y^2)^2} \left(1 - 2v + \frac{4x^2}{x^2 + y^2} \right),$$

(9.1.23)

$$\sigma_y(x, y) = M \frac{\partial}{\partial y} u_{21}(x, y)$$

$$= - \frac{M}{2\pi(1 - v)} \frac{xy}{(x^2 + y^2)^2} \left(3 - 2v - \frac{4y^2}{x^2 + y^2} \right),$$

$$\tau_{xy}(x, y) = M \frac{\partial}{\partial y} u_{31}(x, y)$$

$$= - \frac{M}{4\pi(1 - v)} \frac{1}{(x^2 + y^2)^2} \left[(x^2 - y^2)\left(1 - 2v + \frac{2x^2}{x^2 + y^2} \right) - \frac{4x^2y^2}{x^2 + y^2} \right]. \quad (9.1.23')$$

Let us now consider an elastic plane acted on by a rotational concentrated moment of magnitude M and applied at the origin of the co-ordinate axes (Figure 9.4);

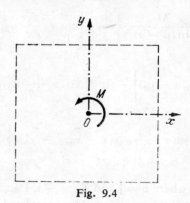

Fig. 9.4

the equivalent load may be written in the form

$$X(x, y) = \frac{1}{2} M\delta_y'(x, y), \quad Y(x, y) = - \frac{1}{2} M\delta_x'(x, y), \quad (9.1.24)$$

while the state of stress is given by

$$\sigma_x(x, y) = \frac{M}{2}\left[\frac{\partial}{\partial y} u_{11}(x, y) - \frac{\partial}{\partial x} u_{12}(x, y)\right] = \frac{M}{\pi} \frac{xy}{(x^2 + y^2)^2},$$

$$\sigma_y(x, y) = \frac{M}{2}\left[\frac{\partial}{\partial y} u_{21}(x, y) - \frac{\partial}{\partial x} u_{22}(x, y)\right] \tag{9.1.25}$$

$$= -\frac{M}{\pi} \frac{xy}{(x^2 + y^2)^2} = -\sigma_x(x, y),$$

$$\tau_{xy}(x, y) = \frac{M}{2}\left[\frac{\partial}{\partial y} u_{31}(x, y) - \frac{\partial}{\partial x} u_{32}(x, y)\right] = -\frac{M}{2\pi} \frac{x^2 - y^2}{(x^2 + y^2)^2}. \tag{9.1.25'}$$

Other different cases of loading may be treated in a similar way.

9.1.3. Dynamic problem

Let there be an elastic plane acted on by the dynamic loads $X = X(x, y; t)$, $Y = Y(x, y; t)$ (Figure 9.1); we replace the body forces by the distributions $\overline{X}(x, y; t)$, $\overline{Y}(x, y; t)$ defined by the formulae (7.2.47) and (7.2.47'). The components of the displacement vector shall be also replaced by the distributions $\overline{u}(x, y; t)$, $\overline{v}(x, y; t)$ specified by the formulae (7.2.48), (7.2.48').

For a solution in terms of displacements we shall use Lamé's equations with initial conditions of the form

$$u(x, y; 0) = u_0(x, y), \quad v(x, y; 0) = v_0(x, y), \tag{9.1.26}$$

$$\dot{u}(x, y; 0) = \dot{u}_0(x, y), \quad \dot{v}(x, y; 0) = \dot{v}_0(x, y), \tag{9.1.26'}$$

corresponding to the initial moment $t = 0$. With the aid of the distributions introduced we obtain the modified form of Lamé's equations given by (7.2.50), where the notation (7.2.40) has been used.

In the following discussion we shall apply the Laplace and Fourier transforms in order to obtain the solution corresponding to arbitrary body forces; we shall also compute the inverse transforms. Using these results we can then consider a few particular cases of loading and determine the fundamental solution.

9.1.3.1. Case of arbitrary body forces

In connection with this we shall first apply to equations (7.2.50), the simple Laplace transform with respect to the time variable, and then the double Fourier transform with respect to the spatial variables; we obtain

$$[\mu(\alpha^2 + \beta^2) + \rho p^2]F[L[\bar{u}(x, y; t)]] + (\lambda + \mu)\alpha(\alpha F[L[\bar{u}(x, y; t)]]$$

$$+ \beta F[L[\bar{v}(x, y; t)]]) = F[L[\bar{X}(x, y; t)]] + \rho p F[u_0(x, y)] + \rho F[\dot{u}_0(x, y)],$$

(9.1.27)

$$[\mu(\alpha^2 + \beta^2) + \rho p^2]F[L[\bar{v}(x, y; t)]] + (\lambda + \mu)\beta(\alpha F[L[\bar{u}(x, y; t)]]$$

$$+ \beta F[L[\bar{v}(x, y; t)]]) = F[L[\bar{Y}(x, y; t)]] + \rho p F[v_0(x, y)] + \rho F[\dot{v}_0(x, y)],$$

where p is a complex variable corresponding to the Laplace transform and α and β are complex variables corresponding to the Fourier transform.

Multiplying the first equation (9.1.27) by α and the second by β, and adding we may write

$$f(\alpha, \beta; p) = \alpha F[L[\bar{u}(x, y; t)]] + \beta F[L[\bar{v}(x, y; t)]]$$

$$= \frac{1}{(\lambda + 2\mu)(\alpha^2 + \beta^2) + \rho p^2} \; [\alpha F[L[\bar{X}(x, y; t)]] + \beta F[L[\bar{Y}(x, y; t)]]$$

$$+ \rho p(\alpha F[u_0(x, y)] + \beta F[v_0(x, y)]) + \rho(\alpha F[\dot{u}_0(x, y)] + \beta F[\dot{v}_0(x, y)])]; \quad (9.1.28)$$

this allows us to solve easily the system of linear algebraic equations (9.1.27), and the integral transforms are given by

$$F[L[\bar{u}(x, y; t)]] = \frac{1}{\mu(\alpha^2 + \beta^2) + \rho p^2} \; [F[L[\bar{X}(x, y; t)]]$$

$$+ \rho p F[u_0(x, y)] + \rho F[\dot{u}_0(x, y)] - (\lambda + \mu)\alpha f(\alpha, \beta; p)],$$

(9.1.29)

$$F[L[\bar{v}(x, y; t)]] = \frac{1}{\mu(\alpha^2 + \beta^2) + \rho p^2} \; [F[L[\bar{Y}(x, y; t)]]$$

$$+ \rho p F[v_0(x, y)] + \rho F[\dot{v}_0(x, y)] - (\lambda + \mu)\beta f(\alpha, \beta; p)].$$

In order to obtain the inverse integral transforms some **preliminary** calculations are required. Thus, we remark that

$$F_y[e^{-a|y|}] = \frac{2a}{\alpha^2 + \beta^2}, \quad \text{Re } a > 0; \tag{9.1.30}$$

putting

$$a = \sqrt{\alpha^2 + k^2 p^2}, \tag{9.1.31}$$

we may write

$$F_y\left[\frac{e^{-|y|\sqrt{\alpha^2 + k^2 p^2}}}{\sqrt{\alpha^2 + k^2 p^2}}\right] = \frac{2}{\alpha^2 + \beta^2 + k^2 p^2}. \tag{9.1.30'}$$

On the other hand

$$\frac{1}{\pi} F_x[K_0(kpr)] = \frac{e^{-|y|\sqrt{\alpha^2 + k^2 p^2}}}{\sqrt{\alpha^2 + k^2 p^2}}, \tag{9.1.32}$$

where we have introduced the radius vector (3.3.35), while $K_0(kpr)$ represents the *modified Bessel function of zero order.*

Taking into account relations (9.1.30') and (9.1.32) we obtain the double Fourier transform

$$\frac{1}{2\pi} F[K_0(kpr)] = \frac{1}{\alpha^2 + \beta^2 + k^2 p^2}, \quad \text{Re } p > 0, \tag{9.1.33}$$

whence

$$F^{-1}\left[\frac{1}{\alpha^2 + \beta^2 + k^2 p^2}\right] = \frac{1}{2\pi} K_0(kpr). \tag{9.1.33'}$$

Taking into account

$$\frac{\lambda + \mu}{[(\lambda + 2\mu)(\alpha^2 + \beta^2) + \rho p^2][\mu(\alpha^2 + \beta^2) + \rho p^2]}$$

$$= \frac{1}{\rho p^2}\left[\frac{1}{\alpha^2 + \beta^2 + \dfrac{\rho p^2}{\lambda + 2\mu}} - \frac{1}{\alpha^2 + \beta^2 + \dfrac{\rho p^2}{\mu}}\right],$$

we may write

$$F^{-1}\left[\frac{\lambda + \mu}{[(\lambda + 2\mu)(\alpha^2 + \beta^2) + \rho p^2][\mu(\alpha^2 + \beta^2) + \rho p^2]}\right]$$

$$= \frac{1}{2\pi\rho p^2}\left[K_0\left(p\frac{r}{c_1}\right) - K_0\left(p\frac{r}{c_2}\right)\right], \tag{9.1.34}$$

where c_1, c_2 are the wave propagation velocities given by the formulae (7.2.6).

Using the formulae (1.3.2.4) and (1.3.27) and the above results, we obtain the inverse Fourier transform

$$L[\bar{u}(x, y; t)] = \frac{1}{2\pi G}\left\{[L[\bar{X}(x, y; t)] + \rho p u_0(x, y) + \rho \dot{u}_0(x, y)] * K_0\left(p\frac{r}{c_2}\right)\right.$$

$$+ c_2^2\left[\frac{\partial}{\partial x}L[\bar{X}(x, y; t)] + \frac{\partial}{\partial y}L[\bar{Y}(x, y; t)]\right.$$

$$\left.\left. + \rho p \varepsilon_p^0(x, y) + \rho \dot{\varepsilon}_p^0(x, y)\right] * \frac{1}{p^2}\frac{\partial}{\partial x}\left[K_0\left(p\frac{r}{c_1}\right) - K_0\left(p\frac{r}{c_2}\right)\right]\right\}, \tag{9.1.35}$$

$$L[\bar{v}(x, y; t)] = \frac{1}{2\pi G}\left\{[L[\bar{Y}(x, y; t)] + \rho p v_0(x, y) + \rho \dot{v}_0(x, y)] * K_0\left(p\frac{r}{c_2}\right)\right.$$

$$+ c_2^2\left[\frac{\partial}{\partial x}L[\bar{X}(x, y; t)] + \frac{\partial}{\partial y}L[\bar{Y}(x, y; t)]\right.$$

$$\left.\left. + \rho p \varepsilon_p^0(x, y) + \rho \dot{\varepsilon}_p^0(x, y)\right] * \frac{1}{p^2}\frac{\partial}{\partial y}\left[K_0\left(p\frac{r}{c_1}\right) - K_0\left(p\frac{r}{c_2}\right)\right]\right\},$$

where we have introduced the notations

$$\varepsilon_p^0(x, y) = \frac{\partial}{\partial x}u_0(x, y) + \frac{\partial}{\partial y}v_0(x, y),$$

$$\tag{9.1.36}$$

$$\dot{\varepsilon}_p^0(x, y) = \frac{\partial}{\partial x}\dot{u}_0(x, y) + \frac{\partial}{\partial y}\dot{v}_0(x, y)$$

and we have observed that $\mu = G$.

Applying the inverse Laplace transform we may write

$$\bar{u}(x, y; t) = L^{-1}[L[\bar{u}(x, y; t)]],$$

$$\bar{v}(x, y; t) = L^{-1}[L[\bar{v}(x, y; t)]].$$

(9.1.37)

For the computation of the inverse transforms we introduce the notation

$$f_0(t; kr) = L^{-1}[K_0(pkr)] = \begin{cases} 0 & \text{for } 0 < t < kr \\ \dfrac{1}{\sqrt{t^2 - k^2 r^2}} & \text{for } t > kr. \end{cases}$$

(9.1.38)

Starting from the above distribution we may compute the inverse Laplace transform

$$f_1(t; kr) = L^{-1}[pK_0(pkr)] = \begin{cases} 0 & \text{for } 0 < t < kr \\ -\dfrac{t}{\sqrt{(t^2 - k^2 r^2)^3}} & \text{for } t > kr, \end{cases}$$

(9.1.39)

together with the inverse Laplace transforms

$$f_{-1}(t; kr) = L^{-1}\left[\frac{1}{p} K_0(pkr)\right] = \begin{cases} 0 & \text{for } 0 < t < kr \\ \log\left(t + \sqrt{t^2 - k^2 r^2}\right) & \text{for } t > kr, \end{cases}$$

(9.1.39')

$$f_{-2}(t; kr) = L^{-1}\left[\frac{1}{p^2} K_0(pkr)\right]$$

$$= \begin{cases} 0 & \text{for } 0 < t < kr \\ t \log\left(t + \sqrt{t^2 - k^2 r^2}\right) - \sqrt{t^2 - k^2 r^2} & \text{for } t > kr, \end{cases}$$

(9.1.39'')

$$f_{-3}(t; kr) = L^{-1}\left[\frac{1}{p^3} K_0(pkr)\right]$$

$$= \begin{cases} 0 & \text{for } 0 < t < kr \\ \dfrac{2t^2 + k^2 r^2}{4}\log\left(t + \sqrt{t^2 - k^2 r^2}\right) - \dfrac{3t}{4}\sqrt{t^2 - k^2 r^2} & \text{for } t > kr. \end{cases}$$

(9.1.39''')

Hence, the generalized displacements (9.1.37) may be written as

$$\bar{u}(x, y; t) = \frac{1}{2\pi G} \left\{ L^{-1}\left[L[\bar{X}(x, y; t)] * K_0\left(p\frac{r}{c_2} \right) \right] \right.$$

$$+ \rho u_0(x, y) * f_1\left(t; \frac{r}{c_2} \right) + \rho \dot{u}_0(x, y) * f_0\left(t; \frac{r}{c_2} \right)$$

$$+ c_2^2 \left\{ L^{-1}\left[L[\bar{X}(x, y; t)] * \frac{1}{p^2} \frac{\partial^2}{\partial x^2}\left[K_0\left(p\frac{r}{c_1} \right) - K_0\left(p\frac{r}{c_2} \right) \right] \right] \right.$$

$$+ L^{-1}\left[L[\bar{Y}(x, y; t)] * \frac{1}{p^2} \frac{\partial^2}{\partial x \partial y}\left[K_0\left(p\frac{r}{c_1} \right) - K_0\left(p\frac{r}{c_2} \right) \right] \right]$$

$$+ \rho \varepsilon_p^0(x, y) * \frac{\partial}{\partial x}\left[f_{-1}\left(t; \frac{r}{c_1} \right) - f_{-1}\left(t; \frac{r}{c_2} \right) \right]$$

$$+ \rho \dot{\varepsilon}_p^0(x, y) * \frac{\partial}{\partial x}\left[f_{-2}\left(t; \frac{r}{c_1} \right) - f_{-2}\left(t; \frac{r}{c_2} \right) \right] \right\} \right\},$$

$$\bar{v}(x, y; t) = \frac{1}{2\pi G} \left\{ L^{-1}\left[L[\bar{Y}(x, y; t)] * K_0\left(p\frac{r}{c_2} \right) \right] \right.$$

$$+ \rho v_0(x, y) * f_1\left(t; \frac{r}{c_2} \right) + \rho \dot{v}_0(x, y) * f_0\left(t; \frac{r}{c_2} \right)$$

$$+ c_2^2 \left\{ L^{-1}\left[L[\bar{X}(x, y; t)] * \frac{1}{p^2} \frac{\partial^2}{\partial x \partial y}\left[K_0\left(p\frac{r}{c_1} \right) - K_0\left(p\frac{r}{c_2} \right) \right] \right] \right.$$

$$+ L^{-1}\left[L[\bar{Y}(x, y; t)] * \frac{1}{p^2} \frac{\partial^2}{\partial y^2}\left[K_0\left(p\frac{r}{c_1} \right) - K_0\left(p\frac{r}{c_2} \right) \right] \right]$$

$$+ \rho \varepsilon_p^0(x, y) * \frac{\partial}{\partial y}\left[f_{-1}\left(t; \frac{r}{c_1} \right) - f_{-1}\left(t; \frac{r}{c_2} \right) \right]$$

$$+ \rho \dot{\varepsilon}_p^0(x, y) * \frac{\partial}{\partial y}\left[f_{-2}\left(t; \frac{r}{c_1} \right) - f_{-2}\left(t; \frac{r}{c_2} \right) \right] \right\} \right\}.$$

$$(9.1.37')$$

For example, in the case of homogeneous initial conditions

$$u_0(x, y) = v_0(x, y) = \varepsilon_p^0(x, y) = 0,$$

$$\dot{u}_0(x, y) = \dot{v}_0(x, y) = \dot{\varepsilon}_p^0(x, y) = 0$$

$$(9.1.40)$$

we obtain

$$\bar{u}(x, y; t) = \frac{1}{2\pi G} \left\{ L^{-1}\left[L[\bar{X}(x, y; t)] * K_0\left(p\frac{r}{c_2} \right) \right] \right.$$

$$+ c_2^2 \left\{ L^{-1}\left[L[\bar{X}(x, y; t)] * \frac{1}{p^2} \frac{\partial^2}{\partial x^2}\left[K_0\left(p\frac{r}{c_1} \right) - K_0\left(p\frac{r}{c_2} \right) \right] \right] \right.$$

$$\left. + L^{-1}\left[L[\bar{Y}(x, y; t)] * \frac{1}{p^2} \frac{\partial^2}{\partial x\,\partial y}\left[K_0\left(p\frac{r}{c_1} \right) - K_0\left(p\frac{r}{c_2} \right) \right] \right] \right\} \right\},$$

$$(9.1.41)$$

$$\bar{v}(x, y; t) = \frac{1}{2\pi G} \left\{ L^{-1}\left[L[\bar{Y}(x, y; t)] * K_0\left(p\frac{r}{c_2} \right) \right] \right.$$

$$+ c_2^2 \left\{ L^{-1}\left[L[\bar{X}(x, y; t)] * \frac{1}{p^2} \frac{\partial^2}{\partial x\partial y}\left[K_0\left(p\frac{r}{c_1} \right) - K_0\left(p\frac{r}{c_2} \right) \right] \right. \right.$$

$$\left. + L^{-1}\left[L[\bar{Y}(x, y; t)] * \frac{1}{p^2} \frac{\partial^2}{\partial y^2}\left[K_0\left(p\frac{r}{c_1} \right) - K_0\left(p\frac{r}{c_2} \right) \right] \right] \right\} \right\}.$$

If the body forces are null and only the initial conditions occur, we have

$$\bar{u}(x, y; t) = \frac{1}{2\pi} \left\{ \frac{1}{c_2^2}\left[u_0(x, y) * f_1\left(t; \frac{r}{c_2} \right) + \dot{u}_0(x, y) * f_0\left(t; \frac{r}{c_2} \right) \right] \right.$$

$$+ \varepsilon_p^0(x, y) * \frac{\partial}{\partial x}\left[f_{-1}\left(t; \frac{r}{c_1} \right) - f_{-1}\left(t; \frac{r}{c_2} \right) \right]$$

$$\left. + \dot{\varepsilon}_p^0(x, y) * \frac{\partial}{\partial x}\left[f_{-2}\left(t; \frac{r}{c_1} \right) - f_{-2}\left(t; \frac{r}{c_2} \right) \right] \right\},$$

$$(9.1.42)$$

$$\bar{v}(x, y; t) = \frac{1}{2\pi} \left\{ \frac{1}{c_2^2}\left[v_0(x, y) * f_1\left(t; \frac{r}{c_2} \right) + \dot{v}_0(x, y) * f_0\left(t; \frac{r}{c_2} \right) \right] \right.$$

$$+ \varepsilon_p^0(x, y) * \frac{\partial}{\partial y}\left[f_{-1}\left(t; \frac{r}{c_1} \right) - f_{-1}\left(t; \frac{r}{c_2} \right) \right]$$

$$\left. + \dot{\varepsilon}_p^0(x, y) * \frac{\partial}{\partial y}\left[f_{-2}\left(t; \frac{r}{c_1} \right) - f_{-2}\left(t; \frac{r}{c_2} \right) \right] \right\}.$$

We further remark that

$$\frac{1}{x}\frac{\partial}{\partial x}f_0(t;kr) = \frac{1}{y}\frac{\partial}{\partial y}f_0(t;kr) = \begin{cases} 0 & \text{for } 0 < t < kr \\ \dfrac{k^2}{\sqrt{(t^2 - k^2r^2)^3}} & \text{for } t > kr, \end{cases} \tag{9.1.43}$$

$$\frac{1}{x}\frac{\partial}{\partial x}f_1(t;kr) = \frac{1}{y}\frac{\partial}{\partial y}f_1(t;kr) = \begin{cases} 0 & \text{for } 0 < t < kr \\ -\dfrac{3k^2t}{\sqrt{(t^2 - k^2r^2)^3}} & \text{for } t > kr, \end{cases} \tag{9.1.44}$$

$$\frac{1}{x}\frac{\partial}{\partial x}f_{-1}(t;kr) = \frac{1}{y}\frac{\partial}{\partial y}f_{-1}(t;kr)$$

$$= \begin{cases} 0 & \text{for } 0 < t < kr \\ \dfrac{k^2}{t^2 - k^2r^2 + t\sqrt{t^2 - k^2r^2}} & \text{for } t > kr, \end{cases} \tag{9.1.45}$$

$$\frac{1}{x}\frac{\partial}{\partial x}f_{-2}(t;kr) = \frac{1}{y}\frac{\partial}{\partial y}f_{-2}(t;kr) = \begin{cases} 0 & \text{for } 0 < t < kr \\ \dfrac{k^2}{t + \sqrt{t^2 - k^2r^2}} & \text{for } t > kr, \end{cases} \tag{9.1.45'}$$

$$\frac{1}{x}\frac{\partial}{\partial x}f_{-3}(t;kr) = \frac{1}{y}\frac{\partial}{\partial y}f_{-3}(t;kr)$$

$$= \begin{cases} 0 & \text{for } 0 < t < kr \\ \dfrac{k^2}{4}\left[2\log(t + \sqrt{t^2 - k^2r^2}) + \dfrac{2t}{t + \sqrt{t^2 - k^2r^2}} + 1\right] & \text{for } t > kr. \end{cases} \tag{9.1.45''}$$

9.1.3.2. Examples

Let us consider an elastic plane acted on by body forces of the form (Figure 9.1)

$$X(x, y; t) = P(t)X(x, y), \quad Y(x, y; t) = 0, \tag{9.1.46}$$

with *homogeneous initial conditions* of the form (9.1.40); we introduce the distribution

$$\overline{P}(t) = \begin{cases} 0 & \text{for } t < 0 \\ \\ P(t) & \text{for } t \geqslant 0. \end{cases} \tag{9.1.46'}$$

The Laplace transforms (9.1.35) of the generalized displacements become

$$L[\bar{u}(x, y; t)] = \frac{1}{2\pi G} X(x, y) * L[\overline{P}(t)] \left\{ K_0 \left(p \frac{r}{c_2} \right) \right.$$

$$+ c_2^2 \frac{1}{p^2} \frac{\partial^2}{\partial x^2} \left[K_0 \left(p \frac{r}{c_1} \right) - K_0 \left(p \frac{r}{c_2} \right) \right] \Bigg\}, \tag{9.1.47}$$

$$L[\bar{v}(x, y; t)] = \frac{1}{2\pi\rho} X(x, y) * \frac{1}{p^2} L[\overline{P}(t)] \frac{\partial^2}{\partial x \partial y} \left[K_0 \left(p \frac{r}{c_1} \right) - K_0 \left(p \frac{r}{c_2} \right) \right].$$

Assuming that the convolution products of the right member exist in distributions and applying the inverse Fourier transform, we obtain

$$\bar{u}(x, y; t) = \frac{1}{2\pi G} X(x, y) * \left\{ \overline{P}(t) \underset{(t)}{*} \left\{ f_0 \left(t; \frac{r}{c_2} \right) \right. \right.$$

$$+ c_2^2 \frac{\partial^2}{\partial x^2} \left[f_{-2} \left(t; \frac{r}{c_1} \right) - f_{-2} \left(t; \frac{r}{c_2} \right) \right] \Bigg\} \Bigg\}. \tag{9.1.48}$$

$$\bar{v}(x, y; t) = \frac{1}{2\pi\rho} X(x, y) * \left\{ \overline{P}(t) \underset{(t)}{*} \frac{\partial^2}{\partial x \partial y} \left[f_{-2} \left(t; \frac{r}{c_1} \right) - f_{-2} \left(t; \frac{r}{c_2} \right) \right] \right\}.$$

For the computations of these generalized displacements we remark that

$$\frac{\partial^2}{\partial x^2} f_{-2}(t; kr) = \begin{cases} 0 & \text{for } 0 < t < kr \\ \\ \dfrac{k^2(t^2 - k^2 y^2 + t\sqrt{t^2 - k^2 r^2})}{2t(t^2 - k^2 r^2) + (2t^2 - k^2 r^2)\sqrt{t^2 - k^2 r^2}} & \text{for } t > kr, \end{cases} \tag{9.1.49}$$

$$\frac{\partial^2}{\partial y^2} f_{-2}(t; kr) = \begin{cases} 0 & \text{for } 0 < t < kr \\ \\ \dfrac{k^2(t^2 - k^2 x^2 + t\sqrt{t^2 - k^2 r^2})}{2t(t^2 - k^2 r^2) + (2t^2 - k^2 r^2)\sqrt{t^2 - k^2 r^2}} & \text{for } t > kr, \end{cases}$$

$$\frac{\partial^2}{\partial x \partial y} f_{-2}(t; kr) = \begin{cases} 0 & \text{for } 0 < t < kr \\ \\ \dfrac{k^4 xy}{2t(t^2 - k^2 r^2) + (2t^2 - k^2 r^2)\sqrt{t^2 - k^2 r^2}} & \text{for } t > kr; \end{cases} \tag{9.1.49'}$$

it should be noted that we have marked by "$*$" the convolution product in distributions with respect to the space variables, and by "$\underset{(t)}{*}$" the convolution product in distributions with respect to the time variable.

In the particular case

$$\bar{P}(t) = P\delta(t), \tag{9.1.50}$$

which correspond to *a shock at the initial moment*, the generalized displacements are given by

$$\bar{u}(x, y; t) = \frac{P}{2\pi G} X(x, y) * \left\{ f_0\left(t; \frac{r}{c_2}\right) + c_2^2 \frac{\partial^2}{\partial x^2}\left[f_{-2}\left(t; \frac{r}{c_1}\right) - f_{-2}\left(t; \frac{r}{c_2}\right)\right]\right\}, \tag{9.1.51}$$

$$\bar{v}(x, y; t) = \frac{P}{2\pi\rho} X(x, y) * \frac{\partial^2}{\partial x \partial y}\left[f_{-2}\left(t; \frac{r}{c_1}\right) - f_{-2}\left(t; \frac{r}{c_2}\right)\right].$$

Also, if in (9.1.46) we put

$$X(x, y) = \delta(x, y) \tag{9.1.52}$$

we obtain

$$\bar{u}(x, y; t) = \frac{1}{2\pi G} \bar{P}(t) \underset{(t)}{*} \left\{ f_0\left(t; \frac{r}{c_2}\right) \right.$$

$$\left. + c_2^2 \frac{\partial^2}{\partial x^2}\left[f_{-2}\left(t; \frac{r}{c_1}\right) - f_{-2}\left(t; \frac{r}{c_2}\right)\right]\right\}, \tag{9.1.53}$$

$$\bar{v}(x, y; t) = \frac{1}{2\pi\rho} \bar{P}(t) \underset{(t)}{*} \frac{\partial^2}{\partial x \partial y}\left[f_{-2}\left(t; \frac{r}{c_1}\right) - f_{-2}\left(t; \frac{r}{c_2}\right)\right].$$

By introducing now in relation (9.1.51) the spatial distribution of body forces (9.1.52), we obtain

$$\bar{u}(x, y; t) = \frac{P}{2\pi G} \left\{ f_0\left(t; \frac{r}{c_2}\right) + c_2^2 \frac{\partial^2}{\partial x^2}\left[f_{-2}\left(t; \frac{r}{c_1}\right) - f_{-2}\left(t; \frac{r}{c_2}\right)\right]\right\}, \tag{9.1.54}$$

$$\bar{v}(x, y; t) = \frac{P}{2\pi\rho} \frac{\partial^2}{\partial x \partial y}\left[f_{-2}\left(t; \frac{r}{c_1}\right) - f_{-2}\left(t; \frac{r}{c_2}\right)\right],$$

which is a state of generalized displacement corresponding to a concentrated force applied at the origin of the co-ordinate axes, in the direction of the Ox axis, under the form of a *shock at the initial moment*.

If in the formulae (9.1.53) we take

$$\bar{P}(t) = P\theta(t), \tag{9.1.55}$$

where $\theta(t)$ is the Heaviside distribution, we obtain the state of generalized displacement

$$\bar{u}(x, y, t) = \frac{P}{2\pi G} X(x, y) * \left\{ f_{-1}\left(t; \frac{r}{c_2}\right) \right.$$

$$+ c_2^2 \frac{\partial^2}{\partial x^2} \left[f_{-3}\left(t; \frac{r}{c_1}\right) - f_{-3}\left(t; \frac{r}{c_2}\right) \right] \right\}, \qquad (9.1.56)$$

$$\bar{v}(x, y; t) = \frac{P}{2\pi\rho} X(x, y) * \frac{\partial^2}{\partial x \partial y} \left[f_{-3}\left(t; \frac{r}{c_1}\right) - f_{-3}\left(t; \frac{r}{c_2}\right) \right],$$

which corresponds to a *load applied suddenly and then maintained constant in time.*

By introducing now the spatial distribution of body forces (9.1.52) we obtain the generalized displacements

$$\bar{u}(x, y; t) = \frac{P}{2\pi G} \left\{ f_{-1}\left(t; \frac{r}{c_2}\right) + c_2^2 \frac{\partial^2}{\partial x^2} \left[f_{-3}\left(t; \frac{r}{c_1}\right) - f_{-3}\left(t; \frac{r}{c_2}\right) \right] \right\},$$

$$\bar{v}(x, y; t) = \frac{P}{2\pi\rho} \frac{\partial^2}{\partial x \partial y} \left[f_{-3}\left(t; \frac{r}{c_1}\right) - f_{-3}\left(t; \frac{r}{c_2}\right) \right], \qquad (9.1.57)$$

which correspond to a *concentrated force applied suddenly and maintained constant in time.*

It is useful to note that

$$\frac{\partial^2}{\partial x^2} f_{-3}(t; kr)$$

$$= \begin{cases} 0 & \text{for } 0 < t < kr \\ \frac{k^2}{2} \left[\log(t + \sqrt{t^2 - k^2 r^2}) - \frac{k^2}{2} \frac{x^2 - y^2}{2t^2 - k^2 r^2 + 2t\sqrt{t^2 - k^2 r^2}} + 1 \right] & \text{for } t > kr, \end{cases} \qquad (9.1.58)$$

$$\frac{\partial^2}{\partial y^2} f_{-3}(t; kr)$$

$$= \begin{cases} 0 & \text{for } 0 < t < kr \\ \frac{k^2}{2} \left[\log(t + \sqrt{t^2 - k^2 r^2}) + \frac{k^2}{2} \frac{x^2 - y^2}{2t^2 - k^2 r^2 + 2t\sqrt{t^2 - k^2 r^2}} + 1 \right] & \text{for } t > kr, \end{cases} \qquad (9.1.58')$$

$$\frac{\partial^2}{\partial x \partial y} f_{-3}(t; kr) = \begin{cases} 0 & \text{for } 0 < t < kr \\ -\frac{k^4}{2} \frac{xy}{2t^2 - k^2 r^2 + 2t\sqrt{t^2 - k^2 r^2}} & \text{for } t > kr. \end{cases} \qquad (9.1.58'')$$

Let us now consider an elastic plane acted on by a concentrated rotational moment of magnitude $M(t)$ at the origin of the co-ordinate axes (Figure 9.4); the equivalent load may be written

$$X(x, y; t) = \frac{1}{2} M(t) \frac{\partial}{\partial y} \delta(x, y),$$

$$\tag{9.1.59}$$

$$Y(x, y; t) = -\frac{1}{2} M(t) \frac{\partial}{\partial x} \delta(x, y).$$

Assuming homogeneous initial conditions of the form (9.1.40) and remarking that

$$\frac{\partial}{\partial x} L[\bar{X}(x, y; t)] + \frac{\partial}{\partial y} L[\bar{Y}(x, y; t)] = 0, \tag{9.1.60}$$

we may write the generalized displacements in the form

$$\bar{u}(x, y; t) = \frac{1}{4\pi G} \bar{M}(t) \underset{(t)}{*} \frac{\partial}{\partial y} f_0\left(t; \frac{r}{c_2}\right),$$

$$\tag{9.1.61}$$

$$\bar{v}(x, y; t) = -\frac{1}{4\pi G} \bar{M}(t) \underset{(t)}{*} \frac{\partial}{\partial x} f_0\left(t; \frac{r}{c_2}\right),$$

where we have introduced the distribution

$$\bar{M}(t) = \begin{cases} 0 & \text{for } t < 0 \\ M(t) & \text{for } t \geqslant 0. \end{cases} \tag{9.1.62}$$

In particular, if

$$\bar{M}(t) = M\delta(t) \tag{9.1.63}$$

we have to deal with a *shock at the initial moment* and we may write

$$\bar{u}(x, y; t) = \frac{M}{4\pi G} \frac{\partial}{\partial y} f_0\left(t; \frac{r}{c_2}\right),$$

$$\tag{9.1.64}$$

$$\bar{v}(x, y; t) = -\frac{M}{4\pi G} \frac{\partial}{\partial x} f_0\left(t; \frac{r}{c_2}\right).$$

Similarly, if

$$\bar{M}(t) = M\theta(t) \tag{9.1.65}$$

we obtain the generalized displacements

$$\bar{u}(x, y; t) = \frac{M}{4\pi G} \frac{\partial}{\partial y} f_{-1}\left(t; \frac{r}{c_2}\right),$$

$$\bar{v}(x, y; t) = -\frac{M}{4\pi G} \frac{\partial}{\partial x} f_{-1}\left(t; \frac{r}{c_2}\right), \tag{9.1.66}$$

which correspond to a *rotational concentrated moment applied suddenly at the origin of the co-ordinate axes and maintained constant in time.*

Let us now consider a tangential load uniformly distributed along a square the side of which is $2a$ and the centre at the origin of the co-ordinate axes; such a load corresponds to a positive rotation in the plane Oxy. If the magnitude of the load is $\dfrac{M(t)}{8a^2}$ the equivalent body forces may be written

$$X(x, y; t) = \frac{M(t)}{8a^2} \frac{\partial}{\partial y} h(x, y),$$

$$Y(x, y; t) = -\frac{M(t)}{8a^2} \frac{\partial}{\partial x} h(x, y), \tag{9.1.67}$$

where $h(x, y)$ is the characteristic function given by (1.2.109).

Remarking that relation (9.1.60) is satisfied and using the distribution (9.1.62), we may write the Laplace transforms of the generalized displacements in the form

$$L[\bar{u}(x, y; t)] = \frac{1}{16\pi Ga^2} h(x, y) * L[\bar{M}(t)] \frac{\partial}{\partial y} K_0\left(p \frac{r}{c_2}\right),$$

$$L[\bar{v}(x, y; t)] = -\frac{1}{16\pi Ga^2} h(x, y) * L[\bar{M}(t)] \frac{\partial}{\partial x} K_0\left(p \frac{r}{c_2}\right). \tag{9.1.68}$$

The generalized displacements are given by

$$\bar{u}(x, y; t) = \frac{1}{16\pi Ga^2} h(x, y) * \left[\bar{M}(t) \underset{(t)}{*} \frac{\partial}{\partial y} f_0\left(t; \frac{r}{c_2}\right)\right],$$

$$\bar{v}(x, y; t) = -\frac{1}{16\pi Ga^2} h(x, y) * \left[\bar{M}(t) \underset{(t)}{*} \frac{\partial}{\partial x} f_0\left(t; \frac{r}{c_2}\right)\right]. \tag{9.1.69}$$

In particular, for (9.1.63) we obtain

$$\bar{u}(x, y; t) = \frac{M}{16\pi Ga^2} h(x, y) * \frac{\partial}{\partial y} f_0\left(t; \frac{r}{c_2}\right),$$

(9.1.70)

$$\bar{v}(x, y; t) = -\frac{M}{16\pi Ga^2} h(x, y) * \frac{\partial}{\partial x} f_0\left(t; \frac{r}{c_2}\right),$$

while for (9.1.65) we may write

$$\bar{u}(x, y; t) = \frac{1}{16\pi Ga^2} h(x, y) * \frac{\partial}{\partial y} f_{-1}\left(t; \frac{r}{c_2}\right),$$

(9.1.71)

$$\bar{v}(x, y; t) = -\frac{1}{16\pi Ga^2} h(x, y) * \frac{\partial}{\partial x} f_{-1}\left(t; \frac{r}{c_2}\right).$$

In the case of the representative δ sequence, specified by formula (1.2.109′) the equivalent body forces (9.1.67) corespond to a rotation centre given by (9.1.59); under these conditions formulae (9.1.69) lead to formulae (9.1.61), formulae (9.1.70) permit us to write formulae (9.1.64), and the formulae (9.1.71) supply formulae (9.1.66).

We introduce the fundamental solution matrix

$$(U) \equiv \begin{pmatrix} u_{11}(x, y; t) & u_{12}(x, y; t) \\ u_{21}(x, y; t) & u_{22}(x, y; t) \end{pmatrix}$$

(9.1.72)

and the matrices

$$(\bar{u}) \equiv \begin{pmatrix} \bar{u} \\ \bar{v} \end{pmatrix}, \quad (Q) \equiv \begin{pmatrix} \bar{X} \\ \bar{Y} \end{pmatrix};$$

(9.1.73)

the state of displacement is given by

$$(\bar{u}) = (U) * (Q).$$

(9.1.74)

Using the formulae (9.1.54) where we put $P = 1$, and considering the formulae corresponding to the case where the concentrated force is acting along the Oy

axis too, we obtain the components of the fundamental solution matrix, namely

$$u_{11} = \frac{1}{2\pi G} \left\{ f_0\left(t; \frac{r}{c_2}\right) + c_2^2 \frac{\partial^2}{\partial x^2} \left[f_{-2}\left(t; \frac{r}{c_1}\right) - f_{-2}\left(t; \frac{r}{c_2}\right) \right] \right\},$$

$$(9.1.75)$$

$$u_{22} = \frac{1}{2\pi G} \left\{ f_0\left(t; \frac{r}{c_2}\right) + c_2^2 \frac{\partial^2}{\partial y^2} \left[f_{-2}\left(t; \frac{r}{c_1}\right) - f_{-2}\left(t; \frac{r}{c_2}\right) \right] \right\},$$

$$u_{12} = u_{21} = \frac{1}{2\pi\rho} \frac{\partial^2}{\partial x \partial y} \left[f_{-2}\left(t; \frac{r}{c_1}\right) - f_{-2}\left(t; \frac{r}{c_2}\right) \right]. \qquad (9.1.75')$$

By a similar procedure one may obtain results for the state of stress.

9.2. Elastic half-plane

Let there be an elastic half-plane $y \geqslant 0$ acted on by a tangential load $p = p(x)$ and by a normal load $q = q(x)$, both expressed in distributions (Figure 9.5). It is assumed that we are in the case of a *state of plane strain* (hence in the case of an elastic half-space subjected, on the separation plane, to constant external loads over a certain direction and having no component along that direction). We shall

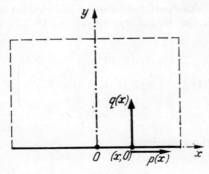

Fig. 9.5

show how to obtain the fundamental solution and then, using the latter, how to obtain the solution corresponding to a case of arbitrary loading.

Also, we shall show how to consider the case of arbitrary body forces.

9.2.1. Fundamental solution in the static case

In the absence of body forces and for a solution in terms of stresses the system of equations (7.2.13), (9.1.1) takes the form

$$\frac{\partial \sigma_x}{\partial x} + \frac{\partial \tau_{xy}}{\partial y} = 0,$$

$$(9.2.1)$$

$$\frac{\partial \tau_{xy}}{\partial x} + \frac{\partial \sigma_y}{\partial y} = 0,$$

$$\Delta(\sigma_x + \sigma_y) = 0 \, ; \qquad (9.2.1')$$

the boundary conditions are set on the separation line

$$\lim_{y \to +0} \tau_{yx} = - p(x), \quad \lim_{y \to +0} \sigma_y = - q(x) \qquad (9.2.2)$$

and at infinity (conditions of regularity)

$$\lim_{y \to \infty} \sigma_x = \lim_{y \to \infty} \sigma_y = \lim_{y \to \infty} \tau_{xy} = 0. \qquad (9.2.2')$$

We shall denote by *fundamental solution* of the system of differential equations (9.2.1) and (9.2.1') with the boundary conditions (9.2.2) and (9.2.2'), the matrix

$$(V) \equiv \begin{pmatrix} v_{11}(x, y) & v_{12}(x, y) \\ v_{21}(x, y) & v_{22}(x, y) \\ v_{31}(x, y) & v_{32}(x, y) \end{pmatrix}, \qquad (9.2.3)$$

whose elements satisfy the equations

$$\frac{\partial v_{11}}{\partial x} + \frac{\partial v_{31}}{\partial y} = 0, \quad \frac{\partial v_{31}}{\partial x} + \frac{\partial v_{21}}{\partial y} = 0, \quad \Delta(v_{11} + v_{21}) = 0, \quad (9.2.4)$$

$$\frac{\partial v_{12}}{\partial x} + \frac{\partial v_{32}}{\partial y} = 0, \quad \frac{\partial v_{32}}{\partial x} + \frac{\partial v_{22}}{\partial y} = 0, \quad \Delta(v_{12} + v_{22}) = 0, \quad (9.2.4')$$

and the boundary conditions

$$\lim_{y \to +0} v_{21} = 0, \qquad \lim_{y \to +0} v_{31} = -\delta(x), \qquad \lim_{y \to \infty} v_{i1} = 0 \, (i = 1,2,3), \qquad (9.2.5)$$

$$\lim_{y \to +0} v_{22} = -\delta(x), \quad \lim_{y \to +0} v_{32} = 0, \qquad \lim_{y \to \infty} v_{i2} = 0 \, (i = 1,2,3); \qquad (9.2.5')$$

the limits are considered in the sense of the theory of distributions.

In order to obtain the fundamental solution we shall apply the simple Fourier transform in terms of distributions,

9.2.1.1. Fourier transform

Applying the Fourier transform with respect to x and considering y as a parameter, the equations (9.2.4) may be written

$$-is\, F_x\,[v_{11}(x, y)] + \frac{\mathrm{d}}{\mathrm{d}y} F_x\,[v_{31}(x, y)] = 0,$$

$$\qquad (9.2.6)$$

$$-is\, F_x\,[v_{31}(x, y)] + \frac{\mathrm{d}}{\mathrm{d}x} F_x\,[v_{21}(x, y)] = 0,$$

$$-s^2\,(F_x\,[v_{11}(x, y)] + F_x\,[v_{21}(x, y)])$$

$$+\frac{\mathrm{d}^2}{\mathrm{d}y^2}\,(F_x[v_{11}(x, y)] + F_x[v_{21}(x, y)]) = 0, \qquad (9.2.6')$$

where s is a complex variable.

Eliminating the Fourier transforms $F_x[v_{11}(x, y)]$ and $F_x[v_{31}(x, y)]$ from these equations, we obtain the differential equation

$$\frac{\mathrm{d}^4}{\mathrm{d}y^4}\, F_x[v_{21}(x, y)] - 2s^2 \frac{\mathrm{d}^2}{\mathrm{d}y^2}\, F_x[v_{21}(x, y)] + s^4 F_x[v_{21}(x, y)] = 0, \qquad (9.2.7)$$

the general integral of which is

$$F_x[v_{21}(x, y)] = (A + By)\,\mathrm{e}^{-|s|y} + (C + Dy)\,\mathrm{e}^{|s|y}, \qquad (9.2.8)$$

where the coefficients A, B, C, D are dependent on s. In order to determine these coefficients we apply the Fourier transform to the boundary conditions (9.2.5) too, thus obtaining

$$\lim_{y \to +0} F_x[v_{21}(x, y)] = 0, \quad \lim_{y \to +0} F_x[v_{31}(x, y)] = -1,$$

$$\lim_{y \to \infty} F_x[v_{i1}(x, y)] = 0 \, (i = 1,2,3).$$

(9.2.9)

The third condition (9.2.9) leads for $i = 2$ to

$$C = D = 0$$

(9.2.10)

and the first condition (9.2.9) to

$$A = 0.$$

(9.2.10')

Substituting the Fourier transform $F_x[v_{21}(x, y)]$, given by relation (9.2.8), in the second equation (9.2.6), and setting the second condition (9.2.9), we obtain

$$B = -is.$$

(9.2.10'')

Thus, it follows that

$$F_x[v_{21}(x, y)] = -isy \, e^{-|s|y}$$

(9.2.11)

and equations (9.2.6) permit us to write also

$$F_x[v_{11}(x, y)] = -2i \frac{|s|}{s} e^{-|s|y} + isy \, e^{-|s|y},$$

$$F_x[v_{31}(x, y)] = -(1 - |s|y) e^{-|s|y} \, ;$$

(9.2.11')

we remark that the third condition (9.2.9) is satisfied for $i = 1,3$, too.

9.2.1.2. Inverse Fourier transform

To compute the inverse Fourier transform we remark that

$$\frac{1}{\pi} F_x\left[\frac{y}{x^2 + y^2} \right] = e^{-|s|y}, \quad y > 0,$$

(9.2.12)

whence

$$\frac{1}{\pi} F_x\left[\frac{d}{dx}\left(\frac{y}{x^2 + y^2} \right) \right] = -\frac{is}{\pi} F_x\left[\frac{y}{x^2 + y^2} \right] = -ise^{-|s|y}, \, y > 0, \quad (9.2.12')$$

and

$$\frac{1}{\pi}\frac{\mathrm{d}}{\mathrm{d}y}F_x\left[\frac{y}{x^2+y^2}\right]=\frac{1}{\pi}F_x\left[\frac{\mathrm{d}}{\mathrm{d}y}\left(\frac{y}{x^2+y^2}\right)\right]$$

$$=\frac{\mathrm{d}}{\mathrm{d}y}e^{-|s|y}=-|s|e^{-|s|y},\ y>0.\qquad(9.2.12'')$$

Also, we can write

$$\frac{\mathrm{d}}{\mathrm{d}x}\arctan\frac{y}{x}=-\frac{y}{x^2+y^2}$$

and, taking into account (9.2.12), it follows that

$$\frac{1}{\pi}F_x\left[\frac{\mathrm{d}}{\mathrm{d}x}\arctan\frac{y}{x}\right]=-e^{-|s|y},\quad y>0,\qquad(9.2.13)$$

whence

$$F_x\left[\arctan\frac{y}{x}\right]=-\frac{\pi\mathrm{i}}{s}e^{-|s|y},\quad y>0\ ;\qquad(9.2.13')$$

differentiating with respect to the parameter y we obtain

$$\frac{\mathrm{d}}{\mathrm{d}y}F_x\left[\arctan\frac{y}{x}\right]=F_x\left[\frac{\mathrm{d}}{\mathrm{d}y}\arctan\frac{y}{x}\right]=\pi\mathrm{i}\frac{|s|}{s}e^{-|s|y},\ y>0,$$

and therefore

$$F_x\left[\frac{x}{x^2+y^2}\right]=\pi\mathrm{i}\frac{|s|}{s}e^{-|s|y},\quad y>0.\qquad(9.2.13'')$$

From formulae (9.2.11) and (9.2.11') and the above results we obtain

$$v_{11}(x,y)=-\frac{2}{\pi}\frac{x}{x^2+y^2}-\frac{1}{\pi}y^2\frac{\mathrm{d}}{\mathrm{d}x}\left(\frac{1}{x^2+y^2}\right)=-\frac{2}{\pi}\frac{x^3}{(x^2+y^2)^2},$$

$$v_{21}(x,y)=\frac{1}{\pi}y\frac{\mathrm{d}}{\mathrm{d}x}\left(\frac{y}{x^2+y^2}\right)=-\frac{2}{\pi}\frac{xy^2}{(x^2+y^2)^2},\qquad(9.2.14)$$

$$v_{31}(x,y)=-\frac{1}{\pi}\frac{y}{x^2+y^2}-\frac{1}{\pi}y\frac{\mathrm{d}}{\mathrm{d}y}\left(\frac{y}{x^2+y^2}\right)=-\frac{2}{\pi}\frac{x^2y}{(x^2+y^2)^2};$$

by proceeding similarly with the equations (9.2.4′) and the boundary conditions (9.2.5′) we find that

$$v_{12}(x, y) = -\frac{2}{\pi} \frac{x^2 y}{(x^2 + y^2)^2},$$

$$v_{22}(x, y) = -\frac{2}{\pi} \frac{y^3}{(x^2 + y^2)^2}, \qquad\qquad (9.2.14')$$

$$v_{32}(x, y) = -\frac{2}{\pi} \frac{xy^2}{(x^2 + y^2)^2},$$

whereby the fundamental solution is specified fully.

9.2.2. Case of arbitrary static loads

By use of the matrix (V) of the fundamental solution we can construct the solution corresponding to arbitrary loads acting on the separation line. With the aid of this mode of expression (convolution product) we shall then give results for a few particular cases of loading. Also, we shall show how to examine the case of internal loads (expressed by body forces).

9.2.2.1. Case of arbitrary loads on the separation line

We introduce the matrices

$$(\sigma) \equiv \begin{pmatrix} \sigma_x(x, \ y) \\ \sigma_y(x, \ y) \\ \tau_{xy}(x, \ y) \end{pmatrix}, \qquad (Q) \equiv \begin{pmatrix} p(x) \\ q(x) \end{pmatrix}; \qquad (9.2.15)$$

we may write

$$(\sigma) = (v) * (Q), \qquad\qquad (9.2.16)$$

and the state of stress corresponding to arbitrary loads on the separation line is expressed in the form

$$\sigma_x(x, y) = v_{11}(x, y) * p(x) + v_{12}(x, y) * q(x),$$

$$\sigma_y(x, y) = v_{21}(x, y) * p(x) + v_{22}(x, y) * q(x), \qquad\qquad (9.2.17)$$

$$\tau_{xy}(x, y) = v_{31}(x, y) * p(x) + v_{32}(x, y) * q(x). \qquad\qquad (9.2.17')$$

This result may be also proved directly, namely: by using the results of section 1.4.2.2. it is found that equations (9.2.1), and (9.2.1′) are satisfied; in order to verify the boundary conditions (9.2.2) we remark that the components of the fundamental solution correspond to representative δ sequences. The verification of the conditions of regularity (9.2.2′) follows immediately.

9.2.2.2. Examples

Let $y \geqslant 0$ be an elastic half-plane acted on by a concentrated force $\mathbf{F}(F_x,\ F_y)$ at the point $A(x_0, 0)$ (Figure 9.6); the load equivalent to the concentrated force may be written

$$p(x) = F_x\,\delta(x - x_0), \quad q(x) = F_y\,\delta(x - x_0). \tag{9.2.18}$$

Using formulae (9.2.17) and (9.2.17′) we obtain the corresponding state of stress in the form

$$\sigma_x(x, y) = F_x v_{11}(x - x_0, y) + F_y v_{12}(x - x_0, y),$$

$$\sigma_y(x, y) = F_x v_{21}(x - x_0, y) + F_y v_{22}(x - x_0, y), \tag{9.2.19}$$

$$\tau_{xy}(x, y) = F_x v_{31}(x - x_0, y) + F_y v_{32}(x - x_0, y). \tag{9.2.19'}$$

Now let $y \geqslant 0$ be the elastic plane acted on at the point $A(x_0, 0)$ by a directed concentrated moment of magnitude M and defined by the unit vectors $\mathbf{u}(1,0)$ and

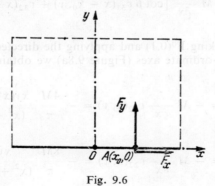

Fig. 9.6

$\mathbf{F}^0(\cos\beta,\ \sin\beta)$, $\beta = 0$, π (Figure 9.7); the load equivalent to the directed concentrated moment is expressed in the form

$$p(x) = -\,M \cot \beta\,\frac{\mathrm{d}}{\mathrm{d}x}\,\delta(x - x_0), \quad q(x) = -\,M\,\frac{\mathrm{d}}{\mathrm{d}x}\,\delta(x - x_0). \tag{9.2.20}$$

In that case the state of stress is given by

$$\sigma_x(x, y) = - M \frac{\partial}{\partial x} [\cot \beta \, v_{11}(x - x_0, y) + v_{12}(x - x_0, y)],$$

$$(9.2.21)$$

$$\sigma_y(x, y) = - M \frac{\partial}{\partial x} [\cot \beta \, v_{21}(x - x_0, y) + v_{22}(x - x_0, y)],$$

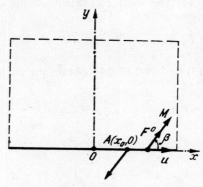

Fig. 9.7

$$\tau_{xy}(x, y) = - M \frac{\partial}{\partial x} [\cot \beta \, v_{31}(x - x_0, y) + v_{32}(x - x_0, y)]. \qquad (9.2.21')$$

In particular, by taking $\mathbf{F}^0(0,1)$ and applying the directed concentrated moment at the origin of the co-ordinate axes (Figure 9.8a) we obtain

$$\sigma_x(x, y) = - M \frac{\partial}{\partial x} v_{12}(x, y) = - \frac{4M}{\pi} \frac{xy(x^2 + y^2)}{(x^2 + y^2)^3},$$

$$(9.2.22)$$

$$\sigma_y(x, y) = - M \frac{\partial}{\partial x} v_{22}(x, y) = - \frac{8M}{\pi} \frac{xy^3}{(x^2 + y^2)^3},$$

$$\tau_{xy}(x, y) = - M \frac{\partial}{\partial x} v_{32}(x, y) = - \frac{2M}{\pi} \frac{y^2(3x^2 - y^2)}{(x^2 + y^2)^3}. \qquad (9.2.22')$$

We remark that in the case of a rotational concentrated moment of magnitude M, acting at the same point (Figure 9.8b), we obtain the same state of stress (9.2.22),

(9.2.22′). This is explained by the fact that the separation line is the principal direction for the dipoles of concentrated forces considered from the point of view of the tensorial properties of concentrated forces emphasized in section 5.3.2.1.

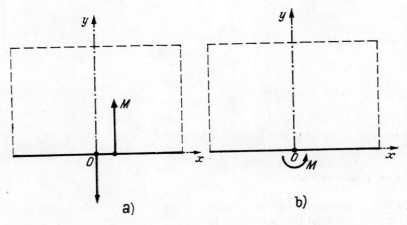

Fig. 9.8

Now let us consider the elastic half-plane $y \geqslant 0$ acted on at the point $A(x_0, 0)$ by a concentrated moment of the linear dipole type, of magnitude D

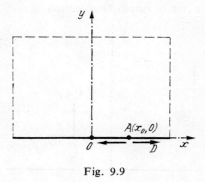

Fig. 9.9

(Figure 9.9); the equivalent load is

$$p(x) = -D \frac{\mathrm{d}}{\mathrm{d}x} \delta(x - x_0), \quad q(x) = 0, \tag{9.2.23}$$

and we obtain the state of stress

$$\sigma_x = - D \frac{\partial}{\partial x} v_{11}(x - x_0, y) = - \frac{2D}{\pi} \frac{(x - x_0)^2}{[(x - x_0)^2 + y^2]^3} [(x - x_0)^2 - 3y^2],$$

$$(9.2.24)$$

$$\sigma_y = - D \frac{\partial}{\partial x} v_{21}(x - x_0, y) = - \frac{2D}{\pi} \frac{y^2}{[(x - x_0)^2 + y^2]^3} [3(x - x_0)^2 - y^2],$$

$$\tau_{xy} = - D \frac{\partial}{\partial x} v_{31}(x - x_0, y) = - \frac{4D}{\pi} \frac{(x - x_0)y}{[(x - x_0)^2 + y^2]^3} [(x - x_0)^2 - y^2]. \quad (9.2.24')$$

Other cases of loading may be treated in a similar way.

9.2.2.3. Case of internal loads

Let $y \geqslant 0$ be an elastic half-plane acted on by the arbitrary body forces $X = X(x, y)$, $Y = Y(x, y)$ (Figure 9.10) expressed in terms of distributions.

To obtain the state of stress corresponding to that case of loading we shall proceed from the problem of the elastic plane acted on by body forces; we have seen that the corresponding state of stress is given by formulae (9.1.17), and (9.1.17'). Using these results we obtain, on the separation line $y = 0$, a state of stress which should be zero; in order to cancel these stresses we shall consider the elastic half-plane $y \geqslant 0$ acted on by loads the magnitudes of which are equal to that of the stresses mentioned above but of opposite direction.

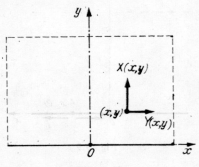

Fig. 9.10

We shall denote these stresses by $\sigma_y^0(x)$ and $\tau_{yx}^0(x)$; they are obtained from the results corresponding to the elastic plane by passing to the limit for $y \to +0$ in the sense of the theory of distributions.

The state of stress which is sought will be expressed in that case in the form

$$\sigma_x(x, y) = u_{11}(x, y) * X(x, y) + u_{12}(x, y) * Y(x, y)$$

$$+ v_{11}(x, y) * \tau_{yx}^0(x) + v_{12}(x, y) * \sigma_y^0(x),$$

$$\sigma_y(x, y) = u_{21}(x, y) * X(x, y) + u_{22}(x, y) * Y(x, y)$$

$$+ v_{21}(x, y) * \tau_{yz}^0(x) + v_{22}(x, y) * \sigma_y^0(x), \tag{9.2.25}$$

$$\tau_{xy}(x, y) = u_{31}(x, y) * X(x, y) + u_{32}(x, y) * Y(x, y)$$

$$+ v_{31}(x, y) * \tau_{yx}^0(x) + v_{32}(x, y) * \sigma_y^0(x). \tag{9.2.25'}$$

Let us consider, for example, the elastic half-plane $y \geqslant 0$ acted on by a concentrated internal force $F(F_x, F_y)$ (Figure 9.11) applied at the point $A(x_0, y_0)$; we may express this concentrated force by equivalent loads (9.1.18).

Taking into account relations (9.1.19) and (9.1.19') we obtain on the boundary

$$\tau_{yx}^0(x) = \lim_{y \to +0} [F_x u_{31}(x - x_0, y - y_0) + F_y u_{32}(x - x_0, y - y_0)],$$

$$\tag{9.2.26}$$

$$\sigma_y^0(x) = \lim_{y \to +0} [F_x u_{21}(x - x_0, y - y_0) + F_y u_{22}(x - x_0, y - y_0)].$$

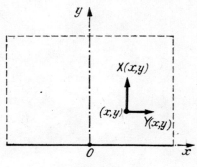

Fig. 9.11

If we introduce the notations

$$u_{2i}^0(x - x_0) = \lim_{y \to +0} u_{2i}(x - x_0, y - y_0) \quad (i = 1, 2),$$

$$\tag{9.2.27}$$

$$u_{3i}^0(x - x_0) = \lim_{y \to +0} u_{3i}(x - x_0, y - y_0) \quad (i = 1, 2),$$

we may express the state of stress corresponding to the elastic half-plane acted on by an internal concentrated force **F**, in the form

$$\sigma_x(x, y) = F_x[u_{11}(x - x_0, y - y_0) + v_{11}(x, y) * u_{31}^0(x - x_0)$$

$$+ v_{12}(x, y) * u_{21}^0(x - x_0)] + F_y[u_{12}(x - x_0, y - y_0)$$

$$+ v_{11}(x, y) * u_{32}^0(x - x_0) + v_{12}(x, y) * u_{22}(x - x_0)],$$

$$\sigma_y(x, y) = F_x[u_{21}(x - x_0, y - y_0) + v_{21}(x, y) * u_{31}^0(x - x_0)$$

$$+ v_{22}(x, y) * u_{21}^0(x - x_0)] + F_y[u_{22}(x - x_0, y - y_0)$$

$$+ v_{21}(x, y) * u_{32}^0(x - x_0) + v_{22}(x, y) * u_{22}^0(x - x_0)],$$

$$\tau_{xy}(x, y) = F_x[u_{31}(x - x_0, y - y_0) + v_{31}(x, y) * u_{31}^0(x - x_0)$$

$$+ v_{32}(x, y) * u_{21}^0(x - x_0)] + F_y[u_{32}(x - x_0, y - y_0)$$

$$+ v_{31}(x, y) * u_{32}^0(x - x_0) + v_{32}(x, y) * u_{22}^0(x - x_0)].$$

(9.2.28)

(9.2.28')

Different cases of internal loading may be treated in a similar way.

10

Three-dimensional problems of the theory of elasticity

In the following discussion we shall show how to apply the methods of the theory of distributions to three-dimensional problems of the theory of elasticity. We shall consider the elastic space and half-space acted on by static or dynamic loads. The methods applied may be also used in the case of different three-dimensional domains. Furthermore we shall also use the representation of Schaefer type given in sections 7.2.4.3 and 7.2.4.4.

It is interesting to note the difficulties (or simplifications) which appear in calculations when we pass from plane to similar three-dimensional problems.

10.1. Elastic space

Let there be an elastic space acted on by body forces $X = X(x, y, z)$, $Y = Y(y, y, z)$, $Z = Z(x, y, z)$ expressed in distributions (Figure 10.1). We shall first show how to obtain the fundamental solution and then we shall use the latter to obtain the solution corresponding to arbitrary body forces.

Then we shall study an elastic space acted on by body forces $X = X(x, y, z; t)$, $Y = Y(x, y, z; t)$, $Z = Z(x, y, z; t)$ variable in time.

10.1.1. Static problem. Fundamental solution

The state of displacement of the elastic space will be expressed by Lamé's equations (7.2.20) where the operations are considered in the sense of theory of distributions. The solution of that system of equations must satisfy certain *conditions of regularity at infinity;* thus, if R is the radius vector expressed by (3.3.35′), the components of the displacement vector must approach zero for $R \to \infty$

$$\lim_{R \to \infty} u(x, y, z) = \lim_{R \to \infty} v(x, y, z) = \lim_{R \to \infty} w(x, y, z) = 0. \tag{10.1.1}$$

The matrix

$$(U) \equiv \begin{pmatrix} u_{11}(x, y, z) & u_{12}(x, y, z) & u_{13}(x, y, z) \\ u_{21}(x, y, z) & u_{22}(x, y, z) & u_{23}(x, y, z) \\ u_{31}(x, y, z) & u_{32}(x, y, z) & u_{33}(x, y, z) \end{pmatrix} \qquad (10.1.2)$$

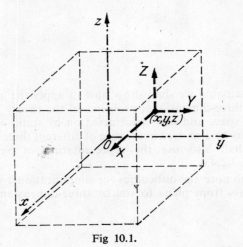

Fig 10.1.

will be termed the *fundamental solution* corresponding to the elastic space, if its elements satisfy the systems of equations

$$\mu \, \Delta u_{11} + (\lambda + \mu) \frac{\partial \varepsilon_v'}{\partial x} + \delta(x, y, z) = 0,$$

$$\mu \, \Delta u_{21} + (\lambda + \mu) \frac{\partial \varepsilon_v'}{\partial y} = 0, \qquad (10.1.3)$$

$$\mu \, \Delta u_{31} + (\lambda + \mu) \frac{\partial \varepsilon_v'}{\partial z} = 0,$$

$$\mu \, \Delta u_{12} + (\lambda + \mu) \frac{\partial \varepsilon_v''}{\partial x} = 0,$$

$$\mu \, \Delta u_{22} + (\lambda + \mu) \frac{\partial \varepsilon_v''}{\partial y} + \delta(x, y, z) = 0, \qquad (10.1.3')$$

$$\mu \, \Delta u_{32} + (\lambda + \mu) \frac{\partial \varepsilon_v''}{\partial z} = 0,$$

$$\mu \, \Delta u_{13} + (\lambda + \mu) \frac{\partial \varepsilon_v'''}{\partial x} = 0,$$

$$\mu \, \Delta u_{23} + (\lambda + \mu) \frac{\partial \varepsilon_v'''}{\partial y} = 0, \qquad (10.1.3'')$$

$$\mu \, \Delta u_{33} + (\lambda + \mu) \frac{\partial \varepsilon_v'''}{\partial z} + \delta(x, y, z) = 0,$$

with the notations

$$\varepsilon_v' = \frac{\partial u_{11}}{\partial x} + \frac{\partial u_{21}}{\partial y} + \frac{\partial u_{31}}{\partial z},$$

$$\varepsilon_v'' = \frac{\partial u_{12}}{\partial x} + \frac{\partial u_{22}}{\partial y} + \frac{\partial u_{32}}{\partial z}, \qquad (10.1.4)$$

$$\varepsilon_v''' = \frac{\partial u_{13}}{\partial x} + \frac{\partial u_{23}}{\partial y} + \frac{\partial u_{33}}{\partial z},$$

as well as the limiting conditions

$$\lim_{R \to \infty} u_{ij}(x, y, z) = 0 \qquad (i, j = 1, 2, 3). \qquad (10.1.5)$$

For the determination of the fundamental solution we shall apply the triple Fourier transform in terms of distributions.

10.1.1.1. Fourier transform

We shall first consider the system of equations (10.1.3); by applying the triple Fourier transform we obtain the system of linear algebraic equations

$$\mu(\alpha^2 + \beta^2 + \gamma^2) \, F[u_{11}(x, y, z)] + (\lambda + \mu) \, \alpha(\alpha F[u_{11}(x, y, z)]$$

$$+ \beta F[u_{21}(x, y, z)] + \gamma F[u_{31}(x, y, z)]) = 1,$$

$$\mu(\alpha^2 + \beta^2 + \gamma^2) \, F[u_{21}(x, y, z)] + (\lambda + \mu) \, \beta[\alpha F[u_{11}(x, y, z)]$$

$$+ \beta F[u_{21}(x, y, z)] + \gamma F[u_{31}(x, y, z)]) = 0, \qquad (10.1.6)$$

$$\mu(\alpha^2 + \beta^2 + \gamma^2) \, F[u_{31}(x, y, z)] + (\lambda + \mu) \, \gamma(\alpha F[u_{11}(x, y, z)$$

$$+ \beta F[u_{21}(x, y, z)] + \gamma F[u_{31}(x, y, z)]) = 0,$$

where α, β and γ are complex variables.

Multiplying these equations by α, β and γ, respectively, and adding member by member, we obtain

$$\alpha F[u_{11}(x, y, z)] + \beta F[u_{21}(x, y, z)] + \gamma F[u_{31}(x, y, z)]$$

$$= \frac{1}{\lambda + 2\mu} \frac{\alpha}{\alpha^2 + \beta^2 + \gamma^2} ; \qquad (10.1.7)$$

substituting this expression in the system (10.1.6) we obtain the Fourier transforms in the form

$$F[u_{11}(x, y, z)] = -\frac{\lambda + \mu}{\mu(\lambda + 2\mu)} \frac{\alpha^2}{(\alpha^2 + \beta^2 + \gamma^2)^2} + \frac{1}{\mu} \frac{1}{\alpha^2 + \beta^2 + \gamma^2} ,$$

$$F[u_{21}(x, y, z)] = -\frac{\lambda + \mu}{\mu(\lambda + 2\mu)} \frac{\alpha\beta}{(\alpha^2 + \beta^2 + \gamma^2)^2} , \qquad (10.1.8)$$

$$F[u_{31}(x, y, z)] = -\frac{\lambda + \mu}{\mu(\lambda + 2\mu)} \frac{\alpha\gamma}{(\alpha^2 + \beta^2 + \gamma^2)^2} ,$$

whence

$$u_{11}(x, y, z) = F^{-1}[F[u_{11}(x, y, z)]]$$

$$= -\frac{\lambda + \mu}{\mu(\lambda + 2\mu)} F^{-1}\left[\frac{\alpha^2}{(\alpha^2 + \beta^2 + \gamma^2)^2} \right] + \frac{1}{\mu} F^{-1}\left[\frac{1}{\alpha^2 + \beta^2 + \gamma^2} \right],$$

$$u_{21}(x, y, z) = F^{-1}[F[u_{21}(x, y, z)]] \qquad (10.1.8')$$

$$= -\frac{\lambda + \mu}{\mu(\lambda + 2\mu)} F^{-1}\left[\frac{\alpha\beta}{(\alpha^2 + \beta^2 + \gamma^2)^2} \right],$$

$$u_{31}(x, y, z) = F^{-1}[F[u_{31}(x, y, z)]] = -\frac{\lambda + \mu}{\mu(\lambda + 2\mu)} F^{-1}\left[\frac{\alpha\gamma}{(\alpha^2 + \beta^2 + \gamma^2)^2} \right] .$$

10.1.1.2. Inverse Fourier transform

Applying the Fourier transform to relation (1.4.52) we obtain

$$\frac{1}{4\pi} F\left[\frac{1}{R} \right] = \frac{1}{\alpha^2 + \beta^2 + \gamma^2} \qquad (10.1.9)$$

Differentiating with respect to α in the sense of the theory of distributions and remarking that this derivative is identical to the derivative in the ordinary sense, we may write

$$\frac{\partial}{\partial \alpha}\left[\frac{1}{\alpha^2 + \beta^2 + \gamma^2}\right] = \frac{1}{4\pi} F\left[i\frac{x}{R}\right] = \frac{i}{4\pi} F\left[\frac{x}{R}\right],$$

whence

$$\frac{1}{8\pi} F\left[\frac{x}{R}\right] = i\frac{\alpha}{(\alpha^2 + \beta^2 + \gamma^2)^2} . \tag{10.1.10}$$

Likewise, we have

$$F\left[\frac{\partial}{\partial x}\left(\frac{x}{R}\right)\right] = F\left[\frac{1}{R}\right] - F\left[\frac{x^2}{R^3}\right] = -i\alpha F\left[\frac{x}{R}\right]$$

$$= 8\pi \frac{\alpha^2}{(\alpha^2 + \beta^2 + \gamma^2)^2} , \tag{10.1.11}$$

$$F\left[\frac{\partial}{\partial y}\left(\frac{x}{R}\right)\right] = -F\left[\frac{xy}{R^3}\right] = -i\beta F\left[\frac{x}{R}\right] = 8\pi \frac{\alpha\beta}{(\alpha^2 + \beta^2 + \gamma^2)^2} , \tag{10.1.11'}$$

$$F\left[\frac{\partial}{\partial z}\left(\frac{x}{R}\right)\right] = -F\left[\frac{xz}{R^3}\right] = -i\gamma F\left[\frac{x}{R}\right] = 8\pi \frac{\alpha\gamma}{(\alpha^2 + \beta^2 + \gamma^2)^2} . \tag{10.1.11''}$$

Taking into account relation (7.1.16), we may write

$$\frac{\lambda + \mu}{\lambda + 2\mu} = \frac{1}{2(1 - v)} , \quad \frac{\lambda + 3\mu}{\lambda + 2\mu} = \frac{3 - 4v}{2(1 - v)} , \tag{10.1.12}$$

where v is Poisson's ratio.

Using the above results we may write the inverse Fourier transforms (10.1.8') in the form

$$u_{11}(x, y, z) = \frac{1}{16\pi(1 - v)G} \frac{1}{R}\left(3 - 4v + \frac{x^2}{R^2}\right),$$

$$u_{21}(x, y, z) = \frac{1}{16\pi(1 - v)G} \frac{xy}{R^3} , \tag{10.1.13}$$

$$u_{31}(x, y, z) = \frac{1}{16\pi(1 - v)G} \frac{xz}{R^3} ,$$

by taking into account that $\mu = G$.

From equations (10.1.3) and (10.1.3'') we obtain, by the same method,

$$u_{12}(x, y, z) = \frac{1}{16\pi(1 - v)G} \frac{yx}{R^3} \ ,$$

$$u_{22}(x, y, z) = \frac{1}{16\pi(1 - v)G} \frac{1}{G}\left(3 - 4v + \frac{y^2}{R^2}\right), \qquad (10.1.13')$$

$$u_{32}(x, y, z) = \frac{1}{16\pi(1 - v)G} \frac{yz}{R^3} \ ,$$

and

$$u_{13}(x, y, z) = \frac{1}{16\pi(1 - v)G} \frac{zx}{R^3} \ ,$$

$$u_{23}(x, y, z) = \frac{1}{16\pi(1 - v)G} \frac{zy}{R^3}, \qquad (10.1.13'')$$

$$u_{33}(x, y, z) = \frac{1}{16\pi(1 - v)G} \frac{1}{R}\left(3 - 4v + \frac{z^2}{R^2}\right).$$

10.1.1.3. Fundamental solution corresponding to a state of stress

Using relations (7.1.2) and (7.1.2') between strains and displacements, Hooke's law (7.1.13), (7.1.13') and relations (7.1.16), we may introduce the matrix

$$(V) \equiv \begin{pmatrix} v_{11}(x, y, z) & v_{12}(x, y, z) & v_{13}(x, y, z) \\ v_{21}(x, y, z) & v_{22}(x, y, z) & v_{23}(x, y, z) \\ v_{31}(x, y, z) & v_{32}(x, y, z) & v_{33}(x, y, z) \\ v_{41}(x, y, z) & v_{42}(x, y, z) & v_{43}(x, y, z) \\ v_{51}(x, y, z) & v_{52}(x, y, z) & v_{53}(x, y, z) \\ v_{61}(x, y, z) & v_{62}(x, y, z) & v_{63}(x, y, z) \end{pmatrix} ; \qquad (10.1.14)$$

this represents the *fundamental solution* of the system of equations (7.1.7), (7.2.1) and (7.2.1') expressed in terms of stress.

The matrix components may be written in the form

$$\frac{1}{x} v_{11}(x, y, z) = \frac{1}{y} v_{61}(x, y, z) = \frac{1}{z} v_{51}(x, y, z)$$

$$= -\frac{1}{8\pi(1 - v)} \frac{1}{R^3} \left(1 - 2v + 3 \frac{x^2}{R^2} \right),$$

$$\frac{1}{x} v_{62}(x, y, z) = \frac{1}{y} v_{22}(x, y, z) = \frac{1}{z} v_{42}(x, y, z)$$

$$= -\frac{1}{8\pi(1 - v)} \frac{1}{R^3} \left(1 - 2v + 3 \frac{y^2}{R^2} \right),$$

$$\frac{1}{x} v_{53}(x, y, z) = \frac{1}{y} v_{43}(x, y, z) = \frac{1}{z} v_{33}(x, y, z)$$

$$= -\frac{1}{8\pi(1 - v)} \frac{1}{R^3} \left(1 - 2v + 3 \frac{z^2}{R^2} \right),$$

(10.1.15)

$$\frac{1}{y} v_{12}(x, y, z) = \frac{1}{z} v_{13}(x, y, z) = \frac{1}{8\pi(1 - v)} \frac{1}{R^3} \left(1 - 2v - 3 \frac{x^2}{R^3} \right),$$

$$\frac{1}{z} v_{23}(x, y, z) = \frac{1}{x} v_{21}(x, y, z) = \frac{1}{8\pi(1 - v)} \frac{1}{R^3} \left(1 - 2v - 3 \frac{y^2}{R^2} \right), \quad (10.1.15')$$

$$\frac{1}{x} v_{31}(x, y, z) = \frac{1}{y} v_{32}(x, y, z) = \frac{1}{8\pi(1 - v)} \frac{1}{R^3} \left(1 - 2v - 3 \frac{z^2}{R^2} \right),$$

$$v_{41}(x, y, z) = v_{52}(x, y, z) = v_{63}(x, y, z) = -\frac{3}{8\pi(1 - v)} \frac{xyz}{R^5}. \quad (10.1.15'')$$

10.1.2. Static problem. Case of arbitrary loads

With the aid of the matrix (U) of the fundamental solution (corresponding to a solution in terms of displacements), or by using the matrix (V) of the fundamental solution (corresponding to a solution in terms of stresses), we may construct the solution corresponding to arbitrary body forces with the help of the convolution product. We shall apply this method and give results for a few particular cases of loading.

10.1.2.1. Case of arbitrary body forces

Introducing the matrices

$$(\sigma) \equiv \begin{pmatrix} \sigma_x \\ \sigma_y \\ \sigma_z \\ \tau_{yz} \\ \tau_{zx} \\ \tau_{xy} \end{pmatrix}, \quad (u) \equiv \begin{pmatrix} u \\ v \\ w \end{pmatrix}, \quad Q \equiv \begin{pmatrix} X \\ Y \\ Z \end{pmatrix}, \tag{10.1.16}$$

we may write

$$(u) = (U) * (Q) \tag{10.1.17}$$

and

$$(\sigma) = (V) * (Q). \tag{10.1.17'}$$

Thus, the state of displacement for arbitrary body forces may be expressed in the form

$$u(x, y, z) = u_{11}(x, y, z) * X(x, y, z) + u_{12}(x, y, z) * Y(x, y, z)$$
$$+ u_{23}(x, y, z) * Z(x, y, z),$$

$$v(x, y, z) = u_{21}(x, y, z) * X(x, y, z) + u_{22}(x, y, z) * Y(x, y, z) \tag{10.1.18}$$
$$+ u_{23}(x, y, z) * Z(x, y, z),$$

$$w(x, y, z) = u_{31}(x, y, z) * X(x, y, z) + u_{32}(x, y, z) * Y(x, y, z)$$
$$+ u_{33}(x, y, z) * Z(x, y, z),$$

and the corresponding state of stress will be given by

$$\sigma_x(x, y, z) = v_{11}(x, y, z) * X(x, y, z) + v_{12}(x, y, z) * Y(x, y, z)$$
$$+ v_{13}(x, y, z) * Z(x, y, z),$$

$$\sigma_y(x, y, z) = v_{21}(x, y, z) * X(x, y, z) + v_{22}(x, y, z) * Y(x, y, z) \tag{10.1.19}$$
$$+ v_{23}(x, y, z) * Z(x, y, z),$$

$$\sigma_z(x, y, z) = v_{31}(x, y, z) * X(x, y, z) + v_{32}(x, y, z) * Y(x, y, z)$$
$$+ v_{33}(x, y, z) * Z(x, y, z),$$

$$\tau_{yz}(x, y, z) = v_{41}(x, y, z) * X(x, y, z) + v_{42}(x, y, z) * Y(x, y, z)$$
$$+ v_{43}(x, y, z) * Z(x, y, z),$$

$$\tau_{zx}(x, y, z) = v_{51}(x, y, z) * X(x, y, z) + v_{52}(x, y, z) * Y(x, y, z)$$
$$+ v_{53}(x, y, z) * Z(x, y, z), \qquad (10.1.19')$$

$$\tau_{xy}(x, y, z) = v_{61}(x, y, z) * X(x, y, z) + v_{62}(x, y, z) * Y(x, y, z)$$
$$+ v_{63}(x, y, z) * Z(x, y, z).$$

Using the above results, we can now verify the complete system of equations of the theory of elasticity by effecting all the necessary operations in the sense of the theory of distributions; to this end, we may apply the results of section 1.4.2.2. However, taking into account the proof given in the preceding section, such a verification is not necessary.

10.1.2.2. Examples

Let us consider an elastic space acted on by a concentrated force $\mathbf{F}(F_x, F_y, F_z)$ at the point $A(x_0, y_0, z_0)$ (Figure 10.2); the load equivalent to the concentrated force may be written

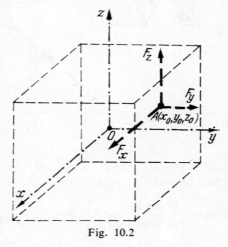

Fig. 10.2

$$X(x, y, z) = F_x \delta(x - x_0, y - y_0, z - z_0),$$
$$Y(x, y, z) = F_y \delta(x - x_0, y - y_0, z - z_0), \qquad (10.1.20)$$
$$Z(x, y, z) = F_z \delta(x - x_0, y - y_0, z - z_0).$$

Using the formulae (10.1.18) the state of displacement is expressed by

$$u(x, y, z) = F_x u_{11}(x - x_0, y - y_0, z - z_0)$$

$$+ F_y u_{12}(x - x_0, y - y_0, z - z_0) + F_z u_{13}(x - x_0, y - y_0, z - z_0),$$

$$v(x, y, z) = F_x u_{21}(x - x_0, y - y_0, z - z_0) \qquad (10.1.21)$$

$$+ F_y u_{22}(x - x_0, y - y_0, z - z_0) + F_z u_{23}(x - x_0, y - y_0, z - z_0),$$

$$w(x, y, z) = F_x u_{31}(x - x_0, y - y_0, z - z_0)$$

$$+ F_y u_{32}(x - x_0, y - y_0, z - z_0) + F_z u_{33}(x - x_0, y - y_0, z - z_0) ;$$

for the state of stress we proceed in the same way.

Let us consider an elastic space acted on at the point $A(x_0, y_0, z_0)$ by a directed concentrated moment of magnitude M, specified by the unit vectors $\mathbf{u}(\cos\alpha_1, \cos\alpha_2, \cos\alpha_3)$

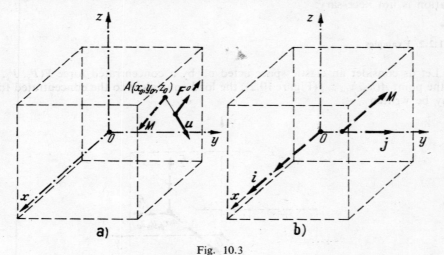

Fig. 10.3

and $\mathbf{F}^0(\cos\beta_1, \cos\beta_2, \cos\beta_3)$ (Figure 10.3a). The load equivalent to the directed concentrated moment may be written, using formulae (5.2.4''), in the form

$$X(x, y, z) = - \frac{M \cos \beta_1}{|\mathbf{u} \times \mathbf{F}^0|} \frac{\partial}{\partial u} \delta (x - x_0, y - y_0, z - z_0),$$

$$Y(x, y, z) = - \frac{M \cos \beta_2}{|\mathbf{u} \times \mathbf{F}^0|} \frac{\partial}{\partial u} \delta (x - x_0, y - y_0, z - z_0), \qquad (10.1.22)$$

$$Z(x, y, z) = - \frac{M \cos \beta_3}{|\mathbf{u} \times \mathbf{F}^0|} \frac{\partial}{\partial u} \delta(x - x_0, y - y_0, z - z_0) ;$$

the directional derivative is given by the formula (5.2.5) and we may write

$$\frac{\partial}{\partial u} = \cos \alpha_1 \frac{\partial}{\partial x} + \cos \alpha_2 \frac{\partial}{\partial y} + \cos \alpha_3 \frac{\partial}{\partial z}.$$ (10.1.23)

Therefore, the state of displacement is expressed by

$$u(x, y, z) = - \frac{M}{|\mathbf{u} \times \mathbf{F}^0|} \left[\cos \beta_1 \frac{\partial}{\partial u} u_{11}(x - x_0, y - y_0, z - z_0) \right.$$

$$+ \cos \beta_2 \frac{\partial}{\partial u} u_{12}(x - x_0, y - y_0, z - z_0)$$

$$\left. + \cos \beta_3 \frac{\partial}{\partial u} u_{13}(x - x_0, y - y_0, z - z_0) \right] = - \frac{M}{|\mathbf{u} \times \mathbf{F}^0|} \frac{\partial}{\partial u} u_{\mathbf{F}^0},$$

$$v(x, y, z) = - \frac{M}{|\mathbf{u} \times \mathbf{F}^0|} \left[\cos \beta_1 \frac{\partial}{\partial u} u_{21}(x - x_0, y - y_0, z - z_0) \right.$$

$$+ \cos \beta_2 \frac{\partial}{\partial u} u_{22}(x - x_0, y - y_0, z - z_0)$$ (10.1.24)

$$\left. + \cos \beta_3 \frac{\partial}{\partial u} u_{23}(x - x_0, y - y_0, z - z_0) \right] = - \frac{M}{|\mathbf{u} \times \mathbf{F}^0|} \frac{\partial}{\partial u} v_{\mathbf{F}^0},$$

$$w(x, y, z) = - \frac{M}{|\mathbf{u} \times \mathbf{F}^0|} \left[\cos \beta_1 \frac{\partial}{\partial u} u_{31}(x - x_0, y - y_0, z - z_0) \right.$$

$$+ \cos \beta_2 \frac{\partial}{\partial u} u_{32}(x - x_0, y - y_0, z - z_0)$$

$$\left. + \cos \beta_3 \frac{\partial}{\partial u} u_{33}(x - x_0, y - y_0, z - z_0) \right] = - \frac{M}{|\mathbf{u} \times \mathbf{F}^0|} \frac{\partial}{\partial u} w_{\mathbf{F}^0}$$

where $u_{\mathbf{F}^0}$, $v_{\mathbf{F}^0}$, $w_{\mathbf{F}^0}$ are the displacements corresponding to the action of a concentrated force $\mathbf{F}^0$, whose magnitude is equal to unity, in accordance with formulae (10.1.21).

In the particular case $\mathbf{F}^0 = -\mathbf{i}$ and $\mathbf{u} = \mathbf{j}$, where $\mathbf{i}$ and $\mathbf{j}$ are the unit vectors of the co-ordinate axes, we obtain a directed concentrated moment, applied at the origin, causing a positive rotation in the plane Oxy (Figure 10.3b), and which is expressed by the equivalent load

$$X(x, y, z) = M \frac{\partial}{\partial y} \delta(x, y, z), \quad Y(x, y, z) = Z(x, y, z) = 0.$$ (10.1.25)

In that case, the state of displacement will be written

$$u(x, y, z) = M \frac{\partial}{\partial y} u_{11}(x, y, z) = - \frac{M}{16\pi(1 - v)G} \frac{y}{R^3} \left(3 - 4v + 3 \frac{x^2}{R^2} \right),$$

$$v(x, y, z) = M \frac{\partial}{\partial y} u_{21}(x, y, z) = \frac{M}{16\pi(1 - v)G} \frac{x}{R^3} \left(1 - 3 \frac{y^2}{R^2} \right), \qquad (10.1.26)$$

$$w(x, y, z) = M \frac{\partial}{\partial y} u_{31}(x, y, z) = - \frac{3M}{16\pi(1 - v)G} \frac{xyz}{R^5}.$$

We remark that the derivatives occurring in the above expressions may be considered in the ordinary sense, since the formal derivatives of the components of the matrix (U) are locally integrable functions.

Let us consider an elastic space acted on at the point $A(x_0, y_0, z_0)$ by a rotational concentrated moment of magnitude M and specified by the unit vector

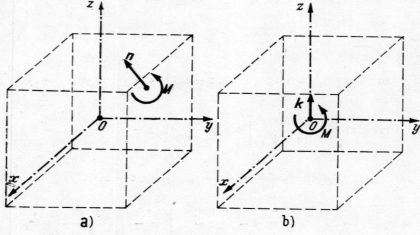

a)   b)

Fig. 10.4

$\mathbf{n}$ $(\cos\gamma_1, \cos\gamma_2, \cos\gamma_3)$, normal to the plane of action (Figure 10.4a); the load equivalent to the rotational concentrated moment may be written in the form

$$X(x, y, z) = - \frac{1}{2} M \left[\cos \gamma_2 \frac{\partial}{\partial z} \delta(x - x_0, y - y_0, z - z_0) \right.$$

$$\left. - \cos \gamma_3 \frac{\partial}{\partial y} \delta(x - x_0, y - y_0, z - z_0) \right],$$

$$Y(x, y, z) = -\frac{1}{2} M\left[\cos \gamma_3 \frac{\partial}{\partial x} \delta(x - x_0, y - y_0, z - z_0)\right.$$

$$(10.1.27)$$

$$\left. -\cos \gamma_1 \frac{\partial}{\partial z} \delta(x - x_0, y - y_0, z - z_0)\right],$$

$$Z(x, y, z) = -\frac{1}{2} M\left[\cos \gamma_1 \frac{\partial}{\partial y} \delta(x - x_0, y - y_0, z - z_0)\right.$$

$$\left. -\cos \gamma_2 \frac{\partial}{\partial x} \delta(x - x_0, y - y_0, z - z_0)\right],$$

and the state of displacement is given by

$$u(x, y, z) = \frac{1}{2} M\left\{\cos \gamma_1 \left[\frac{\partial}{\partial z} u_{12}(x - x_0, y - y_0, z - z_0)\right.\right.$$

$$\left. -\frac{\partial}{\partial y} u_{13}(x - x_0, y - y_0, z - z_0)\right]$$

$$+\cos \gamma_2 \left[\frac{\partial}{\partial x} u_{13}(x - x_0, y - y_0, z - z_0) - \frac{\partial}{\partial z} u_{11}(x - x_0, y - y_0, z - z_0)\right]$$

$$+\cos \gamma_3 \left[\frac{\partial}{\partial y} u_{11}(x - x_0, y - y_0, z - z_0) - \frac{\partial}{\partial x} u_{12}(x - x_0, y - y_0, z - z_0)\right]\right\},$$

$$v(x, y, z) = \frac{1}{2} M\left\{\cos \gamma_1 \left[\frac{\partial}{\partial z} u_{22}(x - x_0, y - y_0, z - z_0)\right.\right.$$

$$\left. -\frac{\partial}{\partial y} u_{23}(x - x_0, y - y_0, z - z_0)\right]$$

$$(10.1.28)$$

$$+\cos \gamma_2 \left[\frac{\partial}{\partial x} u_{23}(x - x_0, y - y_0, z - z_0) - \frac{\partial}{\partial z} u_{21}(x - x_0, y - y_0, z - z_0)\right]$$

$$+ \cos \gamma_3 \left[\frac{\partial}{\partial y} u_{21}(x - x_0, y - y_0, z - z_0) - \frac{\partial}{\partial x} u_{22}(x - x_0, y - y_0, z - z_0) \right] \Bigg\},$$

$$w(x, y, z) = \frac{1}{2} M \Bigg\{ \cos \gamma_1 \left[\frac{\partial}{\partial z} u_{32}(x - x_0, y - y_0, z - z_0) \right. $$

$$\left. - \frac{\partial}{\partial y} u_{33}(x - x_0, y - y_0, z - z_0) \right]$$

$$+ \cos \gamma_2 \left[\frac{\partial}{\partial x} u_{33}(x - x_0, y - y_0, z - z_0) - \frac{\partial}{\partial z} u_{31}(x - x_0, y - y_0, z - z_0) \right]$$

$$+ \cos \gamma_3 \left[\frac{\partial}{\partial y} u_{31}(x - x_0, y - y_0, z - z_0) \right] - \frac{\partial}{\partial x} u_{32}(x - x_0, y - y_0, z - z_0) \Bigg\}.$$

In particular, if $\mathbf{n} = \mathbf{k}$, where $\mathbf{k}$ is the unit vector of the Oz axis, we obtain a rotational concentrated moment, applied at the origin, inducing a positive rotation in the Oxy plane (Figure 10.4b) and expressed by the equivalent load

$$X(x, y, z) = \frac{1}{2} M \frac{\partial}{\partial y} \delta(x, y, z),$$

$$Y(x, y, z) = - \frac{1}{2} M \frac{\partial}{\partial x} \delta(x, y, z), \qquad (10.1.29)$$

$$Z(x, y, z) = 0 ;$$

in that case the state of displacement is obtained in the form

$$u(x, y, z) = \frac{1}{2} M \left[\frac{\partial}{\partial y} u_{11}(x, y, z) - \frac{\partial}{\partial x} u_{12}(x, y, z) \right] = - \frac{M}{8\pi G} \frac{y}{R^3} ,$$

$$v(x, y, z) = \frac{1}{2} M \left[\frac{\partial}{\partial y} u_{21}(x, y, z) - \frac{\partial}{\partial x} u_{22}(x, y, z) \right] = \frac{M}{8\pi G} \frac{x}{R^3} , \qquad (10.1.30)$$

$$w(x, y, z) = \frac{1}{2} M \left[\frac{\partial}{\partial y} u_{31}(x, y, z) - \frac{\partial}{\partial x} u_{32}(x, y, z) \right] = 0.$$

Other cases of loading may be treated in a similar way.

10.1.3. Dynamic problem

We consider an elastic space acted on by the dynamic loads $X = X(x, y, z; t)$, $Y = Y(x, y, z; t)$, $Z = Z(x, y, z; t)$ (Figure 10.1); to solve the problem we replace these body forces by the distributions $\bar{X} = \bar{X}(x, y, z; t)$, $\bar{Y} = \bar{Y}(x, y, z; t)$, $\bar{Z} = \bar{Z}(x, y, z; t)$ defined by the formulae (7.2.41). The components of the displacement vector too, will be replaced by the distributions $\bar{u}(x, y, z; t)$, $\bar{v}(x, y, z; t)$, $\bar{w}(x, y, z; t)$ specified by the formulae (7.2.42).

For a solution in terms of displacements we use Lamé's equations (7.2.23) with initial conditions of the form

$$u(x, y, z; 0) = u_0(x, y, z), \quad v(x, y, z; 0) = v_0(x, y, z),$$

$$w(x, y, z; 0) = w_0(x, y, z), \tag{10.1.31}$$

$$\dot{u}(x, y, z; 0) = \dot{u}_0(x, y, z), \quad \dot{v}(x, y, z; 0) = \dot{v}_0(x, y, z),$$

$$\dot{w}(x, y, z; 0) = \dot{w}_0(x, y, z), \tag{10.1.31'}$$

corresponding to the initial moment $t = 0$. With the aid of the distributions introduced above we obtain Lamé's equations in a modified form as given in equation (7.2.46) and where the notation (7.2.45) has been used.

In the following discussion we shall apply the Laplace and Fourier transforms to obtain the solution corresponding to arbitrary body forces; also, we shall compute the inverse transforms. Using these results, we shall consider a few particular cases of loading.

10.1.3.1. Case of arbitrary body forces

We shall first apply to equations (7.2.46) the simple Laplace transform with respect to the time variable and then the triple Fourier transform with respect to the space variables; thus we obtain

$$[\mu(\alpha^2 + \beta^2 + \gamma^2) + \rho p^2] F[L[\bar{u}(x, y, z; t)]] + (\lambda + \mu)\,\alpha(\alpha F[L[\bar{u}(x, y, z; t)]]$$

$$+ \beta F[L[\bar{v}(x, y, z; t)]] + \gamma F[L[\bar{w}(x, y, z; t)]])$$

$$= F[L[\bar{X}(x, y, z; t)]] + \rho\, p\, F[u_0(x, y, z)] + \rho\, F[\dot{u}_0(x, y, z)],$$

$$[\mu(\alpha^2 + \beta^2 + \gamma^2) + \rho\, p^2]\, F[L[\bar{v}(x, y, z; t)]] + (\lambda + \mu)\, \beta\, (\alpha F[L[\bar{u}(x, y, z; t)]]$$

$$+ \beta F\, [L[\bar{v}(x, y, z; t)]] + \gamma F[L[\bar{w}(x, y, z; t)]]) \qquad (10.1.32)$$

$$= F[L[\bar{Y}[(x, y, z; t)]] + \rho\, p\, F[v_0(x, y, z)] + \rho\, F[\dot{v}_0(x, y, z)],$$

$$[\mu(\alpha^2 + \beta^2 + \gamma^2) + \rho\, p^2]\, F[L[\bar{w}(x, y, z; t)]] + (\lambda + \mu)\, \gamma\, (\alpha F[L[\bar{u}(x, y, z; t)]]$$

$$+ \beta F[L[\bar{v}(x, y, z; t)]] + \gamma F[L[\bar{w}(x, y, z; t)]])$$

$$= F[L[\bar{Z}(x, y, z; t)]] + \rho\, p F[w_0(x, y, z)] + \rho\, F[\dot{w}_0(x, y, z,)],$$

where p is a complex variable corresponding to the Laplace transform and α, β and γ are complex variables corresponding to the Fourier transform.

Multiplying equations (10.1.32) by α, β and γ, respectively, and adding member by member, we may write

$$f(\alpha, \beta, \gamma; p) = \alpha F\, [L[\bar{u}(x, y, z; t)]] + \beta F\, [L[\bar{v}(x, y, z; t)]] + \gamma F\, [L[w(x, y, z; t)]]$$

$$= \frac{1}{(\lambda + 2\mu)(\alpha^2 + \beta^2 + \gamma^2) + \rho p^2}\, [\alpha F[L[\bar{X}(x, y, z; t)]]$$

$$+ \beta F[\bar{L}[Y(x, y, z; t)]] + \gamma F[\bar{L}[Z(x, y, z; t)]] + \rho p(\alpha F[u_0(x, y, z)]$$

$$+ \beta F[v_0(x, y, z)] + \gamma F[w_0(x, y, z)])$$

$$+ \rho(\alpha F[\dot{u}_0(x, y, z)] + \beta F[\dot{v}_0(x, y, z)] + \gamma F[\dot{w}_0(x, y, z)])]\, ; \qquad (10.1.33)$$

thus we may solve easily the system of linear algebraic equations (10.1.32); the integral transforms are given by

$$F[L[\bar{u}(x, y, z; t)]] = \frac{1}{\mu(\alpha^2 + \beta^2 + \gamma^2) + \rho\, p^2}\, [F[L[\bar{X}(x, y, z; t)]]$$

$$+ \rho p F[u_0(x, y, z)] + \rho F[\dot{u}_0(x, y, z)] - (\lambda + \mu)\, \alpha f(\alpha, \beta, \gamma; p)],$$

$$F[L[\bar{v}(x, y, z; t)]] = \frac{1}{\mu(\alpha^2 + \beta^2 + \gamma^2) + \rho\, p^2} \; [F[L[\bar{Y}(x, y, z; t)]]$$

$$\text{(10.1.34)}$$

$$+ \rho\, p\, F[v_0(x, y, z)] + \rho\, F[\dot{v}_0(x, y, z)] - (\lambda + \mu)\, \beta f(\alpha, \beta, \gamma; p)],$$

$$F[L[\bar{w}(x, y, z; t)]] = \frac{1}{\mu(\alpha^2 + \beta^2 + \gamma^2) + \rho\, p^2} \; [F[L[\bar{Z}(x, y, z; t)]]$$

$$+ \rho p F[w_0(x, y, z)] + \rho F[\dot{w}_0(x, y, z)] - (\lambda + \mu)\, \gamma f(\alpha, \beta, \gamma; p)].$$

Taking into account the inverse Fourier transform (3.3.40′), we obtain the following inverse Fourier transform

$$F^{-1}\left[\frac{\lambda + \mu}{[(\lambda + 2\mu)(\alpha^2 + \beta^2 + \gamma^2) + \rho p^2]\,[\mu(\alpha^2 + \beta^2 + \gamma^2) + \rho p^2]} \right]$$

$$= \frac{1}{4\pi\, \rho p^2 R}\, (e^{-p\frac{R}{c_1}} - e^{-p\frac{R}{c_2}}),$$

where we have introduced the *wave propagation velocities* given by relation (7.2.6).

Using the properties of the convolution product as regards differentiation, we may express the inverse Fourier transforms of expressions (10.1.34) in the form

$$L[\bar{u}(x, y, z; t)] = \frac{1}{4\pi\mu}\Bigg\{ [L[\bar{X}(x, y, z; t)]$$

$$+ \rho p u_0(x, y, z) + \rho\dot{u}_0(x, y, z)] * \frac{1}{R} e^{-p\frac{R}{c_2}}$$

$$+ c_2^2\bigg[\frac{\partial}{\partial x} L[\bar{X}(x, y, z; t)] + \frac{\partial}{\partial y} L[\bar{Y}(x, y, z; t)] + \frac{\partial}{\partial z} [L[\bar{Z}(x, y, z; t)]$$

$$+ \rho p \varepsilon_v^0(x, y, z) + \rho\dot{\varepsilon}_v^0(x, y, z)\bigg] * \frac{1}{p^2}\frac{\partial}{\partial x}\bigg[\frac{1}{R}\, (e^{-p\frac{R}{c_1}} - e^{-p\frac{R}{c_2}})\bigg]\Bigg\},$$

$$L[\bar{v}(x, y, z; t)] = \frac{1}{4\pi\mu}\left\{ L[\bar{Y}(x, y, z; t)]\right.$$

$$+ \rho p v_0(x, y, z) + \rho \dot{v}_0(x, y, z)] * \frac{1}{R}e^{-p\frac{R}{c_2}}$$

(10.1.35)

$$+ c_2^2\left[\frac{\partial}{\partial x} L[\bar{X}(x, y, z; t)] + \frac{\partial}{\partial y} L[\bar{Y}(x, y, z; t)] + \frac{\partial}{\partial z} L[\bar{Z}(x, y, z; t)]\right.$$

$$+ \rho p\,\varepsilon_v^0(x, y, z) + \rho\dot{\varepsilon}_v^0(x, y, z)] * \frac{1}{p^2}\frac{\partial}{\partial y}\left[\frac{1}{R}(e^{-p\frac{R}{c_1}} - e^{-p\frac{R}{c_2}})\right]\right\},$$

$$L[\bar{w}(x, y, z; t)] = \frac{1}{4\pi\mu}\left\{ [L[\bar{Z}(x, y, z; t)]\right.$$

$$+ \rho p w_0(x, y, z) + \rho\dot{w}_0(x, y, z)] * \frac{1}{R}e^{-p\frac{R}{c_2}}$$

$$+ c_2^2\left[\frac{\partial}{\partial x} L[\bar{X}(x, y, z; t)] + \frac{\partial}{\partial y} L[\bar{Y}(x, y, z; t)] + \frac{\partial}{\partial z} L[\bar{Z}(x, y, z; t)]\right.$$

$$+ \rho p\varepsilon_v^0(x, y, z) + \rho\dot{\varepsilon}_v^0(x, y, z)] * \frac{1}{p^2}\frac{\partial}{\partial z}\left[\frac{1}{R}(e^{-p\frac{R}{c_1}} - e^{-p\frac{R}{c_2}})\right]\right\},$$

where we have introduced the notations

$$\varepsilon_v^0(x, y, z) = \frac{\partial}{\partial x} u_0(x, y, z) + \frac{\partial}{\partial y} v_0(x, y, z) + \frac{\partial}{\partial z} w_0(x, y, z),$$

(10.1.31'')

$$\dot{\varepsilon}_v(x, y, z) = \frac{\partial}{\partial x} \dot{u}_0(x, y, z) + \frac{\partial}{\partial y} \dot{v}_0(x, y, z) + \frac{\partial}{\partial z} \dot{w}_0(x, y, z).$$

Now, if we apply the inverse Laplace transform, we obtain the generalized displacements

$$
\begin{cases}
\bar{u}(x, y, z; t) = L^{-1}[L[\bar{u}(x, y, z; t)]], \\[2mm]
\bar{v}(x, y, z; t) = L^{-1}[L[\bar{v}(x, y, z; t)]], \\[2mm]
\bar{w}(x, y, z; t) = L^{-1}[L[\bar{w}(x, y, z; t)]].
\end{cases}
\tag{10.1.36}
$$

For the effective computation of these transforms we remark that

$$
L^{-1}[e^{-kpR}] = \delta(t - kR),
\tag{10.1.37}
$$

and

$$
L^{-1}[p\,e^{-kpR}] = \dot{\delta}(t - kR),
\tag{10.1.38}
$$

$$
L^{-1}\left[\frac{1}{p}\,e^{-kpR}\right] = \theta(t - kR),
\tag{10.1.38'}
$$

$$
L^{-1}\left[\frac{1}{p^2}\,e^{-kpR}\right] = (t - kR)\,\theta(t - kR) = (t - kR)_+,
\tag{10.1.38''}
$$

$$
L^{-1}\left[\frac{1}{p^3}\,e^{-kpR}\right] = \frac{1}{2}(t - kR)^2\,\theta(t - kR) = \frac{1}{2}(t - kR)^2_+.
\tag{10.1.38'''}
$$

In that case the generalized displacements (10.1.36) become

$$
\bar{u}(x, y, z; t) = \frac{1}{4\pi G}\left\{ L^{-1}\left[L[\bar{X}(x, y, z; t)] * \frac{1}{R}\,e^{-p\frac{R}{c_2}} \right] \right.
$$

$$
+ \rho u_0(x, y, z) * \frac{1}{R}\,\dot{\delta}\left(t - \frac{R}{c_2}\right) + \rho \dot{u}_0(x, y, z) * \frac{1}{R}\,\delta\left(t - \frac{R}{c_2}\right)
$$

$$
+ c_2^2\left\{ L^{-1}\left[L[\bar{X}(x, y, z; t)] * \frac{1}{p^2}\,\frac{\partial^2}{\partial x^2}\left[\frac{1}{R}\left(e^{-p\frac{R}{c_1}} - e^{-p\frac{R}{c_2}} \right) \right] \right] \right.
$$

$$
+ L^{-1}\left[L[\bar{Y}(x, y, z; t)] * \frac{1}{p^2}\,\frac{\partial^2}{\partial x\,\partial y}\left[\frac{1}{R}\left(e^{-p\frac{R}{c_1}} - e^{-p\frac{R}{c_2}} \right) \right] \right]
$$

$$+ L^{-1}\left[L[\bar{Z}(x, y, z; t)] * \frac{1}{p^2} \frac{\partial^2}{\partial z\, \partial x}\left[\frac{1}{R}\left(e^{-t\frac{R}{c_1}} - e^{-t\frac{R}{c_2}} \right)\right]\right]\right]$$

$$+ \rho\varepsilon_v^0(x, y, z) * \frac{\partial}{\partial x}\left\{ \frac{1}{R}\left[\theta\left(t - \frac{R}{c_1} \right) - \theta\left(t - \frac{R}{c_2} \right)\right]\right\}$$

$$+ \rho\dot{\varepsilon}_v^0(x, y, z) * \frac{\partial}{\partial x}\left\{ \frac{1}{R}\left[\left(t - \frac{R}{c_1} \right)_+ - \left(t - \frac{R}{c_2} \right)_+\right]\right\}\right\}\right\},$$

$$\bar{v}(x, y, z; t) = \frac{1}{4\pi G}\left\{ L^{-1}\left[L[\bar{Y}(x, y, z; t)] * \frac{1}{R} e^{-p\frac{R}{c_2}}\right]\right.$$

$$+ \rho v_0(x, y, z) * \frac{1}{R}\dot{\delta}\left(t - \frac{R}{c_2} \right) + \rho\dot{v}_0(x, y, z) * \frac{1}{R}\delta\left(t - \frac{R}{c_2} \right)$$

$$+ c_2^2\left\{ L^{-1}\left[L[\bar{X}(x, y, z; t)] * \frac{1}{p^2} \frac{\partial}{\partial x\, \partial y}\left[\frac{1}{R}\left(e^{-p\frac{R}{c_1}} - e^{-p\frac{R}{c_2}} \right)\right]\right]\right]$$

$$+ L^{-1}\left[L[\bar{Y}(x, y, z; t)] * \frac{1}{p^2} \frac{\partial^2}{\partial y^2}\left[\frac{1}{R}\left(e^{-p\frac{R}{c_1}} - e^{-p\frac{R}{c_2}} \right)\right]\right]\right] \qquad (10.1.36')$$

$$+ L^{-1}\left[L[\bar{Z}(x, y, z; t)] * \frac{1}{p^2} \frac{\partial^2}{\partial y\, \partial z}\left[\frac{1}{R}\left(e^{-p\frac{R}{c_1}} - e^{-p\frac{R}{c_2}} \right)\right]\right]\right]$$

$$+ \rho\varepsilon_v^0(x, y, z) * \frac{\partial}{\partial y}\left\{ \frac{1}{R}\left[\theta\left(t - \frac{R}{c_1} \right) - \theta\left(t - \frac{R}{c_2} \right)\right]\right\}$$

$$+ \rho\dot{\varepsilon}_v^0(x, y, z) * \frac{\partial}{\partial y}\left\{ \frac{1}{R}\left[\left(t - \frac{R}{c_1} \right)_+ - \left(t - \frac{R}{c_2} \right)_+\right]\right\}\right\}\right\},$$

$$\bar{w}(x, y, z; t) = \frac{1}{4\pi G}\left\{ L^{-1}\left[L[\bar{Z}(x, y, z; t)] * \frac{1}{R} e^{-p\frac{R}{c_2}}\right]\right.$$

$$+ \rho w_0(x, y, z) * \frac{1}{R}\dot{\delta}\left(t - \frac{R}{c_2} \right) + \rho\dot{w}_0(x, y, z) * \frac{1}{R}\delta\left(-\frac{R}{c_2} \right)$$

$$+ c_2^2\left\{ L^{-1}\left[L[\bar{X}(x, y, z; t)] * \frac{1}{p^2} \frac{\partial^2}{\partial z\, \partial x}\left[\frac{1}{R}\left(e^{-p\frac{R}{c_1}} - e^{-p\frac{R}{c_2}} \right)\right]\right]\right]$$

$$+ L^{-1}\left[L\left[\overline{Y}(x, y, z; t)\right] * \frac{1}{p^2} \frac{\partial^2}{\partial y\, \partial z}\left[\frac{1}{R}\left(e^{-p\frac{R}{c_1}} - e^{-p\frac{R}{c_2}}\right)\right]\right]$$

$$+ L^{-1}\left[L\left[\overline{Z}(x, y, z; t)\right] * \frac{1}{p^2} \frac{\partial^2}{\partial z^2}\left[\frac{1}{R}\left(e^{-p\frac{R}{c_1}} - e^{-p\frac{R}{c_2}}\right)\right]\right]$$

$$+ \rho\varepsilon_v^0(x, y, z) * \frac{\partial}{\partial z}\left\{\frac{1}{R}\left[\theta\left(t - \frac{R}{c_1}\right) - \theta\left(t - \frac{R}{c_2}\right)\right]\right\}$$

$$+ \rho\dot{\varepsilon}_v^0(x, y, z) * \frac{\partial}{\partial z}\left\{\frac{1}{R}\left[\left(t - \frac{R}{c_1}\right)_+ - \left(t - \frac{R}{c_2}\right)_+\right]\right\}\bigg\}\bigg\}.$$

10.1.3.2. Examples

Let there be an elastic space acted on by body forces of the form

$$X(x, y, z; t) = P(t)\, X(x, y, z),$$

$$Y(x, y, z; t) = Z(x, y, z; t) = 0,$$

$$(10.1.39)$$

with the homogeneous initial conditions (zero initial conditions)

$$u_0(x, y, z) = v_0(x, y, z) = w_0(x, y, z) = 0,$$

$$\dot{u}_0(x, y, z) = \dot{v}_0(x, y, z) = \dot{w}_0(x, y, z) = 0;$$

$$(10.1.40)$$

taking into account the properties of the convolution product with respect to differentiation, the Laplace transforms (10.1.35) become

$$L[\overline{u}(x, y, z; t)] = \frac{1}{4\pi\mu} X(x, y, z) * L[\overline{P}(t)]\left\{\frac{1}{R}e^{-p\frac{R}{c_2}}\right.$$

$$+ c_2^2 \frac{1}{p^2} \frac{\partial^2}{\partial x^2}\left[\frac{1}{R}\left(e^{-p\frac{R}{c_1}} - e^{-p\frac{R}{c_2}}\right)\right]\bigg\},$$

$$L[\overline{v}(x, y, z; t)] = \frac{1}{4\pi\rho} X(x, y, z) * \frac{1}{p^2} L[\overline{P}(t)] \frac{\partial^2}{\partial x\, \partial y}\left[\frac{1}{R}\left(e^{-p\frac{R}{c_1}} - e^{-p\frac{R}{c_2}}\right)\right],$$

$$L[\overline{w}(x, y, z; t)] = \frac{1}{4\pi\rho} X(x, y, z) * \frac{1}{p^2} L[\overline{P}(t)] \frac{\partial^2}{\partial x\, \partial z}\left[\frac{1}{R}\left(e^{-p\frac{R}{c_1}} - e^{-p\frac{R}{c_2}}\right)\right],$$

where we have introduced the distribution

$$\bar{P}(t) = \begin{cases} 0 & \text{for } t < 0 \\ P(t) & \text{for } t \geqslant 0. \end{cases} \qquad (10.1.41)$$

Applying the inverse Laplace transform and using relation (10.1.38), we obtain the generalized displacements

$$\bar{u}(x, y, z; t) = \frac{1}{4\pi G} X(x, y, z) * \left\{ \frac{1}{R} \bar{P}\left(t - \frac{R}{c_2}\right) \right.$$

$$\left. + c_2^2 \frac{\partial^2}{\partial x^2} \left\{ \frac{1}{R} \left\{ \bar{P}(t) \underset{(t)}{*} \left[\left(t - \frac{R}{c_1}\right)_+ - \left(t - \frac{R}{c_2}\right)_+ \right] \right\} \right\} \right\},$$

$$\qquad (10.1.42)$$

$$\bar{v}(x, y, z; t) = \frac{1}{4\pi\rho} X(x, y, z) * \frac{\partial^2}{\partial x \, \partial y} \left\{ \frac{1}{R} \left\{ \bar{P}(t) \underset{(t)}{*} \left[\left(t - \frac{R}{c_1}\right)_+ - \left(t - \frac{R}{c_2}\right)_+ \right] \right\} \right\},$$

$$\bar{w}(x, y, z; t) = \frac{1}{4\pi\rho} X(x, y, z) * \frac{\partial^2}{\partial x \, \partial z} \left\{ \frac{1}{R} \left\{ \bar{P}(t) \underset{(t)}{*} \left[\left(t - \frac{R}{c_1}\right)_+ - \left(t - \frac{R}{c_2}\right)_+ \right] \right\} \right\}.$$

In particular, it will be assumed that

$$X(x, y, z) = \delta(x, y, z), \qquad (10.1.43)$$

which constitutes the case of a concentrated force variable in time; the generalized displacements become

$$\bar{u}(x, y, z; t) = \frac{1}{4\pi G} \left\{ \frac{1}{R} \bar{P}\left(t - \frac{R}{c_2}\right) + c_2^2 \frac{\partial^2}{\partial x^2} \left\{ \frac{1}{R} \left\{ \bar{P}(t) \underset{(t)}{*} \left[\left(t - \frac{R}{c_1}\right)_+ - \left(t - \frac{R}{c_2}\right)_+ \right] \right\} \right\} \right\},$$

$$\bar{v}(x, y, z; t) = \frac{1}{4\pi\rho} \frac{\partial^2}{\partial x \, \partial y} \left\{ \frac{1}{R} \left\{ \bar{P}(t) \underset{(t)}{*} \left[\left(t - \frac{R}{c_1}\right)_+ - \left(t - \frac{R}{c_2}\right)_+ \right] \right\} \right\}, \quad (10.1.44)$$

$$\bar{w}(x, y, z; t) = \frac{1}{4\pi\rho} \frac{\partial^2}{\partial x \, \partial z} \left\{ \frac{1}{R} \left\{ \bar{P}(t) \underset{(t)}{*} \left[\left(t - \frac{R}{c_1}\right)_+ - \left(t - \frac{R}{c_2}\right)_+ \right] \right\} \right\}.$$

This result could have been obtained also from the representation of Schaefer type given in sections 7.2.4.3 and 7.2.4.4. Formula (7.2.83) leads to

$$\Phi_x = -\frac{1}{4\pi R}\,\overline{P}\!\left(t - \frac{R}{c_2}\right),$$

$$\Phi_y = \Phi_z = 0 \tag{10.1.45}$$

and formula (7.2.85) gives

$$\Omega = \frac{1}{2\pi}\,c_2^2 \overline{P}(t)\underset{(t)}{*}\frac{\partial}{\partial x}\left\{\frac{1}{R}\left[\left(t - \frac{R}{c_1}\right)_+ - \left(t - \frac{R}{c_2}\right)_+\right]\right\}; \tag{10.1.45'}$$

formula (7.2.72) leads to the state of displacement (10.1.44) while the state of stress is given by formulae (7.2.71) and (7.2.71').

Considering $P(t)$ as a function in the usual sense, G. G. Stokes obtained this result in 1849 in the form

$$u(x, y, z; t) = \frac{1}{4\pi\rho R}\left[\frac{1}{c_1^2}\frac{x^2}{R^2}P\!\left(t - \frac{R}{c_1}\right) + \frac{1}{c_2^2}\left(1 - \frac{x^2}{R^2}\right)P\!\left(t - \frac{R}{c_2}\right)\right.$$

$$\left. - \left(1 - 3\frac{x^2}{R^2}\right)\int_{1/c_1}^{1/c_2}\lambda P(t - \lambda R)\,\mathrm{d}\lambda\right],$$

$$\tag{10.1.44'}$$

$$v(x, y, z; t) = \frac{xy}{4\pi\rho R}\left[\frac{1}{c_1^2}P\!\left(t - \frac{R}{c_1}\right) - \frac{1}{c_2^2}P\!\left(t - \frac{R}{c_2}\right) + 3\int_{1/c_1}^{1/c_2}\lambda P(t - \lambda R)\,\mathrm{d}\lambda\right],$$

$$w(x, y, z; t) = \frac{xz}{4\pi\rho R}\left[\frac{1}{c_1^2}P\!\left(t - \frac{R}{c_1}\right) - \frac{1}{c_2^2}P\!\left(t - \frac{R}{c_2}\right) + 3\int_{1/c_1}^{1/c_2}\lambda P(t - \lambda R)\,\mathrm{d}\lambda\right].$$

Also, if we put in formulae (10.1.42)

$$\overline{P}(t) = P\delta(t), \tag{10.1.46}$$

which constitutes a *shock at the initial moment*, we obtain the state of generalized displacement

$$\overline{u}(x, y, z; t) = \frac{P}{4\pi G}\,X(x, y, z) * \left\{\frac{1}{R}\delta\!\left(t - \frac{R}{c_2}\right)\right.$$

$$\left. + c_2^2\frac{\partial^2}{\partial x^2}\left\{\frac{1}{R}\left[\left(t - \frac{R}{c_1}\right)_+ - \left(t - \frac{R}{c_2}\right)_+\right]\right\}\right\},$$

$$\tag{10.1.47}$$

$$\overline{v}(x, y, z; t) = \frac{P}{4\pi\rho}\,X(x, y, z) * \frac{\partial^2}{\partial x\,\partial y}\left\{\frac{1}{R}\left[\left(t - \frac{R}{c_1}\right)_+ - \left(t - \frac{R}{c_2}\right)_+\right]\right\},$$

$$\overline{w}(x, y, z; t) = \frac{P}{4\pi\rho}\,X(x, y, z) * \frac{\partial^2}{\partial x\,\partial z}\left\{\frac{1}{R}\left[\left(t - \frac{R}{c_1}\right)_+ - \left(t - \frac{R}{c_2}\right)_+\right]\right\}.$$

Similarly, in the case of a concentrated force applied in the form of a shock at the initial moment, at the origin of the co-ordinate axes and in the direction of the Ox axis, we obtain

$$\bar{u}(x, y, z; t) = \frac{P}{4\pi G} \left\{ \frac{2}{R} \delta\left(t - \frac{R}{c_2}\right) + c_2^2 \frac{\partial^2}{\partial x^2} \left\{ \frac{1}{R} \left[\left(t - \frac{R}{c_1}\right)_+ - \left(t - \frac{R}{c_2}\right)_+ \right] \right\} \right\},$$

$$\bar{v}(x, y, z; t) = \frac{P}{4\pi\rho} \frac{\partial^2}{\partial x\, \partial y} \left\{ \frac{1}{R} \left[\left(t - \frac{R}{c_1}\right)_+ - \left(t - \frac{R}{c_2}\right)_+ \right] \right\}, \qquad (10.1.48)$$

$$\bar{w}(x, y, z; t) = \frac{P}{4\pi\rho} \frac{\partial^2}{\partial x\, \partial z} \left\{ \frac{1}{R} \left[\left(t - \frac{R}{c_1}\right)_+ - \left(t - \frac{R}{c_2}\right)_+ \right] \right\}.$$

Finally, if we *apply suddenly a concentrated force* at the origin of the co-ordinate axes, in the direction of the Ox axis, *at the initial moment, and then maintain the concentrated force constant in time*, we may write

$$\bar{P}(t) = P\theta(t); \qquad (10.1.49)$$

we thus obtain the generalized displacements

$$\bar{u}(x, y, z; t) = \frac{P}{8\pi G} \left\{ \frac{2}{R} \theta\left(t - \frac{R}{c_2}\right) + c_2^2 \frac{\partial^2}{\partial x^2} \left\{ \frac{1}{R} \left[\left(t - \frac{R}{c_1}\right)_+^2 - \left(t - \frac{R}{c_2}\right)_+^2 \right] \right\} \right\},$$

$$\bar{v}(x, y, z; t) = \frac{P}{8\pi\rho} \frac{\partial^2}{\partial x\, \partial y} \left\{ \frac{1}{R} \left[\left(t - \frac{R}{c_1}\right)_+^2 - \left(t - \frac{R}{c_2}\right)_+^2 \right] \right\}, \qquad (10.1.50)$$

$$\bar{w}(x, y, z; t) = \frac{P}{8\pi\rho} \frac{\partial^2}{\partial x\, \partial z} \left\{ \frac{1}{R} \left[\left(t - \frac{R}{c_1}\right)_+^2 - \left(t - \frac{R}{c_2}\right)_+^2 \right] \right\}.$$

We shall now consider an elastic space acted on by a rotational concentrated moment of magnitude $M(t)$, applied at the origin of the co-ordinate axes, in a plane whose normal is $\mathbf{n}(\cos \gamma_1, \cos \gamma_2, \cos \gamma_3)$; also, we assume homogeneous initial conditions. The load equivalent to the rotational concentrated moment expressed in vectorial notation is

$$\mathbf{Q}(x, y, z; t) = -\frac{1}{2} M(t)\, \mathbf{n} \times \operatorname{grad} \delta(x, y, z) \qquad (10.1.51)$$

and its components are

$$X(x, y, z; t) = \frac{1}{2} M(t) \left[\cos \gamma_3 \frac{\partial}{\partial y} \delta(x, y, z) - \cos \gamma_2 \frac{\partial}{\partial z} \delta(x, y, z) \right],$$

$$Y(x, y, z; t) = \frac{1}{2} M(t) \left[\cos \gamma_1 \frac{\partial}{\partial z} \delta(x, y, z) - \cos \gamma_3 \frac{\partial}{\partial x} \delta(x, y, z) \right], \quad (10.1.51')$$

$$Z(x, y, z; t) = \frac{1}{2} M(t) \left[\cos \gamma_2 \frac{\partial}{\partial x} \delta(x, y, z) - \cos \gamma_1 \frac{\partial}{\partial y} \delta(x, y, z) \right].$$

Remarking that

$$\frac{\partial}{\partial x} L[\bar{X}(x, y, z; t)] + \frac{\partial}{\partial y} L[\bar{Y}(x, y, z; t)] + \frac{\partial}{\partial z} L[\bar{Z}(x, y, z; t)] = 0$$

and still assuming homogeneous initial conditions, we obtain the generalized displacement vector in the form

$$\bar{\mathbf{u}}(x, y, z; t) = -\frac{1}{8\pi G} \mathbf{n} \times \operatorname{grad} \left[\frac{1}{R} \bar{M} \left(t - \frac{R}{c_2} \right) \right]; \quad (10.1.52)$$

its components are given by

$$\bar{u}(x, y, z; t) = \frac{1}{8\pi G} \left\{ \cos \gamma_3 \frac{\partial}{\partial y} \left[\frac{1}{R} \bar{M} \left(t - \frac{R}{c_2} \right) \right] \right.$$

$$\left. - \cos \gamma_2 \frac{\partial}{\partial z} \left[\frac{1}{R} \bar{M} \left(t - \frac{R}{c_2} \right) \right] \right\},$$

$$\bar{v}(x, y, z; t) = \frac{1}{8\pi G} \left\{ \cos \gamma_1 \frac{\partial}{\partial z} \left[\frac{1}{R} \bar{M} \left(t - \frac{R}{c_2} \right) \right] \right.$$

$$\qquad\qquad\qquad\qquad\qquad\qquad\qquad\qquad (10.1.52')$$

$$\left. - \cos \gamma_3 \frac{\partial}{\partial x} \left[\frac{1}{R} \bar{M} \left(t - \frac{R}{c_2} \right) \right] \right\},$$

$$\bar{w}(x, y, z; t) = \frac{1}{8\pi G} \left\{ \cos \gamma_2 \frac{\partial}{\partial x} \left[\frac{1}{R} \bar{M} \left(t - \frac{R}{c_2} \right) \right] \right.$$

$$\left. - \cos \gamma_1 \frac{\partial}{\partial y} \left[\frac{1}{R} \bar{M} \left(t - \frac{R}{c_2} \right) \right] \right\}.$$

In particular, if the moment of magnitude $M(t)$ is acting in the plane Oxy, we obtain the generalized displacements

$$\bar{u}(x, y, z; t) = \frac{1}{8\pi G} \frac{\partial}{\partial y}\left[\frac{1}{R}\bar{M}\left(t - \frac{R}{c_2}\right)\right],$$

$$\bar{v}(x, y, z; t) = -\frac{1}{8\pi G} \frac{\partial}{\partial x}\left[\frac{1}{R}\bar{M}\left(t - \frac{R}{c_2}\right)\right], \qquad (10.1.53)$$

$$\bar{w}(x, y, z; t) = 0.$$

We have introduced the distribution

$$\bar{M}(t) = \begin{cases} 0 & \text{for } t < 0 \\ M(t) & \text{for } t \geqslant 0. \end{cases} \qquad (10.1.54)$$

If

$$\bar{M}(t) = M\delta(t) \qquad (10.1.55)$$

which corresponds to an *initial shock*, we may write

$$\bar{u}(x, y, z; t) = \frac{M}{8\pi G} \frac{\partial}{\partial y}\left[\frac{1}{R}\delta\left(t - \frac{R}{c_2}\right)\right],$$

$$\bar{v}(x, y, z; t) = -\frac{M}{8\pi G} \frac{\partial}{\partial x}\left[\frac{1}{R}\delta\left(t - \frac{R}{c_2}\right)\right], \qquad (10.1.56)$$

$$\bar{w}(x, y, z; t) = 0,$$

while if

$$\bar{M}(t) = M\theta(t), \qquad (10.1.57)$$

which corresponds to a *concentrated moment applied suddenly at the initial moment and then maintained constant in time,* we obtain

$$\bar{u}(x, y, z; t) = \frac{M}{8\pi G} \frac{\partial}{\partial y}\left[\frac{1}{R}\theta\left(t - \frac{R}{c_2}\right)\right],$$

$$\bar{v}(x, y, z; t) = -\frac{M}{8\pi G} \frac{\partial}{\partial x}\left[\frac{1}{R}\theta\left(t - \frac{R}{c_2}\right)\right], \qquad (10.1.58)$$

$$\bar{w}(x, y, z; t) = 0.$$

Let us now consider an elastic space acted on at the origin of the co-ordinate axes by a centre of spatial expansion (a concentrated moment of spatial dipole type), of variable magnitude $D_s(t)$; homogeneous initial conditions are assumed. The load equivalent to this concentrated moment of spatial dipole type may be written

$$\mathbf{Q}(x, y, z; t) = -\frac{1}{3} D_s(t) \operatorname{grad} \delta(x, y, z), \tag{10.1.59}$$

where the components are

$$X(x, y, z; t) = -\frac{1}{3} D_s(t) \frac{\partial}{\partial x} \delta(x, y, z),$$

$$Y(x, y, z; t) = -\frac{1}{3} D_s(t) \frac{\partial}{\partial y} \delta(x, y, z), \tag{10.1.59'}$$

$$Z(x, y, z; t) = -\frac{1}{3} D_s(t) \frac{\partial}{\partial z} \delta(x, y, z).$$

After a few calculations we obtain the generalized displacements

$$\bar{u}(x, y, z; t) = -\frac{1}{12\pi(\lambda + 2\mu)} \frac{\partial}{\partial x} \left[\frac{1}{R} \bar{D}_s\left(t - \frac{R}{c_1}\right) \right],$$

$$\bar{v}(x, y, z; t) = -\frac{1}{12\pi(\lambda + 2\mu)} \frac{\partial}{\partial y} \left[\frac{1}{R} \bar{D}_s\left(t - \frac{R}{c_1}\right) \right], \tag{10.1.60}$$

$$\bar{w}(x, y, z; t) = -\frac{1}{12\pi(\lambda + 2\mu)} \frac{\partial}{\partial z} \left[\frac{1}{R} \bar{D}_s\left(t - \frac{R}{c_1}\right) \right],$$

where we have introduced the distribution

$$\bar{D}_s(t) = \begin{cases} 0 & \text{for } t < 0 \\ \\ D_s(t) & \text{for } t \geqslant 0 ; \end{cases} \tag{10.1.61}$$

in vectorial notation, we have

$$\bar{\mathbf{u}}(x, y, z; t) = -\frac{1}{12\pi(\lambda + 2\mu)} \operatorname{grad} \left[\frac{1}{R} \bar{D}_s\left(t - \frac{R}{c_1}\right) \right]. \tag{10.1.60'}$$

In particular, if

$$\bar{D}_s(t) = D_s \delta(t), \tag{10.1.62}$$

we have to deal with an *initial shock* and there follow the generalized displacements

$$\bar{\mathbf{u}}(x, y, z; t) = -\frac{D_s}{12\pi(\lambda + 2\mu)} \, \text{grad}\left[\frac{1}{R}\delta\left(t - \frac{R}{c_1}\right)\right].$$ (10.1.63)

Also, if

$$\bar{D}_s(t) = D_s\theta(t),$$ (10.1.64)

which corresponds to *a load applied suddenly and then maintained constant in time,* we obtain the generalized displacement

$$\bar{\mathbf{u}}(x, y, z; t) = -\frac{D_s}{12\pi(\lambda + 2\mu)} \, \text{grad}\left[\frac{1}{R}\theta\left(t - \frac{R}{c_1}\right)\right].$$ (10.1.65)

We remark that the case of loading by a centre of spatial dilatation, expressed in the form (10.1.59), is conservative; applying the results obtained in subsection 7.2.4.4, formula (7.2.84) leads to

$$\omega = -\frac{1 - 2v}{12\pi(1 - v)} \frac{1}{R} D\left(t - \frac{R}{c_1}\right),$$ (10.1.66)

while formula (7.2.81) gives the state of displacement (10.1.60′).

Setting in the formulae (10.1.48) the particular value $P = 1$ and writing the formulae corresponding to the cases where the concentrated force is acting in the direction of the Oy axis or of the Oz axis, respectively, we obtain the components of the fundamental solution matrix (7.2.52) in the form

$$u_{11} = \frac{1}{4\pi G}\left\{\frac{1}{R}\delta\left(t - \frac{R}{c_2}\right) + c_2^2\frac{\partial^2}{\partial x^2}\left\{\frac{1}{R}\left[\left(t - \frac{R}{c_1}\right)_+ - \left(t - \frac{R}{c_2}\right)_+\right]\right\}\right\},$$

$$u_{22} = \frac{1}{4\pi G}\left\{\frac{1}{R}\delta\left(t - \frac{R}{c_2}\right) + c_2^2\frac{\partial^2}{\partial y^2}\left\{\frac{1}{R}\left[\left(t - \frac{R}{c_1}\right)_+ - \left(t - \frac{R}{c_2}\right)_+\right]\right\}\right\},$$ (10.1.67)

$$u_{33} = \frac{1}{4\pi G}\left\{\frac{1}{R}\delta\left(t - \frac{R}{c_2}\right) + c_2^2\frac{\partial^2}{\partial z^2}\left\{\frac{1}{R}\left[\left(t - \frac{R}{c_1}\right)_+ - \left(t - \frac{R}{c_2}\right)_+\right]\right\}\right\},$$

$$u_{23} = u_{32} = \frac{1}{4\pi\rho}\frac{\partial^2}{\partial y\,\partial z}\left\{\frac{1}{R}\left[\left(t - \frac{R}{c_1}\right)_+ - \left(t - \frac{R}{c_2}\right)_+\right]\right\},$$

$$u_{31} = u_{13} = \frac{1}{4\pi\rho}\frac{\partial^2}{\partial z\,\partial x}\left\{\frac{1}{R}\left[\left(t - \frac{R}{c_1}\right)_+ - \left(t - \frac{R}{c_2}\right)_+\right]\right\},$$ (10.1.67′)

$$u_{12} = u_{21} = \frac{1}{4\pi\rho}\frac{\partial^2}{\partial x\,\partial y}\left\{\frac{1}{R}\left[\left(t - \frac{R}{c_1}\right)_+ - \left(t - \frac{R}{c_2}\right)_+\right]\right\}.$$

Applying the considerations of subsection 7.2.3.2, we may obtain with the above formulae the state of displacement of an elastic body for an arbitrary dynamic loading. Obviously, for the calculation of the various convolutions which may occur, computers are very useful.

10.2. Elastic half-space

Let $z \geqslant 0$ be an elastic half-space acted on by the tangential loads $p_x(x, y)$, $p_y(x, y)$ and a normal load $q(x, y)$, expressed in terms of distributions (Figure 10.5). We shall

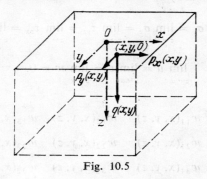

Fig. 10.5

show how to obtain the fundamental solution and then how to use it for obtaining the solution corresponding to the case of an arbitrary loading. Also, we shall consider the corresponding dynamic problem.

10.2.1. Fundamental solution in the static case

In the absence of body forces the system of Lamé's equations (7.2.20) takes the form

$$\mu \Delta u(x, y, z) + (\lambda + \mu) \frac{\partial}{\partial x} \varepsilon_v(x, y, z) = 0,$$

$$\mu \Delta v(x, y, z) + (\lambda + \mu) \frac{\partial}{\partial y} \varepsilon_v(x, y, z) = 0, \tag{10.2.1}$$

$$\mu \Delta w(x, y, z) + (\lambda + \mu) \frac{\partial}{\partial z} \varepsilon_v(x, y, z) = 0;$$

the boundary conditions on the separation plane are

$$\lim_{z \to +0} \tau_{zx} = \mu \lim_{z \to +0} \left(\frac{\partial w}{\partial x} + \frac{\partial u}{\partial z} \right) = - p_x(x, y),$$

$$\lim_{z \to +0} \tau_{xy} = \mu \lim_{z \to +0} \left(\frac{\partial v}{\partial z} + \frac{\partial w}{\partial y} \right) = - p_y(x, y),$$

(10.2.2)

$$\lim_{z \to +0} \sigma_z = \lim_{z \to +0} \left(\lambda \theta + 2\mu \frac{\partial w}{\partial z} \right) = - q(x, y)$$

(10.2.2′)

and the conditions of regularity at infinity are

$$\lim_{z \to \infty} \sigma_x = \lim_{z \to \infty} \sigma_y = \lim_{z \to \infty} \sigma_z = \lim_{z \to \infty} \tau_{yz} = \lim_{z \to \infty} \tau_{zx} = \lim_{z \to \infty} \tau_{xy} = 0,$$

(10.2.3)

$$\lim_{z \to \infty} u = \lim_{z \to \infty} v = \lim_{z \to \infty} w = 0.$$

(10.2.3′)

The matrix

$$(W) \equiv \begin{pmatrix} w_{11}(x, y, z) & w_{12}(x, y, z) & w_{13}(x, y, z) \\ w_{21}(x, y, z) & w_{22}(x, y, z) & w_{23}(x, y, z) \\ w_{31}(x, y, z) & w_{32}(x, y, z) & w_{33}(x, y, z) \end{pmatrix},$$

(10.2.4)

whose elements satisfy the systems of differential equations

$$\mu \Delta w_{11} + (\lambda + \mu) \frac{\partial \varepsilon_v'}{\partial x} = 0,$$

$$\mu \Delta w_{21} + (\lambda + \mu) \frac{\partial \varepsilon_v'}{\partial y} = 0,$$

(10.2.5)

$$\mu \Delta w_{31} + (\lambda + \mu) \frac{\partial \varepsilon_v'}{\partial z} = 0,$$

$$\mu \Delta w_{12} + (\lambda + \mu) \frac{\partial \varepsilon_v''}{\partial x} = 0,$$

$$\mu \Delta w_{22} + (\lambda + \mu) \frac{\partial \varepsilon_v''}{\partial y} = 0,$$

(10.2.5′)

$$\mu \Delta w_{32} + (\lambda + \mu) \frac{\partial \varepsilon_v''}{\partial z} = 0,$$

$$\mu \Delta w_{13} + (\lambda + \mu) \frac{\partial \varepsilon_v'''}{\partial x} = 0,$$

$$\mu \Delta w_{23} + (\lambda + \mu) \frac{\partial \varepsilon_v'''}{\partial y} = 0, \qquad (10.2.5'')$$

$$\mu \Delta w_{33} + (\lambda + \mu) \frac{\partial \varepsilon_v'''}{\partial z} = 0,$$

with the boundary conditions

$$\lim_{z \to +0} \left(\frac{\partial w_{31}}{\partial x} + \frac{\partial w_{11}}{\partial z} \right) = - \frac{1}{\mu} \delta(x, y),$$

$$\lim_{z \to +0} \left(\frac{\partial w_{21}}{\partial z} + \frac{\partial w_{31}}{\partial y} \right) = 0, \qquad (10.2.6)$$

$$\lim_{z \to +0} \left(\lambda \varepsilon_v' + 2\mu \frac{\partial w_{31}}{\partial z} \right) = 0,$$

$$\lim_{z \to +0} \left(\frac{\partial w_{32}}{\partial x} + \frac{\partial w_{12}}{\partial z} \right) = 0,$$

$$\lim_{z \to +0} \left(\frac{\partial w_{22}}{\partial z} + \frac{\partial w_{32}}{\partial y} \right) = - \frac{1}{\mu} \delta(x, y), \qquad (10.2.6')$$

$$\lim_{z \to +0} \left(\lambda \varepsilon_v'' + 2\mu \frac{\partial w_{32}}{\partial z} \right) = 0,$$

$$\lim_{z \to +0} \left(\frac{\partial w_{33}}{\partial x} + \frac{\partial w_{13}}{\partial z} \right) = 0,$$

$$\lim_{z \to +0} \left(\frac{\partial w_{23}}{\partial z} + \frac{\partial w_{33}}{\partial y} \right) = 0, \qquad (10.2.6'')$$

$$\lim_{z \to +0} \left(\lambda \varepsilon_v''' + 2\mu \frac{\partial w_{33}}{\partial z} \right) = - \delta(x, y)$$

on the separation plane, is termed the fundamental solution of the system of equations (10.2.1) with the boundary conditions (10.2.2) — (10.2.3′); the following notations have been introduced

$$\varepsilon'_v = \frac{\partial w_{11}}{\partial x} + \frac{\partial w_{21}}{\partial y} + \frac{\partial w_{31}}{\partial z},$$

$$\varepsilon''_v = \frac{\partial w_{12}}{\partial x} + \frac{\partial w_{22}}{\partial y} + \frac{\partial w_{32}}{\partial z}, \tag{10.2.7}$$

$$\varepsilon'''_v = \frac{\partial w_{13}}{\partial x} + \frac{\partial w_{23}}{\partial y} + \frac{\partial w_{33}}{\partial z},$$

to which must be added the conditions of regularity at infinity

$$\lim_{z \to \infty} w_{ij} = 0 \qquad (i, j = 1, 2, 3). \tag{10.2.8}$$

In order to obtain the fundamental solution we shall apply the double Fourier transform with respect to the variables x and y, expressed in terms of distributions.

10.2.1.1. Fourier transform

Applying the double Fourier transform with respect to the variables x and y and considering z as a parameter, we may write equations (10.2.5) in the form

$$\mu\left[\frac{\mathrm{d}^2}{\mathrm{d}z^2} F[w_{11}(x, y, z)] - (\alpha^2 + \beta^2)F[w_{11}(x, y, z)] \right]$$

$$+ (\lambda + \mu)\alpha\left(-\alpha F[w_{11}(x, y, z)] - \beta F[w_{21}(x, y, z)] - i\frac{\mathrm{d}}{\mathrm{d}z}F[w_{31}(x, y, z)] \right) = 0,$$

$$\mu\left[\frac{\mathrm{d}^2}{\mathrm{d}z^2} F[w_{21}(x, y, z)] - (\alpha^2 + \beta^2)F[w_{21}(x, y, z)] \right]$$

$$\tag{10.2.9}$$

$$+ (\lambda + \mu)\beta\left(-\alpha F[w_{11}(x, y, z)] - \beta F[w_{21}(x, y, z)] - i\frac{\mathrm{d}}{\mathrm{d}z}F[w_{31}(x, y, z)] \right) = 0,$$

$$\mu\left[\frac{\mathrm{d}^2}{\mathrm{d}z^2} F[w_{31}(x, y, z)] - (\alpha^2 + \beta^2)F[w_{31}(x, y, z)] \right]$$

$$+ (\lambda + \mu)\frac{\mathrm{d}}{\mathrm{d}z}\left(-i\alpha F[w_{11}(x, y, z)] - i\beta F[w_{21}(x, y, z)] + \frac{\mathrm{d}}{\mathrm{d}z}F[w_{31}(x, y, z)] \right) = 0,$$

where $F[\ ] = F_y[F_x[\ ]]$ and α and β are complex variables; applying the Fourier transform to the boundary conditions (10.2.6) as well, we obtain

$$\mu \lim_{z \to +0} \left(\frac{\mathrm{d}}{\mathrm{d}z} F[w_{11}(x, y, z)] - \mathrm{i}\,\alpha F[w_{31}(x, y, z)] \right) = -1,$$

$$\lim_{z \to +0} \left(\frac{\mathrm{d}}{\mathrm{d}z} F[w_{21}(x, y, z)] - \mathrm{i}\,\beta F[w_{31}(x, y, z)] \right) = 0,$$

(10.2.10)

$$2\mu \lim_{z \to +0} \frac{\mathrm{d}}{\mathrm{d}z} F[w_{31}(x, y, z)] + \lambda \lim_{z \to +0} \left(-\mathrm{i}\,\alpha F[w_{11}(x, y, z)] \right.$$

$$\left. - \mathrm{i}\,\beta F[w_{21}(x, y, z)] + \frac{\mathrm{d}}{\mathrm{d}z} F[w_{31}(x, y, z)] \right) = 0.$$

Multiplying the first two equations (10.2.9) by α and β, respectively, and adding we have

$$\mu \frac{\mathrm{d}^2}{\mathrm{d}z^2} (\alpha F[w_{11}(x, y, z)] + \beta F[w_{21}(x, y, z)]) - (\lambda + 2\mu)(\alpha^2 + \beta^2)(\alpha F[w_{11}(x, y, z)]$$

$$+ \beta F[w_{21}(x, y, z)]) = \mathrm{i}(\lambda + \mu)(\alpha^2 + \beta^2) \frac{\mathrm{d}}{\mathrm{d}z} F[w_{31}(x, y, z)]; \qquad (10.2.11)$$

on the other hand the third equation (10.2.9) may be written

$$\frac{\mathrm{d}}{\mathrm{d}z} (\alpha F[w_{11}(x, y, z)] + \beta F[w_{21}(x, y, z)])$$

$$= \frac{1}{\mathrm{i}(\lambda + \mu)} \left[(\lambda + 2\mu) \frac{\mathrm{d}^2}{\mathrm{d}z^2} F[w_{31}(x, y, z)] - \mu(\alpha^2 + \beta^2) F[w_{31}(x, y, z)] \right]. \quad (10.2.11')$$

Thus, from equations (10.2.11) and (10.2.11′) we obtain the differential equation which must be satisfied by the Fourier transform $F[w_{31}(x, y, z)]$, namely

$$\frac{\mathrm{d}^4}{\mathrm{d}z^4} F[w_{31}(x, y, z)] - 2(\alpha^2 + \beta^2) \frac{\mathrm{d}^2}{\mathrm{d}z^2} [w_{31}(x, y, z)]$$

$$+ (\alpha^2 + \beta^2)^2 F[w_{31}(x, y, z)] = 0. \qquad (10.2.12)$$

The general solution of the above equation will be of the form

$$F[w_{31}(x, y, z)] = (A + Bz)e^{-\gamma z} + (C + Dz)e^{-\gamma z}, \qquad (10.2.13)$$

where

$$\gamma = \sqrt{\alpha^2 + \beta^2} \qquad (10.2.14)$$

and the coefficients A, B, C, D are dependent on the parameters α and β; the condition of regularity for $z \to \infty$ leads to

$$C = D = 0 \qquad (10.2.15)$$

and therefore expression (10.2.13) becomes

$$F[w_{31}(x, y, z)] = (A + Bz)e^{-\gamma z}. \qquad (10.2.13')$$

Taking into account that

$$\lim_{z \to +0} F[w_{31}(x, y, z)] = A \qquad (10.2.16)$$

and the first two conditions (10.2.10), equation (10.2.11′) gives

$$2\mu(\lambda + \mu)\gamma^2 A = 2\mu(\lambda + 2\mu)\gamma B - i\alpha(\lambda + \mu) \qquad (10.2.16')$$

for $z \to +0$.

Substituting the Fourier transform $F[w_{31}(x, y, z)]$ given by (10.2.13′) into equation (10.2.11′) and integrating, we obtain

$$i(\lambda + \mu)(\alpha F[w_{11}(x, y, z)] + \beta F[w_{21}(x, y, z)])$$

$$= \{2(\lambda + 2\mu)B - (\lambda + \mu)[\gamma A + (1 + \gamma z)B]\}e^{-\gamma z}; \qquad (10.2.17)$$

the integration constant which occurs is zero since

$$\lim_{z \to \infty} (\alpha F[w_{11}(x, y, z)] + \beta F[w_{21}(x, y, z)]) = 0.$$

Passing to the limit for $z \to +0$ in relation (10.2.17) and taking into account the third boundary condition (10.2.10), we obtain

$$\mu B = (\lambda + \mu)\gamma A. \qquad (10.2.16'')$$

Thus, the system of equations (10.2.16′) and (10.2.16″) will give the coefficients A, B in the form

$$A = \frac{i}{2(\lambda + \mu)} \frac{\alpha}{\gamma},$$

$$ \tag{10.2.15′}$$

$$B = \frac{i}{2\mu} \frac{\alpha}{\gamma}.$$

From the first two equations (10.2.9) we obtain the equation

$$\frac{d^2}{dz^2} (\beta F[w_{11}(x, y, z)] - \alpha F[w_{21}(x, y, z)])$$

$$- \gamma^2 (\beta F[w_{11}(x, y, z)] - \alpha F[w_{21}(x, y, z)]) = 0, \tag{10.2.18}$$

whose general solution is of the form

$$\beta F[w_{11}(x, y, z)] - \alpha F[w_{21}(x, y, z)] = Le^{-\gamma z} + Ne^{\gamma z}, \tag{10.2.19}$$

where L and N are the parameters to be determined; from the condition of regularity for $z \to \infty$ we obtain

$$N = 0 \tag{10.2.20}$$

and the first two conditions (10.2.10) give

$$L = \frac{1}{\mu} \frac{\beta}{\gamma}. \tag{10.2.20′}$$

Ultimately, from relations (10.2.17), (10.2.19) and (10.2.13′), and taking into account the coefficients (10.2.15), (10.2.20), (10.2.20′), we may write the Fourier transforms in the form

$$F[w_{11}(x, y, z)] = \frac{1}{\mu\gamma} \left[1 - \frac{\lambda}{2(\lambda + \mu)} \frac{\alpha^2}{\gamma^2} - \frac{1}{2} \frac{\alpha^2}{\gamma} z \right] e^{-\gamma z},$$

$$F[w_{21}(x, y, z)] = -\frac{1}{2\mu} \frac{\alpha\beta}{\gamma^2} \left[\frac{\lambda}{(\lambda + \mu)\gamma} + z \right] e^{-\gamma z}, \tag{10.2.21}$$

$$F[w_{31}(x, y, z)] = \frac{i}{2} \frac{\alpha}{\gamma} \left[\frac{1}{(\lambda + \mu)\gamma} + \frac{1}{\mu} z \right] e^{-\gamma z}.$$

10.2.1.2. Inverse Fourier transform

For the computation of the inverse Fourier transform we remark that

$$F_x\left[\frac{1}{R}\right] = 2K_0(\alpha\sqrt{y^2 + z^2}),$$ (10.2.22)

where $K_0(\;)$ is the modified Bessel function of the second species and zero order; then we have

$$\frac{1}{\pi}F_y[K_0(\alpha\sqrt{y^2 + z^2})] = \frac{1}{\gamma}e^{-\gamma z}, \qquad \text{Re } \alpha > 0,$$ (10.2.22′)

whence

$$\frac{1}{2\pi}F\left[\frac{1}{R}\right] = \frac{1}{\gamma}e^{-\gamma z}, \qquad z > 0.$$ (10.2.23)

On the other hand, we may write

$$F\left[\frac{\partial}{\partial x}\left(\frac{1}{R}\right)\right] = -i\alpha F\left[\frac{1}{R}\right] = -2\pi i\frac{\alpha}{\gamma}e^{-\gamma z}.$$ (10.2.24)

Since

$$\int_z^\infty \frac{1}{\gamma}e^{-\gamma z}\,dz = \frac{1}{\gamma^2}e^{-\gamma z},$$

there follows

$$F\left[\int_z^\infty \frac{\partial}{\partial x}\left(\frac{1}{R}\right)dz\right] = \int_z^\infty F\left[\frac{\partial}{\partial x}\left(\frac{1}{R}\right)\right]dz = -2\pi i\frac{\alpha}{\gamma^2}e^{-\gamma z}.$$ (10.2.25)

In the following we shall express the inverse Fourier transforms with the aid of the functions $\frac{1}{R}$ and $\log(R + z)$, $z > 0$, between which there exists the differential relation

$$\frac{\partial}{\partial z}\log(R + z) = \frac{1}{R};$$ (10.2.26)

also, we may write

$$\int_z^\infty \frac{1}{R}\,dz = -\log(R + z),$$ (10.2.26′)

a relation to which partial differentiation with respect to x and y can be applied.

Thus, we may write the double Fourier transforms

$$\frac{1}{2\pi} F\left[\frac{\partial^2}{\partial x^2} \log(R+z)\right] = \frac{\alpha^2}{\gamma^2} e^{-\gamma z}, \qquad (10.2.27)$$

$$\frac{1}{2\pi} F\left[\frac{\partial^2}{\partial x \partial y} \log(R+z)\right] = \frac{\alpha\beta}{\gamma^2} e^{-\gamma z}, \qquad (10.2.27')$$

whence we deduce also that

$$\frac{1}{2\pi} F\left[\int_z^\infty \frac{\partial^2}{\partial x^2} \log(R+z)\,dz\right] = \frac{\alpha^2}{\gamma^3} e^{-\gamma z}, \qquad (10.2.28)$$

$$\frac{1}{2\pi} F\left[\int_z^\infty \frac{\partial^2}{\partial x \partial y} \log(R+z)\,dz\right] = \frac{\alpha\beta}{\gamma^3} e^{-\gamma z}. \qquad (10.2.28')$$

Finally, we obtain

$$w_{11}(x, y, z) = \frac{1}{4\pi\mu}\left[\frac{2}{R} - \frac{\lambda}{\lambda+\mu}\int_z^\infty \frac{\partial^2}{\partial x^2} \log(R+z)\,dz - z\frac{\partial^2}{\partial x^2} \log(R+z)\right]$$

$$= \frac{1}{4\pi G}\left\{\frac{1}{R}\left(1 + \frac{x^2}{R^2}\right) + (1-2v)\frac{1}{R+z}\left[1 - \frac{x^2}{R(R+z)}\right]\right\},$$

$$w_{21}(x, y, z) = -\frac{1}{4\pi\mu}\left[\frac{\lambda}{\lambda+\mu}\int_z^\infty \frac{\partial^2}{\partial x \partial y} \log(R+z)\,dz + z\frac{\partial^2}{\partial x \partial y} \log(R+z)\right]$$
$$(10.2.29)$$

$$= \frac{1}{4\pi G}\frac{xy}{R}\left[\frac{1}{R^2} - (1-2v)\frac{1}{(R+z)^2}\right],$$

$$w_{31}(x, y, z) = \frac{1}{4\pi\mu}\left[\frac{\mu}{\lambda+\mu}\frac{\partial}{\partial x} \log(R+z) - z\frac{\partial}{\partial x}\left(\frac{1}{R}\right)\right]$$

$$= \frac{1}{4\pi G}\frac{x}{R}\left[\frac{z}{R^2} + (1-2v)\frac{1}{R+z}\right].$$

Likewise, from equations (10.2.5′) and the conditions on the separation plane (10.2.6′) we obtain

$$w_{12}(x, y, z) = -\frac{1}{4\pi\mu}\left[\frac{\lambda}{\lambda + \mu}\int_z^\infty \frac{\partial^2}{\partial x\,\partial y}\log(R + z)\,dz + z\,\frac{\partial^2}{\partial x\,\partial y}\log(R + z)\right]$$

$$= \frac{1}{4\pi G}\frac{xy}{R}\left[\frac{1}{R^2} - (1 - 2v)\frac{1}{(R + z)^2}\right],$$

$$w_{22}(x, y, z) = \frac{1}{4\pi\mu}\left[\frac{2}{R} - \frac{\lambda}{\lambda + \mu}\int_z^\infty \frac{\partial^2}{\partial y^2}\log(R + z)\,dz - z\,\frac{\partial^2}{\partial y^2}\log(R + z)\right]$$

$$\text{(10.2.29′)}$$

$$= \frac{1}{4\pi G}\left\{\frac{1}{R}\left(1 + \frac{y^2}{R^2}\right) + (1 - 2v)\frac{1}{R + z}\left[1 - \frac{y^2}{R(R + z)}\right]\right\},$$

$$w_{32}(x, y, z) = \frac{1}{4\pi\mu}\left[\frac{\mu}{\lambda + \mu}\frac{\partial}{\partial y}\log(R + z) - z\,\frac{\partial}{\partial y}\left(\frac{1}{R}\right)\right]$$

$$= \frac{1}{4\pi G}\frac{y}{R}\left[\frac{z}{R^2} + (1 - 2v)\frac{1}{R + z}\right],$$

while the equations (10.2.5″) and the boundary conditions (10.2.6″) give

$$w_{13}(x, y, z) = -\frac{1}{4\pi\mu}\left[z\,\frac{\partial}{\partial x}\left(\frac{1}{R}\right) + \frac{\mu}{\lambda + \mu}\frac{\partial}{\partial x}\log(R + z)\right]$$

$$= \frac{1}{4\pi G}\frac{x}{R}\left[\frac{z}{R^2} - (1 - 2v)\frac{1}{R + z}\right],$$

$$w_{23}(x, y, z) = -\frac{1}{4\pi\mu}\left[z\,\frac{\partial}{\partial y}\left(\frac{1}{R}\right) + \frac{\mu}{\lambda + \mu}\frac{\partial}{\partial y}\log(R + z)\right] \qquad \text{(10.2.29″)}$$

$$= \frac{1}{4\pi G}\frac{y}{R}\left[\frac{z}{R^2} - (1 - 2v)\frac{1}{R + z}\right],$$

$$w_{33}(x, y, z) = \frac{1}{4\pi\mu}\left[\frac{\lambda + 2\mu}{\lambda + \mu}\frac{1}{R} - z\,\frac{\partial}{\partial z}\left(\frac{1}{R}\right)\right] = \frac{1}{4\pi G}\frac{1}{R}\left[2(1 - v) + \frac{z^2}{R^2}\right].$$

10.2.1.3. Fundamental solution corresponding to a state of stress

As in section 10.1.1.3 we introduce the matrix

$$
(H) \equiv
\begin{pmatrix}
h_{11}(x, y, z) & h_{12}(x, y, z) & h_{13}(x, y, z) \\
h_{21}(x, y, z) & h_{22}(x, y, z) & h_{23}(x, y, z) \\
h_{31}(x, y, z) & h_{32}(x, y, z) & h_{33}(x, y, z) \\
h_{41}(x, y, z) & h_{42}(x, y, z) & h_{43}(x, y, z) \\
h_{51}(x, y, z) & h_{52}(x, y, z) & h_{53}(x, y, z) \\
h_{61}(x, y, z) & h_{62}(x, y, z) & h_{63}(x, y, z)
\end{pmatrix}
\tag{10.2.30}
$$

which represents the fundamental solution corresponding to a formulation in terms of stresses for the system of equations (7.1.7), (7.2.1) and (7.2.1'), in the absence of body forces.

The components of the above matrix may be written in the form

$$
h_{11}(x, y, z) = \frac{1}{2\pi}\left[\frac{3\lambda + 2\mu}{\lambda + \mu} \frac{\partial}{\partial x}\left(\frac{1}{R}\right) \right.
$$

$$
\left. - \frac{\lambda}{\lambda + \mu} \int_z^\infty \frac{\partial^3}{\partial x^3} \log(R + z)\, dz - z \frac{\partial^3}{\partial x^3} \log(R + z) \right]
$$

$$
= \frac{1}{2\pi}\frac{x}{R}\left\{ \frac{1}{R^2}\left(1 - 2v - 3\frac{x^2}{R^2}\right) - (1 - 2v)\frac{1}{(R + z)^2}\left[3 - \frac{x^2(3R + z)}{R^2(R + z)}\right] \right\},
$$

$$
h_{12}(x, y, z) = \frac{1}{2\pi}\left[\frac{\lambda}{\lambda + \mu} \frac{\partial}{\partial y}\left(\frac{1}{R}\right) \right.
$$

$$
\tag{10.2.31}
$$

$$
\left. - \frac{\lambda}{\lambda + \mu} \int_z^\infty \frac{\partial^3}{\partial x^2 \partial y} \log(R + z)\, dz - z \frac{\partial^3}{\partial x^2 \partial y} \log(R + z) \right]
$$

$$
= \frac{1}{2\pi}\frac{y}{R}\left\{ \frac{1}{R^2}\left(1 - 2v - 3\frac{x^2}{R^2}\right) - (1 - 2v)\frac{1}{(R + z)^2}\left[3 - \frac{x^2(3R + z)}{R^2(R + z)}\right] \right\},
$$

$$
h_{13}(x, y, z) = \frac{1}{2\pi}\left[\frac{\lambda}{\lambda + \mu} \frac{\partial}{\partial z}\left(\frac{1}{R}\right) - \frac{\mu}{\lambda + \mu} \frac{\partial^2}{\partial x^2} \log(R + z) - z \frac{\partial^2}{\partial x^2}\left(\frac{1}{R}\right) \right]
$$

$$
= \frac{1}{2\pi}\frac{z}{R}\left\{ \frac{1}{R^2}\left(1 - 2v - 3\frac{x^2}{R^2}\right) - (1 - 2v)\frac{1}{z(R + z)}\left[1 - \frac{x^2(2R + z)}{R^2(R + z)}\right] \right\},
$$

$$h_{21}(x, y, z) = \frac{1}{2\pi}\left[\frac{\lambda}{\lambda + \mu} \frac{\partial}{\partial x}\left(\frac{1}{R}\right)\right.$$

$$-\frac{\lambda}{\lambda + \mu}\int_z^\infty \frac{\partial^3}{\partial x \partial y^2}\log(R + z)\,dz - z\,\frac{\partial^3}{\partial x \partial y^2}\log(R + z)\left.\right]$$

$$= \frac{1}{2\pi}\frac{x}{R}\left\{\frac{1}{R^2}\left(1 - 2v - 3\frac{y^2}{R^2}\right) - (1 - 2v)\frac{2}{(R + z)^2}\left[3 - \frac{y^2(3R + z)}{R^2(R + z)}\right]\right\},$$

$$h_{22}(x, y, z) = \frac{1}{2\pi}\left[\frac{3\lambda + 2\mu}{\lambda + \mu} \frac{\partial}{\partial y}\left(\frac{1}{R}\right)\right.$$

$$(10.2.31')$$

$$-\frac{\lambda}{\lambda + \mu}\int_z^\infty \frac{\partial^3}{\partial y^3}\log(R + z)\,dz - z\,\frac{\partial^3}{\partial y^3}\log(R + z)\left.\right]$$

$$= \frac{1}{2\pi}\frac{y}{R}\left\{\frac{1}{R^2}\left(1 - 2v - 3\frac{y^2}{R^2}\right) - (1 - 2v)\frac{1}{(R + z)^2}\left[3 - \frac{y^2(3R + z)}{R^2(R + z)}\right]\right\},$$

$$h_{23}(x, y, z) = \frac{1}{2\pi}\left[\frac{\lambda}{\lambda + \mu} \frac{\partial}{\partial z}\left(\frac{1}{R}\right) - \frac{\mu}{\lambda + \mu} \frac{\partial^2}{\partial y^2}\log(R + z) - z\,\frac{\partial^2}{\partial y^2}\left(\frac{1}{R}\right)\right]$$

$$= \frac{1}{2\pi}\frac{z}{R}\left\{\frac{1}{R^2}\left(1 - 2v - 3\frac{y^2}{R^2}\right) - (1 - 2v)\frac{1}{z(R + z)}\left[1 - \frac{y^2(2R + z)}{R^2(R + z)}\right]\right\},$$

$$h_{31}(x, y, z) = -\frac{1}{2\pi}z\frac{\partial^2}{\partial x \partial z}\left(\frac{1}{R}\right) = -\frac{3}{2\pi}\frac{xz^2}{R^5},$$

$$h_{32}(x, y, z) = -\frac{1}{2\pi}z\frac{\partial^2}{\partial y \partial z}\left(\frac{1}{R}\right) = -\frac{3}{2\pi}\frac{yz^2}{R^5}, \qquad (10.2.32)$$

$$h_{33}(x, y, z) = \frac{1}{2\pi}\left[\frac{\partial}{\partial z}\left(\frac{1}{R}\right) - z\frac{\partial^2}{\partial z^2}\left(\frac{1}{R}\right)\right] = -\frac{3}{2\pi}\frac{z^3}{R^5},$$

$$h_{41}(x, y, z) = -\frac{1}{2\pi}z\frac{\partial^2}{\partial x \partial y}\left(\frac{1}{R}\right) = -\frac{3}{2\pi}\frac{xyz}{R^5},$$

$$h_{42}(x, y, z) = \frac{1}{2\pi}\left[\frac{\partial}{\partial z}\left(\frac{1}{R}\right) - z\frac{\partial^2}{\partial y^2}\left(\frac{1}{R}\right)\right] = -\frac{3}{2\pi}\frac{y^2 z}{R^5}, \qquad (10.2.33)$$

$$h_{43}(x, y, z) = -\frac{1}{2\pi}z\frac{\partial^2}{\partial y \partial z}\left(\frac{1}{R}\right) = -\frac{3}{2\pi}\frac{yz^2}{R^5},$$

$$h_{51}(x, y, z) = -\frac{1}{2\pi}\left[\frac{\partial}{\partial z}\left(\frac{1}{R}\right) - z\frac{\partial^2}{\partial x^2}\left(\frac{1}{R}\right)\right] = -\frac{3}{2\pi}\frac{x^2 z}{R^5},$$

$$h_{52}(x, y, z) = -\frac{1}{2\pi}z\frac{\partial^2}{\partial x\partial y}\left(\frac{1}{R}\right) = -\frac{3}{2\pi}\frac{xyz}{R^5}, \qquad (10.2.33')$$

$$h_{53}(x, y, z) = -\frac{1}{2\pi}z\frac{\partial^2}{\partial x\partial z}\left(\frac{1}{R}\right) = -\frac{3}{2\pi}\frac{xz^2}{R^5},$$

$$h_{61}(x, y, z) = \frac{1}{2\pi}\left[\frac{\partial}{\partial y}\left(\frac{1}{R}\right)\right.$$

$$\left. -\frac{\lambda}{\lambda + \mu}\int_z^\infty\frac{\partial^3}{\partial x^2\partial y}\log(R + z)\,dz - z\frac{\partial^3}{\partial x^2\partial y}\log(R + z)\right]$$

$$= \frac{1}{2\pi}\frac{y}{R}\left\{\frac{1}{R^2}\left(1 - 3\frac{x^2}{R^2}\right) - (1 - 2v)\frac{1}{(R + z)^2}\left[1 - \frac{x^2(3R + z)}{R^2(R + z)}\right]\right\},$$

$$h_{62}(x, y, z) = \frac{1}{2\pi}\left[\frac{\partial}{\partial x}\left(\frac{1}{R}\right)\right. \qquad (10.2.34)$$

$$\left. -\frac{\lambda}{\lambda + \mu}\int_z^\infty\frac{\partial^3}{\partial x\partial y^2}\log(R + z)\,dz - z\frac{\partial^3}{\partial x\partial y^2}\log(R + z)\right]$$

$$= \frac{1}{2\pi}\frac{x}{R}\left\{\frac{1}{R^2}\left(1 - 3\frac{y^2}{R^2}\right) - (1 - 2v)\frac{1}{(R + z)^2}\left[1 - \frac{y^2(3R + z)}{R^2(R + z)}\right]\right\},$$

$$h_{63}(x, y, z) = -\frac{1}{2\pi}\left[\frac{\mu}{\lambda + \mu}\frac{\partial^2}{\partial x\partial y}\log(R + z) + z\frac{\partial^2}{\partial x\partial y}\left(\frac{1}{R}\right)\right]$$

$$= -\frac{1}{2\pi}\frac{xy}{R^3}\left[3\frac{z}{R^2} - (1 - 2v)\frac{2R + z}{(R + z)^2}\right].$$

10.2.2. Case of arbitrary static loads

With the aid of matrix (W) of the fundamental solution (corresponding to a solution in terms of displacements) or by using matrix (H) of the fundamental solution (corresponding to a solution in terms of stresses), we can construct the solution corresponding to an arbitrary load acting on the separation plane $z = 0$ by applying the convolution product. Proceeding in this way we shall give results for a few particular cases of loading.

10.2.2.1. Case of arbitrary loading on the separation plane

With the aid of the matrices

$$(\sigma) \equiv \begin{pmatrix} \sigma_x(x, y, z) \\ \sigma_y(x, y, z) \\ \sigma_z(x, y, z) \\ \tau_{yz}(x, y, z) \\ \tau_{zx}(x, y, z) \\ \tau_{xy}(x, y, z) \end{pmatrix}, \quad (u) \equiv \begin{pmatrix} u(x, y, z) \\ v(x, y, z) \\ w(x, y, z) \end{pmatrix}, \quad (Q) \equiv \begin{pmatrix} p_x(x, y) \\ p_y(x, y) \\ q(x, y) \end{pmatrix} \quad (10.2.35)$$

we may write

$$(u) = (W) * (Q), \quad (10.2.36)$$

and

$$(\sigma) = (H) * (Q). \quad (10.2.36')$$

Therefore, the state of displacement for an arbitrary loading on the separation plane is given by

$$u(x, y, z) = w_{11}(x, y, z) * p_x(x, y) + w_{12}(x, y, z) * p_y(x, y)$$
$$+ w_{13}(x, y, z) * q(x, y),$$

$$v(x, y, z) = w_{21}(x, y, z) * p_x(x, y) + w_{22}(x, y, z) * p_y(x, y)$$
$$+ w_{23}(x, y, z) * q(x, y), \quad (10.2.37)$$

$$w(x, y, z) = w_{31}(x, y, z) * p_x(x, y) + w_{32}(x, y, z) * p_y(x, y)$$
$$+ w_{33}(x, y, z) * q(x, y),$$

and the corresponding state of stress is expressed in the form

$$\sigma_x(x, y, z) = h_{11}(x, y, z) * p_x(x, y) + h_{12}(x, y, z) * p_y(x, y)$$

$$+ h_{13}(x, y, z) * q(x, y),$$

$$\sigma_y(x, y, z) = h_{21}(x, y, z) * p_x(x, y) + h_{22}(x, y, z) * p_y(x, y)$$

$$+ h_{23}(x, y, z) * q(x, y),$$

$$\sigma_z(x, y, z) = h_{31}(x, y, z) * p_x(x, y) + h_{32}(x, y, z) * p_y(x, y)$$

$$+ h_{33}(x, y, z) * q(x, y),$$

$$\tau_{yz}(x, y, z) = h_{41}(x, y, z) * p_x(x, y) + h_{42}(x, y, z) * p_y(x, y)$$

$$+ h_{43}(x, y, z) * q(x, y),$$

$$\tau_{zx}(x, y, z) = h_{51}(x, y, z) * p_x(x, y) + h_{52}(x, y, z) * p_y(x, y)$$

$$+ h_{53}(x, y, z) * q(x, y),$$

$$\tau_{xy}(x, y, z) = h_{61}(x, y, z) * p_x(x, y) + h_{62}(x, y, z) * p_y(x, y)$$

$$+ h_{63}(x, y, z) * q(x, y).$$

(10.2.38)

(10.2.38′)

Using the above results we can verify the complete system of equations of the theory of elasticity by effecting all the necessary operations in the sense of the theory of distributions; however, considering the proof given in the preceeding section, such a verification is superfluous.

10.2.2.2. Examples

Let $z \geqslant 0$ be an elastic half-space acted on by a concentrated force $\mathbf{F}(F_x, F_y, F_z)$ at the point $A(x_0, y_0, 0)$ on the separation plane (Figure 10.6); the equivalent load to the concentrated force may be written

$$p_x(x, y) = F_x \delta(x - x_0, y - y_0),$$

$$p_y(x, y) = F_y \delta(x - x_0, y - y_0),$$

$$q(x, y) = F_z \delta(x - x_0, y - y_0).$$

(10.2.39)

Using formulae (10.2.37) we obtain the state of displacement

$$u(x, y, z) = F_x w_{11}(x - x_0, y - y_0, z)$$

$$+ F_y w_{12}(x - x_0, y - y_0, z) + F_z w_{13}(x - x_0, y - y_0, z),$$

$$v(x, y, z) = F_x w_{21}(x - x_0, y - y_0, z)$$

$$+ F_y w_{22}(x - x_0, y - y_0, z) + F_z w_{23}(x - x_0, y - y_0, z), \quad (10.2.40)$$

$$w(x, y, z) = F_x w_{31}(x - x_0, y - y_0, z)$$

$$+ F_y w_{32}(x - x_0, y - y_0, z) + F_z w_{33}(x - x_0, y - y_0, z);$$

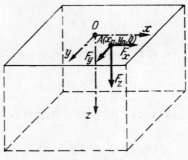

Fig. 10.6

by a similar procedure we obtain the state of stress.

In particular for $F_x = F_y = 0$ we find the solution of J. Boussinesq's problem and for $F_z = 0$ that of V. Cerruti's one.

Let us now consider an elastic half-space acted on at the point $A(x_0, y_0, 0)$ by a directed concentrated moment of magnitude M, specified by the unit vectors $\mathbf{u}(\cos \alpha, \sin \alpha, 0)$ and $\mathbf{F}^0(\cos \beta_1, \cos \beta_2, \cos \beta_3)$ (Figure 10.7a); the equivalent load of the directed concentrated moment is written in the form

$$p_x(x, y) = - \frac{M \cos \beta_1}{|\mathbf{u} \times \mathbf{F}^0|} \frac{\partial}{\partial u} \delta(x - x_0, y - y_0),$$

$$p_y(x, y) = - \frac{M \cos \beta_2}{|\mathbf{u} \times \mathbf{F}^0|} \frac{\partial}{\partial u} \delta(x - x_0, y - y_0), \quad (10.2.41)$$

$$q(x, y) = - \frac{M \cos \beta_3}{|\mathbf{u} \times \mathbf{F}^0|} \frac{\partial}{\partial u} \delta(x - x_0, y - y_0),$$

with the help of the directional derivative given by the operator

$$\frac{\partial}{\partial u} = \cos \alpha \frac{\partial}{\partial x} + \sin \alpha \frac{\partial}{\partial y}. \quad (10.2.42)$$

The state of displacement is expressed in the form

$$u(x, y, z) = - \frac{M}{|\mathbf{u} \times \mathbf{F}^0|} \left[\cos \beta_1 \frac{\partial}{\partial u} w_{11}(x - x_0, y - y_0, z) \right.$$

$$\left. + \cos \beta_2 \frac{\partial}{\partial u} w_{12}(x - x_0, y - y_0, z) + \cos \beta_3 \frac{\partial}{\partial u} w_{13}(x - x_0, y - y_0, z) \right]$$

$$= - \frac{M}{|\mathbf{u} \times \mathbf{F}^0|} \frac{\partial}{\partial u} u_{\mathbf{F}^0},$$

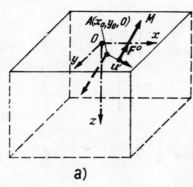 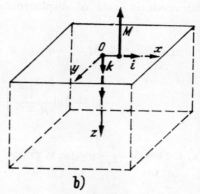

Fig. 10.7

$$v(x, y, z) = - \frac{M}{|\mathbf{u} \times \mathbf{F}^0|} \left[\cos \beta_1 \frac{\partial}{\partial u} w_{21}(x - x_0, y - y_0, z) \right.$$

$$(10.2.43)$$

$$\left. + \cos \beta_2 \frac{\partial}{\partial u} w_{22}(x - x_0, y - y_0, z) + \cos \beta_3 \frac{\partial}{\partial u} w_{23}(x - x_0, y - y_0, z) \right]$$

$$= - \frac{M}{|\mathbf{u} \times \mathbf{F}^0|} \frac{\partial}{\partial u} v_{\mathbf{F}^0},$$

$$w(x, y, z) = - \frac{M}{|\mathbf{u} \times \mathbf{F}^0|} \left[\cos \beta_1 \frac{\partial}{\partial u} w_{31}(x - x_0, y - y_0, z) \right.$$

$$\left. + \cos \beta_2 \frac{\partial}{\partial u} w_{32}(x - x_0, y - y_0, z) + \cos \beta_3 \frac{\partial}{\partial u} w_{33}(x - x_0, y - y_0, z) \right]$$

$$= - \frac{M}{|\mathbf{u} \times \mathbf{F}^0|} \frac{\partial}{\partial u} w_{\mathbf{F}^0},$$

where u_{Fo}, v_{Fo}, w_{Fo} are the displacements corresponding to the action of a concentrated force $\mathbf{F}^0$ of unit magnitude, in accordance with formulae (10.2.40).

In particular if $\mathbf{u} = \mathbf{i}$ and $\mathbf{F}^0 = -\mathbf{k}$, where $\mathbf{i}$ and $\mathbf{k}$ are the unit vectors of the Ox and Oz axes, respectively, we obtain a directed concentrated moment assumed to be concentrated at the origin; the moment will cause a positive rotation in the Ozx plane (Figure 10.7b). The equivalent load is given by

$$p_x(x, y) = 0, \quad p_y(x, y) = 0, \quad q(x, y) = M \frac{\partial}{\partial x} \delta(x, y) \qquad (10.2.44)$$

and the resulting state of displacement is

$$u(x, y, z) = M \frac{\partial}{\partial x} w_{13}(x, y, z) = \frac{M}{4\pi G} \frac{1}{R} \left\{ \frac{z}{R^2} \left(1 - 3 \frac{x^2}{R^2} \right) \right.$$

$$\left. - (1 - 2v) \frac{1}{R + z} \left[1 - \frac{x^2(2R + z)}{R^2(R + z)} \right] \right\},$$

$$\qquad (10.2.45)$$

$$v(x, y, z) = M \frac{\partial}{\partial x} w_{23}(x, y, z) = -\frac{M}{4\pi G} \frac{xy}{R^3} \left[3 \frac{z}{R^2} - (1 - 2v) \frac{2R + z}{(R + z)^2} \right],$$

$$w(x, y, z) = M \frac{\partial}{\partial x} w_{33}(x, y, z) = -\frac{M}{4\pi G} \frac{x}{R^3} \left[2(1 - v) + 3 \frac{z^2}{R^2} \right].$$

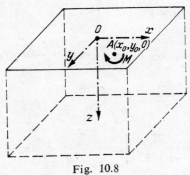

Fig. 10.8

Likewise, let $z \geq 0$ be an elastic half-space acted on at the point $A(x_0, y_0, 0)$ by a rotational concentrated moment of magnitude M, on the separation plane $z = 0$; we assume that the moment causes a positive rotation in that plane (Figure 10.8).

The equivalent load to the rotational concentrated moment may be written in the form

$$p_x(x, y) = \frac{1}{2} M \frac{\partial}{\partial y} \delta(x - x_0, y - y_0),$$

$$(10.2.46)$$

$$p_y(x, y) = -\frac{1}{2} M \frac{\partial}{\partial x} \delta(x - x_0, y - y_0), \quad q(x, y) = 0,$$

and the resulting state of displacement will be

$$u(x, y, z) = \frac{1}{2} M \left[\frac{\partial}{\partial y} w_{11}(x - x_0, y - y_0, z) \right.$$

$$\left. - \frac{\partial}{\partial x} w_{12}(x - x_0, y - y_0, z) \right] = -\frac{M}{4\pi G} \frac{y}{\sqrt{[(x - x_0)^2 + (y - y_0)^2 + z^2]^3}},$$

$$v(x, y, z) = \frac{1}{2} M \left[\frac{\partial}{\partial y} w_{21}(x - x_0, y - y_0, z) \right. \qquad (10.2.47)$$

$$\left. - \frac{\partial}{\partial x} w_{22}(x - x_0, y - y_0, z) \right] = \frac{M}{4\pi G} \frac{x}{\sqrt{[(x - x_0)^2 + (y - y_0)^2 + z^2]^3}},$$

$$w(x, y, z) = \frac{1}{2} M \left[\frac{\partial}{\partial y} w_{31}(x - x_0, y - y_0, z) - \frac{\partial}{\partial x} w_{32}(x - x_0, y - y_0, z) \right] = 0.$$

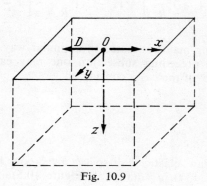

Fig. 10.9

We shall also consider an elastic half-space $z \geqslant 0$ acted on at the origin of the co-ordinate axes on the separation plane by a concentrated moment of the linear dipole type, of magnitude D and having the support in the direction of the Ox axis (Figure 10.9); the equivalent load of the concentrated moment of the linear dipole type is expressed by

$$p_x(x, y) = - D \frac{\partial}{\partial x} \delta(x, y), \; p_y(x, y) = 0, \; q(x, y) = 0 \qquad (10.2.48)$$

and the corresponding state of displacement is obtained in the form

$$u(x, y, z) = - D \frac{\partial}{\partial x} w_{11}(x, y, z)$$

$$= - \frac{D}{4\pi G} \frac{x}{R} \left\{ \frac{1}{R^2} \left(1 - 3 \frac{x^2}{R^2} \right) - (1 - 2v) \frac{1}{(R + z)^2} \left[3 - \frac{x^2(3R + z)}{R^2(R + z)} \right] \right\},$$

$$v(x, y, z) = - D \frac{\partial}{\partial x} w_{21}(x, y, z)$$

$$(10.2.49)$$

$$= - \frac{D}{4\pi G} \frac{y}{R} \left\{ \frac{1}{R^2} \left(1 - 3 \frac{x^2}{R^2} \right) - (1 - 2v) \frac{1}{(R + z)^2} \left[1 - \frac{x^2(3R + z)}{R^2(R + z)} \right] \right\},$$

$$w(x, y, z) = - D \frac{\partial}{\partial x} w_{31}(x, y, z)$$

$$= - \frac{D}{4\pi G} \frac{1}{R} \left\{ \frac{z}{R^2} \left(1 - 3 \frac{x^2}{R^2} \right) + (1 - 2v) \frac{1}{R + z} \left[1 - \frac{x^2(3R + z)}{R^2(R + z)} \right] \right\}.$$

Other cases of loading may be treated in a similar way. Also, by applying the results presented in the preceding subsection one may easily write, in addition, the state of stress in each of the cases considered.

10.2.3. Dynamic case

Let us now consider an elastic half-space acted on by the dynamic loads $p_x = p_x(x, y; t)$, $p_y = p_y(x, y; t)$, $q = q(x, y; t)$ (Figure 10.5); we introduce the functions

$$\bar{p}_x(x, y; t) = \begin{cases} 0 & \text{for } t < 0 \\ p_x(x, y; t) & \text{for } t \geq 0, \end{cases} \qquad (10.2.50)$$

$$\bar{p}_y(x, y; t) = \begin{cases} 0 & \text{for } t < 0 \\ \\ p_y\,(x,y;t) & \text{for } t \geqslant 0, \end{cases} \qquad (10.2.50')$$

$$\bar{q}(x, y; t) = \begin{cases} 0 & \text{for } t < 0 \\ \\ q\,(x,y;t) & \text{for } t \geqslant 0, \end{cases} \qquad (10.2.50'')$$

which lead to the corresponding distributions. The components of the displacement vector too, are replaced by the distributions $\bar{u}(x, y, z; t)$, $\bar{v}(x, y, z; t)$, $\bar{w}(x, y, z; t)$, specified by the formulae (7.2.42).

It is assumed that the body forces vanish

$$X = Y = Z = 0 \qquad (10.2.51)$$

and the initial conditions are homogeneous

$$u_0 = v_0 = w_0 = \dot{u}_0 = \dot{v}_0 = \dot{w}_0 = 0; \qquad (10.2.51')$$

therefore, the derivatives of the generalized displacements in the sense of the theory of distributions are equal to the corresponding derivatives in the ordinary sense.

The general case may be treated in a similar way.

Under the conditions stated, Lamé's equations (7.2.46) become

$$\mu\,\square_2\bar{u} + (\lambda + \mu)\,\frac{\partial\bar{\varepsilon}_v}{\partial x} = 0,$$

$$\mu\,\square_2\bar{v} + (\lambda + \mu)\,\frac{\partial\bar{\varepsilon}_v}{\partial y} = 0, \qquad (10.2.52)$$

$$\mu\,\square_2\bar{w} + (\lambda + \mu)\,\frac{\partial\bar{\varepsilon}_v}{\partial z} = 0,$$

where $\bar{\varepsilon}_v$ is given by formula (7.2.45).

Introducing the generalized stresses $\bar{\sigma}_x$, $\bar{\sigma}_y$, $\bar{\sigma}_z$, $\bar{\tau}_{yz}$, $\bar{\tau}_{zx}$, $\bar{\tau}_{xy}$, by an extension of the form

$$\bar{\sigma}(x, y, z; t) = \begin{cases} 0 & \text{for } t < 0 \\ \\ \sigma\,(x,y,z;t) & \text{for } t \geqslant 0, \end{cases} \qquad (10.2.53)$$

where $\bar{\sigma}$ is an arbitrary component of the generalized stress tensor, we may write the conditions on the separation plane in the form

$$\lim_{z \to +0} \bar{\tau}_{zx} = \mu \lim_{z \to +0} \left(\frac{\partial \overline{w}}{\partial x} + \frac{\partial \overline{u}}{\partial z} \right) = - \bar{p}_x(x, y; t),$$

$$\lim_{z \to +0} \bar{\tau}_{zy} = \mu \lim_{z \to +0} \left(\frac{\partial \overline{w}}{\partial y} + \frac{\partial \overline{v}}{\partial z} \right) = - \bar{p}_y(x, y; t),$$

(10.2.54)

$$\lim_{z \to +0} \bar{\sigma}_z = \lim_{z \to +0} \left(\lambda \bar{\varepsilon}_v + 2\mu \frac{\partial \overline{w}}{\partial z} \right) = - \bar{q}(x, y; t);$$

(10.2.54')

the conditions of regularity at infinity will be

$$\lim_{z \to \infty} (\bar{\sigma}_x, \bar{\sigma}_y, \bar{\sigma}_z, \bar{\tau}_{yz}, \bar{\tau}_{zx}, \bar{\tau}_{xy}) = 0,$$

(10.2.55)

$$\lim_{z \to \infty} (\bar{u}, \bar{v}, \bar{w}) = 0.$$

(10.2.55')

In the following discussion we shall apply the Laplace and Fourier transforms in order to obtain the solution corresponding to the case of an arbitrary loading. Using these results, we shall then consider a particular case of loading.

10.2.3.1. Case of arbitrary surface loads

We shall first apply to equations (10.2.52) the simple Laplace transform with respect to the time variable and then the double Fourier transform with respect to the x and y variables; we obtain

$$\mu \left[\frac{d^2}{dz^2} L[F[\bar{u}(x, y, z; t)]] - (\alpha^2 + \beta^2) L[F[\bar{u}(x, y, z; t)]] \right]$$

$$+ (\lambda + \mu)\alpha \left(-\alpha L[F[\bar{u}(x, y, z; t)]] - \beta L[F[\bar{v}(x, y, z; t)]] \right.$$

$$\left. - i \frac{d}{dz} L[F[\overline{w}(x, y, z; t)]] \right) - \rho p^2 L[F[\bar{u}(x, y, z; t)]] = 0,$$

$$\mu \left[\frac{d^2}{dz^2} L[F[\bar{v}(x, y, z; t)]] - (\alpha^2 + \beta^2) L[F[\bar{v}(x, y, z; t)]] \right]$$

$$+ (\lambda + \mu)\beta \left(-\alpha L[F[\bar{u}(x, y, z; t)]] - \beta L[F[\bar{v}(x, y, z; t)]] \right.$$

(10.2.56)

$$\left. - i \frac{d}{dz} L[F[\overline{w}(x, y, z; t)]] \right) - p p^2 L[F[\bar{v}(x, y, z; t)]] = 0,$$

$$\mu\left[\frac{\mathrm{d}^2}{\mathrm{d}z^2}L[F[\overline{w}(x, y, z; t)]] - (\alpha^2 + \beta^2)L[F[\overline{w}(x, y, z; t)]]\right]$$

$$+ (\lambda + \mu)\frac{\mathrm{d}}{\mathrm{d}z}\left(-i\alpha L[F[\overline{u}(x, y, z; t)]] - i\beta L[F[\overline{v}(x, y, z; t)]]\right.$$

$$\left. + \frac{\mathrm{d}}{\mathrm{d}z}L[F[\overline{w}(x, y, z; t)]]\right) - \rho p^2 L[F[\overline{w}(x, y, z; t)]] = 0,$$

where α, β and p are complex variables corresponding to the variables x, y and t. Also, the boundary conditions (10.2.54) and (10.2.54') become

$$\mu \lim_{z \to +0} (-i\alpha L[F[\overline{w}(x, y, z; t)]] + \frac{\mathrm{d}}{\mathrm{d}z}L[F[\overline{u}(x, y, z; t)]]) = - L[F[\overline{p}_x(x, y; t)]],$$

$$(10.2.57)$$

$$\mu \lim_{z \to +0} \left(-i\beta L[F[\overline{w}(x, y, z; t)]] + \frac{\mathrm{d}}{\mathrm{d}z}L[F[\overline{v}(x, y, z; t)]]\right) = - L[F[\overline{p}_y(x, y; t)]],$$

$$(\lambda + 2\mu)\lim_{z \to +0}\frac{\mathrm{d}}{\mathrm{d}z}L[F[\overline{w}(x, y, z; t)]] - i\lambda \lim_{z \to +0} (\alpha L[F[\overline{u}(x, y, z; t)]]$$

$$+ \beta L[F[\overline{v}(x, y, z; t)]]) = - L[F[\overline{q}(x, y; t)]] \qquad (10.2.57')$$

and the conditions of regularity at infinity become

$$\lim_{z \to \infty} L[F[\overline{\sigma}(x, y, z; t)]] = 0, \qquad (10.2.58)$$

$$\lim_{z \to \infty} L[F[\overline{u}(x, y, z; t)]] = 0, \qquad (10.2.58')$$

where $\overline{\sigma}$ is a component of the stress tensor and $\overline{u}$ is a component of the displacement vector.

Multiplying the first two equations (10.2.56) by α and β, respectively, and adding, we obtain

$$\left\{\mu\frac{\mathrm{d}^2}{\mathrm{d}z^2} - [(\lambda + 2\mu)(\alpha^2 + \beta^2) + \rho p^2]\right\}(\alpha L[F[\overline{u}(x, y, z; t)]]$$

$$+ \beta L[F[\overline{v}(x, y, z; t)]]) = i(\lambda + \mu)(\alpha^2 + \beta^2)\frac{\mathrm{d}}{\mathrm{d}z}L[F[\overline{w}(x, y, z; t)]]; \quad (10.2.59)$$

on the other hand, the third equation of (10.2.56) may be also written

$$i(\lambda + \mu) \frac{d}{dz}(\alpha L[F[\bar{u}(x, y, z; t)]] + \beta L[F[\bar{v}(x, y, z; t)]])$$

$$= \left\{ (\lambda + 2\mu) \frac{d^2}{dz^2} - [\mu(\alpha^2 + \beta^2) + \rho p^2] \right\} L[F[\bar{w}(x, y, z; t)]]. \quad (10.2.59')$$

From the last two equations we obtain an ordinary differential equation

$$\mu(\lambda + 2\mu) \frac{d^4}{dz^4} L[F[\bar{w}(x, y, z; t)]]$$

$$-[2\mu(\lambda + 2\mu)(\alpha^2 + \beta^2) + (\lambda + 3\mu)\rho p^2] \frac{d^2}{dz^2} L[F[\bar{w}(x, y, z; t)]]$$

$$+ [\mu(\alpha^2 + \beta^2) + \rho p^2][(\lambda + 2\mu)(\alpha^2 + \beta^2) + \rho p^2]L[F[\bar{w}(x, y, z; t)]] = 0, \quad (10.2.60)$$

which leads to the general solution

$$L[F[\bar{w}(x, y, z; t)]] = Ae^{-\gamma_1 z} + Be^{-\gamma_2 z} + Ce^{\gamma_1 z} + De^{\gamma_2 z}, \quad (10.2.61)$$

with

$$\gamma_i = \sqrt{\alpha^2 + \beta^2 + \frac{p^2}{c_i^2}} \qquad (i = 1, 2), \quad (10.2.61')$$

where we have introduced the wave propagation velocities (7.2.6).

Using the conditions of regularity (10.2.58′), it follows that

$$C = D = 0 \quad (10.2.62)$$

and we have only to determine the coefficients $A = A(\alpha, \beta; p)$, $B = B(\alpha, \beta; p)$; we remark from (10.2.61) that

$$\lim_{z \to +0} L[F[\bar{w}(x, y, z; t)]] = A + B. \quad (10.2.63)$$

Taking into account (10.2.63), the conditions (10.2.57) lead to

$$i\mu(\alpha^2 + \beta^2)(A + B) - \mu \lim_{z \to +0} \frac{d}{dz}(\alpha L[F[\bar{u}(x, y, z; t)]]$$

$$+ \beta L[F[\bar{v}(x, y, z; t)]]) = \alpha L[F[\bar{p}_x(x, y; t)]] + \beta L[F[\bar{p}_y(x, y; t)]]; \quad (10.2.63')$$

letting $z \to +0$ in equation (10.2.59′) and taking into account conditions (10.2.63) and (10.2.63′), we find that the coefficients A, B must satify the relation

$$(\alpha^2 + \beta^2)A + \gamma_2^2 B = -\frac{i}{2\mu}(\alpha L[F[\bar{p}_x(x, y; t)]] + \beta L[F[\bar{p}_y(x, y; t)]]). \quad (10.2.64)$$

Integrating equation (10.2.59′) with respect to z we obtain

$$i(\lambda + \mu)(\alpha L[F[\bar{u}(x, y, z; t)]] + \beta L[F[\bar{v}(x, y, z; t)]])$$

$$= (\lambda + 2\mu)\frac{d}{dz}L[F[\bar{w}(x, y, z; t)]] + \mu\gamma_2^2\left(\frac{A}{\gamma_1}e^{-\gamma_1 z} + \frac{B}{\gamma_2}e^{-\gamma_2 z}\right), \quad (10.2.65)$$

where the integration constant vanishes owing to the conditions of regularity (10.2.58′).

Condition (10.2.57′) leads to

$$i\lambda \lim_{z \to +0} (\alpha L[F[\bar{u}(x, y, z; t)]] + \beta L[F[\bar{v}(x, y, z; t)]])$$

$$= -(\lambda + 2\mu)(\gamma_1 A + \gamma_2 B) + L[F[\bar{q}(x, y; t)]]; \quad (10.2.57'')$$

for $z \to +0$ in relation (10.2.65) and taking into account relation (10.2.57″), we obtain

$$\frac{1}{\gamma_1}(\alpha^2 + \beta^2 + \gamma_2^2)A + 2\gamma_2 B = \frac{1}{\mu}L[F[\bar{q}(x, y; t)]]. \quad (10.2.64')$$

Thus, we obtain the coefficients A and B from the system of algebraic equations (10.2.64) and (10.2.64′). In general, this entails great computational complications for a general case of loading; in the following we shall consider a particular case of loading where the calculations are considerably simplified.

10.2.3.2. Particular case

We shall consider the case where the loads on the separation plane satisfy the conditions

$$\frac{\partial p_x}{\partial x} + \frac{\partial p_y}{\partial y} = 0, \qquad q = 0; \quad (10.2.66)$$

in that case we obtain

$$\alpha L[F[\overline{p_x}(x, y; t)]] + \beta L[F[\overline{p_y}(x, y; t)]] = 0,$$

$$L[F[\overline{q}(x, y; t)]] = 0 \tag{10.2.66'}$$

and the coefficients A and B vanish

$$A = B = 0, \tag{10.2.67}$$

since the determinant of the system is different from zero.
   This leads to

$$L[F[\overline{w}(x, y, z; t)]] = 0, \tag{10.2.68}$$

whence

$$\overline{w}(x, y, z; t) = 0, \tag{10.2.68'}$$

and the formula (10.2.65) gives

$$\alpha L[F[\overline{u}(x, y, z; t)]] + \beta L[F[\overline{v}(x, y, z; t)]] = 0. \tag{10.2.69}$$

Taking into account relations (10.2.68) and (10.2.69), the first two equations (10.2.56) give

$$\left(\frac{\mathrm{d}^2}{\mathrm{d}z^2} - \gamma_2^2\right) L[F[\overline{u}(x, y, z; t)]] = 0,$$

$$\left(\frac{\mathrm{d}^2}{\mathrm{d}z^2} - \gamma_2^2\right) L[F[\overline{v}(x, y, z; t)]] = 0, \tag{10.2.70}$$

which leads to

$$L[F[\overline{u}(x, y, z; t)]] = A_1 \mathrm{e}^{-\gamma_2 z} + A_2 \mathrm{e}^{\gamma_2 z},$$

$$L[F[\overline{v}(x, y, z; t)]] = B_1 \mathrm{e}^{-\gamma_2 z} + B_2 \mathrm{e}^{\gamma_2 z}; \tag{10.2.70'}$$

the conditions or regularity at infinity give

$$A_2 = B_2 = 0. \tag{10.2.71}$$

The other coefficients A_1 and B_1 are specified by the conditions (10.2.57); thus we obtain

$$L[F[\overline{u}(x, y, z; t)]] = \frac{1}{\mu\gamma_2} L[F[\overline{p_x}(x, y; t)]]\mathrm{e}^{-\gamma_2 z},$$

$$L[F[\overline{v}(x, y, z; t)]] = \frac{1}{\mu\gamma_2} L[F[\overline{p_y}(x, y; t)]]\mathrm{e}^{-\gamma_2 z}, \tag{10.2.72}$$

whence

$$\bar{u}(x, y, z; t) = \frac{1}{2\pi G} L^{-1} \left[L[\bar{p}_x(x, y; t)] * \frac{1}{R} e^{-p\frac{R}{c_2}} \right],$$

$$\bar{v}(x, y, z; t) = \frac{1}{2\pi G} L^{-1} \left[L[\bar{p}_y(x, y; t)] * \frac{1}{R} e^{-p\frac{R}{c_2}} \right], \tag{10.2.73}$$

taking into account that $\mu = G$.

In the case of a rotational concentrated moment $M(t)$ acting at the origin on the separation plane and whose equivalent load is

$$p_x(x, y; t) = \frac{1}{2} M(t) \frac{\partial}{\partial y} \delta(x, y),$$

$$p_y(x, y; t) = - \frac{1}{2} M(t) \frac{\partial}{\partial x} \delta(x, y), \tag{10.2.74}$$

$$q(x, y; t) = 0,$$

the conditions (10.2.66) are satisfied. The resulting generalized state of displacement is

$$\bar{u}(x, y, z; t) = \frac{1}{4\pi G} \frac{\partial}{\partial y} \left[\frac{1}{R} \overline{M} \left(t - \frac{R}{c_2} \right) \right],$$

$$\bar{v}(x, y, z; t) = - \frac{1}{4\pi G} \frac{\partial}{\partial x} \left[\frac{1}{R} \overline{M} \left(t - \frac{R}{c_2} \right) \right], \tag{10.2.75}$$

$$\bar{w}(x, y, z; t) = 0,$$

where we have introduced the distribution $\overline{M}(t)$ specified by formula (10.2.54).

This shows that the particular case considered includes some remarkable cases of loading. A comparison between formulae (10.1.53) and (10.2.75) shows that an elastic space may be considered as being made up of two elastic half-spaces on the separation planes of which rotational concentrated moments of the magnitude $\frac{1}{2} M(t)$ are acting.

Bibliography

1. APPEL, P., *Traité de mécanique rationnelle*. Vol. I — V. 5th ed., Gauthier-Villars et Cie, Paris (1930—1932).

2. ARSAC, J., *Transformation de Fourier et théorie des distributions*. Dunod éd., Paris (1961).

3. ATANASIU, V., Applications de la théorie des distributions à l'étude de la stabilité élastique des barres. *Bul. Şt. Inst. Constr. Bucureşti*, **12**, 161—166 (1969).

4. — Extension des opérations avec des distributions et leur application dans l'étude de la barre droite élastique. *Bul. Şt. Inst. Constr. Bucureşti*, **14** (1971).

5. BANACH, ST., *Mechanics*. Naklad. Polsk. Tow. Mat., Warszawa. Wroclaw (1951).

6. BATEMAN, H. and ERDÉLYI, A., *Tables of Integral Transforms*. Vol. I. McGraw-Hill Book Co. Inc., New York — Toronto — London (1954).

7. BAUŞIC, V., HAVÎRNEANU, TH., PREDA, N., HORBANIUC, D. and AGACHI, P., Contribuţiuni în generalizarea ecuaţiei fibrei elastice a barelor drepte (Romanian) (Contributions to the generalization of the equation of the elastic line of straight bars). *Bul. Inst. Polit. Iaşi, ser. nouă*, **3** (**7**) (1957).

8. BOCHNER, S., *Vorlesungen über Fouriersche Integrale*. Leipzig (1932).

9. BOUIX, M., *Les fonctions généralisées ou distributions*. Masson et Cie, Paris (1964).

10. BREMERMANN, H., *Distributions, Complex Variables, and Fourier Transforms*. Addison-Wesley Publ. Co., Reading, Massachusetts (1965).

11. CONSTANTINESCU, I., Studiul mişcării cu puncte de discontinuitate a unei plăci plane (Romanian) (Study of the motion with points of discontinuity of a plane plate). *An. Univ. Timişoara, ser. şt. mat.-fiz.*, **3**, 101—108 (1965).

12. — Mişcarea firului sub acţiunea forţelor percutante (Romanian) (Motion of the string subjected to forces of percussion). *Lucr. Şt. Inst. Mine Petroşani, ser. şt. cult. gen.*, **6**, 219—229 (1969).

13. CONSTANTINESCU, I. and TOCACI, E., Sur les équations de deuxième espèce de Lagrange pour le cas des systèmes matériaux percutés. *Elektr. mech.-górn. hutnic.*, Kraków, **33**, 11—17 (1969).

14. CRĂCIUNAŞ, P., GRECU, A. and TOCACI, E., Su alcuni resultati concernenti le definizioni attuali delle distribuzioni. *Atti Accad. Naz. Lincei, Rend. Cl. Sci. Fis. Mat. Natur.*, ser. VIII, **40** (1966).

15. CRISTESCU, R., *Elemente de analiză funcţională şi introducere în teoria distribuţiilor* (Romanian) (Elements of Functional Analysis and Introduction to the Theory of Distritions). Ed. tehnică, Bucureşti (1966).

16. CRISTESCU, R. and MARINESCU, G., *Applications of the Theory of Distributions*. Ed. Acad., Bucureşti; J. Wiley and Sons, London — New York — Sydney — Toronto (1973).

17. DARABONT, A., Vibraţiile transversale ale barelor cu mase adiţionale concentrate (Romanian) (Transverse vibrations of bars with additional concentrated masses). *An. Univ. Bucureşti, mat.-mec.*, **20**, 63−90 (1971).

18. — Vibraţiile transversale ale plăcilor plane cu mase concentrate ataşate (Romanian) (Transverse vibrations of plane plates with attached concentrated masses). *Stud. cerc. mec. apl.*, **30**, 1331−1351 (1971).

19. — *Contribuţii la mecanica sistemelor cu caracteristici discontinuu variabile* (Romanian) (Contributions to the mechanics of systems with discontinuous variable characteristics). Doctoral Thesis, Inst. Polit. Bucureşti (1971).

20. DIACONU, M. and PREDA, N., Aplicarea parametrilor iniţiali la calculul săgeţii grinzilor încovoiate, ţinînd seama de forţele tăietoare (Romanian) (Application of the initial parameters to the calculus of the deflection of beams subjected to bending, taking into account the shear forces). *Bul. Inst. Polit. Iaşi, ser. nouă*, **3** (7) (1957).

21. DIRAC, P. A. M., The physical interpretation of the quantum mechanics. *Proc. Roy. Soc., London*, A, 113, 621−641 (1926−1927).

22. — *The Principles of Quantum Mechanics.* 3rd ed., New York (1947).

23. DITKIN, V. A. and PRUDNIKOV, A. P., *Integral Transforms and Operational Calculus.* Pergamon Press, Oxford − London − Edinburgh − New York − Paris − Frankfurt (1965).

24. DOETSCH, G., *Handbuch der Laplace Transformation.* Vol. I−III. Verlag Birkhäuser, Basel (1950, 1955, 1956).

25. FIHTENGOL'C, G. M., *Kurs differencial'nogo i integral'nogo isčislenija* (Russian) (Course of Differential and Integral Calculus). Vol. III. Fizmatgiz, Moscow − Leningrad (1963).

26. FLÜGGE, S. (editor), *Handbuch der Physik.* Vol. III/1, III/2, III/3, IIIa/2, VI. Springer-Verlag, Berlin − Heidelberg − New York (1958−1972).

27. FODOR, G., *Laplace Transforms in Engineering.* Akadémiai Kiado, Budapest (1965).

28. GARSOUX, J., *Espaces vectorielles topologiques et distributions.* Dunod, Paris (1963).

29. GELFAND, I. M. and ŠILOV, G. E., *Obobščennye funkcii* (Russian) (Generalized Functions). Vol. I−IV, Fizmatgiz, Moscow (1958−1961).

30. GERMAIN, P., *Mécanique des milieux continus.* Masson et Cie, Paris (1962).

31. GHICA, A., *Analiza funcţională* (Romanian) (Functional Analysis). Ed. Academiei, Bucureşti (1967).

32. HADAMARD, J., *Le problème de Cauchy et les équations aux dérivées partielles linéaires hyperboliques.* Hermann, Paris (1932).

33. HAIMOVICI, M., *Teoria elasticităţii* (Romanian) (Theory of Elasticity). Ed. didact. ped., Bucureşti (1969).

34. HALPERIN, I., *Introduction to the Theory of Distributions.* Toronto (1952).

35. HEAVISIDE, O., On Operators in Mathematical Physics. *Proc. Roy. Soc., London*, A, **52**, 504−529 (1893).

36. HÖRMANDER, L., *Linear Partial Differential Operators.* Springer-Verlag, Berlin − Göttingen − Heidelberg (1964).

37. IACOB, C., *Mecanica teoretică* (Romanian) (Theoretical Mechanics). Ed. didact. ped., Bucureşti (1970).

38. IVANENKO, D. and SOKOLOV, A., *Klassičeskaja teorija polja (novye problemy)* (Russian) (Classical Theory of Field (New Problems)). Gostehizdat, Moscow − Leningrad (1951).

39. KAMKE, E., *Das Lebesgue-Stieltjes-Integral.* B. G. Teubner Verlagsgesellschaft, Leipzig (1956).

40. Kecs, W., Sur les problèmes concernant le demi-espace élastique. *Bull. Math. Soc. Sci. Math., R. S. Roumanie*, **6** (**54**), 157—173 (1962).

41. — Asupra unor aplicații ale teoriei distribuțiilor în mecanica teoretică și aplicată (Romanian) (On the applications of the theory of distributions in theoretical and applied mechanics). *St. cerc. mat.*, **15**, 83—95 (1664).

42. — Reprezentarea cîmpurilor vectoriale cu ajutorul funcțiilor generalizate (Romanian) (Representation of vector fields with the aid of generalized functions). *St. cerc. mat.*, **15**, 265—273 (1964).

43. — Applications de la théorie des fonctions généralisées concernant les problèmes du plan élastique infini. *Rev. Roumaine Math. Pures Appl.*, **9**, 529—547 (1964).

44. — Extinderea ecuațiilor de echilibru a firelor perfect flexibile și inextensibile în cazul unor sarcini ce se exprimă printr-o funcție continuă, avînd discontinuități de speța întîi) (Romanian) (Generalization of balance equations of perfectly flexible and inextensible strings, in the case of loads expressed by the aid of a continuous function, having discontinuities of the first species). *St. cerc. mat.*, **17**, 367—377 (1665).

45. — Aplicații ale teoriei distribuțiilor privind încovoierea barelor drepte (Romanian) (Applications of the theory of distributions concerning the bending of straight bars). *St. cerc. mat.*, **17**, 1035—1048 (1965).

46. — Représentation mathématique des moments du type dipôle. *Rev. Roumaine Math. Pures Appl.*, **10**, 1391—1402 (1965).

47. — Sur les problèmes concernant le demi-plan élastique. *Bull. Math. Soc. Sci. Math., R. S. Roumanie*, **9** (**57**), 67—81 (1965).

48. — Les solutions généralisées des problèmes concernant l'espace élastique infini. *Rev. Roumaine Math. Pures Appl.*, **11**, 27—41 (1966).

49. — L'espace des chocs plastiques. *Bull. Math. Soc. Sci. Math., R. S. Roumanie*, **11** (**59**), 163—176 (1967).

50. — Asupra soluției fundamentale a problemei Cauchy pentru o clasă de ecuații diferențiale cu derivate parțiale cu coeficienți constanți (Romanian) (On the fundamental solution of Cauchy's problem for a class of partial differential equations with constant coefficients). *An. Univ. București, ser. șt. nat., mat.-mec.*, **17**, 59—68 (1968).

51. — Représentation mathématique de certaines limites de groupes de charges et leurs applications dans la mécanique des corps déformables (I). *Bul. Inst. Polit. Iași, ser. nouă*, **14** (**18**), 443—448 (1968).

52. — Extension de la notion de variation et quelques conséquences qui en dérivent. *Rev. Roumaine Math. Pures Appl.*, **13**, 347—364 (1968).

53. — L'extension du cadre d'applicabilité des équations de Meščerski et Lagrange. *Bull. Math. Soc. Sci. Math., R. S. Roumanie*, **12** (**60**), 127—135 (1968).

54. — Sur les équations d'équilibre de la théorie de l'élasticité. *Bull. Math. Soc. Sci. Math., R. S. Roumanie*, **12** (**60**), 137—143 (1968).

55. — Asupra derivării distribuțiilor compuse și unele aplicații ale lor (Romanian) (On the differentiation of composed distributions and applications). *St. cerc. mat.*, **21**, 223—233 (1969).

56. — Les solutions généralisées de l'espace élastique sous l'action de certaines charges variables. *Rev. Roumaine Math. Pures Appl.*, **15** (1970).

57. — Sur les conditions nécessaires d'extrêmum concernant les intégrales multiples. *Bull. Math. Soc. Sci. Math., R. S. Roumanie*, **15** (**63**), 181—192 (1971).

58. KECS, W., Sur une généralisation de la condition de Legendre du calcul variationnel pour les intégrales dépendant des dérivées d'ordre supérieur. *Rev. Roumaine Math. Pures Appl.*, **18** (1973).

59. — Sur le problème dynamique du demi-espace élastique. *Rev. Roumaine Math. Pures Appl.*, **18** (1973).

60. KECS, W. and TEODORESCU, P. P., On the plane problem of an elastic body acted upon by dynamic loads. I, II. *Bull. Acad. Polon. Sci., Sér. Sci. Techn.*, **19**, 47—55; 57—62 (1971).

61. — Sur le calcul variationnel en distributions. *Bull. Math. Soc. Sci. Math., R. S. Roumanie*, **15** (**63**) (1971).

62. KNESCHKE, A., *Differentialgleichungen und Randwertprobleme*. Vol. I—III. B. G. Teubner Verlagsgesellschaft, Leipzig (1960—1962).

63. KOREVAAR, J., Distributions defined from the point of view of applied Mathematics. *Proc. Kon. Nederl. Akad. Wetenschappen, ser. A*, **58**, 368—383; 463—503; 563—674 (1955).

64. KÖNIG, H., Neue Begründung der Theorie der "Distributionen" von L. Schwartz. *Math. Nachr.*, **9**, 130—448 (1953).

65. KUPRADZE, V. D., GEGELIJA, G. G., BAŠELEIŠVILI, M. O. and BURČULADZE, T. V., *Trehmernye zadači matematičeskoĭ teorii uprugosti* (Russian) (Three-Dimensional Problems of the Mathematical Theory of Elasticity). Tbilisi (1968).

66. LANDAU, L. D. and LIFŠIC, E. M., *Kvantovaia mehanika. Nereljativiskaja teorija* (Russian) (Quantum Mechanics. Non-Relativistic Theory). Fizmatgiz, Moscow (1963).

67. LAVOINE, J., *Calcul symbolique des distributions et des pséudofonctions*. Thèse, Paris (1962).

68. LAVRENTIEV, M. A. and LJUSTERNIK, L. D., *Kurs variacionnogo isčislenija* (Russian) (Course of Variational Calculus). Gostehizdat, Moscow — Leningrad (1950).

69. LAVRENTIEV, M. A. and ŠABAT, B. V., *Metody teorii funkcii kompleksnogo peremennogo* (Russian) (Methods of the Theory of Functions of Complex Variable). Fizmatgiz, Moscow (1958).

70. LEIBENZON, L. S., *Sobranie trudov* (Russian) (Complete Works). Vol. I. Izd. Akad. Nauk SSSR, Moscow (1953).

71. LERAY, J., Un prolongement de la transformation de Laplace, qui transforme la solution unitaire d'un opérateur hyperbolique en sa solution élémentaire (Problème de Cauchy, IV). *Bull. Soc. Math.*, France, **90**, 39—156 (1962).

72. LOVE, A. E. H., *A Treatise on the Mathematical Theory of Elasticity*. 4th ed., Cambridge University Press (1934).

73. LUR'E, A. I., *Teorija uprugosti* (Russian) (Theory of Elasticity). Izd. "Nauka", Moscow (1970).

74. MARINESCU, G., *Espaces vectorielles pseudotopologiques et théorie des distributions*. VEB Deutscher Verlag der Wiss., Berlin (1963).

75. MARINESCU, G. and TUDOR, C., Sur la transformation de Laplace des distributions. *Rev. Roumaine Math. Pures Appl.*, **12** (1967).

76. MIKUSIŃSKI, J., *Operational Calculus*. Pergamon Press-PWN Polish Sci. Publ., Oxford — London — Edinburgh — New York — Toronto — Sydney — Paris — Braunschweig (1959).

77. MIKUSIŃSKI, J. and SIKORSKI, R., The Elementary Theory of Distributions. Rozprawy mat., Warszawa (1957).

78. Moisil Gr. C., *Mecanica mediilor continue deformabile* (Romanian) (Mechanics of Continuous Deformable Solids). Ed. Univ., Bucureşti (1950).

79. Mucichescu, D., Quelques applications de la théorie des distributions dans le calcul du deuxième ordre de la poutre droite. *Bul. Şt. Inst. Constr. Bucureşti*, **12**, 87—94 (1969).

80. — Aplicaţii ale teoriei distribuţiilor în studiul încovoierii plăcilor plane dreptunghiulare simplu rezemate pe contur (Romanian) (Applications of the theory of distributions in the study of simply supported rectangular plane plates). *Bul. Şt. Inst. Constr. Bucureşti*, **13**, 55—64 (1970).

81. — Cu privire la condiţia generalizată de ortogonalitate a funcţiilor proprii în vibraţiile libere ale barei elastice (Romanian) (On the generalized condition of orthogonality of the eigenfunctions for free vibrations of the elastic bar). *Bul. Şt. Inst. Constr. Bucureşti*, **13**, 219—223 (1970).

82. — Sur la propagation des ondes transversales dans une barre élastique infinie. *Rev. Roumaine Math. Pures Appl.*, **15**, 285—292 (1970).

83. — Une généralisation de la notion de charge et ses applications au problème des lignes d'influence. *Rev. Roumaine Sci. Techn., Méc. Appl.*, **15**, 75—90 (1970).

84. — Sur la flexion de la dalle infinie posée sur un milieu élastique. *Rev. Roumaine Sci. Techn., Méc. Appl.*, **16**, 131—138 (1971).

85. — Sur l'application de la théorie des distributions dans la dynamique des plaques planes élastiques. *Bul. Şt. Inst. Constr. Bucureşti*, **14**, 185—194 (1971).

86. Mushelišvili, N. I., *Some Basic Problems of the Mathematical Theory of Elasticity*. P. Noordhoff Ltd., Groningen (1963).

87. Nicolescu, M., *Analiza matematică* (Romanian) (Mathematical Analysis). Vol. I—III. Ed. tehnică, Bucureşti (1957—1960).

88. Noaghi, Tr., Studiul curburii curbelor plane în punctele singulare cu ajutorul teoriei distribuţiilor (Romanian) (Study of the curvature of the plane curves in singular points with the aid of the theory of distributions). *An. Univ. Bucureşti, mat.-mec.*, **17**, 109—121 (1968).

89. — Curburile unei suprafeţe în anumite puncte singulare, exprimate prin distribuţii (Romanian) (Curvatures of a surface in certain singular points, expressed in distributions). *An. Univ. Bucureşti, mat.-mec.*, **17** (1969).

90. Nowacki, W., *Dynamics of Elastic Systems*. Chapman and Hall Ltd., London (1963).

91. — *Teoria sprężystości* (Polish) (Theory of Elasticity). Państ. Wyd. Naukowe, Warszawa (1970).

92. Olariu, V., *Ecuaţii cu derivate parţiale. Soluţii generalizate.* (Romanian) (Partial Derivative Equations. Generalized Solutions). Vol. I, II. Ed. Univ., Bucureşti (1970—1971).

93. Onicescu, O., *Mecanica* (Romanian) (Mechanics). Ed. tehnică, Bucureşti (1969).

94. Pauli, W., *Meson Theory of Nuclear Forces*. New York (1946).

95. Planck, M., *Mechanik Deformierbarer Körper*. 3. Aufl., S. Hirzel, Leipzig (1931).

96. Pol, B. van der and Bremmer, H., *Operational Calculus Based on the Two-Sided Laplace Integral*. Cambridge University Press (1950).

97. Schwartz, L., *Théorie des distributions*. Vol. I, II. Hermann, Paris (1950—1951).

98. — *Méthodes mathématiques pour les sciences physiques*. Hermann, Paris (1961).

99. Sebastiano e Silva, J., Sur une construction axiomatique de la théorie des distributions. *Rev. Fac. Ciêne, Lisboa (A)*, 79—186 (1955).

100. SIKORSKI, R., A definition of the notion of distribution. *Bull. Acad. Polon. Sci.* sér. III, 2, 209—211 (1954).

101. — On substitutions in the Delta Distribution. *Bull. Acad. Polon. Sci., sér. Sci. Math. Astr. Phys.*, 8 (1960).

102. SMIRNOV, V. I., *Kurs vysšei matematiki* (Russian) (Course of High Mathematics). Vol. V. Fizmatgiz, Moscow (1959).

103. SNEDDON, I. N., *Fourier Transforms*. McGraw-Hill Book Co., New York — Toronto — London (1951).

104. SNEDDON, I. N. and BERRY, D. S., *The Classical Theory of Elasticity*. In *Handbuch der Physik*, Vol. VI, Springer-Verlag, Berlin — Göttingen — Heidelberg (1958).

105. SOBOLEV, S. A., Méthode nouvelle pour résoudre le problème de Cauchy pour les équations hyperboliques normales. *Mat. sbornik*, 1, 39—72 (1936).

106. — *Uravnenija matematičeskoi fiziki* (Russian) (Equations of Mathematical Physics). Gostehizdat, Moscow — Leningrad (1960).

107. SOKOLNIKOFF, I. S., *Mathematical Theory of Elasticity*. 2nd ed., McGraw-Hill Book Co., New York — Toronto — London (1956).

108. SOMMERFELD, A., *Partielle Differentialgleichungen der Physik*. 5. Auflage, Akad. Verlagsgesellschaft Geest Protig K.-G., Leipzig (1962).

109. SOÓS, E., Tensor Kelvina-Somigliany dla ciała lepkosprężystego (Polish) (Kelvin-Somigliana Tensor for Viscoelastic Bodies). *Mech. Teor. Štos.*, 2, 31—33 (1965).

110. STEFANIAK, J., Concentrated loads as body forces. *Rev. Roumaine Math. Pures Appl.*, 14, 119—127 (1969).

111. ŞANDRU, N., Asupra acţiunii unei forţe concentrate în spaţiul elastic nemărginit (Romanian) (On the action of a concentrated force in the elastic space). *Com. Acad. R. P. Romảne*, 13, 1119—1021 (1963).

112. — O deistvii peremennyh sil v neogranicennom uprugom prostranstve (Russian) (On the action of a variable force in the elastic space). *Bull. Acad. Polon. Sci., sér. Sci. Techn.*, 12, 45—47, (1964).

113. ŞESAN A., Relations statiques générales pour l'étude de la flexion. *Bul. Inst. Polit. Iaşi*, ser. nouă, 4 (8) (1958).

114. TEODORESCU, N. and OLARIU, V., *Ecuaţiile fizicii matematice* (Romanian) (Equations of Mathematical Physics). Vol. I, II. Ed. didact. ped., Bucureşti; Ed. Univ., Bucureşti (1970—1972).

115. TEODORESCU, P. P., Über die Berechnung des elastischen Halbraumes unter örtlicher Belastung. *Bull. Math. Soc. Sci. Math. Phys.*, R. P. Roumaine, 2 (50), 113—121 (1958).

116. — Über die Berechnung des periodisch belasteten elastischen Halbraumes. *Rev. Méc. Appl.*, 4, 141—148 (1959).

117. — Über einige räumliche Probleme der Elastizitätstheorie. *Apl. Mat.*, 4, 225—238 (1959).

118. — Sur le problème du coin plan élastique. *C. R. Acad. Sci., Paris*, 250, 3446—3448 (1960).

119. — Sur le problème du quart de plan élastique. *Apl. mat.*, 6, 359—378 (1961).

120. — Sur certains problèmes plans de l'élastodynamique. *Rev. Roum. Sci. Techn., Méc. Appl.*, 9, 895—918 (1964).

121. — Sur l'action des charges concentrées dans le problème plan de la théorie de l'élasticité. *Bull. Math. Soc. Sci. Math., R. S. Roumanie*, 8 (56) 243—287 (1964).

122. TEODORESCU P. P., Considerații asupra modului de a defini sarcinile concentrate în problema plană a teoriei elasticității (Romanian) (Considerations on defining concentrated loads in the plane problem of the theory of elasticity). *An. Univ. Timișoara, ser. șt. mat.-fiz.*, **3**, 287—305 (1965).

123. — Über das dreidimensionale Problem der Elastokinetik. *Z. A. M. M.*, **45**, 513—523 (1965).

124. — Schwingungen der elastischen Kontinua. III. *Konferenz über nichtlineare Schwingungen, Berlin*, 25—30 Mai 1964, Teil II, Technische Schwingungsprobleme und Fragen der Regelung und Steurung, *Abh. der deutschen Akad. der Wiss. zu Berlin Kl. für Math., Phys. und Technik*, Akad.-Verlag, Berlin, 29—67 (1965).

125. — Sur quelques problèmes dynamiques de la théorie de l'élasticité. *Rev. Roumaine Math. Pures Appl.*, **11**, 773—786 (1966).

126. — Asupra acțiunii sarcinilor concentrate în teoria elasticității (Romanian) (On the action of concentrated loads in the theory of elasticity). *An. Univ. București, ser. șt. nat., mat.-mec.*, **15**, 15—24 (1966).

127. — Sur un certain caractère tensoriel des charges concentrées. *Atti Accad. Naz. Lincei, Rend. Cl. Sci. Fis. Mat. Natur.*, ser. VIII, **40**, 251—257 (1966).

128. — Sur l'action des charges concentrées dans le problème plan de la mécanique des solides déformables. *Arch. Mech. Stos.*, **18**, 567—579 (1966).

129. — *Probleme plane în teoria elasticității* (Romanian) (Plane Problems in the Theory of Elasticity). Vol. I, II. Ed. Academiei, București (1961—1966).

130. — Sur la notion de moment massique dans le cas des corps du type de Cosserat. *Bull. Acad. Polon. Sci., sér. Sci. Techn.*, **15**, 65—70 (1967).

131. — On the action of concentrated loads in the case of a Cosserat continuum. *Mechanics of Generalized Continua*. IUTAM-Symposium, Freudenstadt — Stuttgart, 28 August — 2 September 1967, Springer — Verlag, Berlin — Göttingen — New York, 120—125 (1968).

132. — Sur les corps du type de Cosserat à élasticité linéaire. *Symposium on Polar Continua*, Roma, 2—5 April 1968, *Ist. Naz. Alta Matem. Symposia Math.*, **1**, 375—409 (1968).

133. — Considérations concernant l'introduction des fonctions potentiel de déplacement dans les problèmes en espace de la théorie de l'élasticité (cas statique). *An. Univ. București, ser. șt. nat., mat.-mec.*, **18**, 127—135 (1969).

134. — Asupra propagării undelor în medii elastice (Romanian) (On the propagation of waves in elastic media). *Bul. Șt. Inst. Polit. Cluj, ser. Constr.*, **13**, 97—104 (1970).

135. — Sur le tenseur de Finzi et sur quelques de ses applications et généralisations. *Ann. Mat. Pura Appl.*, ser. IV, **84**, 225—244 (1970).

136. — Funcții de tensiune în elastodinamică (Romanian) (Stress functions in elastodynamics). *An. Univ. București, mat.-mec.*, **19**, 145—158 (1970).

137. — *Probleme spațiale în teoria elasticității* (Romanian) (Space Problems in the Theory of Elasticity). Ed. Academiei, București (1970).

138. — Considerații în legătură cu utilizarea funcțiilor potențial în teoria elasticității (Romanian) (Considerations concerning the use of potential functions in the theory of elasticity). *Lucr. Șt. Inst. Ped., Galați*, **5**, 27—35 (1971).

139. — Considérations concernant l'introduction des fonctions-potentiel de déplacement dans les problèmes en espace de la théorie de l'élasticité (cas dynamique). *An. Univ. București, mat.-mec.*, **20**, *1*, 125—139 (1971).

140. TEODORESCU, P. P., Sur l'introduction des fonctions-potentiel en élasticité linéaire. *An. Univ. Bucureşti, mat.-mec.*, **20**, *2*, 131—140 (1971).

141. — Asupra acţiunii unor sarcini dinamice concentrate (Romanian) (On the action of certain dynamic concentrated loads). *Bul. Şt. Inst. Polit. Cluj, ser. Constr.*, **14**, 119—126 (1971).

142. — Über ein Analogon der Schaeferschen Darstellung in der Elastokinetik und einige Anwendungen. *Z.A.M.M.*, **52**, T 154—T 156 (1972).

143. — Stress Functions in Three-Dimensional Elastodynamics. *Acta Mech.*, **14**, 103—118 (1972).

144. — Quelques considérations sur le problème de l'espace élastique et sur l'utilisation de sa solution. *An. Univ. Bucureşti, mat.-mec.*, **21** (1972).

145. — Sur une représentation du type de Papkovitch-Neuber dans le cas des corps du type de Cosserat. *Rev. Roumaine Math. Pures Appl.*, **17**, 1097—1106 (1972).

146. — Sur le problème dynamique des corps du type de Cosserat. *Rev. Roumaine Math. Pures Appl.*, **17**, 1097—1106 (1972).

147. — *Dinamica corpurilor liniar elastice* (Romanian) (Dynamics of Linear Elastic Bodies). Ed. Academiei, Bucureşti (1972).

148. TEODORESCU, P. P. and ŞANDRU, N., Sur l'action des charges concentrées en élasticité asymétrique plane. *Rev. Roumaine Math. Pures Appl.*, **12**, 1399—1405 (1967).

149. TIHONOV, A. N. and SAMARSKII, A. A., *Uravnenija matematičeskoi fiziki* (Russian) (Equations of Mathematical Physics). Gostehizdat, Moscow (1953).

150. TOCACI, E., Contribuţii la studiul sistemelor materiale percutate (Romanian) (Contributions to the study of material systems subjected to percussion). *Bul. Inst. Polit. Iaşi, ser. nouă*, **11 (15)** (1965).

151. — Asupra ecuaţiilor mecanicii analitice pentru calculul sistemelor materiale percutate (Romanian) (On the equations of analytical mechanics for the calculus of material systems subjected to percussion). *St. cerc. mat.*, **18**, 1129—1138 (1966).

152. — Ecuaţiile lui Lagrange pentru sisteme materiale de masă discontinuu variabilă (Romanian) (Lagrange's equations for material systems of discontinuous variable mass). *An. Univ. Bucureşti, ser. şt. nat., mat.-mec.*, **15** (1966).

153. — Energia de acceleraţie (Romanian) (Energy of acceleration). *Bul. com. Inst. Mine Petroşani* (1966).

154. — Aspecte ale utilizării teoriei distribuţiilor în mecanică (Romanian) (Aspects of the use of the theory of distributions in mechanics). *St. cerc. mec. apl.*, **25** (1970).

155. — Asupra mecanicii analitice a mişcărilor discontinue ale fluidelor (Romanian) (On analytical mechanics of discontinuous motion of fluids). *Bul. com. Inst. Mine Petroşani* (1970).

156. — *Fenomene discontinue în mecanica şi rezistenţa materialelor* (Romanian) (Discontinuous Phenomena in Mechanics and Strength of Materials). Doctoral Thesis, Inst. Polit. Iaşi (1971).

157. TOCACI, E., SIMA, P. and OLARIU, V., Sur les équations de Mangeron pour les systèmes matériels de masse discontinue variable. *Bul. Inst. Polit. Iaşi, ser. nouă*, **12 (16)** (1966).

158. TONELLI, L., *Fondamenti di calcolo delle variazioni.* Vol. I, II. Bologna (1921).

159. TUDOR, C., Transformarea Laplace a distribuţiilor (Romanian) (Laplace transformation of distributions). St. cerc. mat., **19**, 1521—1536 (1967).

160. UFLJAND, JA. S., *Integral'nye preobrazovanija v zadačah teorii uprugosti* (Russian) (Integral Transforms in the Problems of the Theory of Elasticity). Izd. Akad. Nauk SSSR, Moscow — Leningrad (1963).

161. VALIRON, G., *Équations fonctionnelles. Applications.* IIme éd., Masson et Cie, Paris (1950).

162. VÂLCOVICI, V., BĂLAN, ŞT. and VOINEA, R., *Mecanica teoretică* (Romanian) (Theoretical Mechanics). 3rd ed., Ed. tehnică, Bucureşti (1968).

163. VERMA, G. R., Application of Dirac's delta function in isolated force problems of infinite elastic solid of isotropic and non-isotropic materials. *Z.A.M.M.*, **37** (1957).

164. WLADIMIROW, W. S., *Gleichungen der mathematischen Physik.* VEB Deutscher Verlag der Wiss., Berlin (1972).

165. VODIČKA, V., Ein durch allgemeine Massenkräfte beanspruchtes unendliches elastisches Medium. *Z.A.M.M.*, **39**, 2—8 (1959).

166. VOICULESCU, D., *Aplicarea calculului operaţional şi matricial la probleme de rezistenţa materialelor* (Romanian) (Application of Operational and Matricial Calculus to Problems of Strength of Materials). Doctoral Thesis, Inst. Polit. Bucureşti (1964).

167. — Quelques applications de la théorie des distributions dans la résistance des matériaux. *Bul. Inst. Polit. Bucureşti*, **31**, 77—96 (1969).

168. — Contribution à la détermination des déformations et des efforts par la méthode des paramètres initiaux à la barre droite à section constante et à section variable en degrés. *Bul. Inst. Polit. Bucureşti*, **31**, 83—96 (1969).

169. ZEMANIAN, A. H., *Distribution Theory and Transform Analysis.* McGraw-Hill Book Co., New York — Saint Louis — San Francisco — Toronto — London — Sydney (1965).

170. — *Generalized Integral Transformations.* Interscience Publ., J. Wiley and Sons, New York — London — Sydney — Toronto (1968).